Springer Proceedings in Mathematics & Statistics

Volume 26

For further volumes:
http://www.springer.com/series/10533

Springer Proceedings in Mathematics & Statistics

This book series features volumes composed of select contributions from workshops and conferences in all areas of current research in mathematics and statistics, including OR and optimization. In addition to an overall evaluation of the interest, scientific quality, and timeliness of each proposal at the hands of the publisher, individual contributions are all refereed to the high quality standards of leading journals in the field. Thus, this series provides the research community with well-edited, authoritative reports on developments in the most exciting areas of mathematical and statistical research today.

Miguel Sánchez • Miguel Ortega • Alfonso Romero
Editors

Recent Trends in Lorentzian Geometry

Springer

Editors
Miguel Sánchez
Department of Geometry and Topology
Faculty of Science
University of Granada
Campus Fuentenueva s/n
Granada, Spain

Alfonso Romero
Department of Geometry and Topology
Faculty of Science
University of Granada
Campus Fuentenueva s/n
Granada, Spain

Miguel Ortega
Department of Geometry and Topology
Faculty of Science
University of Granada
Campus Fuentenueva s/n
Granada, Spain

ISSN 2194-1009 ISSN 2194-1017 (electronic)
ISBN 978-1-4614-4896-9 ISBN 978-1-4614-4897-6 (eBook)
DOI 10.1007/978-1-4614-4897-6
Springer New York Heidelberg Dordrecht London

Library of Congress Control Number: 2012950329

Mathematics Subject Classification (2010): 53C50, 53C40, 53C22, 53C30, 53C60, 53C80

Printed on acid-free paper

Springer is part of Springer Science+Business Media (www.springer.com)

GEOMETRY
LORENTZ
GRANADA
'11

September 6~9

Preface

Since the beginning of general relativity, Lorentzian geometry has provided the language and background for such a theory, as well as a mathematical arena where prospective extensions of Riemannian techniques could be sounded out. However, the variety and depth of its developments have consolidated Lorentzian geometry as a branch of differential geometry which is interesting by itself, as it provides applications to many parts of mathematical physics and yields an appealing framework where many mathematical techniques merge: geometric analysis, functional analysis, partial differential equations, Lie groups, and so on.

Ten years ago, a biennial series of meetings focused on Lorentzian Geometry was born in the town of Benalmádena (Spain). In the sixth edition, celebrated in Granada, September 2012, around 120 researchers of 18 countries gathered to discuss on the new trends of this geometry. In fact, the progress along this decade has attracted a renewed interest for many researchers: long-standing open problems have been solved, outstanding Lorentzian spaces and groups have been classified, new applications to mathematical relativity and high energy physics have been found, and further connections with other geometries have been developed.

In this volume, a sampler of the recent progress in Lorentzian geometry is presented. Topics such as geodesics, constant mean curvature submanifolds, trapped surfaces, gravitational collapse, classifications of manifolds with relevant symmetries, connections with Finsler geometry, and applications to mathematical physics are included. The contributions to this volume give a general perspective on these topics and provide new substantial results in some of them.

Let us give a very short overview of the contents.

The first five contributions constitute a block devoted to several problems on notable surfaces (maximal, constant mean curvature, umbilical, trapped) in Lorentzian manifolds. They are studied from different viewpoints, which include connections with other classical parts of differential geometry and mathematical relativity. More precisely, Fujimori, Kawakami, Kokubu, Rossman, Umehara, and Yamada introduce and develop an original notion of extended hyperbolic metric (i.e., a hyperbolic metric with a certain kind of singularities on a Riemann surface). Surfaces endowed with such metrics will be related to surfaces of constant mean curvature one in de Sitter space $\mathbb{S}_1^3$. This relation is developed specifically for catenoids in $\mathbb{S}_1^3$, and the classification of such catenoids will provide a classification of the corresponding moduli space of hyperbolic metrics. Albujer and Alías revisit both the classical Calabi–Bernstein theorem (i.e., the only entire maximal graphs in Lorentz-Minkowski space $\mathbb{L}^3$ are the space-like planes) and quite a few of its extensions. Very recent generalizations to a product spacetime $M^2 \times \mathbb{L}^1$ are specially considered. In particular, a local approach based on a parabolicity criterium is introduced so that a new proof to Calabi–Bernstein result is achieved. Senovilla focuses on umbilical space-like 2-surfaces in a Lorentzian manifold of dimension four. He introduces the notion of *ortho-umbilicity* and provides an original criterion to characterize total umbilicity in terms of the conmutativity of two independent Weingarten operators. Some consequences are analyzed, and extensions to arbitrary signatures and higher dimensions are also discussed. Mars focuses on marginally outer trapped surfaces (MOTS), which play an important role in gravitational theory as indicators of strong gravitational fields and, eventually, of black hole boundaries (*event horizons*). They share some of the properties of minimal hypersurfaces, in particular, the existence of a useful notion of *stability*. The implications of stability on the topology of MOTS, its interplay with spacetime symmetries, and, then, the stability of Killing horizons are carefully analyzed. As a further development, Jaramillo analyzes the existence of a set of inequalities involving the area, angular momentum, and charges of stably outermost marginally trapped surfaces in a generic spacetime under natural hypotheses. These inequalities provide lower bounds for the area of spatial sections that offers quasi-local models of black hole horizons. As in Mars contribution, the extension to a Lorentzian setting of tools employed in minimal surfaces in Riemannian contexts is emphasized.

The next three contributions are devoted to properties of geodesics and gravitational collapse. Caponio makes a thorough analysis of the notion of convexity for a hypersurface in a semi-Riemannian manifold. Classical Bishop's Riemannian result stating that infinitesimal convexity is equivalent to local convexity is reviewed, and its failure for the Lorentzian case is remarked. The analogous problem for the Finsler case has been solved very recently, and the author shows that the techniques for this more general case are also applicable to the semi-Riemannian problem. Applications to geodesic connectivity and further questions are also discussed. Variational methods and techniques of global analysis in manifolds are used by Bartolo, Candela, and Flores in order to study geodesics in spacetimes. After the successful results in the last two decades about the existence of connecting geodesics in causally well-behaved spacetimes, the authors focus on Gödel-type spacetimes. The results on this case are reviewed, and further improvements are obtained. Giambò and Magli analyze the geometry of isotropic fluids under gravitational collapse. Under general assumptions defining the fluid model, the null geodesics and causal structure, as well as the possible formation of horizons and nature of singularities, are discussed, with special attention to the case of bariotropic fluids obeying a linear equation of state.

The next block of three contributions is related to the recently developed connection between the class of (conformally) standard stationary spacetimes and the class of Finsler manifolds of Randers type. Javaloyes gives a general overview of such a stationary-to-Randers correspondence. This includes relations already developed at the levels of light-like geodesics, causality or causal boundaries on the stationary side, with different Finslerian elements (geodesics, convexity/completeness, and Busemann boundaries, respectively), as well as further prospective relations on topics such as isometry groups and curvature. In this framework, Matveev solves a natural question on arbitrary Finsler manifolds. More specifically, he characterizes when a Finsler metric F can be made complete by using a trivial projective change ($F \to F + df$). This question is inspired in a result that can be proved for the class of Randers metrics as a direct consequence of the stationary-to-Randers correspondence. Flores and Herrera contribution has several aims. Firstly, they review both the new redefinition of the classical *causal boundary* of a spacetime and the tools for its computation. These include, on the one hand, Penrose's conformal boundary and, on the other, connections with several boundaries in differential geometry (Cauchy, Gromov, Busemann), which have been developed for Finsler manifolds recently. Secondly, by using such tools, the causal boundary of the stationary part of Kerr spacetime is computed explicitly here.

The last four contributions study different aspects of symmetries of Lorentzian manifolds. Lichtenfelz, Piccione, and Zeghibs contribution provides a critical survey on some topics about the isometry group of a Lorentzian manifold. They revisit carefully the subtleties to endow this group with a Lie group structure. Then, recent results on (compact or not) Lie groups acting on a compact Lorentzian manifold are reviewed. Honda and Tsukada progress towards the local classification of conformally flat homogeneous Lorentzian manifolds. Such a complete classification is available in the Riemannian setting, but in the Lorentzian one has been obtained only in dimension three. As the Schouten tensor determines the curvature in this case, the authors focus on its algebraic structure and obtain the classification for all cases with nontrivial Jordan form, except when a triple root of the minimal polynomial exists. Díaz-Ramos gives an updated review on polar and hyperpolar actions on symmetric spaces, including the discussion of the differences between the compact and the noncompact cases. The study is carried out at the Riemannian level, and the Lorentzian one appears as a prospective challenge. Finally, Gilkey and Nikcevic, after surveying some known results in geometric realizability (including the semi-Riemannian and para-Hermitian settings), provide a new result on Kähler and para-Kähler Weyl structures. Specifically, a decomposition of the corresponding space of curvature tensors (which stresses the differences between dimension 4 and higher) is obtained. Then, every (para-)Kähler algebraic structure is shown to be geometrically realized by a (para-)Kähler manifold.

Summing up, these contributions, as a whole, provide a progress and an updated guide for many of the most interesting topics in present-day research on Lorentzian geometry. Thus, we would like to acknowledge the careful work of all the contributors, as well as of the anonymous referees. We would also like to thank

Springer for this opportunity to spread this branch of knowledge and, specially, the very friendly assistance provided by Meredith Rich during the elaboration of the book.

Finally, we would also like to acknowledge all the participants of the Lorentzian meeting for the excellent scientific level and pleasant atmosphere of this congress http://gigda.ugr.es/gelogra/, as well as the support of its sponsors, the Spanish state grants MTM2010-11595-E and i-MATH SARE-C6-0364, the regional research group FQM-324 and the University of Granada. The editors also acknowledge the support of their work by the Spanish projects with FEDER funds MTM2010-18099 and P09FQM-4496. We also thank Francisco Torralbo for meeting the artwork, including the logo of the conference reproduced in this volume.

Granada, Spain

Miguel Sánchez
Alfonso Romero
Miguel Ortega

Contents

Hyperbolic Metrics on Riemann Surfaces and Space-Like CMC-1 Surfaces in de Sitter 3-Space

Shoichi Fujimori, Yu Kawakami, Masatoshi Kokubu, Wayne Rossman, Masaaki Umehara, and Kotaro Yamada

Mathematics Subject Classification (2010): Primary 53A10; Secondary 53A35, 53C50.

The authors were partially supported by the Grant-in-Aid for Young Scientists (B) No. 21740052 (Fujimori), (B) No. 21740053 (Kawakami), and by the Grant-in-Aid for Scientific Research (C) No. 22540100 (Kokubu), (B) No. 20340012 (Rossman), (A) No. 22244006 (Umehara) and (B) No. 21340016 (Yamada), Japan Society for the Promotion of Science.

S. Fujimori
Department of Mathematics, Okayama University, Okayama 700-8530, Japan
e-mail: fujimori@math.okayama-u.ac.jp

Y. Kawakami
Department of Mathematics, Graduate School of Science and Engineering, Yamaguchi University, Yamaguchi 753-8572, Japan
e-mail: y-kwkami@yamaguchi-u.ac.jp

M. Kokubu
Department of Mathematics, School of Engineering, Tokyo Denki University, 120-8551, Tokyo, Japan
e-mail: kokubu@cck.dendai.ac.jp

W. Rossman
Department of Mathematics, Faculty of Science, Kobe University, 657-8501, Rokko, Kobe, Japan
e-mail: wayne@math.kobe-u.ac.jp

M. Umehara
Department of Mathematical and Computing Sciences, Tokyo Institute of Technology, 152-8552, Tokyo, Japan
e-mail: umehara@is.titech.ac.jp

K. Yamada (✉)
Department of Mathematics, Tokyo Institute of Technology, 152-8551, Tokyo, Japan
e-mail: kotaro@math.titech.ac.jp

M. Sánchez et al. (eds.), *Recent Trends in Lorentzian Geometry*, Springer Proceedings in Mathematics & Statistics 26, DOI 10.1007/978-1-4614-4897-6_1,

Introduction

In this chapter, we give a new notion called the *extended hyperbolic metric* on a Riemann surface (the precise definition is given in Sect. 2), which is a canonical generalization of hyperbolic metrics (i.e. metrics of constant curvature -1): Connecting two hyperbolic planes by identifying their ideal boundaries, we get the *hyperbolic* 2*-sphere* S^2_H, that is, S^2_H is a Riemann sphere $(S^2=)\mathbf{C}\cup\{\infty\}$ with the metric $d\sigma_P^2 := 4|dz|^2/(1-|z|^2)^2$, where $|dz|^2 := dz d\bar{z}$ and z is the canonical complex coordinate of $\mathbf{C}$. We call the metric $d\sigma_P^2$ on S^2_H the *spherical Poincaré metric*. An arbitrarily given extended hyperbolic metric on a Riemann surface can be realized as the pull-back metric of the spherical Poincaré metric $d\sigma_P^2$ on S^2_H by its developing map. Such an object has been discussed mainly as a projective structure with $\mathrm{SL}(2,\mathbf{R})$-monodromy (cf. Goldman [10] and Gallo–Kapovich–Marden [8]) in the study of Teichmüller spaces, but it seems that singularities of such metrics have not been precisely examined yet. (It should be remarked that $\mathrm{SL}(2,\mathbf{R})$ is conjugate to $\mathrm{SU}(1,1)$ in $\mathrm{SL}(2,\mathbf{C})$.) In fact, an extended hyperbolic metric $d\sigma^2$ may have singularities called *ideal boundary points*, denoted also as ∂^∞-points, where any path accumulating to that ideal boundary point has infinite length, and also may have singularities called *proper singularities*, which are isolated singular points of the Schwarzian derivative of $d\sigma^2$ (cf. Sect. 2 and Appendix A). In Sect. 3, we give several important properties of extended hyperbolic metrics.

It is known that constant mean curvature one surfaces (resp. flat surfaces) in hyperbolic 3-space are closely related to spherical metrics (resp. flat metrics) on Riemann surfaces (cf. [18, 19] for spherical metrics, and [12, Theorem 4.4] for flat metrics). Similarly, extended hyperbolic metrics bijectively correspond to space-like surfaces of constant mean curvature one (CMC-1) in de Sitter 3-space S^3_1 with a given hyperbolic Gauss map (cf. Theorems 2.10 and 2.11). In other words, CMC-1 surfaces in de Sitter 3-space S^3_1 can be considered as geometric realizations of extended hyperbolic metrics. In fact, the singular set of a given CMC-1 surface in S^3_1 coincides with that of the associated co-orientable extended hyperbolic metric. Several important examples of CMC-1 surface in S^3_1 are given in Lee-Yang [14] and [1–3,5,6]. In Sect. 4, we classify S^3_1-catenoids, i.e. weakly complete CMC-1 faces in S^3_1 of genus zero with two regular ends whose hyperbolic Gauss map is of degree one.

In Sect. 5, we classify extended hyperbolic metrics with at most two regular singularities on S^2, using the above correspondence, where a proper singularity is called a *regular singularity* if the Schwarzian derivative of the metric has at most a pole of order 2 (cf. Sect. 2).

1 Generalized CMC-1 Faces in de Sitter 3-Space

First, we recall some fundamental facts about CMC-1 faces in de Sitter 3-space. For detailed expressions, see [1, 5].

Generalized CMC-1 Faces

Let $\mathbf{R}^4_1$ be the Lorentz-Minkowski 4-space with the metric $\langle\ ,\ \rangle$ of signature $(-,+,+,+)$. Then de Sitter 3-space is expressed as

$$S^3_1 = \{X \in \mathbf{R}^4_1 ; \langle X, X\rangle = 1\}$$

with metric induced from $\mathbf{R}^4_1$, which is a simply-connected Lorentzian 3-manifold with constant sectional curvature 1. We identify $\mathbf{R}^4_1$ with the set of 2×2 Hermitian matrices Herm(2) by

$$\mathbf{R}^4_1 \ni (x_0,x_1,x_2,x_3) \longleftrightarrow \begin{pmatrix} x_0+x_3 & x_1+\mathrm{i}x_2 \\ x_1-\mathrm{i}x_2 & x_0-x_3 \end{pmatrix} \in \mathrm{Herm}(2), \tag{1}$$

where $\mathrm{i}=\sqrt{-1}$. Then de Sitter 3-space is represented as

$$\begin{aligned} S^3_1 &= \{X \in \mathrm{Herm}(2)\,;\, \det X = -1\} \\ &= \{ae_3a^*\,;\, a \in \mathrm{SL}(2,\mathbf{C})\} = \mathrm{SL}(2,\mathbf{C})/\mathrm{SU}(1,1) \qquad \left(e_3 := \begin{pmatrix} 1 & 0 \\ 0 & -1\end{pmatrix}\right), \end{aligned} \tag{2}$$

where $a^* := {}^t\overline{a}$ is the transposed conjugate matrix of a, and

$$\mathrm{SU}(1,1) := \{a \in \mathrm{SL}(2,\mathbf{C})\,;\, ae_3a^* = e_3\}.$$

We consider the projection

$$\pi_S : \mathrm{SL}(2,\mathbf{C}) \ni a \longmapsto ae_3a^* \in S^3_1 = \mathrm{SL}(2,\mathbf{C})/\mathrm{SU}(1,1). \tag{3}$$

The group $\mathrm{SL}(2,\mathbf{C})$ acts isometrically on S^3_1 as

$$S^3_1 \ni X \longmapsto aXa^* \in S^3_1 \qquad (a \in \mathrm{SL}(2,\mathbf{C})). \tag{4}$$

In fact, $\mathrm{PSL}(2,\mathbf{C}) = \mathrm{SL}(2,\mathbf{C})/\{\pm\text{identity matrix}\}$ can be considered as the connected component of the identity of the isometry group of S^3_1.

[1] introduced the notion of CMC-1 faces in S^3_1, which corresponds to maxfaces (i.e., maximal surfaces with admissible singularities, see [20]) in the Lorentz-Minkowski 3-space $\mathbf{R}^3_1$. As a generalization of CMC-1 faces, we define generalized CMC-1 faces as follows. (Later, we show that an extended hyperbolic metric induces a generalized CMC-1 face with a given hyperbolic Gauss map, see Theorems 2.10 and 2.11.) We fix a Riemann surface M.

Definition 1.1. A C^∞-map $f : M \to S^3_1$ is called a *generalized CMC-1 face* if there exists an open dense subset W of M such that the restriction $f|_W$ of f on W gives a conformal (space like) immersion of constant mean curvature one.

A *singular point* of f is a point at which f is not an immersion. A singular point p satisfying $df(p)=0$ is called a *branch point* of f. Moreover, f is called a *CMC-1 face* if f does not have any branch points. (A CMC-1 face may have singular points in general).

Remark 1.2. The above definition of CMC-1 face is simpler than the definition given in [1,5]. However, as seen in the following Proposition 1.4, the new definition is equivalent to the previous one. Similarly, the definition of 'maxface' given in [20] can be simplified as follows: *A C^∞-map $f: M \to \mathbf{R}^3_1$ is a maxface if and only if there exists an open dense subset W of M such that the restriction $f|_W$ of f on W gives a conformal (space like) maximal immersion and df has no zeros on M.* The proof is easier than for the case of CMC-1 faces.

To state the Weierstrass-type representation formula, we prepare some notions:

Definition 1.3. A pair (G,Q) of a meromorphic function G and a holomorphic 2-differential Q on M is said to be *definite* (resp. *semi-definite*) if

$$ds^2_\# = \left(1+|G|^2\right)^2 \left|\frac{Q}{dG}\right|^2 \tag{5}$$

is a positive definite (resp. positive semi-definite) metric on M.

We denote by $\widetilde{M}$ the universal cover of M and by

$$\pi: \widetilde{M} \longrightarrow M$$

its covering projection.

Proposition 1.4. *Let (G,Q) be a semi-definite pair on M and let $F=(F_{ij}): \widetilde{M} \to \mathrm{SL}(2,\mathbf{C})$ be a holomorphic map of $\widetilde{M}$ such that*

$$(dF)F^{-1} = \Psi \qquad \left(\Psi := \begin{pmatrix} G & -G^2 \\ 1 & -G \end{pmatrix} \frac{Q}{dG}\right). \tag{6}$$

Then F is a null holomorphic map, that is, F is a holomorphic map such that $\det(dF/dz)$ vanishes identically for each local complex coordinate z on M. And

$$f := Fe_3F^* : \widetilde{M} \longrightarrow S^3_1 \tag{7}$$

is a generalized CMC-1 face if $|g|$ is not identically 1, where g is a meromorphic function on $\widetilde{M}$ (called the secondary Gauss map) defined by

$$g := -\frac{dF_{12}}{dF_{11}} = -\frac{dF_{22}}{dF_{21}}. \tag{8}$$

The induced metric ds^2 *and the second fundamental form* II *are expressed as*

$$ds^2 = (1-|g|^2)^2 \left|\frac{Q}{dg}\right|^2, \qquad \mathrm{II} = Q + \overline{Q} + ds^2, \tag{9}$$

respectively. (The meromorphic function G *and the holomorphic* 2*-differential* Q *are called the hyperbolic Gauss map and the Hopf differential, respectively.)*

Moreover, f *is a CMC-*1 *face if and only if* F *is an immersion, that is,* (G,Q) *is a definite pair on* M*.*

*Conversely, any non-totally-umbilical generalized CMC-*1 *face defined on* $\widetilde{M}$ *with hyperbolic Gauss map* G *and Hopf differential* Q *is obtained in this manner. (Later, we give a necessary and sufficient condition for* f *to be single-valued on* M*; see Proposition 1.9.)*

Proof. When the pair (G,Q) is definite, the assertion has been proved in [6, Proposition 4.2]. So we assume (G,Q) is semi-definite. Since the solution F of the ordinary differential Equation (6) is analytic, the local existence of F implies the existence of the solution on $\widetilde{M}$. Thus, it is sufficient to show that Ψ (as in Eq. (6)) is a holomorphic matrix-valued 1-form on M. In fact, under the assumption that Ψ is holomorphic, one can directly check that $f = Fe_3F^*$ is a conformal CMC-1 immersion at p if $|g(p)| \neq 1$ and $ds^2_\#$ as in (5) is positive definite at p (cf. [6, Proposition 4.2]).

We fix a point $p \in M$ arbitrarily. If G is holomorphic at p, then the boundedness of $ds^2_\#$ at p implies that Q/dG is holomorphic at p, and so is Ψ. So we consider the case that G has a pole at p. Then $\hat{G} = 1/G$ is holomorphic at p. Since

$$ds^2_\# = (1+|\hat{G}|^2)^2 \left|\frac{Q}{d\hat{G}}\right|^2$$

holds, we can conclude that $Q/d\hat{G}$ is holomorphic at p. Moreover, we have the following expression

$$\Psi = \begin{pmatrix} -\hat{G} & 1 \\ -\hat{G}^2 & \hat{G} \end{pmatrix} \frac{Q}{d\hat{G}},$$

which implies that Ψ is holomorphic at p also in this case. Here, we have shown that Ψ is holomorphic even when (G,Q) is semi-definite. Moreover, it holds that

$$(\partial f)f^{-1} = \partial(Fe_3F^*)((F^*)^{-1}e_3F^{-1}) = (dF)F^{-1} = \Psi, \tag{10}$$

which implies that f is a CMC-1 face if and only if (G,Q) is a definite pair on M, unless Q vanishes identically making f totally umbilical, where we consider S^3_1 a set of 2×2 matrices as in Eq. (2) and f a matrix-valued function. Now, one can prove all of the remaining assertions except the converse part by imitating the proof of [6, Proposition 4.2].

So we now prove the converse assertion: Let $f : M \to S^3_1$ be a generalized CMC-1 face. By definition, there exists an open dense subset W of M such that $f|_W$ is a conformal space-like CMC-1 immersion. It is well-known that f can be lifted to a null holomorphic immersion $F : \widetilde{W} \to \mathrm{SL}(2,\mathbf{C})$, where $\widetilde{W}$ is the universal covering of W. By Eq. (10), the identity

$$(\partial f)f^{-1} = (dF)F^{-1}$$

holds on W. Since W is dense, the $\mathrm{SL}(2,\mathbf{C})$-valued 1-form $(\partial f)f^{-1}$ is holomorphic on M. Then there exists a holomorphic map $\widetilde{F} : \widetilde{M} \to \mathrm{SL}(2,\mathbf{C})$ such that $(d\widetilde{F})\widetilde{F}^{-1}$ is equal to $(\partial f)f^{-1}$. Since df vanishes if ∂f does as well, $\widetilde{F}$ is an immersion if and only if df never vanishes on M. □

A generalized CMC-1 face is totally umbilical if and only if its image lies in an S^3_1-horosphere (cf. [5]). To avoid this exceptional case, we assume that the Hopf differential Q of the generalized CMC-1 face does not vanish identically, in Sect. 1–3.

Definition 1.5. For a generalized CMC-1 face f obtained as in Proposition 1.4, the null holomorphic map F in Eq. (7) is called a *null holomorphic lift* of f. The metric $ds^2_\#$ as in Eq. (5) is called the *dual metric* of f.

Remark 1.6. Let f be a generalized CMC-1 face obtained from given (G,Q) using Proposition 1.4. Then

1. The Eq. (6) should be regarded as an equation on the universal cover $\widetilde{M}$ (see the appendix in [6]). However, for simplicity, we use the notation here.
2. For each $a \in \mathrm{SL}(2,\mathbf{C})$, $f_a := afa^*$ gives a generalized CMC-1 face which is congruent to f [cf. Eq. (4)]. For a null holomorphic lift F of f, $F_a := aF$ is a null holomorphic lift of f_a. In particular, the hyperbolic Gauss map and the Hopf differential of f_a are $a \star G$ and Q, respectively, where

$$a \star G := \frac{a_{11}G + a_{12}}{a_{21}G + a_{22}} \qquad \left(a = (a_{ij})\right)$$

denotes the Möbius transformation (cf. [18]).

3. The choice of a null holomorphic lift F of f has the ambiguity $F \mapsto Fb^{-1}$ for $b \in \mathrm{SU}(1,1)$. Under this change, the secondary Gauss map g is transformed as

$$b \star g = \frac{b_{11}g + b_{12}}{b_{21}g + b_{22}} \qquad \left(b = (b_{ij})\right).$$

4. If $f : M \to S^3_1$ is a generalized CMC-1 face, the singular set of f is given by

$$\Sigma_f := \{\pi(p) \in M\,;\, |g(p)| = 1 \ (p \in \widetilde{M})\} \cup \{q \in M\,;\, df(q) = 0\}. \tag{11}$$

The condition in Proposition 1.4 that $|g|$ is not identical to 1 is necessary to avoid the example all of whose points are singular points of f. Such an example is unique up to isometry, whose image is a light-like line in S^3_1 (see [5, Remark 1.3]).

5. The metric

$$d\sigma^2 := \frac{4\,dg\,d\bar{g}}{(1-|g|^2)^2}$$

has the expression (cf. [5, Remark 1.10])

$$d\sigma^2 := K ds^2 \quad (K\text{: Gaussian curvature of } ds^2),$$

and has constant Gaussian curvature -1. It is the pull-back of the spherical Poincaré metric by $g : \widetilde{M} \to \mathbf{C} \cup \{\infty\}$ (cf. the Introduction and Sect. 2). In other words, $d\sigma^2$ is an extended hyperbolic metric whose developing map is g (see Sect. 3 for details). The identity

$$d\sigma^2 ds^2 = 4Q\overline{Q} = d\sigma_\#^2 ds_\#^2 \tag{12}$$

holds, where

$$d\sigma_\#^2 := \frac{4\,dG\,d\overline{G}}{(1+|G|^2)^2}. \tag{13}$$

These holomorphic data are related by

$$S(g) - S(G) = 2Q, \tag{14}$$

where

$$S(h) := S_z(h) dz^2, \qquad S_z(h) := \left(\frac{h''}{h'}\right)' - \frac{1}{2}\left(\frac{h''}{h'}\right)^2 \qquad \left({}' = \frac{d}{dz}\right). \tag{15}$$

Here z is a local complex coordinate on M and $S(\cdot)$ is the Schwarzian derivative. The Schwarzian derivative has the property that $S(a \star h) = S(h)$ for each $a \in \mathrm{SL}(2,\mathbf{C})$. The difference $S(g) - S(G)$ of the Schwarzian derivatives of two meromorphic functions g and G on a given Riemann surface M does not depend on a choice of local complex coordinate z. Fundamental properties of the Schwarzian derivative are given in the appendix of [17].

On the other hand, the null holomorphic lift $\pm F$ can be expressed using G and g by

$$F = \begin{pmatrix} G\dfrac{da}{dG} - a & G\dfrac{db}{dG} - b \\ \dfrac{da}{dG} & \dfrac{db}{dG} \end{pmatrix}, \quad a := \sqrt{\frac{dG}{dg}}, \quad b = -ga, \tag{16}$$

which is called the *Bianchi-Small formula*. (Although a (and so F also) has $\pm$-ambiguity, the CMC-1 face $f = \pi_S \circ F = Fe_3F^*$ is uniquely determined.)

Proposition 1.7. *The set of zeros of the Hopf differential Q of a given generalized CMC-1 face $f : M \to S^3_1$ coincides exactly with the union of the set of umbilics and the set of branch points of f.*

Proof. Let p be a fixed point of M. By a suitable motion of S^3_1, we may assume that the hyperbolic Gauss map G is holomorphic at p. Then, p is a branch point of f if and only if Q/dG vanishes at p, in particular $Q(p) = 0$. It is well-known that p is an umbilic point of f if and only if Q vanishes and Q/dG does not vanish at p. So we get the assertion. □

Here, we give the following additional result on branch points.

Proposition 1.8. *Let $f : M \to S^3_1$ be a generalized CMC-1 face, then the following three conditions are equivalent:*

1. *$p \in M$ is a branch point of f.*
2. *$F : \widetilde{M} \to \mathrm{SL}(2, \mathbf{C})$ is not an immersion at p.*
3. *$ds^2_\#$ degenerates at p.*

Proof. The equivalency of the first two conditions is obvious. Since $ds^2_\#$ is the Hermitian metric induced by the inverse matrix F^{-1}, so the assertion follows from the fact that F^{-1} is an immersion if and only if F is as well. □

The Monodromy Representation of Generalized CMC-1 Faces

For a semi-definite pair (G, Q) on M, there exists a representation $\rho_F : \pi_1(M) \to \mathrm{SL}(2, \mathbf{C})$ associated to the solution F of Eq. (6) as in the appendix in [6]:

$$F \circ T = F \rho_F(T), \tag{17}$$

where $T \in \pi_1(M)$ is an element of the fundamental group considered as a covering transformation of the universal cover $\widetilde{M}$. Let $f : \widetilde{M} \to S^3_1$ be a generalized CMC-1 face and $g : \widetilde{M} \to \mathbf{C} \cup \{\infty\}$ the secondary Gauss map of f as in Eq. (8). By Eq. (14), $S(g)$ is a projective connection on M (cf. Appendix A), since (G, Q) are defined on M. As a consequence, there exists a group representation

$$\rho_g : \pi_1(M) \longrightarrow \mathrm{PSL}(2, \mathbf{C}) := \mathrm{SL}(2, \mathbf{C})/\{\pm e_0\}, \qquad e_0 := \begin{pmatrix} 1 & 0 \\ 0 & 1 \end{pmatrix}$$

such that

$$g \circ T^{-1} = \rho_g(T) \star g, \tag{18}$$

where $T \in \pi_1(M)$ is a covering transformation. Since $a \star g = (-a) \star g$ for $a \in \mathrm{SL}(2, \mathbf{C})$, $a \star g$ is well-defined for $\mathrm{PSL}(2, \mathbf{C})$. Let

$$\hat{\pi} : \mathrm{SL}(2, \mathbf{C}) \longrightarrow \mathrm{PSL}(2, \mathbf{C}) \tag{19}$$

be the double covering as a group homomorphism. Then it holds that

$$\hat{\pi} \circ \rho_F = \rho_g \tag{20}$$

(see [5, (1.12) and (1.13)]). One can easily prove the following criteria for a given generalized CMC-1 face $f : \widetilde{M} \to S^3_1$ as in Proposition 1.4 to be single-valued on M:

Proposition 1.9. *Let $f : \widetilde{M} \to S^3_1$ be a generalized CMC-1 face associated to a semi-definite pair (G, Q) on M. Then the following conditions are mutually equivalent:*

1. *f is single-valued on M.*
2. *$\rho_F(\pi_1(M)) \subset \mathrm{SU}(1,1)$.*
3. *$\rho_g(\pi_1(M)) \subset \mathrm{PSU}(1,1) := \mathrm{SU}(1,1)/\{\pm e_0\}$.*

Completeness and Reducibility of CMC-1 Faces

We define completeness and weak completeness of CMC-1 faces.

Definition 1.10 ([5, Definitions 1.2 and 1.3]). A CMC-1 face $f : M \to S^3_1$ is called *complete* if there exists a symmetric 2-tensor T which vanishes outside some compact set in M, such that $ds^2 + T$ is a complete Riemannian metric on M, where ds^2 is the induced metric by f as in Eq. (9). On the other hand, f is called *weakly complete* if the metric $ds^2_\#$ in Eq. (5) is complete.

The following assertion is useful:

Fact 1.11 ([21, 22] and Okuyama–Yamanoi [15]). *Let $f : M \to S^3_1$ be a weakly complete CMC-1 face. Then f is complete if and only if there exist a closed Riemann surface $\overline{M}$ and a finite number of points $p_1, \dots, p_n \in \overline{M}$ such that M is conformally equivalent to $\overline{M} \setminus \{p_1, \dots, p_n\}$, and the set of singular points Σ_f is compact.*

Remark 1.12. This fact was proved in [22] under the assumption that the Hopf differential of f is meromorphic. The last two authors pointed out in [21, 22] that the assumption of the Hopf differential can be removed if one can establish a certain generalization of the completeness lemma in minimal surface theory. Recently, Okuyama-Yamanoi [15] accomplished this, and as a consequence Fact 1.11 is obtained.

Let f be a complete CMC-1 face. Each point p_j is called an *end* of f. A complete end p_j is said to be *regular* if the hyperbolic Gauss map G has at most a pole at p_j, which is equivalent to the order of the Hopf differential Q at p_j being at least -2.

Next, we recall the reducibility of CMC-1 faces:

Definition 1.13. A CMC-1 face f is called *irreducible* (resp. *reducible*) if the image of the representation ρ_g is not (resp. is) an abelian subgroup of $\mathrm{PSU}(1,1)$. When f is reducible, it is called 3-*reducible* if ρ_g is a trivial representation (i.e., the image of ρ_g coincides with $\{\pm e_0\}$), and f is called 1-*reducible* if the image of ρ_g is abelian and not equal to $\{\pm e_0\}$. In particular, ρ_g is 3-reducible if and only if g is single-valued on M.

In the case of CMC-1 surfaces in the hyperbolic 3-space H^3, 1-reducible (resp. 3-reducible) corresponds to the terminology H^1-reducible (resp. H^3-reducible). If f is a 1-reducible (resp. 3-reducible) CMC-1 surface in S^3_1, then f has a 1-parameter family (resp. 3-parameter family) of deformations of f preserving the hyperbolic Gauss map G and the Hopf differential Q. The numbers 1 and 3 for reducibility come from the numbers of these freedoms, as follows:

Theorem 1.14. *Let $f : M \to S^3_1$ be a CMC-*1 *face with a given hyperbolic Gauss map G and Hopf differential Q. Then f is uniquely determined if f is irreducible. On the other hand, if f is* 3*-reducible (resp.* 1*-reducible), then there is a* 3*-parameter family (resp.* 1*-parameter family) of CMC-*1 *faces (as mappings of M into S^3_1) having the same hyperbolic Gauss map G and Hopf differential Q as f.*

Proof. If we replace F by Fa for $a \in \mathrm{SL}(2,\mathbf{C})$, then f changes to $Fae_3a^*F^*$, and this preserves G and Q. On the other hand, a CMC-1 face with the same hyperbolic Gauss map G and Hopf differential Q as f is of the form $Fae_3a^*F^*$ for some $a \in \mathrm{SL}(2,\mathbf{C})$. Furthermore, $Fae_3a^*F^*$ is single-valued on M if and only if the monodromy matrix of Fa belongs to $\mathrm{SU}(1,1)$. Since

$$(Fa)\circ T = F\rho_F(T)a = (Fa)(a^{-1}\rho_F(T)a),$$

$Fae_3a^*F^*$ is single-valued on M if and only if $\hat{\pi}(a) \in \mathrm{PSL}(2,\mathbf{C})$ belongs to C_Γ, where

$$C_\Gamma := \{\sigma \in \mathrm{PSL}(2,\mathbf{C})\,;\, \sigma\Gamma\sigma^{-1} \in \mathrm{PSU}(1,1)\} \qquad (\Gamma := \rho_g(\pi_1(M))).$$

If $\sigma \in C_\Gamma$, it is obvious that $a\sigma \in C_\Gamma$ for all $a \in \mathrm{PSU}(1,1)$. Then the left quotient space $I_\Gamma := \mathrm{PSU}(1,1)\backslash C_\Gamma$ can be considered as a subset of S^3_1 which parametrizes the CMC-1 faces with given (G,Q). As shown in Appendix B, I_Γ is a point if f is irreducible. I_Γ coincides with S^3_1 if f is 3-reducible, and is a geodesic line of S^3_1 if f is 1-reducible, which proves the assertion. □

Remark 1.15. The deformation of CMC-1 faces preserving (G,Q) as in Theorem 1.14 is not an isometric deformation in general. However, it gives the same image of a CMC-1 face in special cases. For example, if $f : \mathbf{C} \to S^3_1$ is an S^3_1-horosphere [cf. Eq. (29)], then it is 3-reducible, since $\mathbf{C}$ is simply connected. In this case, $f = Fe_3F^*$ and $Fae_3a^*F^*$ ($a \in \mathrm{SL}(2,\mathbf{C})$) are both totally umbilical, and thus they are congruent. Similarly, an S^3_1-catenoid $f : \mathbf{C}\setminus\{0\} \to S^3_1$ is reducible which admits a deformation which fixes the image of the surface (see Sect. 4).

For any real number t and $\varepsilon \in \{1,-1\}$, we set

$$E(t) := \begin{pmatrix} e^{it} & 0 \\ 0 & e^{-it} \end{pmatrix}, \quad P(\varepsilon) := \begin{pmatrix} 1+i\varepsilon & -i\varepsilon \\ i\varepsilon & 1-i\varepsilon \end{pmatrix}, \quad H(t) := \begin{pmatrix} \cosh t & \sinh t \\ \sinh t & \cosh t \end{pmatrix}.$$

A matrix in $\mathrm{SU}(1,1)$ is called

- *Elliptic* if it is conjugate to $E(t)$ for some $t \in (-\pi, \pi]$ in $\mathrm{SU}(1,1)$.
- *Parabolic* if it is conjugate to $\pm P(\pm 1)$ in $\mathrm{SU}(1,1)$.
- *Hyperbolic* if it is conjugate to $\pm H(t)$ for some $t > 0$ in $\mathrm{SU}(1,1)$.

Any matrix in $\mathrm{SU}(1,1)$ is one of these three types (see the appendix in [6]). Note that the parabolic matrices $P(1)$ and $P(-1)$ are not conjugate in $\mathrm{SU}(1,1)$. Since a and $-a$ have the same properties for each $a \in \mathrm{SU}(1,1)$, the ellipticity, the parabolicity, and the hyperbolicity are also well-defined for each element of $\mathrm{PSU}(1,1)$.

Singularities

Let $f\colon M \to S^3_1$ be a CMC-1 face whose secondary Gauss map and Hopf differential are g and Q, respectively.

Fact 1.16 ([4, Theorem 3.4], [6, Lemma 2.4]). *Define two meromorphic functions by*

$$\alpha := \frac{dg}{g^2\omega}, \qquad \beta := g\frac{d\alpha}{dg} \qquad \left(\omega = \frac{Q}{dg}\right).$$

Then

- *$p \in M$ is a singular point if and only if $|g(p)| = 1$. Moreover, p is a nondegenerate singular point if and only if $dg(p) \neq 0$.*
- *f gives a front (i.e., wave front; see [12] for the definition of front) on a neighborhood of p if and only if $\mathrm{Re}\,\alpha \neq 0$ at p, where $\mathrm{Re}\,\alpha$ denotes the real part of α.*
- *p is a cuspidal edge if and only if $\mathrm{Re}\,\alpha \neq 0$ and $\mathrm{Im}\,\alpha \neq 0$ hold at p, where $\mathrm{Im}\,\alpha$ denotes the imaginary part of α.*
- *p is a swallowtail if and only if $\mathrm{Re}\,\alpha \neq 0$, $\mathrm{Im}\,\alpha = 0$ and $\mathrm{Re}\,\beta \neq 0$ hold at p.*
- *p is a cuspidal cross cap if and only if $\mathrm{Re}\,\alpha = 0$, $\mathrm{Im}\,\alpha \neq 0$ and $\mathrm{Im}\,\beta \neq 0$ hold at p.*
- *The singular set Σ_f consisting of nondegenerate singular points is a cone-like singularity if and only if Σ_f is compact and $\mathrm{Im}\,\alpha = 0$ holds on Σ_f.*

Remark 1.17. Though Lemma 2.4 in [6] gives a criteria for cone-like singularities of maximal surfaces in Lorentz-Minkowski 3-space $\mathbf{R}^3_1$, one can easily show that it is also a criteria for CMC-1 faces.

As shown in [4], cuspidal edges, swallowtails and cuspidal cross caps are the generic singularities of CMC-1 faces in the compact open C^∞-topology.

HMC-1 Surfaces as Unit Normal Vector Fields of CMC-1 Faces

At the end of this section, we discuss the behavior of unit normal vector fields of CMC-1 faces. Let $f : M \to S^3_1$ be a CMC-1 face and $F : \widetilde{M} \to \mathrm{SL}(2,\mathbf{C})$ its null holomorphic lift. Then the unit normal vector field ν has the following expression (see [1, Remark 1.2]):

$$\nu := \frac{1}{|g|^2-1} F \begin{pmatrix} 1+|g|^2 & 2g \\ 2\bar{g} & 1+|g|^2 \end{pmatrix} F^* : M \setminus \Sigma_f \longrightarrow H^3_+ \cup H^3_-, \tag{21}$$

where g is the secondary Gauss map of f, Σ_f is the set of singular points, and

$$H^3_\pm := \{X = (x_0,x_1,x_2,x_3) \in \mathbf{R}^4_1 \,;\, \langle X,X\rangle = -1, \pm x_0 > 0\}$$

are the two components of a two-sheeted hyperboloid in $\mathbf{R}^4_1$. As pointed out in [11], when the unit normal vector field ν of a CMC-1 face f meets the singular set $\Sigma_f = \{|g| = 1\}$, the image of ν moves into the other sheet of the hyperboloid $H^3_+ \cup H^3_-$. Moreover, it was shown in [11, Theorem 4.2] that ν is smooth at the singular set under a certain compactification of $H^3_+ \cup H^3_-$ as follows: The *hyperbolic* 3*-sphere* S^3_H is a 3-dimensional manifold diffeomorphic to the 3-sphere:

$$S^3_H := \mathbf{R}^3 \cup \{\infty\} \cong S^3$$

endowed with the metric $4|dx|^2/(1-|x|^2)^2$ on $S^3 \setminus \{\text{the equator}\}$, where $x := (x_1,x_2,x_3) \in \mathbf{R}^3 \cup \{\infty\}$. We consider the stereographic projection

$$\varphi : H^3_+ \cup H^3_- \ni (x_0,x_1,x_2,x_3) \longmapsto \frac{(x_1,x_2,x_3)}{1-x_0} \in S^3_H, \tag{22}$$

which is an isometric embedding, and S^3_H can be considered as a compactification of $H^3_+ \cup H^3_-$. Thus, the unit normal vector field as in Eq. (21) induces a smooth map:

$$\nu : M \longrightarrow S^3_H.$$

We now give the following definition:

Definition 1.18. A C^∞-map $\nu : M \to S^3_H$ is called an *HMC*-1 *face* (i.e. harmonic-mean curvature 1 face) if it is a unit normal vector field of a CMC-1 face $f : M \to S^3_1$.

As pointed out in [11], such a ν actually has the property that the harmonic mean

$$HM := \left(\frac{(\lambda_1)^{-1}+(\lambda_2)^{-1}}{2}\right)^{-1} = \frac{2\lambda_1\lambda_2}{\lambda_1+\lambda_2}$$

of the two principal curvatures λ_1, λ_2 of ν is identically equal to 1. In Sect. 4, we classify S^3_1-catenoids, and then we will also comment on the associated HMC-1 faces that are their unit normal vector fields (see Fig. 7 in Sect. 4).

2 Extended Hyperbolic Metrics on Riemann Surfaces

Let M be a Riemann surface.

Definition 2.1. A C^∞-metric $d\sigma^2$ defined on an open dense subset W of M is called an *extended hyperbolic metric* on M (or sometimes just called a *hyperbolic metric* for the sake of brevity) if it satisfies the following two properties:

1. $d\sigma^2$ is a Hermitian metric of constant curvature -1 on W.
2. There exists a discrete subset S of M such that for each local complex coordinate (U,z) of $M\setminus S$, $e^{-\omega}$ gives a smooth function on $U\setminus S$, where we use a local expression
$$d\sigma^2 = e^{\omega}|dz|^2 \qquad (|dz|^2 := dz\,d\bar{z})$$
on $U\cap W$. Moreover,
$$h(z) := \omega_{zz} - \frac{(\omega_z)^2}{2} \tag{23}$$
can be extended to a holomorphic function on $U\setminus S$.

The word "extended" expresses that the hyperbolic metric might have not only isolated singularities, but also singularities consisting of curves. We choose this terminology, since there are already a number of notions of "generalized hyperbolic metrics".

The first condition is independent of the second condition. In fact, if $e^{\omega}|dz|^2$ is positive definite, the holomorphicity of $h(z)$ implies that $e^{\omega}|dz|^2$ has constant curvature, but one cannot specify that constant. The holomorphic 2-differential $h(z)\,dz^2$ defined on each local complex coordinate induces a projective connection (cf. Appendix A) on $U\setminus S$. The smoothness of $e^{-\omega}$ is required since $U\setminus W$ may be disconnected and the extended hyperbolic metric associated to a given projective connection may not be uniquely determined in general, because of the ambiguity of Möbius transformations of the developing map [cf. Eq. (24) and Appendix A].

Let p be a point in the discrete subset S as above. Then by our definition of extended hyperbolic metric, p is an isolated singularity of the function $h(z)$ [cf. Eq. (23)] defined on a local complex coordinate (U,z) around p. If p is not a removable singular point of $h(z)$, then p is called a *proper singular point*. Moreover, if $h(z)$ has at most a pole of order 2 at $z=p$, then p is called a *regular singular point* of $d\sigma^2$. An extended hyperbolic metric is *proper singularity free* (or *PS-free*) if it has no proper singular points (i.e. $h(z)$ is holomorphic).

Definition 2.2. Two extended hyperbolic metrics on a Riemann surface M are *isometric* if one is obtained as the pull-back of the other by a holomorphic or anti-holomorphic automorphism on M.

It is well-known that there is a unique Hermitian metric of constant curvature -1 defined on an arbitrary closed Riemann surface M of genus greater than one, which can then be considered as a PS-free extended hyperbolic metric. General

PS-free extended hyperbolic metrics on closed Riemann surfaces are discussed in Goldman [10].

We now consider a metric on $S^2 = \mathbf{C} \cup \{\infty\}$, called a *spherical Poincaré metric*, defined by

$$d\sigma_P^2 := \frac{4|dz|^2}{(1-|z|^2)^2},$$

which is a Hermitian metric of constant Gaussian curvature -1 defined on $\{z \in \mathbf{C} \cup \{\infty\};\ |z| \neq 1\}$. The metric $d\sigma_P^2$ is a PS-free extended hyperbolic metric defined on $\mathbf{C} \cup \{\infty\}$. In fact, if we set

$$\omega = \log\left(\frac{4}{(1-|z|^2)^2}\right),$$

then $\omega_{zz} - (\omega_z)^2/2$ vanishes identically. We call the pair

$$S_H^2 := (\mathbf{C} \cup \{\infty\}, d\sigma_P^2)$$

the *hyperbolic sphere*, which was introduced in [11], but has already appeared in Goldman [10, Sect. 2] and [20]. The hyperbolic sphere S_H^2 can be considered as an attachment of two hyperbolic planes at their ideal boundaries, as discussed in the Introduction. The orientation-preserving isometry group $\mathrm{Isom}_+(S_H^2)$ of $d\sigma_P^2$ is generated by $\mathrm{PSU}(1,1)$ and the matrices $\pm\begin{pmatrix} 0 & \mathrm{i} \\ \mathrm{i} & 0 \end{pmatrix}$.

We fix an extended hyperbolic metric $d\sigma^2$ on a connected Riemann surface M. Let $\{(U_\lambda, z_\lambda)\}_{\lambda \in \Lambda}$ be a covering of M consisting of local complex coordinates such that $d\sigma^2 = \exp(\omega^\lambda)|dz_\lambda|^2$, where $\omega^\lambda\ (\lambda \in \Lambda)$ is a C^∞-function on U_λ. We set

$$P_\lambda := \left(\omega^\lambda_{z_\lambda z_\lambda} - \frac{(\omega^\lambda_{z_\lambda})^2}{2}\right)(dz_\lambda)^2 \qquad (\lambda \in \Lambda), \tag{24}$$

which gives a projective connection (cf. Appendix A) defined on $M \setminus S$, where S is the set of proper singular points of $d\sigma^2$. We call $P = \{P_\lambda\}_{\lambda \in \Lambda}$ the *projective connection* induced by $d\sigma^2$, and denote it by

$$S(d\sigma^2) := \{P_\lambda\}_{\lambda \in \Lambda}.$$

In fact, the projective connection P can be considered as a Schwarzian derivative of $d\sigma^2$ because of the identity

$$S_\lambda(g)(dz_\lambda)^2 = P_\lambda \qquad (\lambda \in \Lambda),$$

where g is a developing map of $d\sigma^2$ (see Theorem 2.3).

Theorem 2.3. *Let $d\sigma^2$ be an extended hyperbolic metric on a connected Riemann surface M whose proper singular set is S. Then there exists a meromorphic function g defined on the universal covering space $\pi : \widetilde{M \setminus S} \to M \setminus S$ such that*

$$\pi^* d\sigma^2 = g^* d\sigma_P^2.$$

(Such a meromorphic function g is called a developing map of $d\sigma^2$.)

Conversely, let S be a discrete subset of M and g a meromorphic function on $\widetilde{M \setminus S}$ such that

$$d\sigma^2 := \frac{4\,dg\,d\bar{g}}{(1-|g|^2)^2}$$

gives a positive definite metric defined on an open dense subset W of M. Then $d\sigma^2$ is an extended hyperbolic metric on M, whose developing map is g.

To prove the assertion, we need the following:

Lemma 2.4. *Let g_1 and g_2 be two nonconstant meromorphic functions on a Riemann surface M such that there exists a matrix $a \in \mathrm{SL}(2,\mathbf{C})$ satisfying $g_2 = a \star g_1$. If there exists a neighborhood U of a point $p \in M$ such that*

$$\{q \in U\,;\, |g_1(q)| = 1\} = \{q \in U\,;\, |g_2(q)| = 1\},$$

and if p is not a branch point of g_1 and $|g_1(p)| = 1$, then $\hat{\pi}(a) \in \mathrm{Isom}_+(S^2_H)$.

Proof. Since p is not a branch point of g_1, we may assume that $z = g_1$ gives a local complex coordinate on U. Then we have that

$$\{z \in U\,;\, |z| = 1\} = \{z \in U\,;\, |a \star z| = 1\},$$

which implies that $\hat{\pi}(a) \in \mathrm{Isom}_+(S^2_H)$. □

Proof of Theorem 2.3. We take the maximal open dense subset W of $M \setminus S$ so that $d\sigma^2$ is a Hermitian metric of constant curvature -1 on W. Let $\{W_\lambda\}_{\lambda \in \Lambda}$ be the connected components of W. We take a reference point $p_\lambda \in W_\lambda$, and fix a simply connected local complex coordinate (U, z) centered at p_λ such that $U \subset W_\lambda$. Then there exists a holomorphic function

$$g_\lambda : U \longrightarrow \mathcal{D} = \{w \in \mathbf{C}\,;\, |w| < 1\}$$

such that $g_\lambda^* d\sigma_P^2$ coincides with $d\sigma^2$ on U. Then it can be easily checked that

$$S_z(g_\lambda) = \omega_{zz} - \frac{(\omega_z)^2}{2}$$

holds, where we set $d\sigma^2 = e^{\omega}|dz|^2$ on U.

On the other hand, let

$$g_0 : \widetilde{M \setminus S} \to \mathbf{C} \cup \{\infty\}$$

be a developing map of the associated projective connection of $d\sigma^2$ (cf. Fact A.3 of Appendix A). By definition, it holds that

$$S_z(g_0) = \omega_{zz} - \frac{(\omega_z)^2}{2} \; (= S_z(g_\lambda)) \, .$$

Then there exists a matrix $a_\lambda \in \mathrm{SL}(2, \mathbf{C})$ such that $g_\lambda = a_\lambda \star g_0$ on U. So if we set

$$\tilde{g}_\lambda := a_\lambda \star g_0,$$

then $\tilde{g}_\lambda$ is a meromorphic function defined on $\widetilde{M \setminus S}$. Regarding (U, z) as a local coordinate of $\widetilde{M \setminus S}$, $\tilde{g}_\lambda$ is a meromorphic extension of $g_\lambda \circ \pi$, where $\pi : \widetilde{M \setminus S} \to M \setminus S$ is the covering projection. Since $1/d\sigma^2$ is smooth on $M \setminus S$ as a differential of type $(-1, -1)$ (cf. the property 2 in Definition 2.1 of extended hyperbolic metrics), the real analyticity of $d\sigma^2$ on W_λ implies that

$$1/d\sigma^2 = 1/\left(\tilde{g}_\lambda^* d\sigma_P^2\right) \tag{25}$$

holds on $\overline{W_\lambda}$ as a differential of type $(-1, -1)$. If $W_\lambda = W$, then $\tilde{g}_\lambda$ is the desired developing map of $d\sigma^2$. So we consider the case that W has at least two connected components. In this case, ∂W_λ is not discrete, and we can find a point $p \in \partial W_\lambda \setminus S$. Since the branch point of $\tilde{g}_\lambda$ is discrete, we may also assume that p is not a branch point of $\tilde{g}_\lambda$. If we take a sufficiently small neighborhood U of p, then $U \cap \partial W_\lambda$ is the set of points satisfying $|\tilde{g}_\lambda| = 1$. If $p \in \overline{W}_\lambda \cap \overline{W}_\mu$, then $|\tilde{g}_\lambda| = |\tilde{g}_\mu| = 1$ holds on $U \cap \partial W_\lambda$. By Lemma 2.4,

$$1/(\tilde{g}_\lambda^* d\sigma_P^2) = 1/(\tilde{g}_\mu^* d\sigma_P^2)$$

holds on $\overline{W_\lambda \cup W_\mu}$. Thus Eq. (25) holds on all of $M \setminus S$, and each $\tilde{g}_\lambda$ $(\lambda \in \Lambda)$ gives the desired developing map of $d\sigma^2$. The second statement of the theorem can be proved easily. □

Let

$$g : \widetilde{M \setminus S} \longrightarrow \mathbf{C} \cup \{\infty\}$$

be a developing map of an extended hyperbolic metric $d\sigma^2$ defined on a Riemann surface M. Since $\pi^* d\sigma^2 = g^* d\sigma_P^2$, g induces a group homomorphism

$$\rho_g : \pi_1(M \setminus S) \longrightarrow \mathrm{Isom}_+\left(S_H^2\right) (\subset \mathrm{PSL}(2, \mathbf{C}))$$

such that $g \circ T^{-1} = \rho_g(T) \star g$ holds for each $T \in \pi_1(M \setminus S)$. We call the group homomorphism ρ_g a *monodromy representation* of the extended hyperbolic metric

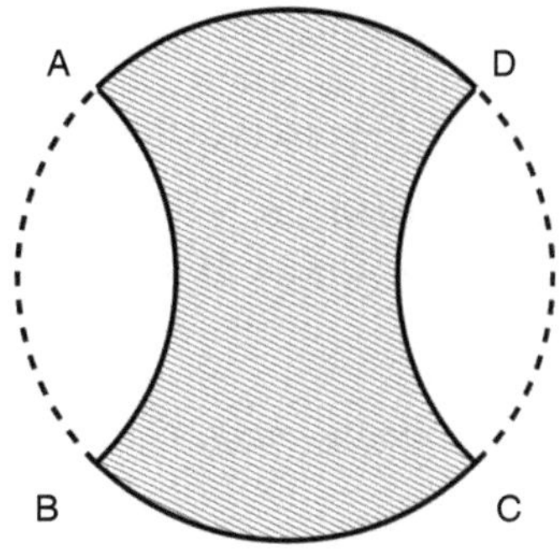

Fig. 1 Example 2.5

$d\sigma^2$. Since the developing map g is not uniquely determined, the representation ρ_g has an ambiguity of the conjugate actions in $\mathrm{Isom}_+(S^2_H)$. On the other hand, the subset

$$\partial^\infty(d\sigma^2) := \left\{\pi(p) \in M\setminus S;\, |g(p)| = 1 \;\left(p \in \widetilde{M\setminus S}\right)\right\} \tag{26}$$

does not depend on the choice of g, which is called the *ideal boundary set* of the metric $d\sigma^2$. Each point in $\partial^\infty(d\sigma^2)$ is called an *ideal boundary point*, or ∂^∞-point. By definition, $d\sigma^2$ is not defined at ∂^∞-points, and also not at proper singular points. The representation ρ_g canonically induces a new representation

$$\hat{\rho}_g : \pi_1(M\setminus S) \longrightarrow \mathbf{Z}_2 := \mathrm{Isom}_+(S^2_H)/\mathrm{PSU}(1,1).$$

If the induced representation $\hat{\rho}_g$ is trivial, that is, if

$$\rho_g\bigl(\pi_1(M\setminus S)\bigr) \subset \mathrm{PSU}(1,1)$$

holds, then the extended hyperbolic metric $d\sigma^2$ is called *co-orientable*. An extended hyperbolic metric which is not co-orientable is said to be *non-co-orientable*. By taking a double covering, a non-co-orientable extended hyperbolic metric becomes co-orientable (see Remark 2.12).

Example 2.5. Consider a 4-gon ABCD in the closed unit disk $\overline{\mathcal{D}} \subset \mathbf{C}$ as in Fig. 1 whose two edges AB and CD are complete geodesics in $\mathcal{D}$ as the Poincaré disk, and BC and DA lie in the ideal boundary of $\mathcal{D}$. Gluing each pair of edges, we get a PS-free extended hyperbolic metric $d\sigma_1^2$ on a torus. This metric $d\sigma_1^2$ is non-co-orientable, since the ideal boundary set is connected (see Proposition 2.18). Similarly, considering a $4k$-gon in $\overline{\mathcal{D}}$, we get a non-co-orientable extended hyperbolic metric defined on a closed Riemann surface of genus $k > 1$ whose ideal boundary set is connected. Moreover, taking its double covering, we also get a co-orientable PS-free hyperbolic metric defined on a closed Riemann surface of positive genus.

In this chapter, we mainly restrict our attention to co-orientable metrics. The following assertion can be proved easily by using the fact that the action of $\mathrm{PSU}(1,1)$ preserves the interior domain of the unit disk.

Proposition 2.6. *Let $d\sigma^2$ be a co-orientable extended hyperbolic metric on a Riemann surface M and*

$$g : \widetilde{M \setminus S} \longrightarrow \mathbf{C} \cup \{\infty\}$$

its developing map. Then there exist two disjoint subsets M_+ and M_- of $M \setminus S$ such that

$$M \setminus S = M_+ \cup \partial^\infty(d\sigma^2) \cup M_-,$$
$$M_+ = \left\{\pi(p) \in M \setminus S;\, |g(p)| < 1 \ \left(p \in \widetilde{M \setminus S}\right)\right\},$$
$$M_- = \left\{\pi(p) \in M \setminus S;\, |g(p)| > 1 \ \left(p \in \widetilde{M \setminus S}\right)\right\}.$$

Moreover, M_+ and M_- are independent of changes of the developing map $g \mapsto a \star g$ for $a \in \mathrm{PSU}(1,1)$, and the action of $\mathrm{Isom}_+(S^2_H) \setminus \mathrm{PSU}(1,1)$ exchanges M_+ and M_-.

If $d\sigma^2$ is non-co-orientable, the sets M_+ and M_- are not well defined.

Parallel to the case of CMC-1 faces, we define the reducibility of hyperbolic metrics.

Definition 2.7. A co-orientable extended hyperbolic metric $d\sigma^2$ (with a developing map g) is called *irreducible* (resp. *reducible*) if the image of the representation ρ_g is not (resp. is) an abelian subgroup of $\mathrm{PSU}(1,1)$. When $d\sigma^2$ is reducible, it is called 3-*reducible* if ρ_g is a trivial representation (i.e. the image of ρ_g coincides with $\{\pm e_0\}$), and $d\sigma^2$ is called 1-*reducible* if the image of ρ_g is abelian and not equal to $\{\pm e_0\}$. In particular, $d\sigma^2$ is 3-reducible if and only if g is single-valued on M.

The *ellipticity*, *parabolicity*, and *hyperbolicity* for a given regular singularity of the co-orientable extended hyperbolic metric are canonically defined as the property of the monodromy matrix.

Remark 2.8. The induced projective connection $S(d\sigma^2)$ on $M \setminus S$ with respect to a given co-orientable extended hyperbolic metric $d\sigma^2$ has a developing map g with $\mathrm{PSU}(1,1)$-monodromy representation, where S is the set of proper singular points of $d\sigma^2$. Conversely, a projective connection with only regular singularities whose monodromy representation ρ is conjugate to a subgroup in $\mathrm{PSU}(1,1)$ is induced from a co-orientable extended hyperbolic metric. Moreover, if the image of ρ is non-abelian, then the hyperbolic metric is uniquely determined from the projective connection. In particular, hyperbolic metrics can be treated as *projective connections with* $\mathrm{PSU}(1,1)$*-monodromy representation* (cf. [8, 10]). Projective connections without regular singularities on a closed Riemann surface M are called $\mathbf{C}P^1$-*structures* on M. Thus, several properties of co-orientable PS-free extended hyperbolic metrics on closed Riemann surfaces of genus greater than 1 are described in terms of $\mathbf{C}P^1$-structures. $\mathbf{C}P^1$-structures on closed surfaces with fixed Fuchsian representations as the holonomy homomorphism are precisely investigated in [10].

We now give a canonical correspondence between generalized CMC-1 faces and co-orientable extended hyperbolic metrics:

Definition 2.9. Let $f : M \to S^3_1$ be a generalized CMC-1 face defined on a Riemann surface M. Then f is called *admissible* if there exists a closed Riemann surface $\overline{M}$ satisfying the following properties:

- M is bi-holomorphic to $\overline{M} \setminus \{p_1, \ldots, p_n\}$, where $p_1, \ldots, p_n$ are distinct points in $\overline{M}$.
- The dual metric $ds^2_\#$ of f is complete at each p_j $(j = 1, \ldots, n)$.
- f is g-regular at each p_j in the sense of [5], namely, the Schwarzian derivative $S(g)$ of the secondary Gauss map g has regular singularities at $p_1, \ldots, p_n$ as a projective connection. In other words, it has at most a pole of order 2 at each p_j $(j = 1, \ldots, n)$.

We call $\overline{M}$ the *compactification* of M. The point p_j is called an *end* of f. And p_j is called *regular* (resp. *irregular*) if the hyperbolic Gauss map G of f has at most a pole at p_j. On the other hand, an end p_j is called a *complete* end if the first fundamental form ds^2_f of f is complete at p_j. Furthermore, an end p_j is called an *embedded* end if it is complete and f is properly embedded at p_j.

As shown in [5, Lemma 3.1], a complete CMC-1 face is admissible. By definition, an admissible generalized CMC-1 face is weakly complete if it has no branch points. In [19], the last two authors showed that the set of CMC-1 surfaces in hyperbolic 3-space H^3 with prescribed hyperbolic Gauss maps corresponds bijectively to the set of conformal metrics of constant curvature one with conical singularities. The following two theorems are an analogue of this correspondence for CMC-1 surfaces in S^3_1:

Theorem 2.10. *Let $f : M \to S^3_1$ be an admissible generalized CMC-1 face which is not totally umbilical, and let $\overline{M}$ be its compactification such that $M = \overline{M} \setminus \{p_1, \ldots, p_n\}$. We denote by ds^2_f the first fundamental form of f and K the Gaussian curvature of ds^2_f. Then*

$$d\sigma^2_f := K ds^2_f$$

gives a co-orientable extended hyperbolic metric on $\overline{M}$, whose developing map is the secondary Gauss map g. Moreover, it has the following properties:

1. *Each umbilical point p of f is a regular singular point of the hyperbolic metric $d\sigma^2_f$.*
2. *Moreover, the set of regular singular points of $d\sigma^2_f$ consists of umbilical points, branch points of f and $\{p_1, \ldots, p_n\}$.*

3. *The monodromy of* $d\sigma_f^2$ *at umbilical points and branch points are trivial, and the monodromy of each end* p_j *of* f *is elliptic, parabolic, or hyperbolic if* p_j *is a regular singular point of* $d\sigma_f^2$ *with elliptic, parabolic, or hyperbolic monodromy, respectively.*
4. *The non-branched singular points of* f *correspond to the ideal boundary points of* $d\sigma_f^2$.
5. $S(d\sigma_f^2) = S(g)$.

Proof. The property 5 of Remark 1.6 yields that $d\sigma_f^2$ is a Hermitian metric of constant curvature -1 on an open dense set of $\overline{M}$. Moreover, the secondary Gauss map g is a developing map of $d\sigma_f^2$ giving a co-orientable extended hyperbolic metric on $\overline{M}$.

Let p be an umbilical point of f. By a suitable motion in S_1^3, we may assume that G is holomorphic at p. By Proposition 1.7, p is a zero of Q. Since $ds_\#^2$ is positive definite, dG vanishes at p. So p is a branch point of G, and thus, $S(G)$ has a pole of order 2 at p. Thus by Eq. (14), $S(g)$ also has a pole of order 2 at p. Hence, p is a regular singularity of $d\sigma_f^2$, which proves the property 1.

Let $p \notin \{p_1, \dots, p_n\}$ be a regular singular point of $S(d\sigma_f^2) = S(g)$. Since Q is holomorphic at p, the formula (14) implies that $S(G)$ also has a pole at p, that is, p is a branch point of G. By a suitable motion in S_1^3, we may assume that G is holomorphic at p. Since $ds_\#^2$ given in Eq. (5) is bounded at p, Q must vanish at p, which proves the property 2. The property 3 follows from the fact that the monodromy representation of f coincides with that of the hyperbolic metric $d\sigma_f^2$, since the developing map of $d\sigma_f^2$ is the secondary Gauss map of f. The property 4 is obvious from Eq. (11). The property 5 is a consequence of the fact that the developing map of the metric $d\sigma_f^2$ is the secondary Gauss map g. □

We can prove the converse assertion under the following situation:

Theorem 2.11. *Let* $d\sigma^2$ *be a co-orientable extended hyperbolic metric on a closed Riemann surface* $\overline{M}$ *with regular singularities. Let* G *be a non-constant meromorphic function on* $\overline{M}$ *and set*

$$Q := \frac{S(d\sigma^2) - S(G)}{2},$$

which gives a meromorphic 2-differential of $\overline{M}$. *Suppose that* Q *does not vanish identically. Then there exist a finite set of points* $p_1, \dots, p_n$ *and a unique admissible generalized CMC-1 face*

$$f : \overline{M} \setminus \{p_1, \dots, p_n\} \longrightarrow S_1^3$$

such that:

1. G *and* Q *are the hyperbolic Gauss map and the Hopf differential of* f.
2. $d\sigma_f^2$ *coincides with the initially given metric* $d\sigma^2$. *In particular, the secondary Gauss map of* f *is the developing map of* $d\sigma^2$.

Proof. Let $p_1,\dots,p_n$ be points where the metric

$$ds^2_{\#} := (1+|G|^2)^2 \left|\frac{Q}{dG}\right|^2$$

diverges. (In fact, such points must be discrete.) Since G and Q are meromorphic, $ds^2_{\#}$ is complete at each p_j $(j=1,\dots,n)$. Let $f : \widetilde{M}_{p_1,\dots,p_n} \to S^3_1$ be a generalized CMC-1 face associated to (G,Q), where we set

$$M_{p_1,\dots,p_n} := \overline{M} \setminus \{p_1,\dots,p_n\},$$

and denote by $\pi : \widetilde{M}_{p_1,\dots,p_n} \to M_{p_1,\dots,p_n}$ its universal covering space. Then the monodromy representation ρ_F with respect to f coincides with the lift of the monodromy representation of $d\sigma^2$, and ρ_F takes values in $\mathrm{SU}(1,1)$. Thus $f = Fe_3F^*$ is single valued on $M \setminus \{p_1,\dots,p_n\}$, which proves the assertion. □

Remark 2.12. One can define *projective de Sitter space* $P^3_1 := S^3_1/\{\pm\}$ by identifying the de Sitter 3-space via the antipodal involution. Then non-co-orientable extended hyperbolic metrics are related to constant mean curvature one surfaces in P^3_1, namely, one can generalize Theorems 2.10 and 2.11 for non-co-orientable extended hyperbolic metrics by suitable modifications.

Definition 2.13. Let $\overline{M}$ be a closed Riemann surface, and let $d\sigma^2$ be a co-orientable extended hyperbolic metric on M whose proper singular points are $p_1,\dots,p_n$. Let

$$\tilde{\rho}_g : \pi_1(\overline{M} \setminus \{p_1,\dots,p_n\}) \longrightarrow \mathrm{SU}(1,1)$$

be the lift of the monodromy representation of $d\sigma^2$ as above. We denote by T_j $(j=1,\dots,n)$ the deck transformations corresponding to the anti-clockwise loops about the ends p_j. We call $\tilde{\rho}_g(T_j)$ the *monodromy matrix* of $d\sigma^2$ at p_j. Since $\tilde{\rho}_g(T_j)$ belongs to $\mathrm{SU}(1,1)$, it is either elliptic, parabolic, or hyperbolic.

Definition 2.14 ([17]). Let $\overline{M}$ be a closed Riemann surface. Let $d\sigma^2$ be a Hermitian metric on $\overline{M} \setminus \{p_1,\dots,p_n\}$, where $p_1,\dots,p_n$ are distinct points. Then $d\sigma^2$ has a *conical singularity* of order μ_j at p_j if $\mu_j > -1$ and $d\sigma^2/|z|^{2\mu_j}$ is positive definite near p_j, where z is a local complex coordinate so that $z=0$ at p_j. The number $2\pi(1+\mu_j)$ is called the *conical angle* of $d\sigma^2$ at p_j.

Example 2.15. Let $g : M \to \mathbf{C} \cup \{\infty\}$ be a non-constant meromorphic function on a Riemann surface M. Then the pull-back metric $g^* d\sigma^2_P$ gives a co-orientable extended hyperbolic metric with ideal boundary set $g^{-1}(\{z \in \mathbf{C} \cup \{\infty\}\,;\, |z|=1\})$. Each proper singular point $p \in M$ coincides with a branch point of the meromorphic function g such that the branching order at p coincides with the order of the conical singularity of $g^* d\sigma^2_P$ at p.

Example 2.16. Let $\triangle\mathrm{ABC}$ be a hyperbolic geodesic triangle on the Poincaré disk $\mathcal{D}$. Taking a copy of the triangle, and gluing two congruent triangles at the corresponding edges, we get a co-orientable extended hyperbolic metric $d\sigma^2$ on

S^2 with three regular singular points having elliptic monodromy matrices; more precisely, $d\sigma^2$ can be considered as a hyperbolic metric on $\mathbf{C}\setminus\{0,1\}$, whose conical angles are $2A$, $2B$, and $2C$, where A,B,C are the interior angles of the triangle $\triangle$ABC. An explicit description of this metric $d\sigma^2$ in terms of hypergeometric functions is given in Kraus-Roth-Sugawa [13].

Example 2.17. Lee-Yang [14] gave an explicit description of many CMC-1 trinoids (i.e. 3-noid, as defined in the beginning of Sect. 4) in S^3_1 in terms of hypergeometric functions. Using the correspondence given in Theorem 2.10, we get many extended hyperbolic metrics with five regular singularities, where two of them, which are the umbilic points, have trivial monodromy matrices. The other three regular singularities consist of elliptic or hyperbolic singularities.

There are many PS-free hyperbolic metrics on $\mathbf{C}$. For example, let $h(z)$ be an entire function on $\mathbf{C}$. Then the pull-back of the spherical Poincaré metric by the holomorphic function $g(z) = a * \int \exp(h(z))dz$ gives a PS-free hyperbolic metric on $\mathbf{C}$, where $a \in PSL(2,\mathbf{C})$. In particular, if we set $h(z) = z$ and choose the matrix a appropriately, then we get a PS-free hyperbolic metric on $\mathbf{C}$ given by

$$d\sigma^2_{\mathbf{C}} := \frac{|dz|^2}{\sin^2(\operatorname{Im} z)}.$$

The metric $d\sigma^2_{\mathbf{C}}$ has the ideal boundary set

$$\partial^\infty(d\sigma^2_{\mathbf{C}}) = \{z \in \mathbf{C}; \operatorname{Im} z \in \pi\mathbf{Z}\},$$

and can be characterized as a PS-free metric on $\mathbf{C}$ such that $S(d\sigma^2_{\mathbf{C}})/dz^2$ is a nonzero constant. We call $(\mathbf{C}, d\sigma^2_{\mathbf{C}})$ the *Poincaré plane*. Any two such hyperbolic metrics on $\mathbf{C}$ are congruent by a suitable coordinate change $z \mapsto cz$ for $c \in \mathbf{C}\setminus\{0\}$. The following assertion holds:

Proposition 2.18. *Any PS-free extended hyperbolic metric on a torus is obtained as a quotient of the Poincaré plane by a lattice generated by* $(\xi,0)$ *and* $(\eta,\pi m)$ *for suitable* $\xi,\eta \in \mathbf{R}$ *and* $m \in \mathbf{Z}_+$. *The metric is co-orientable (resp. non-co-orientable) if* m *is even (resp. odd).*

Proof. We fix a torus $M = \mathbf{C}/\Gamma$, where Γ is a lattice generated by $v_1, v_2 \in \mathbf{C}$. Since any non-co-orientable metric will become co-orientable when taking a double covering, we may assume that $d\sigma^2$ is a co-orientable PS-free extended hyperbolic metric on M. Since $d\sigma^2$ is PS-free, its Schwarzian derivative $S(d\sigma^2)$ is a holomorphic 2-differential on $\mathbf{C}$, that is, there exists a complex number $c \in \mathbf{C}$ such that

$$S(d\sigma^2) = c\,dz^2.$$

We know that the fundamental group of M is $\Gamma = \mathbf{Z}v_1 \oplus \mathbf{Z}v_2$. Let

$$\rho : \Gamma \to \mathrm{SL}(2,\mathbf{C})$$

be the monodromy representation induced from $d\sigma^2$. First, we consider the case $c=0$. Without loss of generality, we may assume that $v_1=1$. Then the developing map of $d\sigma^2$ should be

$$g=a\star z \qquad \left(a:=\begin{pmatrix} a_{11} & a_{12} \\ a_{21} & a_{22}\end{pmatrix}\in \mathrm{SL}(2,\mathbf{C})\right).$$

Then we have that

$$\rho(v_j)=a\begin{pmatrix} 1 & v_j \\ 0 & 1\end{pmatrix}a^{-1}=\begin{pmatrix} 1-a_{11}a_{21}v_j & (a_{11})^2v_j \\ -(a_{21})^2v_j & a_{11}a_{21}v_j+1\end{pmatrix},$$

where we have used the relation $a_{12}a_{12}=a_{11}a_{22}-1$. Then $\rho(v_j)$ belongs to $\mathrm{SU}(1,1)$ if and only if $\rho(v_j)e_3=e_3(\rho(v_j)^{-1})^*$, which is equivalent to the condition

$$a_{11}a_{21}v_j=-\overline{a_{11}a_{21}v_j},\quad (a_{11})^2v_j=-\overline{(a_{21})^2v_j} \qquad (j=1,2). \tag{27}$$

Since $v_1=1$, we have

$$\overline{a_{11}a_{21}}=-a_{11}a_{21}, \qquad (a_{11})^2=-\overline{(a_{21})^2}.$$

Since a is a regular matrix, the second identity implies that $a_{11}a_{21}\neq 0$. Then Eq. (27) for v_2 reduces to the condition $v_2=\overline{v_2}$, which contradicts the fact that v_2 is linearly independent of v_1. So this case does not occur, which implies that the projective connection $S(d\sigma^2)$ associated to $d\sigma^2$ is not trivial, when $d\sigma^2$ is PS-free.

Next, we consider the case that $c\neq 0$. Since a homothetic change of lattice Γ does not affect the complex structure of torus $\mathbf{C}/\Gamma$, we may assume that $v_1>0$ and $v_2=r+\pi\mathrm{i}$ where $r\in\mathbf{R}$.

If we set $t:=\pm\sqrt{-c/2}(\neq 0)$, then $S_z(e^{2tz})=c$ holds. The developing map of $d\sigma^2$ should be

$$g=a\star e^{2tz} \qquad (a\in \mathrm{SL}(2,\mathbf{C})).$$

Then we have that

$$\rho(v_j)=a\begin{pmatrix} e^{-tv_j} & 0 \\ 0 & e^{tv_j}\end{pmatrix}a^{-1} \qquad (j=1,2).$$

Since the eigenvalues of $\mathrm{SU}(1,1)$ are real or unit complex numbers, the fact $v_1>0$ yields that $t\in\mathbf{R}$ or $t\in\mathrm{i}\mathbf{R}$. Moreover, the eigenvalues of

$$\rho(mv_1+nv_2)=a\begin{pmatrix} e^{-t(mv_1+nv_2)} & 0 \\ 0 & e^{t(mv_1+nv_2)}\end{pmatrix}a^{-1} \qquad (m,n\in\mathbf{Z})$$

must also be real numbers or unit complex numbers. This implies $t\in\mathbf{Z}$. Since

$$g = a \star e^{2tz} = a \star \begin{pmatrix} 0 & \mathrm{i} \\ \mathrm{i} & 0 \end{pmatrix} \star e^{-2tz}$$

holds, we may assume that t is a positive integer.

One can choose $a \in \mathrm{SL}(2,\mathbf{C})$ so that $\rho(v_1), \rho(v_2) \in \mathrm{SU}(1,1)$. In fact, such an a can be expressed as

$$a = \frac{1}{\sqrt{2\mathrm{i}}} b \begin{pmatrix} 1 & -\mathrm{i} \\ 1 & \mathrm{i} \end{pmatrix} \begin{pmatrix} e^{\mathrm{i}\alpha} & 0 \\ 0 & e^{-\mathrm{i}\alpha} \end{pmatrix} \qquad (b \in \mathrm{SU}(1,1),\ \alpha \in \mathbf{R}).$$

Then we have that

$$g = a \star e^{2tz} = b \star \frac{e^{2(\mathrm{i}\alpha + tz)} - \mathrm{i}}{e^{2(\mathrm{i}\alpha + tz)} + \mathrm{i}}.$$

By setting $w := 2(\mathrm{i}\alpha + tz)$, the torus is generated by the lattice $\{2tv_1, 2t(r + \pi\mathrm{i})\}$ on the w-plane, and

$$d\sigma^2 = \frac{4|dg|^2}{(1 - |g|^2)^2} = \frac{|dw|^2}{\sin^2(\operatorname{Im} w)},$$

which proves the assertion. □

There are three simply connected Riemann surfaces $\mathcal{D}$, $\mathbf{C}$ and $\mathbf{C} \cup \{\infty\}$, where $\mathcal{D}$ is the unit disk in $\mathbf{C}$. Each of them has a canonical PS-free hyperbolic metric, that is, the usual hyperbolic metric, the planar Poincaré metric $d\sigma_{\mathbf{C}}^2$ and the spherical Poincaré metric, respectively. By the coordinate change $z \mapsto tz$, the planar Poincaré metric changes to

$$d\sigma_t^2 := \frac{t^2 |dz|^2}{\sin^2(t \operatorname{Im} z)},$$

whose fundamental region is a ribbon defined by $0 < \operatorname{Im} z < 2\pi/t$. If $t \to 0$, then

$$\lim_{t \to 0} d\sigma_t^2 = \frac{|dz|^2}{(\operatorname{Im} z)^2} \lim_{t \to 0} \left(\frac{t \operatorname{Im} z}{\sin(t \operatorname{Im} z)} \right)^2 = \frac{|dz|^2}{(\operatorname{Im} z)^2},$$

which implies that the canonical hyperbolic metric on the upper half-plane is a limiting case of the planar Poincaré metric.

3 Fundamental Properties of Co-orientable Extended Hyperbolic Metrics

First of all, we divide the regular singularities of co-orientable extended hyperbolic metrics into the following six types E_0, E_1, P_0, P_1, H_0 and H_1. Such a grouping of regular singularities is useful to describe the behavior of ideal boundary sets near the singularities.

Definition 3.1. Let p be a regular singular point of an extended hyperbolic metric $d\sigma^2$. If the monodromy matrix at p is elliptic, we call p an *elliptic singular point* or an E-singular point. Moreover, if the monodromy matrix at p is $\pm e_0$ (resp. not $\pm e_0$), we call p an *E_0-singular point* (resp. *E_1-singular point*).

Definition 3.2. Let p be a regular singular point of a co-orientable extended hyperbolic metric $d\sigma^2$. If the monodromy matrix at p is parabolic, we call p a *parabolic singular point* or a P-singular point. Moreover, if $S(d\sigma^2) - dz^2/(2z^2)$ has at most a pole of order 1, we call p a *P_0-singular point*, where z is an arbitrarily given local complex coordinate centered at p. On the other hand, if p is not a P_0-singular point, it is called a *P_1-singular point.*

Definition 3.3. Let p be a regular singular point of a co-orientable extended hyperbolic metric $d\sigma^2$. If the monodromy matrix at p is hyperbolic, we call p a *hyperbolic singular point* or an H-singular point. Moreover, if there exists a real number μ (necessarily greater than $1/2$) such that $S(d\sigma^2) - \mu dz^2/z^2$ has at most a pole of order 1, we call p an *H_0-singular point.* On the other hand, if p is not an H_0-singular point, it is called an *H_1-singular point.*

The corresponding concepts for generalized CMC-1 faces are as follows:

Definition 3.4. Let $f : M \to S^3_1$ be an admissible generalized CMC-1 face defined on a Riemann surface M, and let p be an end of f. Then p is called an *elliptic* (resp. *parabolic, hyperbolic*) end if the associated extended hyperbolic metric has an elliptic (resp. parabolic, hyperbolic) singular point at p. Moreover, p is called an end of type E_0, E_1, P_0, P_1, H_0, H_1 if it is a regular singular point of type E_0, E_1, P_0, P_1, H_0, H_1, respectively.

The following three assertions are shown in Lemmas P, E1, E2, and H in [5], although the statements in [5] apply to CMC-1 faces rather than hyperbolic metrics. However, Theorems 2.10 and 2.11 give us a way to translate those results in [5] into the following:

Proposition 3.5 (Lemma P and Lemma E1 in [5]). *Let p be a P_0-singular or an E_1-singular point of a co-orientable extended hyperbolic metric $d\sigma^2$. Then the ideal boundary set of $d\sigma^2$ does not accumulate at the point p.*

Proposition 3.6 (Lemma P and Lemma E2 in [5]). *Let p be an E_0-singular or a P_1-singular point of a co-orientable extended hyperbolic metric $d\sigma^2$. Then the following assertions hold:*

1. *One possibility is that the ideal boundary set $\partial^\infty(d\sigma^2)$ does not accumulate at the point p (this case happens when p is an E_0-singular point).*
2. *The other possibility is that there exists a positive integer m such that the ideal boundary set consists of m regular curves passing through p and dividing a neighborhood of p into $2m$-sectors of angle π/m.*
3. *If p is a P_1-singular point, then $\partial^\infty(d\sigma^2)$ accumulates at the point p.*

Proposition 3.7 (Lemma H in [5]). *Let p be a hyperbolic singular point of a co-orientable extended hyperbolic metric $d\sigma^2$. Then the following assertions hold:*

1. *If p is an H_0-singular point, then there exists a local complex coordinate (U,z) centered at p such that the ideal boundary set on U is a union of countably many circles:*
$$U \cap \partial^\infty(d\sigma^2) \setminus \{p\} = \bigcup_{n\in\mathbf{Z}} \left\{ z \in U\,;\, \log|z| = \frac{\pi n}{\tau} \right\},$$
where $\tau > 0$.
2. *If p is an H_1-singular point, then there exist a positive integer m and a local complex coordinate (U,z) centered at p such that the ideal boundary set on U is the union of $2m$ logarithmic spirals:*
$$U \cap \partial^\infty(d\sigma^2) \setminus \{p\} = \bigcup_{j=0}^{2m-1} \left\{ re^{\mathrm{i}\theta} \in U\,;\, \log r = \frac{\pi j - m\theta}{\tau}, r > 0,\ \theta \in \mathbf{R} \right\},$$
where $\tau \in \mathbf{R} \setminus \{0\}$.

Moreover m and τ are uniquely determined from the coefficient of $1/z^2$ in $S(d\sigma^2)$.

In [5, Theorems 2.1 and 3.1] the following important property of extended hyperbolic metrics was shown:

Theorem 3.8 (Regularity of isolated singularities). *Let $d\sigma^2$ be a conformal hyperbolic metric defined on the unit punctured disk $\mathcal{D}^* := \{z \in \mathbf{C}\,;\, 0 < |z| < 1\}$. Then $z = 0$ is a removable singular point or a regular singular point of $d\sigma^2$, that is, $d\sigma^2$ extends as an extended hyperbolic metric such that $z = 0$ is a non-singular point, a P_0-singular point or an E-singular point.*

Remark 3.9. In [9], a somewhat similar removable singularity theorem for spherical metrics is given, although the method and the situation are different from ours.

A point $p \in M$ is called a *divergent point* of an extended hyperbolic metric $d\sigma^2$ if there exists a sequence $\{q_k\}_{k=1,2,3,\dots}$ on the universal cover $\widetilde{M}$ of M such that $\{\pi(q_k)\}$ converges to p and
$$\lim_{k\to\infty} |g(q_k)| = 1$$
holds, where $\pi\colon \widetilde{M} \to M$ is the covering projection. This definition does not depend on the choice of the developing map g.

Proposition 3.10. *The set of divergent points of (co-orientable or non-co-orientable) $d\sigma^2$ consists of the closure of the ideal boundary set $\partial^\infty(d\sigma^2)$ and P_0-singular points.*

Proof. Without loss of generality, we may assume that $d\sigma^2$ is co-orientable by taking a double covering if necessary. We denote by Δ the divergent set of $d\sigma^2$. It is obvious that the closure of the ideal boundary set is contained in Δ. So we suppose that $p \in \Delta$ does not belong to the closure of the ideal boundary set. Then

$d\sigma^2$ gives a conformal hyperbolic metric defined on a punctured neighborhood of p. By Theorem 3.8, p must be a regular singular point of type P_0 or type E. If p is an E_1-singular point, then we may set $g = z^{\mu}\varphi(z)$, where $\varphi(z)$ is a meromorphic function with $\mu \in (0,1)$. Then $\lim_{z\to 0}|g(z)|$ is 0 or ∞. So the divergent points will not accumulate at the origin, which is a contradiction. If p is an E_0-singular point, then we may set $g = \varphi(z)$, where $\varphi(z)$ is a meromorphic function. Since p is a divergent point, we have that $|\varphi(0)| = 1$, then the set $|\varphi(z)| = 1$ consists of a union of regular curves passing through the origin, which contradicts our assumption that p does not belong to the closure of the ideal boundary set. Thus, p must be a P_0-singular point. On the other hand, by the proof of [5, Lemma P], we may set

$$g(z) = \frac{\log(z) - 1 - \mathrm{i}\varphi(z)}{\log(z) + 1 - \mathrm{i}\varphi(z)},$$

where $\varphi(z)$ is a holomorphic function at p. Then $\lim_{k\to\infty} g(e^{-k}) = 1$ when k is a positive integer. Hence, p is a divergent point if it is a P_0-singular point. □

Next, we show that an extended hyperbolic metric is complete at the divergent points, as follows:

Theorem 3.11 (Completeness at the divergent points). *Let $d\sigma^2$ be a (co-orientable or non-co-orientable) extended hyperbolic metric defined on a Riemann surface M which admits only regular singularities. Let $\gamma : [0,\infty) \to M$ be a piecewise smooth curve which does not pass through ideal boundary points nor proper singular points, and accumulates at a divergent point. Then γ has infinite length.*

Proof. Without loss of generality, we may assume that $d\sigma^2$ is co-orientable by taking a double covering if necessary. Let p be a point in the closure of $\partial^{\infty}(d\sigma^2)$. If p is an ideal boundary point then the assertion is obvious. So we may assume that p is one of the regular singular points of $d\sigma^2$. First, we consider the case that p is an elliptic singular point. By Propositions 3.5 and 3.6, p is an E_0-singular point. Then, by the proof of Lemma E2 in [5], there exists a local complex coordinate z centered at p such that the developing map of the metric at p is expressed by

$$g(z) = \exp z^m = \exp\bigl(r^m(\cos m\theta + \mathrm{i}\sin m\theta)\bigr) \qquad (z = re^{\mathrm{i}\theta},\ r > 0,\ \theta \in \mathbf{R}),$$

where $2\pi(m-1)$ $(m \neq 1)$ is the conical angle of the metric $d\sigma^2$. Since

$$|g| = \exp(r^m \cos m\theta), \qquad |dg| = mr^{m-1}\exp(r^m\cos m\theta)|dz|,$$

we have that

$$\begin{aligned} d\sigma^2 &= \frac{4m^2 r^{2(m-1)}}{\bigl(1-\exp(2r^m\cos m\theta)\bigr)^2}\exp(2r^m\cos m\theta)|dz|^2 \\ &= \frac{m^2 r^{2(m-1)}}{\sinh^2(r^m\cos m\theta)}\left(dr^2 + r^2 d\theta^2\right) \geq \frac{m^2 r^{2(m-1)}}{\sinh^2 r^m}dr^2 \geq \frac{cm^2}{r^2}dr^2 \end{aligned}$$

for $0 < r < 1$, where $0 < c < 1$ is a suitable constant. Since dr^2/r^2 is a complete metric on $\mathbf{R} \setminus \{0\}$ at the origin, an arbitrarily given path γ accumulating at p not passing through any ideal boundary points has infinite length.

Next, we consider the case that p is a hyperbolic or parabolic singular point. Then the developing map g at p can be taken to be of the form

$$g = \frac{g_0 - \mathrm{i}}{g_0 + \mathrm{i}},$$

as in (2) and (3) of [5, Proposition 2.1]. Then it holds that

$$d\sigma^2 = \frac{4\,|dg|^2}{(1-|g|^2)^2} = \frac{|dg_0|^2}{(\mathrm{Im}\, g_0)^2}. \tag{28}$$

We consider the case that p is hyperbolic. In this case, there exists a local complex coordinate z centered at p such that (cf. the proof of Lemma H in [5, p. 409])

$$g_0 := z^{m+\mathrm{i}\tau} = \exp(m \log r - \tau\theta - \mathrm{i}\delta) \qquad (m \in \mathbf{Z},\ \tau > 0),$$

where

$$\delta := \tau \log r + m\theta.$$

We remark that the zeros of $\sin\delta$ coincide with the divergence set. Then, it holds that

$$d\sigma^2 = \frac{m^2+\tau^2}{(r\sin\delta)^2}\left(dr^2 + (r\,d\theta)^2\right) \geq \frac{\tau^2}{r^2}dr^2.$$

Since the right-hand side gives a complete metric on $\mathbf{R} \setminus \{0\}$ at the origin, we get the assertion in this case.

Finally, we consider the case that p is parabolic. Then there exists a local complex coordinate centered at p so that

$$g_0 := \mathrm{i}\log(z) + \varphi(z),$$

where $\varphi(z)$ is a meromorphic function at $z = 0$. Suppose that p is a P_0-singular point. Then $\varphi(z)$ is holomorphic at p. By the coordinate change z to $ze^{-\mathrm{i}\varphi(z)}$, we may set $g_0 = \mathrm{i}\log z$. Then we have that

$$d\sigma^2 = \frac{dr^2 + (r\,d\theta)^2}{(r\log r)^2} \geq \frac{dr^2}{(r\log r)^2}.$$

The integral

$$\int_{\varepsilon}^{r_0} \frac{dr}{r\log r} = \log(-\log r_0) - \log(-\log\varepsilon)$$

diverges to ∞ when ε tends to 0. Then the assertion follows.

On the other hand, suppose that p is a P_1-singular point. Then p is a pole of $\varphi(z)$. Let m be the order of the pole at p. (If $\varphi(z)$ is holomorphic at p, we set $m=0$.) Then there exist two positive numbers c_1 and c_2 such that

$$|g_0'| \geq \frac{c_1}{r^{m+1}}, \qquad |\operatorname{Im}(g_0)| \leq \frac{c_2}{r^m} \qquad \left({}' = \frac{d}{dz}\right)$$

hold on a sufficiently small neighborhood of p. Thus

$$d\sigma^2 = \frac{|g_0'(z)|^2|dz|^2}{|\operatorname{Im}(g_0)|^2} \geq \left(\frac{c_1}{c_2}\right)^2 \frac{dr^2}{r^2},$$

which implies completeness. □

The following assertion follows immediately from Theorem 3.11.

Corollary 3.12 ([10, Theorem 2.2]). *A (co-orientable or non-co-orientable) extended hyperbolic metric defined on a Riemann surface M whose singular set consists of divergent points is a complete metric on $M \setminus \Delta$, where Δ is the divergent set. In particular, a PS-free extended hyperbolic metric is a complete metric on $M \setminus \partial^\infty(d\sigma^2)$.*

Corollary 3.13. *Let $d\sigma^2$ be a (co-orientable or non-co-orientable) extended hyperbolic metric defined on a Riemann surface M whose proper singularities are all regular. Then for each $p \in M$, there exists a neighborhood U of p such that either $d\sigma^2$ or $1/d\sigma^2$ is continuous on U.*

Proof. Let S be the set of regular singularities of $d\sigma^2$. We know that S is a discrete subset of M. Since $d\sigma^2$ gives a Hermitian metric on $M \setminus (S \cup \partial^\infty(d\sigma^2))$, Theorem 3.11 implies that $1/d\sigma^2$ is continuous on $M \setminus S$. So it is sufficient to check the continuity of $d\sigma^2$ or $1/d\sigma^2$ on the set S. We fix a point $p \in S$ arbitrarily. If p is a divergent point, then $\lim_{z\to p} 1/d\sigma^2 = 0$ holds by Theorem 3.11. So we may assume that p is not a divergent point. Then by Lemma 3.14 below, there exists a real number $\mu(>-1)$ $(\mu \neq 0)$ such that p is a conical singularity of order μ, and then $\lim_{z\to p} d\sigma^2 = 0$ if $\mu > 0$, and $\lim_{z\to p} 1/d\sigma^2 = 0$ if $(-1<)\mu<0$. □

Lemma 3.14. *Let p be a regular singular point of $d\sigma^2$ which is not a divergent point. Then there exists a real number $\mu(>-1)$ $(\mu \neq 0)$ such that $d\sigma^2$ has a conical singularity at p of order μ. In particular, p is an elliptic singular point.*

Proof. By Propositions 3.5–3.7 and 3.10, we can conclude that p must be a regular singularity of $S(d\sigma^2)$ with elliptic monodromy. By [5, (1) of Proposition 2.1], we can take a developing map of $d\sigma^2$ of the form

$$g(z) = a \star (z^{\mu+1}h(z)) \qquad (a \in \mathrm{SL}(2,\mathbf{C}),\ \mu \in (-1,0)\cup(0,\infty)),$$

where z is a local complex coordinate centered at p and $h(z)$ is a locally defined holomorphic function satisfying $h(0) \neq 0$. Since p is not a divergent point, $|g(0)| \neq$

1. Hence, there exists $b \in \mathrm{Isom}_+(S^2_H)$ such that $b \star g(0) = 0$. This implies that

$$b \star g = \frac{c_{11} z^{\mu+1} h(z)}{c_{12} z^{\mu+1} h(z) + c_{22}}$$

also gives a developing map of $d\sigma^2$, where $c_{11}c_{22} \neq 0$ and $ba = (c_{ij})_{i,j=1,2}$. Thus, $d\sigma^2$ has a conical singularity of order μ at p. □

4 Classification of de Sitter Catenoids

Let

$$f : \mathbf{C} \cup \{\infty\} \setminus \{p_1, \ldots, p_n\} \longrightarrow S^3_1$$

be a CMC-1 face satisfying the following properties.

- f is a weakly complete CMC-1 face.
- $S(g)$ is a projective connection having only regular singularities, where g is the secondary Gauss map.
- The hyperbolic Gauss map G is a meromorphic function of degree $n-1$ defined on $\mathbf{C} \cup \{\infty\}$, and does not branch at each p_j $(j = 1, \ldots, n)$.

When f is complete, each end is properly embedded (In fact, the equality of the Osserman-type inequality in [5, Theorem 0.2] holds.). We know that $n \geq 1$, since there are no compact CMC-1 faces of genus zero (cf. [5, p. 420]). Each point p_j $(j = 1, 2, \ldots, n)$ is called an end of f. We call such an f an *n-noid*.

Proposition 4.1. *Any* 1*-noid is congruent to the totally umbilical CMC*-1 *immersion called the S^3_1-horosphere given by*

$$f_0 = F e_3 F^* = \begin{pmatrix} 2\,\mathrm{Re}(z) + |z|^2 & -1 + \bar{z} + |z|^2 \\ -1 + z + |z|^2 & -1 + |z|^2 \end{pmatrix} : \mathbf{C} \longrightarrow S^3_1, \qquad (29)$$

where

$$F = \begin{pmatrix} 1+z & 1 \\ z & 1 \end{pmatrix}.$$

Proof. We consider a CMC-1 face $f : \mathbf{C} \to S^3_1$ with a regular embedded end at $z = \infty$. Since $z = \infty$ is a regular end, the Hopf differential Q is holomorphic on $\mathbf{C}$ and has a pole of order at most 2. Since the total order of Q is -4 by the Riemann–Roch theorem, we can conclude that Q must be identically zero. Then f is totally umbilical and must be locally congruent to the S^3_1-horosphere f_0. Since f is defined on $\mathbf{C}$, f must be congruent to f_0 on $\mathbf{C}$. □

Next, we consider the case $n = 2$. It should be remarked that there are many weakly complete CMC-1 faces $f : \mathbf{C} \setminus \{0\} \to S^3_1$ whose hyperbolic Gauss maps are meromorphic functions on $\mathbf{C} \cup \{\infty\}$ of degree greater than 1. The degree 1 case (i.e. the case of that f is a 2-noid) is important for the following two reasons:

- Such f correspond bijectively to co-orientable extended hyperbolic metrics with two regular singularities defined on S^2.
- If f is complete, then all ends of f are embedded if and only if G is of degree 1 (cf. [5, Theorem 0.2]).

Definition 4.2. A 2-noid $f : \mathbf{C} \setminus \{0\} \to S^3_1$ is called a *de Sitter catenoid* or an *S^3_1-catenoid.*

Lee and Yang [14] gave a rough classification of S^3_1-catenoids, where S^3_1-catenoids of hyperbolic type are called surfaces with abnormal ends. However, S^3_1-catenoids of parabolic type were omitted there, and included in [5]. Since $\pi_1(\mathbf{C} \setminus \{0\})$ is abelian, S^3_1-catenoids are all reducible (cf. Remark 1.15). In particular, 3-reducible S^3_1-catenoids are recognized in [14], but are not classified there. In this section, we shall give an explicit classification of congruent classes of S^3_1-catenoids in S^3_1 (The corresponding classification of immersed CMC-1 catenoids of genus zero in hyperbolic 3-space is substantially easier, which is given in [16].).

Let $f \colon \mathbf{C} \setminus \{0\} \to S^3_1$ be an S^3_1-catenoid. Then, the Hopf differential Q of the surface has poles of order at most two at $z = 0, \infty$. Since Q is holomorphic on $\mathbf{C} \setminus \{0\}$, the Riemann–Roch theorem yields that Q is of the form

$$Q = c\frac{dz^2}{z^2} \qquad (c \in \mathbf{C} \setminus \{0\}).$$

On the other hand, since G is of degree 1, and the hyperbolic Gauss map is determined up to Möbius transformations, we may set $G = z$. Since the dual metric

$$ds^2_{\#} = (1 + |G|^2)^2 \left|\frac{Q}{dG}\right|^2 = |c|^2 \left(\frac{1+|z|^2}{|z|^2}\right)^2 |dz|^2$$

gives a complete Riemannian metric on $\mathbf{C} \setminus \{0\}$, each CMC-1 face induced from the pair (G, Q) is weakly complete. This implies that we have determined the definite pair for S^3_1-catenoids. However, as mentioned in Remark 1.6, the determination of (G, Q) does not imply the classification of S^3_1-catenoids. The secondary Gauss map g of an S^3_1-catenoid must satisfy Eq. (14), namely, $S(g) = 2c dz^2/z^2$. Since

$$S(z^{\alpha}) = \frac{1-\alpha^2}{2z^2}dz^2 \qquad (\alpha \in \mathbf{C} \setminus \{0\})$$

and

$$S(\log z) = \frac{1}{2z^2}dz^2,$$

such a g is of the form $g = a \star z^{\alpha}$ or $g = a \star \log z$, where $a \in \mathrm{PSL}(2, \mathbf{C})$. Then it induces a representation

$$\rho_g : \pi_1(\mathbf{C} \setminus \{0\}) \cong \mathbf{Z} \longrightarrow \mathrm{PSL}(2, \mathbf{C}).$$

Recall that elliptic SU$(1,1)$-matrices $E(t)$, parabolic SU$(1,1)$-matrices $P(\varepsilon)$, and hyperbolic SU$(1,1)$-matrices $H(t)$ were all defined in Sect. 1. An S^3_1-catenoid is called *elliptic*, *parabolic*, or *hyperbolic* if the image of a generator of $\pi_1(\mathbf{C} \setminus \{0\})$ is elliptic, parabolic, or hyperbolic, respectively. One can easily check the following assertion:

Lemma 4.3. *An S^3_1-catenoid f is elliptic (resp. hyperbolic) if and only if its secondary Gauss map g satisfies*

$$(2Q =)S(g) = \frac{(1-\alpha^2)dz^2}{2z^2}$$

and $\alpha \in \mathbf{R} \setminus \{0, 1, -1\}$ (resp. $\alpha \in \mathbf{C} \setminus \mathbf{R}$). On the other hand, f is parabolic if and only if

$$(2Q =)S(g) = \frac{dz^2}{2z^2}$$

holds, where Q is the Hopf differential of f.

Using this, we prove the following assertion (This result is also found in Example 5.6 of [5].).

Proposition 4.4 (The classification of S^3_1-catenoids of parabolic type). *An S^3_1-catenoid f of parabolic type is congruent to the CMC-1 face (given in [5, Example 5.2]) $f_P = Fe_3F^*$, where*

$$F = \frac{\mathrm{i}}{2\sqrt{2}} \begin{pmatrix} \sqrt{z} & 0 \\ 0 & 1/\sqrt{z} \end{pmatrix} \begin{pmatrix} 3-\log(z) & -1+\log(z) \\ 1+\log(z) & -3-\log(z) \end{pmatrix}. \tag{30}$$

(see Fig. 2.) Moreover, f_P is complete with embedded ends. The end $z = 0$ lies in the upper ideal boundary $\partial_+ S^3_1$ and the end $z = \infty$ lies in the lower ideal boundary $\partial_- S^3_1$ (see [5] for the definition of $\partial_\pm S^3_1$).

Proof. We set

$$g_P := \frac{\log z + 1}{\log z - 1} = X \star \log z, \qquad X := \frac{\mathrm{i}}{\sqrt{2}} \begin{pmatrix} 1 & 1 \\ 1 & -1 \end{pmatrix}, \tag{31}$$

which has PSU$(1,1)$-monodromy. The above CMC-1 face f_P has hyperbolic Gauss map $G(= z)$, secondary Gauss map g_P, and Hopf differential $Q = dz^2/(4z^2)$. Since the fundamental group of $\mathbf{C} \setminus \{0\}$ is abelian, f_P is reducible. However, since g_P is

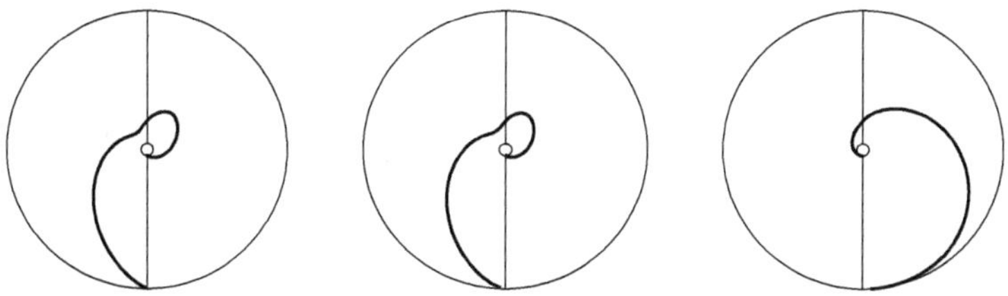

Fig. 2 Profile curves of parabolic and 1-reducible elliptic catenoids: They are shown in the *hollow ball model*, as explained in [1, 2]. In this chapter, for surface graphics in de Sitter 3-space, we always use the *hollow ball model*. The entire surfaces are obtained by rotating the curves around the vertical axis. Cone-like singularities appear at the points where the profile curves meet the *vertical line*

not single valued on $\mathbf{C}\setminus\{0\}$, f_P is 1-reducible. Thus the subgroup $\Gamma := \rho_{g_P}(\pi_1(\mathbf{C}\setminus\{0\}))$ of $\mathrm{PSU}(1,1)$ is generated by a matrix

$$X\begin{pmatrix} 1 & -2\pi\mathrm{i} \\ 0 & 1 \end{pmatrix}X^{-1} = \begin{pmatrix} 1-\pi\mathrm{i} & \pi\mathrm{i} \\ -\pi\mathrm{i} & 1+\pi\mathrm{i} \end{pmatrix}.$$

Replacing the local coordinate z with $e^t z$ ($t \in \mathbf{R}$), we have the expression

$$g_t = \frac{\log z + t + 1}{\log z + t - 1} = X_t \star \log z, \qquad \left(X_t := \frac{\mathrm{i}}{\sqrt{2}}\begin{pmatrix} 1 & t+1 \\ 1 & t-1 \end{pmatrix}\right)$$

which has $\mathrm{PSU}(1,1)$-monodromy for all $t \in \mathbf{R}$. In particular,

$$\sigma_t := X_t X^{-1} = \begin{pmatrix} 1+t/2 & -t/2 \\ t/2 & 1-t/2 \end{pmatrix}$$

satisfies

$$\sigma_t \Gamma (\sigma_t)^{-1} \subset \mathrm{PSU}(1,1),$$

that is, $\sigma_t \in C_\Gamma$ holds, where C_Γ is defined in Appendix B. Moreover, it can be easily checked that $\sigma_t\sigma_s = \sigma_{t+s}$ $(t,s \in \mathbf{R})$, and $\sigma_t \notin \mathrm{SU}(1,1)$ for $t \in \mathbf{R}\setminus\{0\}$ which implies that $\{\tilde{\varphi}(\sigma_t)\}_{t\in\mathbf{R}}$ gives a geodesic in S^3_1, where $\tilde{\varphi}$ is defined in Appendix B (More precisely, one can directly check that the geodesic is light like.). Then Theorem 1.14 and Proposition B.1 in Appendix B yield that any S^3_1-catenoid f of parabolic monodromy is congruent to an S^3_1-catenoid f_t of parabolic type associated to the data $(G,g) = (z, g_t)$ $(t \in \mathbf{R})$. However, we know that f_t is obtained via f by the coordinate change z to $e^t z$, and f_t is congruent to $f_P = f_0$ in S^3_1. Since f_P is complete and attains equality in the Osserman-type inequality [5, Theorem 0.2], the two ends of f are embedded. Moreover, the secondary Gauss map g_P of f_P satisfies

$|g_P(z)| < 1$ if $|z| < 1$ and $|g_P(z)| > 1$ if $|z| > 1$. Then the last assertion follows from [5, Proposition 4.2]. □

Proposition 4.5 (The classification of 1-reducible S^3_1-catenoids of elliptic type). *For each 1-reducible S^3_1-catenoid f of elliptic type, there exists a positive number $\alpha \notin \mathbf{Z}$ such that f is congruent to the CMC-1 face $f_E := Fe_3F^*$, where*

$$F = \frac{1}{2\sqrt{\alpha}} \begin{pmatrix} (\alpha+1)z^{(-\alpha+1)/2} & (\alpha-1)z^{(\alpha+1)/2} \\ (\alpha-1)z^{-(\alpha+1)/2} & (\alpha+1)z^{(\alpha-1)/2} \end{pmatrix}.$$

Moreover, f_E is complete with embedded ends. The end $z = 0$ lies in $\partial_+ S^3_1$ and the end $z = \infty$ lies in $\partial_- S^3_1$ (see Fig. 2.).

Proof. The CMC-1 face $f_E = Fe_3F^*$ has the secondary Gauss map

$$g_E := z^\alpha (= \exp(\alpha \log z)). \tag{32}$$

Since the fundamental group of $\mathbf{C} \setminus \{0\}$ is abelian and g is not single-valued on $\mathbf{C} \setminus \{0\}$, we can conclude that f_E is 1-reducible, which implies that $\tilde{\varphi}(I_\Gamma)$ is 1-dimensional, where $\Gamma := \rho_{g_E}(\pi_1(\mathbf{C} \setminus \{0\}))$ (see Appendix B for the definition of I_Γ.). The generator of Γ is the matrix $\begin{pmatrix} e^{-i\pi\alpha} & 0 \\ 0 & e^{i\pi\alpha} \end{pmatrix}$. Replacing the coordinate z with $e^t z$ $(t \in \mathbf{R})$, we have the expression

$$g_t = e^{\alpha t} z^\alpha = \sigma_t \star z^\alpha, \qquad \sigma_t := \begin{pmatrix} e^{\alpha t/2} & 0 \\ 0 & e^{-\alpha t/2} \end{pmatrix}.$$

Obviously, it satisfies $\sigma_t \Gamma (\sigma_t)^{-1} \subset \mathrm{PSU}(1,1)$ that is $\sigma_t \in C_\Gamma$ holds. Since $\sigma_t \sigma_s = \sigma_{t+s}$ $(t, s \in \mathbf{R})$, and $\sigma_t \notin \mathrm{SU}(1,1)$ for $t \in \mathbf{R} \setminus \{0\}$, we can conclude that $\{\tilde{\varphi}(\sigma_t)\}_{t \in \mathbf{R}}$ gives a geodesic in S^3_1 (More precisely, the geodesic is time like.). Then Theorem 1.14 and Proposition B.1 yield that any S^3_1-catenoid f of elliptic monodromy is congruent to one of these f_t associated to the data $(G, g) = (z, g_t)$ $(t \in \mathbf{R})$. However, we know that f_t is obtained via f_E by the coordinate change z to $e^t z$ $(t \in \mathbf{R})$, and f_t is congruent to f_E in S^3_1. The last assertion follows from [5, Proposition 4.2]. □

Remark 4.6. The singular sets of the S^3_1-catenoid f_P of parabolic type and the S^3_1-catenoid f_E of elliptic type are both equal to the unit circle $\{z \in \mathbf{C}; |z| = 1\}$, which consists of cone-like singularities (cf. Fact 1.16).

Proposition 4.7 (The classification of S^3_1-catenoids of hyperbolic type). *For each S^3_1-catenoid f of hyperbolic type, there exists a complex number $\alpha := m + i\tau$ $(m \in \mathbf{Z}$, $\tau > 0)$ such that f is congruent to the CMC-1 face $f_H = FRe_3R^*F^*$, where (see Fig. 3)*

$$F = \frac{1}{2\sqrt{\alpha}} \begin{pmatrix} (\alpha+1)z^{(-\alpha+1)/2} & (\alpha-1)z^{(\alpha+1)/2} \\ (\alpha-1)z^{-(\alpha+1)/2} & (\alpha+1)z^{(\alpha-1)/2} \end{pmatrix}$$

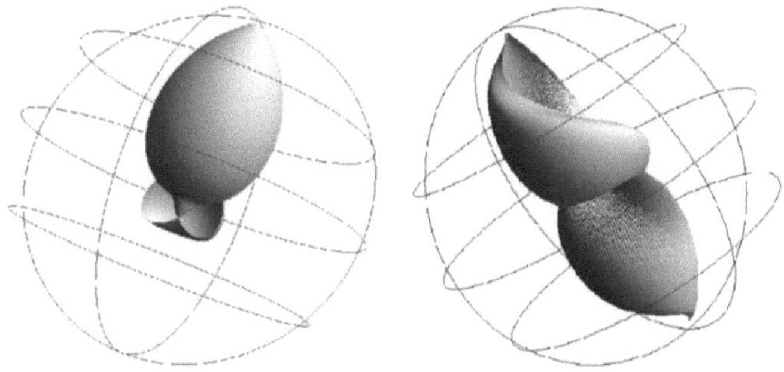

Fig. 3 Catenoids of hyperbolic type. The first surface has $m = 0$ and $\tau = 0.5$ where the domain chosen for producing this graphic is $z = e^s e^{i\theta}$ for $-4 < s < 4$ and $0 < \theta < \pi$. For all graphics shown in this chapter, we use the notation $z = e^s e^{i\theta}$, and in each case we will specify the range of s and θ we use to determine the domain. The second surface has $m = 1$ and $\tau = 1.5$, with $-4 < s < 4$ and $0 < \theta < 2\pi$. The ends of surfaces of hyperbolic type converge to both components of the ideal boundary of de Sitter space, so if we were to draw larger portions of these surfaces, we would expect to see wild oscillation between the two ideal boundary components

$$\text{and} \qquad R = \frac{e^{-3\pi i/4}}{\sqrt{2}} \begin{pmatrix} 1 & 1 \\ i & -i \end{pmatrix}.$$

Proof. The CMC-1 face f_H has the secondary Gauss map

$$g_H = \frac{g_0 - i}{g_0 + i} = R^{-1} \star g_0, \qquad g_0 = z^\alpha (= \exp((m + i\tau)\log z)). \tag{33}$$

Since m is an integer, g_H has a PSU(1,1)-monodromy. Since g_H is not single-valued on $\mathbf{C} \setminus \{0\}$, f_H is 1-reducible. Moreover, the subgroup $\Gamma := \rho_{g_H}(\pi_1(\mathbf{C} \setminus \{0\}))$ is generated by the matrix $\begin{pmatrix} \cosh(\pi\tau) & \sinh(\pi\tau) \\ \sinh(\pi\tau) & \cosh(\pi\tau) \end{pmatrix}$. If we replace z with $e^t z$, the secondary Gauss map g_H of f_H turns out to be

$$g_t = \sigma_t \star g_H, \qquad \sigma_t := R^{-1} \begin{pmatrix} e^{\alpha t/2} & 0 \\ 0 & e^{-\alpha t/2} \end{pmatrix} R,$$

which satisfies $\sigma_t \Gamma (\sigma_t)^{-1} \subset \mathrm{PSU}(1,1)$, that is, $\sigma_t \in C_\Gamma$ holds, where C_Γ is defined as in Appendix B. Moreover, it holds that $\sigma_t \sigma_s = \sigma_{t+s}$ $(t, s \in \mathbf{R})$, and $\sigma_t \notin \mathrm{SU}(1,1)$ for $t \in \mathbf{R} \setminus \{0\}$, which implies that $\{\tilde{\varphi}(\sigma_t)\}_{t \in \mathbf{R}}$ gives a geodesic in S^3_1. (More precisely, the geodesic is space like.) Then by the same argument as in the proofs of Propositions 4.4 and 4.5, we get the conclusion. □

Remark 4.8. The singular set of the S^3_1-catenoid f_H is the union of $2m$ logarithmic spirals

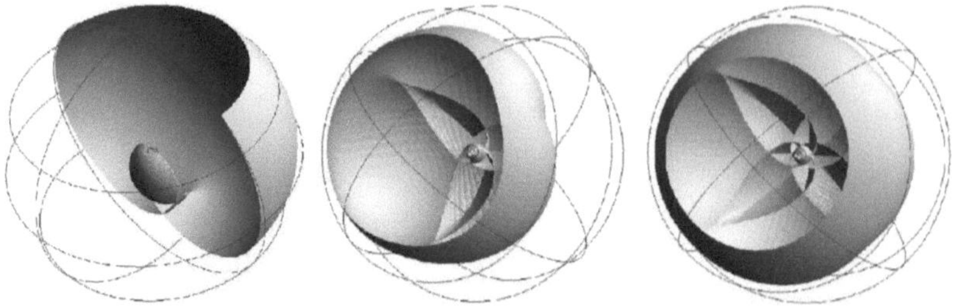

Fig. 4 3-reducible catenoids f_c for $c = 0.91$ that have connected singular sets, first with $m = 1$, then with $m = 2$, and then with $m = 3$. The domain used for the first graphic is $-3 < s < 3$ and $-\pi/2 < \theta < \pi/2$. The domain for the second and third graphics is $-1.5 < s < 1.65$ and $0 < \theta < 2\pi$

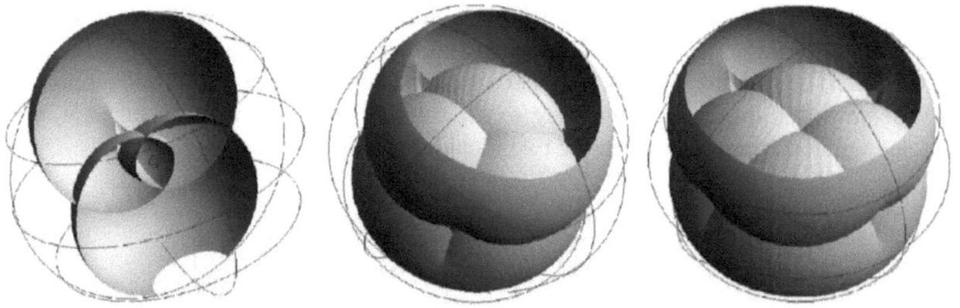

Fig. 5 3-reducible catenoids f_c with $c = 1.05$ that have disconnected singular sets, first with $m = 1$, then with $m = 2$, and then with $m = 3$. The domain used for the first graphic is $-3 < s < 3$ and $-\pi/2 < \theta < \pi/2$. The domain for the second graphic is $-1.5 < s < 1.75$ and $0 < \theta < 2\pi$. The domain for the third graphic is $-1.1 < s < 1.35$ and $0 < \theta < 2\pi$

$$\left\{ z = re^{i\theta} \in \mathbf{C}\,; \quad \log r = \frac{\pi j - m\theta}{\tau} \right\} \qquad (j = 0, 1, \ldots, 2m-1),$$

when $m \neq 0$, namely, $d\sigma^2_{f_H}$ has H_1-singular points at $z = 0, \infty$ (cf. Proposition 3.7). In this case, all singular points are cuspidal edges.

On the other hand, when $m = 0$, the singular set is the union of the circles

$$\left\{ z = re^{i\theta} \in \mathbf{C}\,; \quad \log r = \frac{\pi j}{\tau} \right\} \qquad (j \in \mathbf{Z}),$$

namely, $d\sigma^2_{f_H}$ has H_0-singular points at $z = 0, \infty$ (cf. Proposition 3.7). Each circle corresponds to a cone-like singularity.

Proposition 4.9 (The classification of 3-reducible $\mathbf{S}^3_1$-catenoids with elliptic ends). *Let f be a 3-reducible S^3_1-catenoid. Then one of the following three cases occurs (see Figs. 4 and 5):*

1. *f has two complete ends. In this case, there exist a positive integer m and a non-negative real number c ($\neq 1$) such that f is congruent to $f_c = F_c e_3 F_c^*$, where*

$$F_c := \frac{z^{-(m+2)/2}}{2\sqrt{1+m}} \begin{pmatrix} -(m+2)z & \left(c(m+2) - mz^{m+1}\right)z \\ -m & cm - (m+2)z^{m+1} \end{pmatrix}. \tag{34}$$

The end $z = 0$ lies in $\partial_+ S^3_1$ and the end $z = \infty$ lies in $\partial_- S^3_1$. If $c = 0$, then f_c is rotationally symmetric.

2. *One of the ends of f is complete and the other is incomplete. In this case, there exists a positive integer m such that f is congruent to $f_1 = F_1 e_3 F_1^*$, where*

$$F_1 = \frac{z^{-(m+2)/2}}{2\sqrt{1+m}} \begin{pmatrix} -(m+2)z & \left((m+2) - mz^{m+1}\right)z \\ -m & m - (m+2)z^{m+1} \end{pmatrix}, \tag{35}$$

that is, the case $c = 1$ in Eq. (34).

3. *Both of the ends of f are incomplete, and there exists a positive integer m such that f is congruent to $f_\infty = F_\infty e_3 F_\infty^*$, where*

$$F_\infty = \frac{z^{-(m+2)/2}}{2\sqrt{2(m+1)}} \begin{pmatrix} z\left(m\left(z^{m+1} - 1\right) - 2\right) & \left(-(m+2) - mz^{m+1}\right)z \\ 2z^{m+1} + m\left(z^{m+1} - 1\right) & -m - (m+2)z^{m+1} \end{pmatrix}. \tag{36}$$

(For the sake of convenience, we used the notations F_∞ and f_∞ although they are not the limits of F_c and f_c as $c \to \infty$, respectively.)

Proof. The secondary Gauss map is of the form

$$g = \frac{b_{11}z^{m+1} + b_{12}}{b_{21}z^{m+1} + b_{22}} \qquad (b_{11}b_{22} - b_{12}b_{21} \neq 0). \tag{37}$$

First, we consider the case that the end $z = 0$ is complete. Then the secondary Gauss map $g(z)$ satisfies $|g(0)| \neq 1$. If $|g(0)| > 1$, then we replace g by $1/g$, and may assume that $|g(0)| < 1$. Since $\mathrm{SU}(1,1)$ acts on the unit disk as an isometry of the Poincaré disk transitively, there exists an $a \in \mathrm{SU}(1,1)$ such that $a \star g(0) = 0$. By replacing g by $a \star g$, we may assume that $g(0) = 0$, which implies that $b_{12} = 0$ and $b_{22} \neq 0$. Then, by replacing g with $1/g$, we may set

$$g = \frac{b_{21}}{b_{11}} + \frac{b_{22}}{b_{11}} z^{-m-1}.$$

By replacing the coordinate z with $1/z$, we have that

$$g = \frac{b_{21}}{b_{11}} + \frac{b_{22}}{b_{11}} z^{m+1}.$$

Again, we replace the coordinate z with $(b_{22}/b_{11})^{1/(m+1)}z$, and we get the expression

$$g = z^{m+1} + c_0 \qquad \left(c_0 = \frac{b_{21}}{b_{11}}\right).$$

Now, we express

$$c_0 = ce^{\mathrm{i}s} \qquad (c \geq 0,\ s \in \mathbf{R}).$$

The replacement g by $e^{-\mathrm{i}s}g$ does not change the resulting CMC-1 face. This yields the expression $g = e^{-\mathrm{i}s}z^{m+1} + c$. Finally, we change the coordinate z to $e^{-\mathrm{i}s/(1+m)}z$, giving that

$$g_c = z^{m+1} + c \qquad (c \geq 0). \tag{38}$$

Using Eq. (16), we get Eq. (34) if $c \neq 1$ and Eq. (35) if $c = 1$.

Next, we consider the case that the ends $z = 0, \infty$ are both incomplete. Since the action $\mathrm{SU}(1,1)$ on the boundary of the Poincaré disk is two-point homogeneous, there exists an $a \in \mathrm{SU}(1,1)$ such that $a \star g(0) = -1$ and $a \star g(\infty) = 1$. So we may assume that $g(0) = -1$ and $g(\infty) = 1$ by replacing g with $a \star g$, which implies that $b_{11}/b_{21} = 1$ and $b_{12}/b_{22} = -1$. Since

$$g = \frac{(b_{11}/b_{22})z^{m+1} + (b_{12}/b_{22})}{(b_{21}/b_{22})z^{m+1} + 1},$$

replacing the coordinate z with $(b_{11}/b_{22})^{1/(1+m)}z$, we have that

$$g_\infty = \frac{z^{m+1} - 1}{z^{m+1} + 1}. \tag{39}$$

Finally, we get Eq. (36) by applying Eq. (16). The last assertion follows from [5, Proposition 4.2]. □

The singularities of 3-reducible S^3_1-catenoids in Proposition 4.9 are described as follows:

Proposition 4.10 (Singularities of 3-reducible $\mathbf{S}^3_1$-catenoids). *Let $f_c = F_c e_3 F_c^*$ $(c > 0)$ be a 3-reducible S^3_1-catenoid as in Proposition 4.9. Then the singular set of f_c is (cf. Eq. (38))*

$$\Sigma_c := \left\{ z = re^{\mathrm{i}\theta}\,;\, -1 + c^2 + r^{2(m+1)} + 2cr^{m+1}\cos((m+1)\theta) = 0, r > 0 \right\}. \tag{40}$$

On the other hand, the singular set of f_∞ is (cf. Eq. (39))

$$\Sigma_\infty := \left\{ z = re^{\mathrm{i}\theta}\,;\, \cos\big((m+1)\theta\big) = 0, r > 0 \right\}.$$

In particular (see Fig. 6),

- *When $0 < c < 1$, Σ_c consists of a simple closed regular curve on $\mathbf{C} \setminus \{0\}$ surrounding 0.*
- *When $1 < c (< \infty)$, Σ_c consists of a union of $(m+1)$ disjoint simple closed regular curves.*

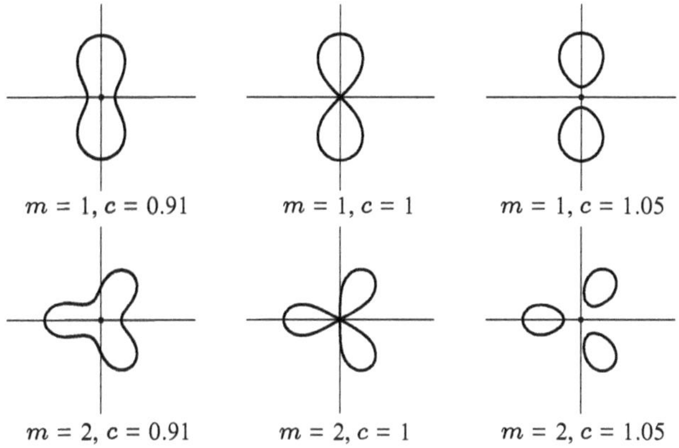

Fig. 6 The singular set Σ_c of f_c (Proposition 4.10)

- *When* $c = 1$, Σ_c *consists of* $(m+1)$ *connected components crossing at the origin.*

All points in Σ_c *are nondegenerate singularities, and*

- *If* $0 < c < 1/\sqrt{2}$, *then the* $2(m+1)$*-points where* $g = z^{m+1} + c \in \{\pm 1\}$ *in* Σ_c *correspond to swallowtails, and other singular points are cuspidal edges.*
- *If* $c = 1/\sqrt{2}$, *the* $2(m+1)$*-points* z *satisfying* $(g =)z^{m+1} + c \in \{\pm 1\}$ *in* Σ_c *correspond to swallowtails. There are* $4(m+1)$ *special singular points where* $(g =)z^{m+1} + c \in \{e^{\pm i\pi/4}\}$, *which are neither cuspidal edges, swallowtails nor cuspidal cross caps. The other singular points are cuspidal edges.*
- *If* $1/\sqrt{2} < c < 1$, *the* $2(m+1)$*-points* z *satisfying* $(g =)z^{m+1} + c \in \{\pm 1\}$ *in* Σ_c *correspond to swallowtails. The* $4(m+1)$*-points* z *satisfying*

$$(g =)z^{m+1} + c \in \{e^{\pm i\varphi}\}, \qquad \varphi = \arccos\left(\frac{1}{2c}\left(1 \pm \sqrt{2c^2 - 1}\right)\right) \tag{41}$$

correspond to cuspidal cross caps, where arccos *is the inverse of the cosine function which takes values in* $[0, \pi]$. *The other singularities are cuspidal edges.*
- *If* $c = 1$, *the end* 0 *of* f_1 *is incomplete, and in each connected component of* Σ_1, *there is one swallowtail at the points* z *satisfying* $z^{m+1} = -2$, *and there are two cuspidal cross caps where* $(g =)z^{m+1} + 1 \in \{\pm i\}$.
- *If* $1 < c (< \infty)$, *each connected component of* Σ_c *includes four swallowtails at* z *where* $(g =)z^{m+1} + c \in \{\pm 1, e^{\pm i\varphi}\}$ $(\varphi = \arccos(1/c))$, *and four cuspidal cross caps at* z *where* $g(z)$ *satisfies Eq. (41). The other singular points are cuspidal edges.*

See Table 1 (The singularities of f_∞ *are not checked yet).*

Proof. Since $g = z^{m+1} + c$, the singular set $\Sigma_c = \{|g| = 1\}$ is expressed as in Eq. (40). Then one can easily show that Σ_c is a simple closed regular curve when $0 < c < 1$,

Table 1 Singular points of f_c (Proposition 4.10)

Range of c	$z=0$ is complete or not	$z=\infty$ is complete or not	Number of comp. of Σ_c	Number of cone-like singularities	Number of swallowtails	Number of cuspidal cross caps	Number of non-generic singularities
0	Yes	Yes	1	1	–	–	–
$\left(0,\frac{1}{\sqrt{2}}\right)$	Yes	Yes	1	–	$2(m+1)$	–	–
$\frac{1}{\sqrt{2}}$	Yes	Yes	1	–	$2(m+1)$	–	$4(m+1)$
$\left(\frac{1}{\sqrt{2}},1\right)$	Yes	Yes	1	–	$2(m+1)$	$4(m+1)$	–
1	No	Yes	$m+1$	–	$m+1$	$2(m+1)$	–
$(1,\infty)$	Yes	Yes	$m+1$	–	$4(m+1)$	$4(m+1)$	–
∞	No	No	$2(m+1)$	–	–	–	–

$m+1$ simple closed regular curves when $c>1$, and $m+1$ simple closed regular curves crossing at 0 when $c=1$.

When $0<c<1$, the restriction of the secondary Gauss map on Σ_c $g|_{\Sigma_c}:\Sigma_c\to S^1\subset \mathbf{C}$ is an $m+1$ to 1 mapping, because the degree of the secondary Gauss map g is equal to $m+1$. Moreover, since $dg\neq 0$ on Σ_c, $g|_{\Sigma_c}$ is a local diffeomorphism. Thus, Σ_c is parametrized by φ such that

$$(g=)z^{m+1}+c=e^{\mathrm{i}\varphi} \qquad (0\leq\varphi<2(m+1)\pi). \tag{42}$$

Using this parametrization, α and β in Fact 1.16 are expressed as

$$\alpha=A(1-2ce^{-\mathrm{i}\varphi}+c^2e^{-2\mathrm{i}\varphi}), \qquad \beta=cB(e^{-\mathrm{i}\varphi}-ce^{-2\mathrm{i}\varphi}),$$

where A and B are nonzero real numbers. Thus, Fact 1.16 yields the conclusion for $0<c<1$.

When $c>1$, denote by Σ' a connected component of Σ_c. Then $g|_{\Sigma'}:\Sigma'\to S^1$ is a diffeomorphism because g is of degree 1 and $dg\neq 0$. Thus, one can parametrize Σ' by φ as

$$(g=)z^{m+1}+c=e^{\mathrm{i}\varphi} \qquad (0\leq\varphi<2\pi), \tag{43}$$

and get the conclusion.

When $c=1$, a connected component of the singular set is parametrized by

$$(g=)z^{m+1}+c=e^{\mathrm{i}\varphi} \qquad (-\pi<\varphi<\pi), \tag{44}$$

($\varphi=\pm\pi$ correspond to the origin), and we have the conclusion. □

Remark 4.11. The two ends of the S^3_1-catenoid in Case 2 of Proposition 4.9 are not congruent to each other, because the end $z=\infty$ is complete and $z=0$ is not. Other S^3_1-catenoids are all symmetric with respect to the two ends. In fact, let f be an S^3_1-catenoid as in Propositions 4.4, 4.5, 4.7 and Case 3 of Proposition 4.9. If we set $\varphi: z\mapsto z^{-1}$, then one can easily check that the extended hyperbolic metric $d\sigma^2$ and the Hopf differential Q associated to f are invariant under φ, which implies that f is

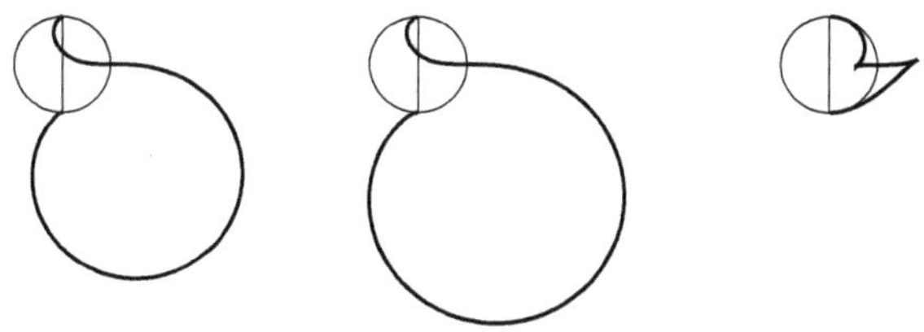

Fig. 7 Profile curves of HMC-1 surfaces corresponding to parabolic and 1-reducible elliptic catenoids (see Fig. 2): In these graphics, the Poincaré ball model for the hyperbolic sphere $S^3_H = \overline{H^3_+ \cup H^3_-}$ is used. The entire surfaces are obtained by rotating the curves around the vertical axis. The ideal boundary at infinity is shown as a *black circle*, and two copies $H^3_\pm$ of H^3 appear, one inside the ideal boundary and one outside the ideal boundary

congruent to $f \circ \varphi$. Also $\psi^* d\sigma^2 = d\sigma^2$ and $\psi^* Q = Q$ hold in Case 1 of Proposition 4.9, where $\psi \colon z \mapsto (c^2-1)^{1/(m+1)} z^{-1}$.

As pointed out in Sect. 1, the unit normal vector fields of CMC-1 faces induce harmonic-mean curvature 1 surfaces (i.e., HMC-1 surfaces) in the hyperbolic sphere (cf. Definition 1.18); see Fig. 7.

In [7], the authors gave a full classification of immersed CMC-1 trinoids in hyperbolic 3-space. The corresponding classification in de Sitter 3-space seems to be an interesting problem. Several examples of CMC-1 trinoids in de Sitter 3-space are given in Lee–Yang [14] (cf. Example 2.17).

5 Hyperbolic Metrics with At Most Two Regular Singularities

In [17], Troyanov gave a classification of spherical metrics on S^2 with two conical singularities, which are related to the classification of immersed CMC-1 catenoids in the hyperbolic 3-space H^3. In this section, we give the corresponding classification of extended hyperbolic metrics on S^2 with two regular singularities.

Proposition 5.1. *An extended hyperbolic metric on the 2-sphere which has at most one regular singular point is isometric to the spherical Poincaré metric on S^2_H. In particular, it is co-orientable. As a consequence, there are no extended hyperbolic metrics on the 2-sphere having exactly one regular singular point.*

The non-existence of spherical metrics on the 2-sphere having exactly one regular singular point was pointed out in [17]. The above assertion is an analogue of this fact.

Proof. Let $d\sigma^2$ be an extended hyperbolic metric on S^2 having at most one proper singular point. Then $S(d\sigma^2)$ is a meromorphic 2-differential having only a pole of order 2 at its regular singular point. So if $d\sigma^2$ has no regular singularity, then $S(d\sigma^2)$ gives a holomorphic 2-differential on $S^2 = \mathbf{C} \cup \{\infty\}$. Then, it must vanish identically and we can conclude that $d\sigma^2$ coincides with the spherical Poincaré metric on S^2_H. We next consider that $d\sigma^2$ has exactly one regular singularity at $p \in S^2$. Then $S(d\sigma^2)$ is a meromorphic 2-differential on $S^2 = \mathbf{C} \cup \{\infty\}$ which has only one pole of order 2. However, this is impossible, since the total sum of signed orders of poles and zeros is equal to -4 by the Riemann–Roch theorem. □

We denote by $\mathrm{Hyp}_2(S^2)$ the set of isometric equivalence classes (cf. Definition 2.2) of the co-orientable extended hyperbolic metrics with two regular singularities on S^2. On the other hand, we denote by $\mathrm{Cat}(S^3_1)$ the set of congruence classes of S^3_1-catenoids. For each $f \in \mathrm{Cat}(S^3_1)$, we denote by K the Gaussian curvature with respect to the first fundamental form ds^2_f of f. The following assertion holds:

Theorem 5.2. *The map*

$$\mathrm{Cat}(S^3_1) \ni f \longmapsto d\sigma^2_f := K ds^2_f \in \mathrm{Hyp}_2(S^2)$$

is well-defined and bijective. In this correspondence, the secondary Gauss map of f coincides with the developing map of the metric $d\sigma^2_f := K\,ds^2_f$. In particular, the monodromy representations of f and $d\sigma^2_f$ are common. Moreover, the ideal boundary set of $d\sigma^2_f$ coincides with the singular set of f.

Proof. It can be easily checked that the map is well-defined. The inverse map is given as follows: We set $G = z$, then the pair $(G, S(d\sigma^2))$ is definite. Let g be a developing map of $d\sigma^2$. Then the CMC-1 face constructed from the data (G, g) via Eq. (16) gives an S^3_1-catenoid. Since $G = z$ has no branch points, this surface has two complete ends at the regular singular points of $d\sigma^2$. □

Corollary 5.3. *Co-orientable extended hyperbolic metrics on the 2-sphere with two regular singularities are obtained from the pull-back of the spherical Poincaré metric by one of the meromorphic functions g_P, g_E, g_H, g_c, and g_∞ given in Eqs. (31)–(33), (38) and (39), defined on the universal covering of $\mathbf{C} \setminus \{0\}$.*

Since we gave a classification of S^3_1-catenoids, the theorem induced a classification of co-orientable extended hyperbolic metrics on the 2-sphere with two regular singularities. As a corollary, we can also get a classification of non-co-orientable extended hyperbolic metrics on the 2-sphere:

Corollary 5.4. *Non-co-orientable extended hyperbolic metrics on the 2-sphere with two regular singularities are obtained from the pullback of the spherical Poincaré metric by the following meromorphic functions, defined on the universal covering of $\mathbf{C} \setminus \{0\}$,*

$$g = \frac{\zeta^{(2m-1+i\tau)/2} - i}{\zeta^{(2m-1+i\tau)/2} + i} \qquad (m \in \mathbf{Z},\ \tau \geq 0), \tag{45}$$

where ζ is the canonical coordinate of $\mathbf{C}$. *If $\tau = 0$ (resp. $\tau \neq 0$), then the double covering of the metric has elliptic (resp. hyperbolic) monodromy. In particular, there are no non-co-orientable extended hyperbolic metrics on S^2 with two regular singularities with parabolic monodromy.*

An outline of the proof: The maps defined by $\pi(z) := z^2$ and $\varphi(z) := -z$ are a double covering and a covering transformation of $\mathbf{C} \setminus \{0\}$, respectively. Let $d\sigma^2$ be a non-co-orientable metric as in the statement. Since the double covering of $\mathbf{C} \setminus \{0\}$ is topologically unique, $\pi^* d\sigma^2$ gives a co-orientable metric. Since $d\sigma^2$ is non-co-orientable, $\varphi(M_+) = M_-$ holds. By Theorem 5.1, there exists $f \in \mathrm{Cat}(S_1^3)$ such that $d\sigma_f^2 = \pi^* d\sigma^2$. If f is as given in Propositions 4.4, 4.5, or Case 1 of Proposition 4.9 with $0 \le c < 1$, then $0 \in M_+$ and $\infty \in M_-$, contradicting the fact that $\varphi(M_+) = M_-$. On the other hand, if f is as given in Cases 1 and 2 of Proposition 4.9 with $c \ge 1$, then the number of components of M_+ and M_- are different, which again contradicts $\varphi(M_+) = M_-$. So f must be as in Proposition 4.7 or Case 3 of Proposition 4.9. By the reducibility of $d\sigma^2$, we may assume that our complex coordinate z coincides with the one used in Propositions 4.7 and 4.9. By setting $\zeta = z^2$, we can prove the assertion. □

Appendix A: Projective Connections

To define extended hyperbolic metrics, we recall the definition of projective connections on Riemann surfaces: Throughout this section, we fix a connected Riemann surface M.

Definition A.1 (cf. [17]). Let $\{(U_\lambda, z_\lambda)\}_{\lambda \in \Lambda}$ be a covering of M consisting of local complex coordinates. A family of meromorphic sections $P := \{h_\lambda\,(dz_\lambda)^2\}_{\lambda \in \Lambda}$ is called a *projective connection* if it satisfies

$$h_\lambda (dz_\lambda)^2 - h_\mu (dz_\mu)^2 = S_{z_\lambda}(z_\mu)(dz_\lambda)^2 \tag{46}$$

on $U_\lambda \cap U_\mu$. Here h_λ is a meromorphic function in z_λ for each $\lambda \in \Lambda$, and $S_{z_\lambda}(z_\mu)$ denotes the Schwarzian derivative of $z_\mu = z_\mu(z_\lambda)$ with respect to the coordinate z_λ.

We now give a typical example.

Example A.2. Let G be a meromorphic function on M and $\{(U_\lambda, z_\lambda)\}_{\lambda \in \Lambda}$ a covering of M consisting of local complex coordinates. We set

$$S(G) := \{S_{z_\lambda}(G)(dz_\lambda)^2\}_{\lambda \in \Lambda}.$$

Then $S(G)$ gives a projective connection on M. In this sense, our definition of the Schwarzian derivative $S(G)$ as in Eq. (15) should be considered as a projective connection.

The difference between two projective connections is a meromorphic 2-differential of M. Let $P := \{h_\lambda\,(dz_\lambda)^2\}_{\lambda \in \Lambda}$ be a projective connection on M. A point $p \in M$ is called a *singularity* of P if it is a pole of $h_\lambda (dz_\lambda)^2$ for λ satisfying

$p \in U_\lambda$. If P has no singularities, it is called a *non-singular projective connection* or *holomorphic projective connection*.

The order m of the pole of $h_\lambda\,(dz_\lambda)^2$ at a singularity p is independent of the choice of the indices λ because of Eq. (46). The integer $m(\geq 1)$ is called the *order* of the singularity at p. If $m \leq 2$, p is called a *regular singularity* of P.

The following fact is well known:

Fact A.3 ([8]). *Let $P := \{h_\lambda(dz_\lambda)^2\}_{\lambda\in\Lambda}$ be a projective connection on M, which is free of singularities. Then there exists a meromorphic function g on $\widetilde{M}$ such that $S_{z_\lambda}(g) = h_\lambda$ holds on U_λ for each index $\lambda \in \Lambda$, where $\widetilde{M}$ is the universal covering of M. Moreover, there exists a group representation $\rho : \pi_1(M) \to \mathrm{PSL}(2,\mathbf{C})$ such that*

$$g \circ T^{-1} = \rho(T) \star g \qquad \bigl(T \in \pi_1(M)\bigr).$$

The map g is called a developing map of P. The representation ρ is called the monodromy representation of P.

A developing map g of a given projective connection is not uniquely determined. For each $a \in \mathrm{SL}(2,\mathbf{C})$, $a \star g$ is also a developing map. The developing maps of P are determined up to such an ambiguity of the action of $\mathrm{SL}(2,\mathbf{C})$.

Appendix B: A Property of Subgroups in $\mathbf{PSU}(1,1)$

This appendix is an analogue of the appendix of [19], where the $\mathrm{PSU}(2)$ case was treated. Let Γ be a subgroup of $\mathrm{PSU}(1,1) = \mathrm{SU}(1,1)/\{\pm 1\}$. We prove a property of a set of groups conjugate to Γ in $\mathrm{PSL}(2,\mathbf{C})$ defined by

$$C_\Gamma := \{\sigma \in \mathrm{PSL}(2,\mathbf{C})\,;\,\sigma\Gamma\sigma^{-1} \subset \mathrm{PSU}(1,1)\}.$$

If $\sigma \in C_\Gamma$, it is obvious that $a\sigma \in C_\Gamma$ for all $a \in \mathrm{PSU}(1,1)$. So if we consider the left quotient space

$$I_\Gamma := \mathrm{PSU}(1,1)\backslash C_\Gamma,$$

the structure of the set C_Γ is completely determined. Define a map $\tilde{\varphi} : C_\Gamma \to S^3_1$ by

$$\tilde{\varphi}(\sigma) := \sigma^* e_3 \sigma,$$

where S^3_1 is the de Sitter 3-space expressed by $S^3_1 := \{a e_3 a^*\,;\, a \in \mathrm{PSL}(2,\mathbf{C})\}$. Then it induces an injective map $\varphi : I_\Gamma \to S^3_1$ such that $\varphi \circ \pi = \tilde{\varphi}$, where $\pi : C_\Gamma \to I_\Gamma$ is the canonical projection. So we can identify I_Γ with a subset $\varphi(I_\Gamma) = \tilde{\varphi}(C_\Gamma)$ of the de Sitter 3-space S^3_1. The following assertion holds.

Proposition B.1. *The subset $\varphi(I_\Gamma)$ is a point, a geodesic line, or all of S^3_1.*

Proof. For each $\gamma \in \Gamma$, we set

$$C_\gamma := \{\sigma \in \mathrm{PSL}(2,\mathbf{C})\,;\,\sigma\gamma\sigma^{-1} \in \mathrm{PSU}(1,1)\}.$$

Then we have

$$C_\Gamma := \bigcap_{\gamma\in\Gamma} C_\gamma. \tag{47}$$

The condition $\sigma\gamma\sigma^{-1} \in \mathrm{PSU}(1,1)$ is rewritten as $e_3\sigma^* e_3 \cdot \gamma = \gamma e_3 \cdot \sigma^* e_3 \cdot \sigma$, that is $e_3\sigma^* e_3\sigma \in Z_\gamma$. So we have

$$\tilde{\varphi}(C_\gamma) = S_1^3 \cap e_3 Z_\gamma, \tag{48}$$

where Z_γ is the center of $\gamma \in \Gamma$ in $\mathrm{PSL}(2,\mathbf{C})$. In the following discussions, Γ can be considered as a subgroup of $\mathrm{SU}(1,1)$ by ignoring the $\pm$-ambiguity.

If $\Gamma \subset \{\pm e_0\}$, then obviously

$$\varphi(I_\Gamma) = S_1^3.$$

So we may assume that $\Gamma \not\subset \{\pm e_0\}$. Take an element $\gamma \in \Gamma$ such that $\gamma \neq \pm e_0$. Then γ is conjugate to Λ_e or Λ_h or Λ_p in $\mathrm{SU}(1,1)$.

Firstly, we consider the case that Γ is abelian. In these three cases, Lemma B.2 implies that Γ must be a subgroup of Z_{Λ_e} or Z_{Λ_h} or Z_{Λ_p}. Then $\varphi(I_\Gamma)$ consists of a geodesic.

Next we suppose that Γ is not abelian. Then there exists an element $\gamma' \in \Gamma$ such that $\gamma\gamma' \neq \gamma'\gamma$. We take $a \in Z_\gamma \cap Z_{\gamma'}$ arbitrarily and suppose that $a \neq \pm e_0$. Since $a \in Z_\gamma$, a belongs to Z_{Λ_e} or Z_{Λ_h} or Z_{Λ_p}. Then Lemma B.2 yields that $Z_a = Z_\gamma$. Moreover, since $a\gamma' = \gamma' a$, we have $\gamma' \in Z_a$. Since $Z_a = Z_\gamma$, we have $\gamma\gamma' = \gamma'\gamma$, a contradiction. Thus $a = \pm e_0$ and $\varphi(I_\Gamma)$ consists of a point. □

In the proof of Proposition B.1, we applied the following assertion, which can be proved easily.

Lemma B.2. *We set*

$$E := \begin{pmatrix} \alpha & 0 \\ 0 & \alpha^{-1} \end{pmatrix} \qquad (\alpha \in \mathbf{C} \setminus \{1,0\}),$$

$$H := \begin{pmatrix} \alpha & \beta \\ \beta & \alpha \end{pmatrix} \qquad (\alpha^2 - \beta^2 = 1,\ \alpha,\beta \in \mathbf{C}),$$

$$P := \begin{pmatrix} 1+\mathrm{i}\beta & \beta \\ \beta & 1-\mathrm{i}\beta \end{pmatrix} \qquad (\beta \in \mathbf{C} \setminus \{0\}).$$

Then the centers of them in $\mathrm{SL}(2,\mathbf{C})$ *are given as follows:*

$$Z_E = \left\{ \begin{pmatrix} e^z & 0 \\ 0 & e^{-z} \end{pmatrix} ; z \in \mathbf{C} \right\},$$

$$Z_H = \left\{ \pm \begin{pmatrix} \cosh z & \sinh z \\ \sinh z & \cosh z \end{pmatrix} ; z \in \mathbf{C} \right\},$$

$$Z_P = \left\{ \begin{pmatrix} 1+\mathrm{i}z & z \\ z & 1-\mathrm{i}z \end{pmatrix} ; z \in \mathbf{C} \right\}.$$

Acknowledgements The authors thank Sadayoshi Kojima and Shingo Kawai for their valuable comments.

References

1. Fujimori, S.: Spacelike CMC 1 surfaces with elliptic ends in de Sitter 3-Space. Hokkaido Math. J. **35**, 289–320 (2006)
2. Fujimori, S.: Spacelike CMC 1 surfaces of genus 1 with two ends in de Sitter 3-space. Kyushu J. Math. **61**, 1–20 (2007)
3. Fujimori, S., Rossman, W.: Higher genus mean curvature 1 catenoids in hyperbolic and de Sitter 3-spaces. Kyushu J. Math. **64**, 169–180 (2010)
4. Fujimori, S., Saji, K., Umehara, M., Yamada, K.: Singularities of maximal surfaces. Math. Z. **259**, 827–848 (2008)
5. Fujimori, S., Rossman, W., Umehara, M., Yamada, K., Yang, S.-D: Spacelike mean curvature one surfaces in de Sitter 3-space. Comm. Anal. Geom. **17**, 383–427 (2009)
6. Fujimori, S., Rossman, W., Umehara, M., Yamada, K., Yang, S.-D: New maximal surfaces in Minkowski 3-space with arbitrary genus and their cousins in de Sitter 3-space. Results Math. **56**, 41–82 (2009)
7. Fujimori, S., Kawakami, Y., Kokubu, M., Rossman, W., Umehara, M., Yamada, K.: CMC-1 trinoids in hyperbolic 3-space and metrics of constant curvature one with conical singularities on the 2-sphere. Proc. Japan Acad. Ser. A. **87**, 144–149 (2011)
8. Gallo, D., Kapovich, M., Marden, A.: The monodromy group of Schwarzian equations on closed Riemann surfaces. Ann. of Math. **151**, 625–704 (2000)
9. Gálvez, J.A., Hauswirth, L., Mira, P.: Surfaces of constant curvature in R^3 with isolated singularities. preprint, arXiv:1007.2523.
10. Goldman, W.M., Projective structures with Fuchsian holonomy. J. Differ. Geom. **25**, 297–326 (1987)
11. Kokubu, M., Umehara, M.: Orientability of linear Weingarten surfaces, spacelike CMC-1 surfaces and maximal surfaces. Math. Nachr. **284**, 1903–1918 (2011)
12. Kokubu, M., Rossman, W., Umehara, M., Yamada, K.: Flat fronts in hyperbolic 3-space and their caustics. J. Math. Soc. Japan **59**, 265–299 (2007)
13. Kraus, D., Roth, O., Sugawa, T.: Metrics with conical singularities on the sphere and sharp extensions of the theorems of Landau and Schottky. Math. Z. **267**, 851–868 (2011)
14. Lee, S., Yang, S.-D.: Spacelike constant mean curvature 1 trinoids in de Sitter space. Osaka J. Math. **43**, 641–663 (2006)
15. Okuyama, Y., Yamanoi, K.: A generalization of completeness lemma in minimal surface theory. unpublished work.
16. Rossman, W., Umehara, M., Yamada, K.: Mean curvature 1 surfaces in hyperbolic 3-space with low total curvature I. Hiroshima Math. J. **34**, 21–56 (2004)
17. Troyanov, M.: Metrics of constant curvature on a sphere with two conical singularities. In: Lecture Notes in Mathematics, vol. 1410, pp. 296–306. Springer, Berlin (1991)
18. Umehara, M., Yamada, K.: Surfaces of constant mean curvature c in $H^3(-c^2)$ with prescribed hyperbolic Gauss map. Math. Ann. **304**, 203–224 (1996)
19. Umehara, M., Yamada, K.: Metrics of constant curvature one with three conical singularities on the 2-sphere. Illinois J. Math. **44**, 72–94 (2000)

20. Umehara, M., Yamada, K.: Maximal surfaces with singularities in Mikowski space. Hokkaido Math. J. **35**, 13–40 (2006)
21. Umehara, M., Yamada, K.: Applications of a completeness lemma in minimal surface theory to various classes of surfaces. Bull. London Math. Soc. **43**, 191–199 (2011)
22. Umehara, M., Yamada, K.: CORRIGENDUM: Applications of a completeness lemma in minimal surface theory to various classes of surfaces. to appear in Bull. London Math. Soc. **44**, 617–618 (2012)

Calabi–Bernstein Results and Parabolicity of Maximal Surfaces in Lorentzian Product Spaces

Alma L. Albujer and Luis J. Alías

Dedicated to Professor Ángel Ferrández on the occasion of his 60th birthday.

1 Introduction

A *maximal surface* in a 3-dimensional Lorentzian manifold is a space-like surface with zero mean curvature. By *space-like*, we mean that the induced metric on the surface from the Lorentzian metric of the ambient space is a Riemannian metric. The terminology maximal comes from the fact that maximal surfaces locally maximize area among all nearby surfaces having the same boundary [12,22]. These surfaces have nice physical and mathematical properties.

From a physical point of view, maximal surfaces and, more generally, constant mean curvature space-like hypersurfaces in a Lorentzian space of arbitrary dimension, have a great importance. In fact, these surfaces are used as initial data for solving the Cauchy problem for the Einstein equations on General Relativity (see [31]). Specifically, in 1944 Lichnerowicz [31] proved that the Cauchy problem with initial conditions on a maximal surface is reduced to a first-order linear differential system, and to a second-order nonlinear elliptic differential equation.

From a mathematical point of view, space-like surfaces with constant mean curvature, and in particular maximal surfaces, present interesting Bernstein-type

A.L. Albujer
Departamento de Matemáticas, Universidad de Córdoba, Campus Universitario de Rabanales, 14071 Córdoba, Spain
e-mail: alma.albujer@uco.es

L.J. Alías (✉)
Departamento de Matemáticas, Universidad de Murcia, Campus de Espinardo, 30100 Espinardo, Murcia, Spain
e-mail: ljalias@um.es

M. Sánchez et al. (eds.), *Recent Trends in Lorentzian Geometry*, Springer Proceedings in Mathematics & Statistics 26, DOI 10.1007/978-1-4614-4897-6_2,

properties. A mathematical property or result is said to be of Bernstein type when, under some suitable assumptions, it allows us to conclude some uniqueness results in a given ambient space. In fact, one of the most relevant results in the context of global geometry of space-like surfaces is the well-known Calabi–Bernstein theorem. This result was established in 1970 by Calabi [14], and, in its nonparametric version, it asserts that the only entire maximal graphs in the 3-dimensional Lorentz–Minkowski space, $\mathbb{R}^3_1$, are the space-like planes. The Calabi–Bernstein theorem can also be formulated in a parametric version. In that case, it states that the only complete maximal surfaces in the 3-dimensional Lorentz–Minkowski space, $\mathbb{R}^3_1$, are the space-like planes.

In Sect. 2 we review the Calabi–Bernstein theorem, and we summarize some of the different extensions and generalizations of it. Specifically, we will sketch three different proofs of the theorem. The first of them is a very neat and simple proof of Romero [39] which only requires the Liouville theorem for harmonic functions on the Euclidean plane $\mathbb{R}^2$. The second one is a proof of Alías and Palmer [9] via a duality between minimal graphs in $\mathbb{R}^3$ and maximal graphs in $\mathbb{R}^3_1$. Finally, the third proof is a local approach of Alías and Palmer [10] based on a local integral inequality for the Gaussian curvature of a maximal surface. The reader can consult [7] for a survey on the classical Calabi–Bernstein theorem in $\mathbb{R}^3_1$ and related results from another point of view.

The aim of the rest of this chapter is to describe some recent results obtained by the authors in [3–6] (see also [2]) for maximal surfaces immersed in a Lorentzian product space of the form $M^2 \times \mathbb{R}_1$ (see Sect. 3 for the details). We will also survey some related results obtained by several authors during these last years.

Our first main result is Theorem 5, which states that any complete maximal surface Σ immersed into a Lorentzian product $M^2 \times \mathbb{R}_1$, where M is a (necessarily complete) Riemannian surface with nonnegative Gaussian curvature, must be totally geodesic. Moreover, if M is non-flat, we conclude that Σ must be a slice $M \times \{t_0\}$, $t_0 \in \mathbb{R}$. Here, by *complete*, it is meant, as usual, that the induced Riemannian metric on Σ from the ambient Lorentzian metric is complete. In Theorem 7 we prove that the same happens if Σ is complete with respect to the metric induced from the Riemannian product $M^2 \times \mathbb{R}$. This allows us to give a nonparametric version of our Calabi–Bernstein result, Theorem 8, where we prove that any entire maximal graph in $M^2 \times \mathbb{R}_1$ must be totally geodesic and that the only entire solutions to the maximal surface equation on any complete, non-flat, Riemannian surface M with nonnegative Gaussian curvature are the constant functions.

It is worth pointing out that, in all the above results, the assumption on the Gaussian curvature of M is necessary as shown by the fact that, when $M = \mathbb{H}^2$ is the hyperbolic plane, there exist examples of complete maximal surfaces in $\mathbb{H}^2 \times \mathbb{R}_1$ which are not totally geodesic, as well as examples of nontrivial entire maximal graphs over $\mathbb{H}^2$. We construct these examples in two different ways. Examples 1 and 2 are a consequence of a simple but nice duality result between solutions to the minimal surface equation in a Riemannian product $M^2 \times \mathbb{R}$ and solutions to the maximal surface equation in a Lorentzian product $M^2 \times \mathbb{R}_1$, Theorem 9, that extends [9, Theorem 3]. On the other hand, Examples 3 and 4 are explicit examples of nontrivial entire maximal graphs in $\mathbb{H}^2 \times \mathbb{R}_1$, which have been found by looking for explicit solutions of the maximal surface equation on $\mathbb{H}^2$. Particularly interesting is the fact that the entire maximal graphs given in Examples 2 and 3 are not complete. As is well known, such circumstance cannot occur in the Lorentz–Minkowski space $\mathbb{R}^3_1$ since, by a result of Cheng and Yau [15], closed surfaces in $\mathbb{R}^3_1$ with constant mean curvature are necessarily complete.

We also introduce a local approach to our Calabi–Bernstein results, which is based on a parabolicity criterium for maximal surfaces with nonempty smooth boundary in $M^2 \times \mathbb{R}_1$, Theorem 10. In particular, we derive that every maximal graph over a star-like domain $\Omega \subseteq M$ is parabolic. This allows us to give an alternative proof of the nonparametric version of the Calabi–Bernstein result, Theorem 8, for maximal entire graphs in $M^2 \times \mathbb{R}_1$.

Finally, in Theorem 11 we present another local approach based on a local integral inequality for the squared norm of the second fundamental form of a maximal surface in $M^2 \times \mathbb{R}_1$. This approach generalizes a previous result of Alías and Palmer, [10, Theorem 1], where they gave a local integral inequality for the Gaussian curvature of a maximal surface in $\mathbb{R}^3_1$. This result only involved the local geometry of the surface and the image of its Gauss map. Under completeness assumption, and as a consequence of Theorem 11, we provide an alternative proof of the parametric version of our Calabi–Bernstein result, Theorem 5, in the case where M^2 is analytic.

2 The Classical Calabi–Bernstein Theorem in $\mathbb{R}^3_1$

Let $\mathbb{R}^3_1$ denote the 3-dimensional Lorentz–Minkowski space, that is, the real vector space $\mathbb{R}^3$ endowed with the Lorentzian metric

$$\langle , \rangle = (dx_1)^2 + (dx_2)^2 - (dx_3)^2,$$

where (x_1, x_2, x_3) are the canonical coordinates in $\mathbb{R}^3$. A smooth immersion $f : \Sigma^2 \to \mathbb{R}^3_1$ of a connected surface Σ is said to be a *space-like surface* if the induced metric via f is a Riemannian metric on Σ, which, as usual, is also denoted by $\langle , \rangle$. A space-like surface Σ is said to be *complete* if the Riemannian induced metric is a complete metric on Σ.

Let us observe that every space-like surface in $\mathbb{R}^3_1$ is orientable. In fact, observe that $(0,0,1)$ is a unit time-like vector field globally defined on $\mathbb{R}^3_1$, which determines a time orientation on $\mathbb{R}^3_1$. Therefore, it allows us to choose the unique time-like unit normal field N on Σ which is in the same time orientation as $(0,0,1)$, and hence we may assume that Σ is oriented by N. We will refer to N as the *future-directed Gauss map* of the surface Σ. The field N can be regarded as a map $N : \Sigma \to \mathbb{H}^2$, where $\mathbb{H}^2$ denotes the 2-dimensional hyperbolic space

$$\mathbb{H}^2 = \{x = (x_1, x_2, x_3) \in \mathbb{R}^3_1 \mid \langle x, x \rangle = -1, x_3 \geq 1\}.$$

The image $N(\Sigma)$ will be called the *hyperbolic image* of the space-like surface.

Let $\overline{\nabla}$ and ∇ be the Levi–Civita connections of $\mathbb{R}^3_1$ and Σ, respectively. Then, the Gauss and Weingarten formulas for Σ in $\mathbb{R}^3_1$ are given, respectively, by

$$\overline{\nabla}_X Y = \nabla_X Y - \langle AX, Y\rangle N \tag{1}$$

and

$$AX = -\overline{\nabla}_X N, \tag{2}$$

for all tangent vector fields $X, Y \in \mathfrak{X}(\Sigma)$, where $A : \mathfrak{X}(\Sigma) \to \mathfrak{X}(\Sigma)$ stands for the shape operator of Σ in $\mathbb{R}^3_1$ with respect to its future-directed Gauss map N. The *mean curvature* of the surface associated to A is defined by

$$H = -\frac{1}{2}\mathrm{tr}(A).$$

A space-like surface is said to be *maximal* if H vanishes on Σ, $H \equiv 0$.

As is well known, the Gaussian curvature of the space-like surface Σ, K, is described in terms of the shape operator of Σ by the Gauss equation, [36, Theorem 4.20],

$$K = -\det(A).$$

Besides, if Σ is maximal then

$$A^2 = \frac{1}{2}\|A\|^2 I,$$

where I denotes the identity map on $\mathfrak{X}(\Sigma)$ and $\|A\|^2 = tr(A^2) = -2\det(A)$. Therefore, the Gauss equation becomes

$$K = \frac{1}{2}\|A\|^2. \tag{3}$$

On the other hand, the Codazzi equation of the surface is given by

$$(\nabla_Y A)X = (\nabla_X A)Y, \tag{4}$$

where $(\nabla_Y A)X = \nabla_Y(AX) - A(\nabla_Y X)$, [36, Corollary 4.34].

Let $f : \Sigma^2 \to \mathbb{R}^3_1$ be a maximal surface oriented by its future-directed Gauss map N. For each vector $a \in \mathbb{R}^3_1$ we can consider the smooth function on Σ given by $\langle N, a\rangle$. With a straightforward computation we get from Eq. (2) that its gradient is given by

$$\nabla\langle N, a\rangle = -A(a^\top), \tag{5}$$

where $a^\top$ denotes the tangential component of a along Σ; that is,

$$a^\top = a + \langle N, a\rangle N \in \mathfrak{X}(\Sigma). \tag{6}$$

On the other hand, $\overline{\nabla}_X a = 0$ for any $X \in \mathfrak{X}(\Sigma)$. Thus, from the Gauss and Weingarten formulas (1) and (2) we easily get from Eq. (6) that

$$\nabla_X a^\top = -\langle N, a\rangle AX$$

for every $X \in \mathfrak{X}(\Sigma)$, and using the Codazzi equation (4) we obtain

$$\nabla_X(\nabla\langle N, a\rangle) = -(\nabla_{a^\top} A)X + \langle N, a\rangle A^2 X.$$

Therefore, the Laplacian of $\langle N, a\rangle$ is given by

$$\Delta\langle N, a\rangle = \|A\|^2\langle N, a\rangle. \tag{7}$$

From Eq. (6) we have $\|a^\top\|^2 = \langle a, a\rangle + \langle N, a\rangle^2$, where for any $X \in \mathfrak{X}(\Sigma)$ $\|X\|$ stands for its norm with respect to the induced metric on Σ from the metric of $\mathbb{R}^3_1$. Thus, from Eq. (5) we obtain

$$\|\nabla\langle N, a\rangle\|^2 = \frac{1}{2}\|A\|^2\left(\langle N, a\rangle^2 + \langle a, a\rangle\right). \tag{8}$$

2.1 Space-Like Graphs and the Calabi–Bernstein Theorem

Let $\Omega \subseteq \mathbb{R}^2$ be an open subset of $\mathbb{R}^2$, then any smooth function $u = u(x_1, x_2) \in \mathscr{C}^\infty(\Omega)$ determines a *graph* over Ω given by

$$\Sigma(u) = \{(x_1, x_2, u(x_1, x_2)) \,|\, (x_1, x_2) \in \Omega\} \subset \mathbb{R}^3_1.$$

The metric induced on Ω from the metric of $\mathbb{R}^3_1$ is given by

$$\langle,\rangle = \langle,\rangle_o - du^2,$$

where $\langle,\rangle_o$ stands for the Euclidean metric on $\mathbb{R}^2$. Therefore, $\Sigma(u)$ is a space-like surface in $\mathbb{R}^3_1$ if and only if $|D_o u|_o^2 < 1$ everywhere on Ω, where $D_o u = (\frac{\partial u}{\partial x_1}, \frac{\partial u}{\partial x_2})$ denotes the Euclidean gradient of u in Ω and $|D_o u|_o$ its Euclidean norm. On the other hand, it is not difficult to see that the vector field

$$N(x) = \frac{1}{\sqrt{1 - |D_o u(x)|_o^2}}(D_o u(x), 1), \quad x \in \Omega,$$

defines the future-pointing Gauss map of $\Sigma(u)$ and that the shape operator with respect to N is given by

$$AX = -\frac{1}{\sqrt{1 - |D_o u|_o^2}} D_{oX} D_o u - \frac{\langle D_{oX} D_o u, D_o u\rangle_o}{(1 - |D_o u|_o^2)^{3/2}} D_o u$$

for every tangent vector field X on Ω, where D_o also denotes the Levi–Civita connection in Ω with respect to the metric $\langle,\rangle_o$. If follows from here that the mean curvature $H(u)$ of a space-like graph $\Sigma(u)$ is given by

$$2H(u) = \mathrm{Div}_o\left(\frac{D_o u}{\sqrt{1-|D_o u|_o^2}}\right),$$

where Div_o stands for the Euclidean divergence operator on Ω. In particular, $\Sigma(u)$ is a maximal graph if and only if the function u satisfies the following partial differential equation on the domain Ω:

$$\mathrm{Div}_o\left(\frac{D_o u}{\sqrt{1-|D_o u|_o^2}}\right) = 0, \quad |D_o u|_o^2 < 1. \tag{9}$$

The graph is said to be *entire* when $\Omega = \mathbb{R}^2$.

An interesting remark on the topology of complete space-like surfaces in $\mathbb{R}^3_1$ is the fact that they are necessarily entire space-like graphs over the space-like plane $x_3 = 0$, in the sense that the projection $\Pi : \Sigma \to \mathbb{R}^2$ of Σ onto the plane $x_3 = 0$ is a diffeomorphism. In fact, it is not difficult to see that since Σ is space like, $\Pi : \Sigma \to \mathbb{R}^2$ is a local diffeomorphism which increases the distance between the Riemannian surfaces Σ and $\mathbb{R}^2$, that is, $\Pi^*(\langle,\rangle_o) \geq \langle,\rangle$. The completeness of Σ implies that $\Pi(\Sigma) = \mathbb{R}^2$ and that Π is a covering map [28, Chap. VIII, Lemma 8.1]. In particular, there exists no compact (without boundary) space-like surface in $\mathbb{R}^3_1$. However, it is worth pointing out that no converse of this fact is true in general. More precisely, there exist examples of space like entire graphs in $\mathbb{R}^3_1$ which are not complete. For instance, let $\phi : \mathbb{R} \to \mathbb{R}$ be a real function defined by

$$\phi(x) = \int_0^{|x|} \sqrt{1-e^{-t}}dt$$

when $|x| \geq 1$, and $\phi(x) = f(x)$ when $|x| < 1$, where $f \in \mathscr{C}^\infty(\mathbb{R})$ is a smooth extension satisfying $f'(x)^2 < 1$ for all $x \in (-1,1)$. Then, it is not difficult to prove that the entire graph of $\mathbb{R}^3_1$ given by $x_3 = \phi(x_1)$ defines a space-like surface which is not complete. This fact represents an important difference between the behavior of surfaces in the Euclidean space $\mathbb{R}^3$ and that of space-like surfaces in the Lorentz–Minkowski space $\mathbb{R}^3_1$. Actually, let us recall that every embedded surface in the Euclidean space which is a closed subset in $\mathbb{R}^3$ is necessarily complete, while there exist examples of complete embedded surfaces in $\mathbb{R}^3$ which are not closed.

In this context, the classical Calabi–Bernstein theorem asserts in its parametric version that

Theorem 1. *The only complete maximal surfaces in the Lorentz–Minkowski space $\mathbb{R}^3_1$ are the space-like planes.*

By the comments above, this result can also be stated in a nonparametric version, in terms of entire maximal graphs.

Theorem 2. *The only entire maximal graphs in the Lorentz–Minkowski space* $\mathbb{R}^3_1$ *are the space-like planes. That is, the only entire solutions to the maximal surface equation (9) on the Euclidean plane* $\mathbb{R}^2$ *are affine functions.*

This theorem was firstly proved by Calabi [14] in 1970, and extended later to the general n-dimensional case by Cheng and Yau [15] as an application of a Simons-type formula in the context of space-like hypersurfaces in a Lorentz ambient space. After that, several authors have approached the Calabi–Bernstein theorem for maximal surfaces ($n = 2$) from different points of view, providing diverse extensions and new proofs of it, both in parametric and nonparametric versions. For instance, Kobayashi introduced in [27] an appropriate Enneper-Weierstrass representation of a maximal surface, providing a new approach to the theorem. McNertey [32] and Estudillo and Romero [18–20] have also considered the study of similar problems in terms of a local complex representation of the surface. We will sketch here three more recent proofs given by Romero [39] and by the second author jointly with Palmer [9, 10]. The choice of these proofs is due to the fact that they have inspired some of the results presented by the authors for maximal surfaces in a Lorentzian product space $M^2 \times \mathbb{R}_1$.

2.2 Romero's Proof Based on the Liouville Theorem for Harmonic Functions on $\mathbb{R}^2$

The proof by Romero of the nonparametric version of the classical Calabi–Bernstein theorem is based on the Liouville theorem for harmonic functions on $\mathbb{R}^2$. It is inspired in a simple proof of the classical Bernstein theorem given by Chern [16].

Let $\Sigma(u)$ be an entire maximal graph in $\mathbb{R}^3_1$ over the space-like plane $x_3 = 0$ and consider $a \in \mathbb{R}^3_1$ a light-like vector ($a \neq 0$, $\langle a,a\rangle = 0$) with $\langle N,a\rangle > 0$. From Eqs. (7) and (8), we get

$$\Delta\left(\frac{1}{\langle N,a\rangle}\right) = -\frac{\Delta\langle N,a\rangle}{\langle N,a\rangle^2} + \frac{2\|\nabla\langle N,a\rangle\|^2}{\langle N,a\rangle^3} = 0,$$

that is, $1/\langle N,a\rangle$ is a positive harmonic function globally defined on Σ.

On the other hand, if we choose $b = (0,0,-1) \in \mathbb{R}^3_1$ then

$$\langle N,b\rangle = \frac{1}{\sqrt{1-|D_o u|_o^2}} \geq 1.$$

And, with a similar computation, we obtain

$$\Delta \log(1+\langle N,b\rangle) = \frac{\Delta\langle N,b\rangle}{\langle N,b\rangle+1} - \frac{\|\nabla\langle N,b\rangle\|^2}{(\langle N,b\rangle+1)^2} = K. \tag{10}$$

Let us consider now on $\Sigma(u)$ the conformal metric

$$\hat{g} = (1+\langle N,b\rangle)^2 g,$$

where for simplicity $g = \langle , \rangle$ is the induced Riemannian metric on $\Sigma(u)$ from $\mathbb{R}^3_1$. As is well known, the Gaussian curvature $\hat{K}$ of $(\Sigma, \hat{g})$ is given by

$$(1 + \langle N, b \rangle)^2 \hat{K} = K - \Delta \log(1 + \langle N, b \rangle).$$

Thus, we can conclude from Eq. (10) that $\hat{g}$ is a flat metric on $\Sigma(u)$.

On the other hand, observe that for every $X \in \mathfrak{X}(\Sigma(u))$ it holds

$$\begin{aligned}\hat{g}(X,X) &\geq \langle N,b\rangle^2 g(X,X) = \langle N,b\rangle^2 \left(|X^*|_o^2 - \langle D_o u, X^*\rangle_o^2\right) \\ &\geq \langle N,b\rangle^2 \left(1 - |D_o u|_o^2\right) |X^*|_o^2 = |X^*|_o^2 = \frac{1}{2}\left(g(X,X) + g'(X,X)\right) \\ &\geq \frac{1}{2} g'(X,X),\end{aligned}$$

where X^* is the projection of X onto the plane $x_3 = 0$ and g' stands for the Riemannian metric induced in $\Sigma(u)$ from the Euclidean 3-dimensional space $\mathbb{R}^3$. This implies that $\hat{L} \geq 1/\sqrt{2}L'$, where $\hat{L}$ and L' denote the length of a curve on $\Sigma(u)$ with respect to the Riemannian metrics $\hat{g}$ and g', respectively. Moreover, since $\Sigma(u)$ is closed in $\mathbb{R}^3_1$ the metric g' is complete on $\Sigma(u)$. As a consequence, $\hat{g}$ is also complete on $\Sigma(u)$.

Summing up, $(\Sigma(u), \hat{g})$ is a flat and complete surface. Therefore, using Cartan's theorem we have a global isometry from the Euclidean plane $\mathbb{R}^2$ onto the entire graph $\Sigma(u)$ endowed with the metric $\hat{g}$. The invariance of harmonic functions under conformal changes of the metric and this isometry allow us to induce $1/\langle N, a \rangle$ on a positive harmonic function on $\mathbb{R}^2$, which must be constant by Liouville theorem. Therefore, the function $\langle N, a \rangle$ is constant on $\Sigma(u)$, which implies by Eq. (7) that the maximal graph is totally geodesic, and so it is a space-like plane.

2.3 Alías and Palmer's Proof Based on a Duality Result

Here we present a duality result given by the second author and Palmer in [9] between minimal entire graphs in the Euclidean space $\mathbb{R}^3$, that is, entire graphs with vanishing mean curvature function and maximal entire graphs in the Lorentz–Minkowski space $\mathbb{R}^3_1$. Specifically, they show how the nonparametric version of the Calabi–Bernstein theorem in $\mathbb{R}^3_1$ can be seen as a consequence of the classical Bernstein theorem on minimal surfaces in the Euclidean space $\mathbb{R}^3$, and viceversa. Before giving this duality result, let us recall that the classical Bernstein theorem states that the only entire minimal graphs in $\mathbb{R}^3$ are the planes. Equivalently, the only entire solutions to the minimal surface equation

$$\mathrm{Div}_o\left(\frac{D_o u}{\sqrt{1 + |D_o u|_o^2}}\right) = 0 \tag{11}$$

are affine functions.

Theorem 3 ([9, Theorem 3]). *Let $\Omega \subseteq \mathbb{R}^2$ be a simply connected domain. There exists a non-affine $\mathscr{C}^2$ solution to the minimal surface equation on Ω (11) if and only if there exists a non-affine $\mathscr{C}^2$ solution to the maximal surface equation on Ω (9).*

Proof. Assume that u is a non-affine solution of the minimal surface equation (11). Recall that for a vector field X on $\mathbb{R}^2$ it holds that

$$(\mathrm{Div}_o X)dx_1 \wedge dx_2 = d\omega_{J_oX},$$

where J_o denotes the positive $\pi/2$-rotation in the plane and ω_{J_oX} denotes the 1-form on $\mathbb{R}^2$ which is metrically equivalent to the field J_oX, that is, ω_{J_oX} satisfies

$$\omega_{J_oX}(Y) = \langle J_oX, Y\rangle$$

for every $Y \in \mathfrak{X}(\mathbb{R}^2)$. Then Eq. (11) is equivalent to the fact that ω_{J_oU} is closed on Ω, where U is the field on Ω given by

$$U = \frac{D_ou}{\sqrt{1+|D_ou|_o^2}}.$$

Then, since the domain Ω is simply connected, we can write

$$J_oU = D_ow \tag{12}$$

for a certain $\mathscr{C}^2$ function w on Ω. Since J_o is an isometry, there follows

$$|D_ow|_o^2 = \frac{|D_ou|_o^2}{1+|D_ou|_o^2} < 1, \tag{13}$$

and also

$$1+|D_ou|_o^2 = \frac{1}{1-|D_ow|_o^2}. \tag{14}$$

From Eq. (13) we conclude that w defines a space-like graph. Besides, using that $J_o^2 = -\mathrm{id}$, we obtain from Eqs. (12) and (14) that

$$J_o\left(\frac{D_ow}{\sqrt{1-|D_ow|_o^2}}\right) = \sqrt{1+|D_ou|_o^2}J_o(D_ow) = D_o(-u).$$

Thus, the maximal surface equation (9) holds on Ω.

Suppose that the function w that we have obtained is affine, then D_ow is a constant vector. Thus, by Eq. (14) $|D_ou|_o$ is also constant, and so it is D_ou because of Eq. (12), contradicting the assumption that u is non-affine.

A similar argument, starting with a non-affine solution of Eq. (9), produces a non-affine solution of Eq. (11) on Ω. □

In the particular case where we consider entire graphs, that is when $\Omega = \mathbb{R}^2$, it follows from Theorem 3 the equivalence between the Bernstein theorem in $\mathbb{R}^3$ and the nonparametric version of the Calabi–Bernstein theorem in $\mathbb{R}^3_1$.

2.4 *Alías and Palmer's Proof Based on a Local Integral Inequality for the Gaussian Curvature*

Finally, we present the proof of the parametric version of the classical Calabi–Bernstein theorem given again by the second author jointly with Palmer [10]. The proof is a consequence of the following local bound for the total curvature of geodesic discs in a maximal surface in $\mathbb{R}^3_1$.

Theorem 4 ([10, Theorem 1]). *Let $f : \Sigma \to \mathbb{R}^3_1$ be a maximal surface in the Lorentz–Minkowski space. Let p be a point of Σ, and let $R > 0$ be a positive real number such that the geodesic disc of radius R about p satisfies $D(p,R) \subset\subset \Sigma$. Then, for all $0 < r < R$, the total curvature of the geodesic disc $D(p,r)$ of radius r centered at p satisfies*

$$0 \leq \int_{D(p,r)} K dA \leq c_r \frac{L(r)}{r \log(R/r)}, \tag{15}$$

where $L(r)$ denotes the length of $\partial D(p,r)$, the geodesic disc of radius r centered at p, and

$$c_r = \frac{\pi^2}{8} \frac{(1+\cosh^2 \rho_r)^2}{\cosh \rho_r \arctan(\cosh \rho_r)} > 0.$$

Here, ρ_r denotes the radius of a geodesic disc in $\mathbb{H}^2$ containing the hyperbolic image of $D(p,r)$.

Before giving the proof of Theorem 4, let us observe how this result implies the parametric version of the classical Calabi–Bernstein theorem, Theorem 1. Indeed, if Σ is complete, for any arbitrary fixed $p \in \Sigma$ and a fixed r, R can approach to infinity in inequality (15), which taking limits gives

$$\int_{D(p,r)} K dA = 0.$$

Taking into account that from Eq. (3) the Gaussian curvature of a maximal surface in $\mathbb{R}^3_1$ is always nonnegative, this yields $K = 0$ on Σ, and Σ must be a space-like plane.

The proof of Theorem 4 follows from the following (intrinsic) local integral inequality on an analytic Riemannian metric with nonnegative Gaussian curvature.

Lemma 1 ([10, Lemma 3]). *Let Σ be an analytic surface endowed with an analytic Riemannian metric with nonnegative Gaussian curvature $K \geq 0$. Let u be a smooth function on Σ which satisfies*

$$u \Delta u \geq 0$$

on Σ. Then, for $0 < r < R$,

$$\int_{D_r} u \Delta u \leq \frac{2L(r)}{r \log(R/r)} \sup_{D_R} u^2,$$

where D_r denotes the geodesic disc of radius r centered at a fixed point in Σ, $D_r \subset D_R \subset\subset \Sigma$, and $L(r)$ denotes the length of ∂D_r, the geodesic disc of radius r.

The proof of Lemma 1 follows from [8, Lemma 2.1 and inequality (2.4)].

Proof of Theorem 4. Let us assume that the hyperbolic image of $D(p,r)$ is contained in a geodesic disc $\tilde{D}(a,\rho_r)$ in $\mathbb{H}^2$ of radius ρ_r and centered at the point $a \in \mathbb{H}^2$. That is,

$$N(D(p,r)) \subseteq \tilde{D}(a,\rho_r) = \{x \in \mathbb{H}^2 \,|\, 1 \leq -\langle x,a\rangle \leq \cosh\rho_r\},$$

so that

$$1 \leq -\langle N(q),a\rangle \leq \cosh\rho_r$$

for all $q \in D(p,r)$.

Observe that, since $f : \Sigma \to \mathbb{R}^3_1$ is a maximal surface, Σ is an analytic Riemannian surface with nonnegative Gaussian curvature, so we may apply Lemma 1 to an appropriate smooth function u. The idea of the proof is to apply it to the smooth function on Σ given by $u = \arctan(-\langle N,a\rangle)$. From Eqs. (7) and (8) it can be easily seen that

$$\Delta u = -\frac{1}{1+\langle N,a\rangle^2}\Delta\langle N,a\rangle + \frac{2\langle N,a\rangle}{(1+\langle N,a\rangle^2)^2}\|\nabla\langle N,a\rangle\|^2 = \frac{-4K\langle N,a\rangle}{(1+\langle N,a\rangle^2)^2},$$

and therefore

$$u\Delta u = \frac{-4\langle N,a\rangle \arctan(-\langle N,a\rangle)}{(1+\langle N,a\rangle^2)^2}K = \phi(-\langle N,a\rangle)K \geq 0, \tag{16}$$

where $\phi : \mathbb{R} \to \mathbb{R}$ is given by

$$\phi(t) = \frac{4t\arctan t}{(1+t^2)^2}. \tag{17}$$

It is immediate to check that the function $\phi(t)$ is strictly decreasing for $t \geq 1$. As a consequence, for $t \in [1,\cosh\rho_r]$ $\phi(t)$ is bounded from below by

$$\phi(t) \geq \phi(\cosh\rho_r) = \frac{4\cosh\rho_r \arctan(\cosh\rho_r)}{(1+\cosh^2\rho_r)^2}.$$

Hence, at each point $q \in D(p,r)$ we obtain, from Eq. (16),

$$u(q)\Delta u(q) \geq \frac{4\cosh\rho_r \arctan(\cosh\rho_r)}{(1+\cosh^2\rho_r)^2}K(q) \geq 0.$$

Integrating now this inequality over $D(p,r)$, and using Lemma 1, we conclude that

$$0 \leq \frac{4\cosh\rho_r \arctan(\cosh\rho_r)}{(1+\cosh^2\rho_r)^2}\int_{D(p,r)} K dA \leq \int_{D(p,r)} u\Delta u dA \leq \frac{\pi^2}{2}\frac{L(r)}{r\log(R/r)}.$$

The theorem follows easily from this inequality. □

3 Some Preliminaries on Lorentzian Product Spaces

Let $(M^2, \langle,\rangle_M)$ be a connected Riemannian surface and consider the topological product space $M^2 \times \mathbb{R}$ endowed with the Lorentzian metric

$$\langle,\rangle = \pi_M^*(\langle,\rangle_M) - \pi_{\mathbb{R}}^*(dt^2),$$

where π_M and $\pi_{\mathbb{R}}$ denote the projections from $M \times \mathbb{R}$ onto each factor. For simplicity, we will write simply

$$\langle,\rangle = \langle,\rangle_M - dt^2,$$

and we will denote by $M^2 \times \mathbb{R}_1$ the 3-dimensional product manifold $M^2 \times \mathbb{R}$ endowed with that Lorentzian metric.

The basic notions on maximal surfaces in a Lorentzian product space $M^2 \times \mathbb{R}_1$ that we need along this chapter are similar to the preliminaries stated in Sect. 2 for maximal surfaces in $\mathbb{R}^3_1$. However, for the sake of completeness, we will present them again in our new context.

A smooth immersion $f : \Sigma^2 \to M^2 \times \mathbb{R}_1$ of a connected surface Σ^2 is said to be a *space-like surface* if f induces a Riemannian metric on Σ, which as usual is also denoted by $\langle,\rangle$. In that case, since

$$\partial_t = (\partial/\partial_t)_{(x,t)}, \quad x \in M, t \in \mathbb{R}$$

is a unitary time-like vector field, globally defined on the ambient spacetime $M^2 \times \mathbb{R}_1$, then there exists a unique unitary time-like normal field N, globally defined on Σ, which is in the same time orientation as ∂_t, so that

$$\langle N, \partial_t \rangle \leq -1 < 0 \quad \text{on} \quad \Sigma.$$

We will refer to N as the *future-pointing Gauss map* of Σ, and we will denote by $\Theta : \Sigma \to (-\infty, -1]$ the smooth function on Σ given by $\Theta = \langle N, \partial_t \rangle$. Observe that the function Θ measures the hyperbolic angle θ between the future-pointing vector fields N and ∂_t along Σ. Indeed, they are related by $\cosh\theta = -\Theta$.

Let $\overline{\nabla}$ and ∇ denote the Levi–Civita connections in $M^2 \times \mathbb{R}_1$ and Σ, respectively. Then the Gauss and Weingarten formulae for the space-like surface $f : \Sigma^2 \to M^2 \times \mathbb{R}_1$ are given by

$$\overline{\nabla}_X Y = \nabla_X Y - \langle AX, Y \rangle N \tag{18}$$

and

$$AX = -\overline{\nabla}_X N, \tag{19}$$

for any tangent vector fields $X, Y \in \mathfrak{X}(\Sigma)$. Here $A : \mathfrak{X}(\Sigma) \to \mathfrak{X}(\Sigma)$ stands for the *shape operator* (or *second fundamental form*) of Σ with respect to its future-pointing Gauss map N. As is well known, the Gaussian curvature K of the surface Σ is described in terms of A and the curvature of the ambient spacetime by the Gauss equation, which is given by

$$K = \overline{K} - \det A, \tag{20}$$

where $\overline{K}$ denotes the sectional curvature in $M^2 \times \mathbb{R}_1$ of the tangent plane to Σ. On the other hand, it is not difficult to see that $\overline{K}$ can be written in terms of the Gaussian curvature of M as

$$\overline{K} = \kappa_M \left(1 + \|\partial_t^\top\|^2\right), \tag{21}$$

where, for simplicity, κ_M stands for the Gaussian curvature of M along the surface Σ; that is, $\kappa_M = K_M \circ \Pi \in \mathscr{C}^\infty(\Sigma)$, where $\Pi = \pi_M \circ f : \Sigma \to M$ denotes the projection of Σ onto M and K_M is the Gaussian curvature of M. Here and in what follows, if Z is a vector field along the immersion $f : \Sigma^2 \to M^2 \times \mathbb{R}_1$, then $Z^\top \in \mathfrak{X}(\Sigma)$ denotes the tangential component of Z along Σ, that is, $Z = Z^\top - \langle N, Z\rangle N$. In particular, $\partial_t^\top = \partial_t + \Theta N$ and then

$$-1 = \|\partial_t^\top\|^2 - \Theta^2. \tag{22}$$

Therefore, Eq. (21) becomes $\overline{K} = \kappa_M \Theta^2$, and the Gauss equation (20) can be written as

$$K = \kappa_M \Theta^2 - \det A.$$

On the other hand, the Codazzi equation of the space-like surface Σ describes the tangent component of $\overline{R}(X,Y)N$ ($\overline{R}$ being the curvature tensor of $M^2 \times \mathbb{R}_1$), for any tangent vector fields $X, Y \in \mathfrak{X}(\Sigma)$, in terms of the derivative of the shape operator. It is given by

$$(\overline{R}(X,Y)N)^\top = (\nabla_X A)Y - (\nabla_Y A)X, \tag{23}$$

where $(\nabla_X A)Y = \nabla_X(AY) - A(\nabla_X Y)$. It is not difficult to see that the curvature tensor of $M^2 \times \mathbb{R}_1$ can be written in terms of κ_M, so that the Codazzi equation (23) becomes

$$(\nabla_X A)Y = (\nabla_Y A)X + \kappa_M \Theta \left(\langle X, \partial_t^\top\rangle Y - \langle Y, \partial_t^\top\rangle X\right).$$

For a space-like surface $f : \Sigma^2 \to M^2 \times \mathbb{R}_1$, we will call the *height function* of Σ h to the projection of Σ onto $\mathbb{R}$. That is, $h \in \mathscr{C}^\infty(\Sigma)$ is the smooth function given by $h = \pi_{\mathbb{R}} \circ f$. Observe that the gradient of $\pi_{\mathbb{R}}$ on $M^2 \times \mathbb{R}_1$ is $\overline{\nabla}\pi_{\mathbb{R}} = -\langle \overline{\nabla}\pi_{\mathbb{R}}, \partial_t\rangle \partial_t = -\partial_t$. Therefore, the gradient of h on Σ is

$$\nabla h = (\overline{\nabla}\pi_{\mathbb{R}})^\top = -\partial_t^\top.$$

Observe that from Eq. (22) we get

$$\|\nabla h\|^2 = \Theta^2 - 1. \tag{24}$$

Since ∂_t is parallel on $M^2 \times \mathbb{R}_1$ we have that

$$\overline{\nabla}_X \partial_t = 0 \tag{25}$$

for any $X \in \mathfrak{X}(\Sigma)$. Writing $\partial_t = -\nabla h - \Theta N$ along Σ and using Gauss (18) and Weingarten (19) formulae, we easily get from Eq. (25) that

$$\nabla_X \nabla h = \Theta AX$$

for every $X \in \mathfrak{X}(\Sigma)$. Thus, the Laplacian of Σ of the height function is given by

$$\Delta h = \Theta \mathrm{tr} A = -2H\Theta, \tag{26}$$

where $H = -\frac{1}{2}\mathrm{tr}(A)$ is the mean curvature function of Σ.

On the other hand, we can compute the gradient and the Laplacian of the function Θ in a similar way as we have done in Sect. 2 for the function $\langle N, a \rangle$. We obtain that

$$\nabla \Theta = A \nabla h, \tag{27}$$

and

$$\Delta \Theta = -2\langle \nabla H, \nabla h \rangle + \Theta \left(\kappa_M (\Theta^2 - 1) + \|A\|^2 \right). \tag{28}$$

Recall that $f : \Sigma \to M^2 \times \mathbb{R}_1$ is said to be *maximal* if $H \equiv 0$. In this case Eq. (26) implies that h is a harmonic function on Σ. Besides, if Σ is maximal then

$$A^2 = \frac{1}{2}\|A\|^2 I = -\det A\, I, \tag{29}$$

where I denotes the identity map on $\mathfrak{X}(\Sigma)$. Therefore, the Gauss equation (20) becomes

$$K = \kappa_M \Theta^2 + \frac{1}{2}\|A\|^2. \tag{30}$$

On the other hand, from Eqs. (24), (27) and (29) we also obtain for a maximal surface that

$$\|\nabla \Theta\|^2 = \frac{1}{2}\|A\|^2 \|\nabla h\|^2 = \frac{1}{2}\|A\|^2 (\Theta^2 - 1). \tag{31}$$

A space-like surface Σ is said to be a *slice* if its height function is constant. Equivalently, by Eq. (24), if $\Theta \equiv -1$ on Σ. The family of slices constitutes a foliation of $M^2 \times \mathbb{R}_1$ by totally geodesic surfaces.

4 A Parametric Version of a Calabi–Bernstein Result

We start by stating the following remarkable property.

Lemma 2 ([5, Lemma 3.1]). *Let M^2 be a Riemannian surface. If $M^2 \times \mathbb{R}_1$ admits a complete space-like surface $f : \Sigma^2 \to M^2 \times \mathbb{R}_1$, then M is necessarily complete and the projection $\Pi = \pi_M \circ f : \Sigma \to M$ is a covering map.*

Proof. The proof follows the ideas of the proof of [11, Lemma 3.1] and is a generalization of the argument given in Sect. 2 for the case of the 3-dimensional Lorentz–Minkowski space $\mathbb{R}^3_1$. We briefly sketch it here. Let $f : \Sigma^2 \to M^2 \times \mathbb{R}_1$ be a space-like surface and consider $\Pi = \pi_M \circ f : \Sigma \to M$ its projection on M. It is not difficult to see that $\Pi^*(\langle , \rangle_M) \geq \langle , \rangle$, where $\langle , \rangle$ stands for the Riemannian metric on Σ induced from the Lorentzian ambient space. This means that Π is a local diffeomorphism which increases the distance between the Riemannian surfaces Σ and M. Then, the proof finishes by recalling that if a map, from a connected complete Riemannian manifold M_1 into another connected Riemannian manifold M_2 of the same dimension, increases the distance, then it is a covering map and M_2 is complete [28, Chap. VIII, Lemma 8.1]. □

In particular, if $M^2 \times \mathbb{R}_1$ admits a compact space-like surface, then M is necessarily compact (see [11, Proposition 3.2 (i)]. An immediate consequence of Eq. (26) in the compact case is the following.

Proposition 1 ([5, Proposition 3.2]). *Let M^2 be a Riemannian surface. If $f : \Sigma^2 \to M^2 \times \mathbb{R}_1$ is a compact space-like surface in $M^2 \times \mathbb{R}_1$ whose mean curvature H does not change sign, then it must be a slice $M \times \{t_0\}$, $t_0 \in \mathbb{R}$. In particular, the only compact maximal surfaces in $M^2 \times \mathbb{R}_1$ are the slices.*

Proof. Observe that, since $\Theta < 0$ and H does not change sign, then Eq. (26) says that the height function h must be either subharmonic or superharmonic on Σ, according to the sign of H. But the compactness of Σ implies that h must be constant, so Σ is a slice. □

Under completeness assumption, we prove the following parametric version of a Calabi–Bernstein result in $M^2 \times \mathbb{R}_1$.

Theorem 5 ([5, Theorem 3.3]). *Let M^2 be a (necessarily complete) Riemannian surface with nonnegative Gaussian curvature, $K_M \geq 0$. Then, any complete maximal surface Σ^2 in $M^2 \times \mathbb{R}_1$ is totally geodesic. In addition, if $K_M > 0$ at some point on M, then Σ is a slice $M \times \{t_0\}$, $t_0 \in \mathbb{R}$.*

Observe first that the assumptions we have made for the Gaussian curvature of M are necessary. In fact, if $M^2 = \mathbb{R}^2$ is the flat Euclidean plane, then $M^2 \times \mathbb{R}_1 = \mathbb{R}^3_1$ is the 3-dimensional Lorentz–Minkowski space, and any space-like affine plane in $\mathbb{R}^3_1$ which is not horizontal determines a complete totally geodesic surface which is not a slice. On the other hand, the assumption $K_M \geq 0$ is necessary as shown by the fact that there exist examples of non-totally geodesic complete maximal surfaces in $\mathbb{H}^2 \times \mathbb{R}_1$, where $\mathbb{H}^2$ is the hyperbolic plane (see Examples 1 and 4 in Sect. 6).

In the proof of Theorem 5, the concept of parabolicity is fundamental.

Definition 1. A Riemannian surface Σ is said to be *parabolic* if any nonpositive subharmonic function on Σ is constant.

In general, it is not easy to decide whether a given surface is or not parabolic. Thus, some parabolicity criteria are usually needed. For our purpose, we find very useful the following classical result due to Ahlfors [1] and Blanc–Fiala–Huber [25].

Proposition 2. *Any complete Riemannian surface Σ with nonnegative Gaussian curvature is parabolic.*

Proof of Theorem 5. Since Σ is maximal and $\kappa_M \geq 0$, Eq. (30) implies that $K \geq 0$ on Σ. Then, as Σ is complete by assumption, from Proposition 2 we deduce that Σ is parabolic.

Recall that $\Theta \leq -1 < 0$. From Eqs. (28) and (31) we can compute

$$\Delta\left(\frac{1}{\Theta}\right) = -\frac{\Delta\Theta}{\Theta^2} + \frac{2\|\nabla\Theta\|^2}{\Theta^3} = -\frac{1}{\Theta}\left(\kappa_M(\Theta^2 - 1) + \frac{\|A\|^2}{\Theta^2}\right) \geq 0. \quad (32)$$

That is, $1/\Theta$ is a negative subharmonic function on the parabolic surface Σ, and hence it must be constant. Then, $\Theta = \Theta_o \leq -1$ is also constant on Σ and, by Eq. (32), we also get that $\|A\|^2 = 0$ and $\kappa_M(\Theta_o^2 - 1) = 0$ on Σ. Therefore, Σ is totally geodesic in $M^2 \times \mathbb{R}_1$ and if $\kappa_M > 0$ at some point on Σ, then it must be $\Theta_o = -1$, so Σ is a slice. Finally, observe that since the projection $\Pi : \Sigma \to M$ is onto (Lemma 2), then $\kappa_M > 0$ at some point on Σ if and only if $K_M > 0$ at some point on M. □

It is worth pointing out that, recently, Caballero et al. [13] have generalized Theorem 5 in the context of generalized Robertson–Walker spacetimes, which are defined as follows. Given $(M^2, \langle,\rangle_M)$ a connected Riemannian surface, a *generalized Robertson–Walker spacetime* is defined as the topological product $M^2 \times I$ endowed with the Lorentzian metric

$$\langle,\rangle = \rho^2(t)\langle,\rangle_M - dt^2,$$

where I is an open subset of $\mathbb{R}$ and $\rho \in \mathscr{C}^\infty(I)$ is a positive smooth function known as *warping function*. This Lorentzian manifold is usually denoted by $M^2_\rho \times I_1$.

It is usual to assume some energy conditions on a generalized Robertson–Walker spacetime or, more generally, in a Lorentzian manifold in order to obtain nice mathematical results in a realistic physical context. In that sense, a Lorentzian manifold obeys the *time-like convergence condition* (TCC) if its Ricci tensor, $\overline{\mathrm{Ric}}$, satisfies

$$\overline{\mathrm{Ric}}(Z,Z) \geq 0,$$

for all time-like vector Z. The TCC condition is the mathematical way to express that gravity, on average, attracts. A weaker energy condition is the *null convergence condition* (NCC) which reads

$$\overline{\mathrm{Ric}}(Z,Z) \geq 0,$$

for any light-like vector Z. Clearly, a continuity argument shows that TCC implies NCC.

Observe that, in the particular case where the warping function is constant, we recover the Lorentzian product spaces. With a straightforward computation we can check in this case that NCC is equivalent to ask the Gaussian curvature of M to be nonnegative. Taking into account these considerations, the generalization of Caballero, Romero and Rubio of Theorem 5 can be stated as follows.

Theorem 6 ([13, Corollary 5.4 and Theorem 5.9]). *Let $M^2_\rho \times I_1$ be a generalized Robertson–Walker spacetime obeying NCC. Let Σ be a complete maximal surface in M such that $\kappa_M \geq -\rho'(h)^2$ on Σ, then Σ is totally geodesic. Moreover, if NCC holds strictly at some point of Σ then Σ is a totally geodesic space-like slice.*

The proof of Theorem 6 follows the ideas of the proof of our parametric version of the Calabi–Bernstein result, Theorem 5.

5 A Nonparametric Version of a Calabi–Bernstein Result

We need to expose some notions about graphs in $M^2 \times \mathbb{R}_1$. These are analogous to the ones presented in Sect. 2.1 for graphs in $\mathbb{R}^3_1$.

Let $\Omega \subseteq M^2$ be a connected domain. Every smooth function $u \in \mathscr{C}^\infty(\Omega)$ determines a graph over Ω given by $\Sigma(u) = \{(x, u(x)) \,|\, x \in \Omega\} \subset M^2 \times \mathbb{R}_1$. The metric induced on Ω from the Lorentzian metric on the ambient space via $\Sigma(u)$ is given by

$$\langle , \rangle = \langle , \rangle_M - du^2.$$

Therefore, $\Sigma(u)$ is a space-like surface in $M^2 \times \mathbb{R}_1$ if and only if $|Du|^2 < 1$ on Ω, where Du denotes the gradient of u in Ω and $|Du|$ denotes its norm, both with respect to the original metric $\langle , \rangle_M$ on Ω. If $\Sigma(u)$ is a space-like graph over a domain Ω, then the future-pointing Gauss map of $\Sigma(u)$ is given by

$$N(x) = \frac{1}{\sqrt{1 - |Du(x)|^2}} \left(Du(x) + \partial_t|_{(x,u(x))} \right), \quad x \in \Omega.$$

We can also compute the mean curvature $H(u)$ of the space-like graph $\Sigma(u)$ with respect to N, and it is given by

$$2H(u) = \mathrm{Div}\left(\frac{Du}{\sqrt{1 - |Du|^2}} \right),$$

where Div stands for the divergence operator on Ω with respect to the metric $\langle , \rangle_M$. In particular, $\Sigma(u)$ is a maximal graph if and only if the function u satisfies the following partial differential equation on the domain Ω, usually known as the maximal surface equation:

$$\mathrm{Div}\left(\frac{Du}{\sqrt{1 - |Du|^2}} \right) = 0, \quad |Du|^2 < 1. \tag{33}$$

A graph is said to be entire if $\Omega = M$. As a direct consequence of Lemma 2, it follows that, when M is a complete Riemannian surface which is simply connected, every complete space-like surface in $M^2 \times \mathbb{R}_1$ is an entire graph. In fact, since M is simply connected, then the projection Π is a diffeomorphism between Σ and M, and hence Σ can be written as the graph over M determined by the function

$u = h \circ \Pi^{-1} \in \mathscr{C}^\infty(M)$. However, as we have pointed out in Sect. 2 for the case $M = \mathbb{R}^2$, an entire space-like graph is not necessarily complete (see also Examples 2 and 3 in Sect. 6 for the case when $M^2 = \mathbb{H}^2$). For that reason, Theorem 5 does not imply in principle that, under the same hypothesis on M, any entire maximal graph in $M^2 \times \mathbb{R}_1$ must be totally geodesic. This is certainly true for entire maximal graphs in $\mathbb{R}^3_1$ as it states the classical Calabi–Bernstein theorem, for entire maximal graphs in Robertson-Walker spaces of the form $\mathbb{R}^2_\rho \times \mathbb{R}_1$ under certain assumptions on the warping function [29] and for generalized Robertson–Walker spaces satisfying certain assumptions on the geometry of the fiber M and on the warping function [13]. However, none of this cases can be applied to Lorentzian product spaces. We will see now that, although we cannot establish a similar result for entire maximal graphs in $M^2 \times \mathbb{R}_1$ just as a direct consequence of our Theorem 5, we can obtain it as a consequence of the following result.

Theorem 7 ([5, Theorem 4.1]). *Let M^2 be a (not necessarily complete) Riemannian surface with nonnegative Gaussian curvature, $K_M \geq 0$. Then, any maximal surface Σ^2 in $M^2 \times \mathbb{R}_1$ which is complete with respect to the metric induced from the Riemannian product $M^2 \times \mathbb{R}$ is totally geodesic. In addition, if $K_M > 0$ at some point on Σ, then M is necessarily complete and Σ is a slice.*

In particular, if M^2 is complete and $f : \Sigma^2 \to M^2 \times \mathbb{R}_1$ is a space-like surface which is properly immersed in $M^2 \times \mathbb{R}_1$, then the metric induced on Σ from the Riemannian product $M^2 \times \mathbb{R}$ is complete. This happens, for instance, when $\Sigma \subset M^2 \times \mathbb{R}_1$ is a closed embedded maximal surface. Thus, it happens for entire maximal graphs, and it yields the following nonparametric version of our Calabi–Bernstein theorem.

Theorem 8 ([5, Theorem 4.3]). *Let M^2 be a complete Riemannian surface with nonnegative Gaussian curvature, $K_M \geq 0$. Then any entire maximal graph $\Sigma(u)$ in $M^2 \times \mathbb{R}_1$ is totally geodesic. In addition, if $K_M > 0$ at some point on M, then u is constant.*

The examples of Sect. 6 show that the assumption $K_M \geq 0$ is necessary.

Proof of Theorem 7. The proof follows the ideas of Romero's proof for the classical Calabi–Bernstein theorem for maximal surfaces in $\mathbb{R}^3_1$ (see Sect. 2.2). We will briefly sketch it here.

Let $f : \Sigma \to M^2 \times \mathbb{R}_1$ be a maximal surface. For simplicity, we denote by $g = \langle , \rangle$ the Riemannian metric induced on Σ from the Lorentzian product $M^2 \times \mathbb{R}_1$. Since $1 - \Theta \geq 2 \geq 0$, we may introduce on Σ the conformal metric

$$\hat{g} = (1 - \Theta)^2 g.$$

The longest part of the proof is to observe that $(\Sigma, \hat{g})$ is a complete surface with nonnegative Gaussian curvature, and this is done just by adapting the proof of Romero to our more general situation. We are not going to develop here all the computations, just recall that, in order to prove that $(\Sigma, \hat{g})$ has nonnegative Gaussian curvature, we have to compare the Gaussian curvatures of Σ with respect to the conformal metrics g and $\hat{g}$. And the proof of the completeness is done by comparison between the metric $\hat{g}$ and the metric induced on Σ from the Riemannian product $M^2 \times \mathbb{R}$, which is complete by assumption.

Once we know that $(\Sigma, \hat{g})$ is a complete surface with nonnegative Gaussian curvature, we conclude that $(\Sigma, \hat{g})$ is parabolic by the same classical parabolicity criterium of Ahlfors and Blanc, Fiala and Huber used in the proof of Theorem 5. The Laplacian Δ on Σ with respect to g and the Laplacian $\hat{\Delta}$ on Σ with respect to the conformal metric $\hat{g}$ are related by

$$\Delta = (1-\Theta)^2 \hat{\Delta},$$

which implies that (Σ, g) is also parabolic. The proof then follows as in the proof of Theorem 5. □

Remark 1. Observe that Theorem 5 can also be seen as a consequence of Theorem 7, since every complete space-like surface Σ in $M^2 \times \mathbb{R}_1$ is also complete with respect to the metric induced from the Riemannian product $M^2 \times \mathbb{R}$. This follows from the fact that $g' \geq g$, where g and g' stand for the metrics induced on Σ from the Lorentzian and the Riemannian product, respectively.

6 Some Nontrivial Entire Maximal Graphs in $\mathbb{H}^2 \times \mathbb{R}_1$

Let us see in this section examples of complete and noncomplete nontrivial entire maximal graphs in $\mathbb{H}^2 \times \mathbb{R}_1$. To do it, we will consider the half-plane model of the hyperbolic plane $\mathbb{H}^2$. That is,

$$\mathbb{H}^2 = \{x = (x_1, x_2) \in \mathbb{R}^2 \,|\, x_2 > 0\}$$

endowed with the complete metric

$$\langle , \rangle_{\mathbb{H}^2} = \frac{1}{x_2^2}(dx_1^2 + dx_2^2),$$

which is conformal to the flat Euclidean metric. Observe that given a smooth function $u = u(x) \in \mathscr{C}^\infty(\mathbb{H}^2)$, its hyperbolic gradient Du in $\mathbb{H}^2$ and its Euclidean gradient $D_o u$ in $\mathbb{R}^2$ are related by

$$Du(x) = x_2^2 D_o u(x), \quad x = (x_1, x_2), \tag{34}$$

and then

$$|Du(x)|^2 = x_2^2 |D_o u(x)|_o^2. \tag{35}$$

On the other hand, the divergence operators of the hyperbolic metric Div and of the Euclidean metric Div_o are related by

$$\mathrm{Div} = \mathrm{Div}_o - \frac{2}{x_2} dx_2.$$

By Eqs. (34) and (35), this implies that the graph determined by u, $\Sigma(u)$, is a maximal surface in $\mathbb{H}^2 \times \mathbb{R}_1$ if and only if

$$\left(1-x_2^2|D_o u|_o^2\right)x_2^2\Delta_o u + x_2^2\left(x_2 u_{x_2}|D_o u|_o^2 + x_2^2 Q(u)\right) = 0 \quad \text{and} \quad x_2^2|D_o u|_o^2 < 1, \tag{36}$$

where Δ_o stands for the Euclidean Laplacian operator and $Q(u) = u_{x_1}^2 u_{x_1x_1} + 2u_{x_1}u_{x_2}u_{x_1x_2} + u_{x_2}^2 u_{x_2x_2}$.

Analogously, it can be seen that $\Sigma(u)$ determines a minimal graph in $\mathbb{H}^2 \times \mathbb{R}$ if and only if

$$\left(1+x_2^2|D_o u|_o^2\right)x_2^2\Delta_o u - x_2^2\left(x_2 u_{x_2}|D_o u|_o^2 + x_2^2 Q(u)\right) = 0. \tag{37}$$

Equations (36) and (37) will help us to construct entire maximal graphs in $\mathbb{H}^2 \times \mathbb{R}_1$ in two different ways: via a duality result between minimal graphs in $\mathbb{H} \times \mathbb{R}$ and maximal graphs in $\mathbb{H} \times \mathbb{R}_1$, and in a direct way by looking for concrete families of solutions of the maximal surface equation (36).

6.1 Duality Between Minimal and Maximal Graphs

In Sect. 2.3 we have presented a duality between minimal graphs in the Euclidean space $\mathbb{R}^3$ and maximal graphs in the Lorentz–Minkowski space $\mathbb{R}_1^3$. Following the ideas of the proof of Alías and Palmer, we observe here that the same duality holds in general between solutions to the minimal surface equation in a Riemannian product space $M \times \mathbb{R}$ and solutions to the maximal surface equation in a Lorentzian product space $M \times \mathbb{R}_1$. First of all, recall that a smooth function u on a connected domain $\Omega \subseteq M^2$ defines a minimal graph $\Sigma(u)$ in $M \times \mathbb{R}$ if and only if u satisfies the following partial differential equation on Ω:

$$\operatorname{Div}\left(\frac{Du}{\sqrt{1+|Du|^2}}\right) = 0, \tag{38}$$

where, as in Eq. (33), Div and Du stand for the divergence operator and the gradient of u in Ω with respect to the metric $\langle,\rangle_M$, respectively. In this context our more general duality result reads as follows.

Theorem 9 ([5, Theorem 5.1]). *Let $\Omega \subseteq M^2$ be a simply connected domain of a Riemannian surface M^2. There exists a nontrivial solution u to the minimal surface equation on Ω (38) if and only if there exists a nontrivial solution w to the maximal surface equation on Ω (33).*

Here by a *nontrivial solution* we mean a solution with nonparallel gradient. With a straightforward computation, it can be seen that nontrivial solutions to either the minimal or maximal surface equation correspond to non-totally geodesic either minimal or maximal graphs.

Proof. Since Ω is simply connected, it is orientable and can be endowed with a globally defined area form $d\Omega$ and an almost complex structure J. The key of the proof is to observe that, as happened in the case $M = \mathbb{R}^2$, in this more general case it also holds that

$$\mathrm{Div}X d\Omega = d\omega_{JX},$$

where ω_{JX} denotes the 1-form in Ω which is metrically equivalent to the field JX, that is,

$$\omega_{JX}(Y) = \langle JX, Y\rangle_M.$$

From that point, the proof follows exactly as in Sect. 2.3. □

Remark 2. Following the philosophy of our result, Lee [30, Theorem 2] has very recently constructed an analogous duality result between graphs of constant mean curvature H in the Bianchi–Cartan–Vranceanu space $\mathbb{R}^3(\kappa,\tau)$ and space-like graphs of constant mean curvature τ in the Lorentzian Bianchi–Cartan–Vranceanu space $\mathbb{R}^3_1(\kappa,H)$. These spaces are defined as

$$\mathbb{R}^3(\kappa,\tau) = \left(V, \frac{dx_1^2+dx_2^2}{\left(1+\frac{\kappa}{4}\left(x_1^2+x_2^2\right)\right)^2} + \left(\tau\frac{x_2dx_1 - x_1dx_2}{1+\frac{\kappa}{4}\left(x_1^2+x_2^2\right)} + dx_3\right)^2 \right)$$

and

$$\mathbb{R}^3_1(\kappa,\tau) = \left(V, \frac{dx_1^2+dx_2^2}{\left(1+\frac{\kappa}{4}\left(x_1^2+x_2^2\right)\right)^2} - \left(\tau\frac{x_2dx_1 - x_1dx_2}{1+\frac{\kappa}{4}\left(x_1^2+x_2^2\right)} + dx_3\right)^2 \right),$$

where $V = \{(x_1,x_2,x_3) \in \mathbb{R}^3 \,|\, 1+\frac{\kappa}{4}(x^2+y^2) > 0\}$.

In particular, observe that $\mathbb{R}^3(0,0) = \mathbb{R}^3$, $\mathbb{R}^3(1,0) = \mathbb{S}^2\times\mathbb{R}$, $\mathbb{R}^3(-1,0) = \mathbb{H}^2\times\mathbb{R}$ and $\mathbb{R}^3_1(0,0) = \mathbb{R}^3_1$, $\mathbb{R}^3_1(1,0) = \mathbb{S}^2\times\mathbb{R}_1$, $\mathbb{R}^3_1(-1,0) = \mathbb{H}^2\times\mathbb{R}_1$. Therefore, the classical duality between minimal graphs in $\mathbb{R}^3$ and maximal graphs in $\mathbb{R}^3_1$ and the duality of Theorem 9 when $M^2 = \mathbb{S}^2$ or $M^2 = \mathbb{H}^2$ are particular cases of Lee's duality.

The interest of Theorem 9 resides on the fact that it allows us to construct new solutions to the maximal surface equation from known solutions to the minimal surface equation and viceversa. In particular, as an application of it, we are able to construct counterexamples which show that our Calabi–Bernstein results are no longer true without the assumption $K_M \geq 0$.

It is immediate to check that the functions

$$u(x_1,x_2) = \log(x_1^2+x_2^2) \tag{39}$$

and

$$u(x_1,x_2) = \frac{x_1}{x_1^2+x_2^2} \tag{40}$$

satisfy the minimal surface equation (37), so that they define nontrivial entire minimal graphs in the Riemannian product space $\mathbb{H}^2 \times \mathbb{R}$. As far as we know, these examples, which are due to Montaldo and Onnis [34, Examples 2.3 and 2.4] (see also [37]), are the first explicit examples of nontrivial entire minimal graphs in $\mathbb{H}^2 \times \mathbb{R}$. Previous existence results for such minimal graphs have been given in [17,35].

Example 1 ([5, Example 5.2]). From Theorem 9 and the entire minimal graph defined by the function (39), we know that there exists a smooth function $w \in \mathscr{C}^\infty(\mathbb{H}^2)$ which determines a nontrivial entire maximal graph in $\mathbb{H}^2 \times \mathbb{R}_1$. This shows that the assumption $K_M \geq 0$ in Theorem 5 is necessary. Moreover, we claim that the entire maximal graph determined by w is also complete, in the sense that the induced metric on $\mathbb{H}^2$ via the graph is complete. To see this, denote by $\langle , \rangle$, as usual, the metric given by

$$\langle , \rangle = \langle , \rangle_{\mathbb{H}^2} - dw^2.$$

Then, for every $X \in \mathfrak{X}(\mathbb{H}^2)$ we have

$$\langle X,X \rangle = \langle X,X \rangle_{\mathbb{H}^2} - X(w)^2 = \langle X,X \rangle_{\mathbb{H}^2} - \langle X, Dw \rangle^2_{\mathbb{H}^2},$$

and using the Cauchy–Schwarz inequality we get

$$\langle X,X \rangle \geq \langle X,X \rangle_{\mathbb{H}^2}(1 - |Dw|^2). \tag{41}$$

Because of the duality between $\Sigma(u)$ and $\Sigma(w)$, the norm of the gradients of u and w satisfies the analogous relation to Eq. (13). That is,

$$|Dw|^2 = \frac{|Du|^2}{1+|Du|^2} = \frac{x_2^2|D_o u|_o^2}{1+x_2^2|D_o u|_o^2}.$$

Therefore, we easily compute from Eq. (39) that

$$1 - |Dw|^2 = \frac{x_1^2 + x_2^2}{x_1^2 + 5x_2^2} \geq \frac{1}{5},$$

which jointly with Eq. (41) implies

$$\langle , \rangle \geq \frac{1}{5} \langle , \rangle_{\mathbb{H}^2}.$$

Since $\langle , \rangle_{\mathbb{H}^2}$ is complete, we conclude that the metric $\langle , \rangle$ is also complete on $\mathbb{H}^2$, as claimed. This shows that the assumption $K_M \geq 0$ in Theorem 8 is also necessary.

It is even possible to get w explicitly. In fact, in a similar way as Eq. (12), the duality allows us to express the gradient of w in terms of the gradient of u as

$$D_o w = (w_{x_1}, w_{x_2}) = \frac{1}{\sqrt{1+x_2^2|D_o u|_o^2}} J_o(D_o u),$$

where, as before, J_o denotes the positive $\pi/2$-rotation on the plane. Then, we can compute

$$w_{x_1} = -\sqrt{\frac{x_1^2+x_2^2}{x_1^2+5x_2^2}}\,u_{x_2} = -\frac{2x_2}{\sqrt{(x_1^2+x_2^2)(x_1^2+5x_2^2)}}, \tag{42}$$

and

$$w_{x_2} = \sqrt{\frac{x_1^2+x_2^2}{x_1^2+5x_2^2}}\,u_{x_1} = \frac{2x_1}{\sqrt{(x_1^2+x_2^2)(x_1^2+5x_2^2)}}. \tag{43}$$

We can explicitly integrate Eqs. (42) and (43) obtaining

$$w(x_1,x_2) = i\frac{2}{\sqrt{5}}F\left(\arcsin\left(i\frac{x_1}{x_2}\right),\frac{1}{\sqrt{5}}\right)+c,$$

where c is a real constant, i stands for the imaginary unit and $F(\phi,k)$ stands for the elliptic integral of the first kind with elliptic modulus k and Jacobi amplitude ϕ. See Fig. 1 (left) for a picture of the graph in the case $c=0$.

Example 2 ([5, Example 5.3]). Similarly, the nontrivial entire minimal graph in $\mathbb{H}^2\times\mathbb{R}$ defined by the function (40) gives rise, via Theorem 9, to another nontrivial entire maximal graph in the Lorentzian product $\mathbb{H}^2\times\mathbb{R}_1$. In contrast to Example 1, this example is not complete. To see this, let $w\in\mathscr{C}^\infty(\mathbb{H}^2)$ stand for the smooth function defining this entire maximal graph, which we denote as usual by $\Sigma(w)$. In an analogous way as in Example 1 we can compute

$$w_{x_1} = -\frac{x_1^2+x_2^2}{\sqrt{(x_1^2+x_2^2)^2+x_2^2}}\,u_{x_2} = \frac{2x_1x_2}{(x_1^2+x_2^2)\sqrt{(x_1^2+x_2^2)^2+x_2^2}}, \tag{44}$$

and

$$w_{x_2} = \frac{x_1^2+x_2^2}{\sqrt{(x_1^2+x_2^2)^2+x_2^2}}\,u_{x_1} = \frac{x_2^2-x_1^2}{(x_1^2+x_2^2)\sqrt{(x_1^2+x_2^2)^2+x_2^2}}. \tag{45}$$

Let $\alpha:(0,1)\to\Sigma(w)$ be the divergent curve in $\Sigma(w)$ given by

$$\alpha(s) = (0,s,w(0,s)).$$

Then $\alpha'(s) = (0,1,w_{x_2}(0,s))$ and

$$\|\alpha'(s)\|^2 = \frac{1}{s^2} - w_{x_2}(0,s)^2 = \frac{1}{1+s^2},$$

which implies that α has finite length since

$$\int_0^1 \|\alpha'(s)\|ds = \int_0^1 \frac{ds}{\sqrt{1+s^2}} = \operatorname{arcsinh}(1) = \log(1+\sqrt{2}).$$

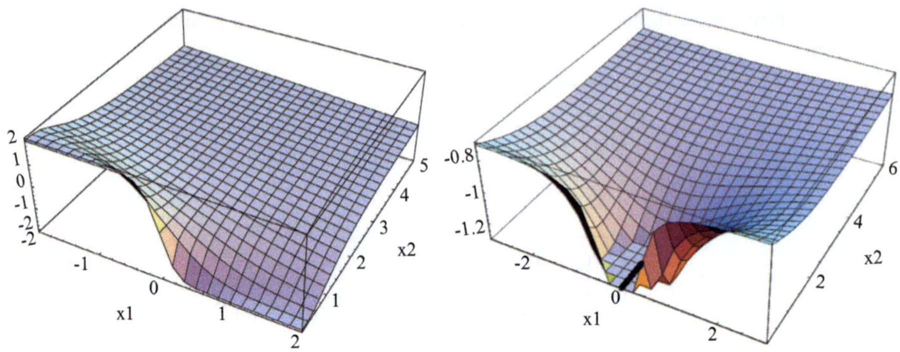

Fig. 1 The complete entire maximal graph of Example 1 and the noncomplete one of Example 2 with its divergent curve of finite length

As a consequence, $\Sigma(w)$ is not complete. Let us recall that such fact cannot occur in the Lorentz–Minkowski space $\mathbb{R}^3_1$, since by a result of Cheng and Yau [15], closed surfaces in $\mathbb{R}^3_1$ with constant mean curvature are necessarily complete.

Again, we can explicitly integrate Eqs. (44) and (45), getting

$$w(x_1,x_2) = \log\left(\frac{x_1^2+x_2^2}{2\left(x_2+\sqrt{x_2^2+(x_1^2+x_2^2)^2}\right)}\right)+c,$$

where c is a real constant. See Fig. 1 (right) for a picture of the graph in the case $c=0$.

Remark 3. Recently, Gálvez and Rosenberg [23] have showed the existence of nontrivial entire minimal graphs in a Riemannian product space $M^2\times\mathbb{R}$, where M^2 is a complete, simply connected, Riemannian surface with not necessarily constant Gaussian curvature bounded from above by a negative constant. That is, $K_M \le c < 0$. By our duality result, Theorem 9, we deduce the existence of nontrivial entire maximal graphs in the corresponding Lorentzian product space $M^2\times\mathbb{R}_1$. Therefore, the nonparametric version of our Calabi–Bernstein theorem is no longer true in such ambient spaces.

6.2 More Examples

Let us show now some explicit entire solutions to the maximal surface equation (36) by solving it for particular types of smooth functions.

Example 3 ([3, Example 3.1]). First, we will look for solutions depending only on one variable. If we consider solutions of the type $u(x_1,x_2)=u(x_1)$, Eq. (36)

reduces to $u''(x_1) = 0$, so u must be of the form $u(x_1,x_2) = ax_1 + b$, $a,b \in \mathbb{R}$. By the space-like condition, we get $x_2^2 u'(x_1)^2 = a^2 x_2^2 < 1$. Thus, as we are looking for entire graphs, the inequality is only satisfied in $\mathbb{H}^2$ for $a = 0$. Consequently, we only obtain slices.

More interesting is the case where u only depends on the second variable, $u(x_1,x_2) = u(x_2)$. Then, u will determine a maximal surface if and only if it satisfies

$$u''(x_2) + x_2 u'(x_2)^3 = 0 \tag{46}$$

and

$$x_2^2 u'(x_2)^2 < 1. \tag{47}$$

Integrating Eq. (46) we get

$$u(x_1,x_2) = u(x_2) = \log\left(x_2 + \sqrt{a + x_2^2}\right) + b, \quad a,b \in \mathbb{R},\ a \geq 0.$$

Moreover, from Eq. (47), u must satisfy

$$x_2^2 u'(x_2)^2 = \frac{x_2^2}{a + x_2^2} < 1,$$

which holds if and only if $a > 0$. Observe that we can assume $b = 0$ up to a translation, which is an isometry of the ambient space. Thus, we have obtained a family of entire maximal graphs, $\Sigma_a(u)$ over $\mathbb{H}^2$ for all $a > 0$. Let us see that, as in Example 2 these graphs are not complete. In fact, consider $\alpha : (1,\infty) \to \Sigma_a(u)$ the divergent curve in $\Sigma_a(u)$ given by

$$\alpha(s) = (0, s, u(s)).$$

Then, $\alpha'(s) = (0, 1, u'(s))$ and

$$\|\alpha'(s)\|^2 = \frac{1}{s^2} - \frac{1}{a + s^2} = \frac{a}{s^2(a + s^2)}.$$

Therefore, α has finite length since

$$\int_1^\infty \|\alpha'(s)\| ds = \int_1^\infty \sqrt{\frac{a}{s^2(a+s^2)}} ds = \operatorname{arcsinh}(\sqrt{a}),$$

being $\Sigma_a(u)$ noncomplete. See Fig. 2 (left) for a picture of this graph in the case $a = 1/2$ and $b = 0$.

Example 4 ([3, Example 3.2]). Let us consider now radial solutions to the maximal surface equation. That is, solutions of type

$$u(x_1,x_2) = f(x_1^2 + x_2^2).$$

In this case, Eq. (36) becomes

$$f'(z) + zf''(z) = 0, \tag{48}$$

and

$$4x_2^2 z f'(z)^2 < 1, \tag{49}$$

where $z = x_1^2 + x_2^2 > 0$. From Eq. (48)

$$f(z) = a\log z + b, \quad a,b \in \mathbb{R},$$

and then Eq. (49) results

$$\frac{4a^2x_2^2}{x_1^2+x_2^2} < 1$$

in $\mathbb{H}^2$, so $a^2 < 1/4$. Therefore, the functions

$$u(x_1,x_2) = a\log(x_1^2+x_2^2) + b, \quad a,b \in \mathbb{R},\ -1/2 < a < 1/2,$$

define a new family of entire maximal graphs in $\mathbb{H}^2 \times \mathbb{R}_1$. We can assume again that, up to an isometry, $b = 0$.

Let us see that these graphs are complete. Observe that, for every $X \in \mathfrak{X}(\mathbb{H}^2)$, we obtain as in Eq. (41)

$$\langle X,X\rangle \geq \langle X,X\rangle_{\mathbb{H}^2}(1-|Du|^2). \tag{50}$$

As $a^2 < 1/4$, we easily get

$$1-|Du|^2 = 1 - \frac{4a^2x_2^2}{x_1^2+x_2^2} \geq 1-4a^2 > 0,$$

which jointly with Eq. (50) implies

$$\langle X,X\rangle \geq (1-4a^2)\langle X,X\rangle_{\mathbb{H}^2},$$

so we conclude that $\langle,\rangle$ is complete following a similar argument as the one in Example 1. See Fig. 2 for a picture of this graph when $a = 1/4$ and $b = 0$.

7 Relative Parabolicity of Maximal Surfaces

Along this section, we will consider maximal surfaces with nonempty smooth boundary. For such surfaces, we can give the following definition related to the concept of parabolicity given in Definition 1 in Sect. 4.

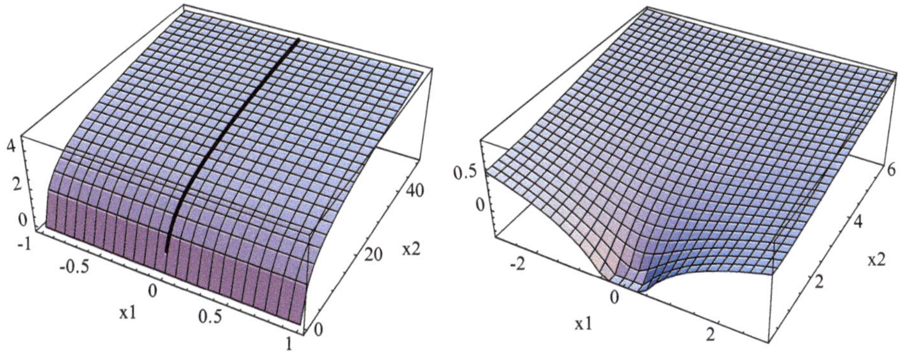

Fig. 2 The noncomplete entire maximal graph of Example 3 with its divergent curve of finite length, and the complete one of Example 4

Definition 2. A Riemannian surface Σ with nonempty smooth boundary, $\partial\Sigma \neq \emptyset$, is said to be *relatively parabolic* if every bounded harmonic function on Σ is determined by its boundary values.

It is interesting to observe that both concepts of parabolicity: parabolicity and relative parabolicity are related in the following way, (see [33] and [24, Theorem 5.1]).

Proposition 3. *A Riemannian surface Σ without boundary is parabolic if and only if for every nonempty open set $O \subset \Sigma$ with smooth boundary, $\Sigma \setminus O$ is relatively parabolic.*

Example 5. It is well known that $\mathbb{R}^2$ is a parabolic surface and $\mathbb{H}^2$ is not parabolic (see for instance [26]). Therefore, by Proposition 3, any open subset with nonempty smooth boundary of $\mathbb{R}^2$ is a relatively parabolic surface, and it exists at least a nonempty open subset $O \subset \mathbb{H}^2$ with smooth boundary such that $\mathbb{H}^2 \setminus O$ is not relatively parabolic. In fact, it is not difficult to prove, with a similar argument as the one in [26], that any connected, unbounded subset $O \subset \mathbb{H}^2$ with nonempty smooth boundary is not a relatively parabolic surface.

It is not easy to decide again whether a given surface with nonempty smooth boundary is or not relatively parabolic. The following criterium is very useful.

Lemma 3 ([6, Lemma 2]). *Let Σ^2 be a Riemannian surface with nonempty smooth boundary, $\partial\Sigma \neq \emptyset$. If there exists a proper continuous function $\psi : \Sigma \to \mathbb{R}$ which is eventually positive and superharmonic, then Σ is relatively parabolic.*

As is usual, by eventually we mean here a property that is satisfied outside a compact set.

The proof of Lemma 3 follows the ideas of the proof of an analogous criterium for proper smooth functions given by Meeks and Pérez [33, 38]. In fact, we only have to observe that the proof of Meeks and Pérez only depends on the minimum principle for superharmonic functions and that this principle holds for continuous functions (not necessarily smooth).

Fernández and López [21, Sect. 4] have recently proved that properly immersed maximal surfaces with nonempty boundary in the Lorentz–Minkowski spacetime $\mathbb{R}^3_1$ are relatively parabolic if the Lorentzian norm on the maximal surface in $\mathbb{R}^3_1$ is eventually positive and proper. Motivated by this work, we study some relative parabolicity criteria for maximal surfaces in Lorentzian product spaces. Observe that a natural generalization of the Lorentzian norm on a surface in $\mathbb{R}^3_1$ to the Lorentzian product $M^2 \times \mathbb{R}_1$ consists in considering the function $\phi = r^2 - h^2$, where the function r measures the distance on the factor M to a fixed point $x_o \in M$

Before describing our main result in this section, we need the following technical observation. Given any function $\hat{\psi} \in \mathscr{C}^\infty(M)$, we can consider its lifting $\bar{\psi} \in \mathscr{C}^\infty(M^2 \times \mathbb{R}_1)$ defined by

$$\bar{\psi}(x,t) = \hat{\psi}(x).$$

Let $f : \Sigma^2 \to M^2 \times \mathbb{R}_1$ be a space-like surface. Then, we can also associate to $\hat{\psi} \in \mathscr{C}^\infty(M)$ the function $\psi \in \mathscr{C}^\infty(\Sigma)$ given by $\psi = \bar{\psi} \circ f$. In this context, the Laplacian on Σ of ψ can be expressed in terms of the Laplacian $\bar{\Delta}$ of $\bar{\psi}$ and the differential operators of $\hat{\psi}$ as follows.

Lemma 4 ([6, Lemma 1]). *Along a space-like surface $f : \Sigma^2 \to M^2 \times \mathbb{R}_1$ we have that*

$$\Delta \psi = \bar{\Delta} \bar{\psi} + 2H \langle N^*, D\hat{\psi} \rangle_M + D^2 \hat{\psi}(N^*, N^*),$$

where $N^* = \pi_M^*(N) = N + \Theta \partial_t$, and D and D^2 denote the gradient and the Hessian operators on M, respectively.

Proof. Since $\bar{\nabla} \bar{\psi} = \nabla \psi - \langle \bar{\nabla} \bar{\psi}, N \rangle N$, we get from the Gauss (18) and Weingarten (19) formulae that the Hessian operators of $\bar{\psi}$ and ψ satisfy

$$\bar{\nabla}^2 \bar{\psi}(X,X) = \nabla^2 \psi(X,X) + \langle AX, X \rangle \langle \bar{\nabla} \bar{\psi}, N \rangle$$

for every $X \in \mathfrak{X}(\Sigma)$. Therefore, it can be easily seen that

$$\bar{\Delta} \bar{\psi} = \Delta \psi - 2H \langle \bar{\nabla} \bar{\psi}, N \rangle - \bar{\nabla}^2 \bar{\psi}(N,N). \tag{51}$$

Observe now that, as the function $\bar{\psi}$ does not depend on t, then $\bar{\nabla} \bar{\psi}(x,t) = D\hat{\psi}(x)$. Thus, $\bar{\nabla}_N \bar{\nabla} \bar{\psi} = D_{N^*} D\hat{\psi}$ and

$$\bar{\nabla}^2 \bar{\psi}(N,N) = D^2 \hat{\psi}(N^*, N^*),$$

so that Lemma 4 follows directly from Eq. (51). □

Following the notation above, consider the function $\hat{r} : M^2 \to \mathbb{R}$ defined by $\hat{r}(x) = \mathrm{dist}_M(x, x_o)$, where $x_o \in M$ is a fixed point. Observe that $\hat{r} \in \mathscr{C}^\infty(M)$ almost everywhere. Actually, $\hat{r}$ is smooth on $M^2 \setminus \mathrm{Cut}(x_o)$, where $\mathrm{Cut}(x_o)$ stands for the cut locus of x_o. As is well known, $\dim \mathrm{Cut}(x_o) < 2$ and $\mathrm{Cut}(x_o)$ is a null set. Let

$\bar{r}(x,t) = \hat{r}(x)$ denote the lifting of $\hat{r}$ to $M^2 \times \mathbb{R}_1$, and for a given space-like surface $f : \Sigma^2 \to M^2 \times \mathbb{R}_1$, let r stand for the restriction of $\bar{r}$ to Σ, $r = \bar{r} \circ f$. By Lemma 2 Π is a covering map. Therefore, $\dim \Pi^{-1}(\mathrm{Cut}(x_o)) = \dim \mathrm{Cut}(x_o) < 2$ and the function r is smooth almost everywhere in Σ.

Our main result in this section is the following relative parabolicity criterium.

Theorem 10 ([6, Theorem 3]). *Let M^2 be a complete Riemannian surface with nonnegative Gaussian curvature. Consider $f : \Sigma^2 \to M^2 \times \mathbb{R}_1$ a maximal surface with nonempty smooth boundary, $\partial\Sigma \neq \emptyset$, and assume that the function $\phi : \Sigma \to \mathbb{R}$ defined by*

$$\phi(p) = r^2(p) - h^2(p)$$

is eventually positive and proper. Then Σ is relatively parabolic.

It is worth pointing out that the assumption on the nonnegativity of the Gaussian curvature of M is necessary. Actually, let $M^2 = \mathbb{H}^2$ and consider $\Omega \subset \mathbb{H}^2$ an unbounded connected domain with smooth boundary. Then, for a fixed $t_o \in \mathbb{R}$, the slice $\Sigma_{t_o} = \{(x,t_o) \in \mathbb{H}^2 \times \mathbb{R}_1 : x \in \Omega\}$ is a non-relatively parabolic maximal surface in $\mathbb{H}^2 \times \mathbb{R}_1$ on which ϕ is eventually positive and proper.

Proof. Let $a > 1$ and consider $K = \{p \in \Sigma : \phi(p) \leq a\} \subseteq \Sigma$. Since ϕ is eventually positive and proper, K is a compact set. As is well known, relative parabolicity is not affected by adding or removing compact subsets, so that Σ is relatively parabolic if and only if $\Sigma \setminus K$ is relatively parabolic.

The function $\log\phi : \Sigma \setminus K \to \mathbb{R}$ is a proper positive function on $\Sigma \setminus K$. Therefore, in order to prove that $\Sigma \setminus K$ is relatively parabolic, it suffices to see that $\log\phi$ is superharmonic on the dense subset $\Sigma' \subset \Sigma \setminus K$ where it is smooth. In what follows, we will work on that subset Σ'. From Eqs. (24) and (26) we get

$$\Delta h^2 = 2h\Delta h + 2\|\nabla h\|^2 = 2(\Theta^2 - 1). \tag{52}$$

On the other hand, applying Lemma 4 to $\psi = r^2$ we get

$$\Delta r^2 = \bar{\Delta}\bar{r}^2 + \hat{\nabla}^2\hat{r}^2(N^*,N^*). \tag{53}$$

As the function $\bar{r}$ does not depend on t, then $\bar{\nabla}\bar{r}(x,t) = D\hat{r}(x)$ and $\bar{\Delta}\bar{r}(x,t) = \hat{\Delta}\hat{r}(x)$, where $\hat{\Delta}$ denotes the Laplacian operator with respect to the metric $\langle , \rangle_M$. Therefore,

$$\bar{\Delta}\bar{r}^2(x,t) = 2\bar{r}(x,t)\bar{\Delta}\bar{r}(x,t) + 2\|\bar{\nabla}\bar{r}(x,t\|^2 = 2\left(\hat{r}(x)\hat{\Delta}\hat{r}(x) + 1\right),$$

since, as is well known, $\|\bar{\nabla}\bar{r}\|^2 = |D\hat{r}|^2 = 1$. And, after a long but straightforward computation, Eq. (53) becomes

$$\Delta r^2 = 2(r\hat{\Delta}\hat{r}(1 + \langle N^*, \tau\rangle_M^2) + 1 + \langle N^*, D\hat{r}\rangle_M^2), \tag{54}$$

for $\tau \perp_M D\hat{r}$ and $|\tau| = 1$.

Now, from Eqs. (52) and (54) we get that

$$\begin{aligned}\frac{1}{2}\Delta\phi &= \frac{1}{2}\Delta r^2 - \frac{1}{2}\Delta h^2\\ &= r\hat{\Delta}\hat{r}(1+\langle N^*,\tau\rangle_M^2) + \langle N^*, D\hat{r}\rangle_M^2 + 2 - \Theta^2. \end{aligned} \tag{55}$$

As M^2 is complete and has nonnegative Gaussian curvature, by the Laplacian comparison theorem we have that $D\hat{r} \leq 1/\hat{r}$, so that

$$r\hat{\Delta}\hat{r} \leq 1$$

on Σ'. Using this in Eq. (55), we obtain that

$$\frac{1}{2}\Delta\phi \leq \|N^*\|^2 + 3 - \Theta^2 = 2, \tag{56}$$

since $\|N^*\|^2 = \Theta^2 - 1$.

On the other hand, $\nabla\phi = 2r\nabla r - 2h\nabla h$, and so we easily obtain

$$\begin{aligned}\|\nabla\phi\|^2 &= 4r^2\|\nabla r\|^2 - 8rh\langle\nabla r, \nabla h\rangle + 4h^2\|\nabla h\|^2\\ &= 4r^2(1+\langle\bar{\nabla}\bar{r}, N\rangle^2) + 8rh\Theta\langle\bar{\nabla}\bar{r}, N\rangle + 4h^2(\Theta^2 - 1)\\ &= 4\phi + 4(r\langle\bar{\nabla}\bar{r}, N\rangle + h\Theta)^2 \geq 4\phi. \end{aligned} \tag{57}$$

Therefore, from Eqs. (56) and (57) we finally get

$$\Delta \log\phi = \frac{1}{\phi^2}(\phi\Delta\phi - \|\nabla\phi\|^2) \leq 0,$$

which means that $\log\phi$ is a superharmonic function on Σ', as we wanted to prove. As a consequence, $\Sigma \setminus K$ is relatively parabolic, so Σ is also relatively parabolic as observed in the beginning of the proof. □

From the existence relation between parabolicity and relative parabolicity, we can state the following consequence of Theorem 10.

Corollary 1 ([6, Corollary 6]). *Let M^2 be a complete Riemannian surface with nonnegative Gaussian curvature, and let Σ be a maximal surface in $M^2 \times \mathbb{R}_1$ without boundary, $\partial\Sigma = \emptyset$. If the function $\phi = r^2 - h^2$ is eventually positive and proper on Σ, then Σ is parabolic.*

Proof. The proof follows from Proposition 3 and the observation that, for any nonempty open set $O \subset \Sigma$ with smooth boundary, the function ϕ restricted to $\Sigma \setminus O$ is also eventually positive and proper on $\Sigma \setminus O$. Therefore, the maximal surface with boundary $\Sigma \setminus O$ is relatively parabolic by Theorem 10, and Σ is a parabolic surface.

7.1 Relative Parabolicity and Entire Maximal Graphs

Consider $\Omega \subseteq M$ a connected domain and let $x_o \in \mathrm{int}(\Omega)$. We will say that Ω is *star-like with respect to* x_o if for every $x \in \Omega$ there exists a (not necessarily unique) minimizing geodesic segment from x_o to x which is contained in Ω. Obviously, if M is a complete Riemannian surface, then M itself is star-like with respect to any of its points.

Let us see that we can apply the relative parabolicity criterium, Theorem 10, to space-like graphs over a star-like domain.

Proposition 4 ([6, Proposition 3]). *Let M^2 be a complete Riemannian surface and let $\Sigma(u)$ be a space-like graph over a domain Ω which is star-like with respect to some point $x_o \in \mathrm{int}(\Omega)$. Then the function $\phi = r^2 - h^2$ is eventually positive and proper on $\Sigma(u)$.*

Proof. We may assume, without loss of generality, that $u(x_o) = 0$. Since $\Sigma(u)$ is homeomorphic to Ω (via the standard embedding $x \in \Omega \hookrightarrow (x, u(x)) \in \Sigma(u)$), and the thesis of our result is topological, it suffices to prove that the function $\varphi = \pi_M \circ f \circ \phi = \hat{r}^2 - u^2$ is eventually positive and proper on Ω.

Firstly, we will prove that φ is positive for every $x \in \Omega - \{x_o\}$. For a given $x \neq x_o$, consider $\gamma : [0, l] \to \Omega$ a minimizing geodesic segment such that $\gamma(0) = x_o$, $\gamma(l) = x$ and $l = \mathrm{dist}_M(x_o, x) = \hat{r}(x) > 0$. Let $\alpha(s) = (\gamma(s), u(s)) \in \Sigma(u)$, where $u(s) := u(\gamma(s))$. Since $\Sigma(u)$ is a space-like surface, $\alpha'(s) = (\gamma'(s), u'(s)) \neq (0,0)$ is a nonvanishing space-like vector, that is,

$$\langle \alpha'(s), \alpha'(s) \rangle = |\gamma'(s)|^2 - u'(s)^2 = 1 - u'(s)^2 > 0.$$

Therefore, $-1 < u'(s) < 1$ for every $0 \leq s \leq l = \hat{r}(x)$, and integrating we get

$$-\hat{r}(x) < u(x) < \hat{r}(x).$$

Consequently, $\varphi(x) > 0$ for every $x \in \Omega$, $x \neq x_o$.

It remains to prove that φ is proper. Let us define

$$\mathscr{W} := \{(x,t) \in \Omega \times \mathbb{R} : \hat{r}^2(x) - t^2 \geq 0\},$$

and consider on $M^2 \times \mathbb{R}$ the standard Riemannian metric, $\langle , \rangle_M + dt^2$. Let us denote by $\mathrm{dist}_+(,)$ the distance related to such Riemannian metric. Let us see now that

$$\mathrm{dist}_+((x,t), \partial \mathscr{W}) = \frac{1}{\sqrt{2}} \min\{\hat{r}(x) - t, \hat{r}(x) + t\} = \frac{1}{\sqrt{2}} (\hat{r}(x) - |t|) \qquad (58)$$

for every $(x,t) \in \mathscr{W}$. Observe that $\partial \mathscr{W} = \partial \mathscr{W}^+ \cup \partial \mathscr{W}^-$, where

$$\partial \mathscr{W}^+ = \{(x, \hat{r}(x)) : x \in \Omega\} \quad \text{and} \quad \partial \mathscr{W}^- = \{(x, -\hat{r}(x)) : x \in \Omega\}.$$

Therefore,

$$\mathrm{dist}_+((x,t),\partial\mathscr{W}) = \min\{\mathrm{dist}_+((x,t),\partial\mathscr{W}^+),\mathrm{dist}_+((x,t),\partial\mathscr{W}^-)\}.$$

Expression (58) is trivial for $x = x_o$ (and necessarily $t = 0$). For a given $x \neq x_o$, let $\gamma : [0,\hat{r}(x)] \to \Omega$ be a minimizing geodesic segment such that $\gamma(0) = x_o$ and $\gamma(\hat{r}(x)) = x$. We will compute first $\mathrm{dist}_+((x,t),\partial\mathscr{W}^+)$. Since γ is minimizing, for every $s \in [0,\hat{r}(x)]$ it holds that $\hat{r}(\gamma(s)) = s$, so that $(\gamma(s),s) \in \partial\mathscr{W}^+$ and

$$\mathrm{dist}_+((x,t),(\gamma(s),s))^2 = \mathrm{dist}_M(x,\gamma(s))^2 + (t-s)^2 = (\hat{r}(x)-s)^2 + (t-s)^2.$$

Observe that this expression attains its minimum at $s_o = (\hat{r}(x)+t)/2$, and

$$\mathrm{dist}_+((x,t),(\gamma(s_o),s_o)) = \frac{1}{\sqrt{2}}(\hat{r}(x)-t). \tag{59}$$

We claim that $\mathrm{dist}_+((x,t),\partial\mathscr{W}^+)$ is given by Eq. (59). In fact, for every $y \in \Omega$ we have that

$$\mathrm{dist}_+(x,y) \geq |\mathrm{dist}_M(x_o,x) - \mathrm{dist}_\mathrm{M}(\mathrm{x_o},\mathrm{y})| = |\hat{r}(x) - \hat{r}(y)|,$$

and so

$$\begin{aligned}
\mathrm{dist}_+((x,t),(y,\hat{r}(y)))^2 &= \mathrm{dist}_+(x,y)^2 + (t-\hat{r}(y))^2 \\
&\geq (\hat{r}(x)-\hat{r}(y))^2 + (t-\hat{r}(y))^2 \\
&\geq \min_{s\geq 0}\left((\hat{r}(x)-s)^2 + (t-s)^2\right) = \frac{1}{2}\left(\hat{r}(x)-t\right)^2.
\end{aligned}$$

Therefore,

$$\mathrm{dist}_+((x,t),\partial\mathscr{W}^+) = \frac{1}{\sqrt{2}}(\hat{r}(x)-t). \tag{60}$$

Analogously, we can see that for $\partial\mathscr{W}^-$ it holds

$$\mathrm{dist}_+((x,t),\partial\mathscr{W}^-) = \frac{1}{\sqrt{2}}(\hat{r}(x)+t). \tag{61}$$

Then Eq. (58) follows from Eqs. (60) and (61).

Let $x \in \Omega$, $x \neq x_o$, and consider again $\gamma : [0,\hat{r}(x)] \to \Omega$ a minimizing geodesic segment such that $\gamma(0) = x_o$ and $\gamma(\hat{r}(x)) = x$. Let us write $u(s) = u(\gamma(s))$. Then $(\gamma(s),u(s)) \in \mathscr{W}$, and Eq. (58) yields

$$\mathrm{dist}_+((\gamma(s),u(s)),\partial\mathscr{W}) = \frac{1}{\sqrt{2}}(s-|u(s)|).$$

In particular, $\mathrm{dist}_+((\gamma(s),u(s)),\partial\mathscr{W})$ is a positive increasing function for $0 < s \leq \hat{r}(x)$. Thus, if we choose $\delta > 0$ such that the geodesic disc verifies $D_\delta = D(x_o,\delta) \subset\subset \Omega$, then it follows that

$$\mathrm{dist}_+((x,u(x)),\partial\mathscr{W}) \geq \varepsilon > 0 \quad \text{for every } x \in \Omega \setminus D_\delta, \tag{62}$$

where $\varepsilon = \min_{x\in\partial D_\delta} \mathrm{dist}_+((x,u(x)),\partial\mathscr{W}) > 0$.

Finally, we are ready to prove that φ is proper on Ω. Since $\Omega = \overline{D}_\delta \cup (\Omega \setminus D_\delta)$ with $\overline{D}_\delta$ compact, it suffices to prove that $\varphi|_{\Omega\setminus D_\delta}$ is proper on $\Omega \setminus D_\delta$. Let $\psi : \Omega \to \Omega \times \mathbb{R}$ be the standard embedding, $\psi(x) = (x,u(x))$, and let $\bar{\phi} : \Omega \times \mathbb{R} \to \mathbb{R}$ be the lifting of ϕ to $\Omega \times \mathbb{R}$; that is, $\bar{\phi}(x,t) = \hat{r}^2(x) - t^2$. Observe that ψ is trivially proper. In fact, if $A \subset \Omega \times \mathbb{R}$ is compact, then

$$\psi^{-1}(A) \subset \psi^{-1}(\pi_M(A) \times \pi_{\mathbb{R}}(A)) = \pi_M(A) \cap u^{-1}(\pi_{\mathbb{R}}(A))$$

is also compact. On the other hand, from Eq. (62) we have that

$$\psi(\Omega \setminus D_\delta) \subset \mathscr{U} = \{(x,t) \in \mathscr{W} \,|\, \mathrm{dist}_+((x,t),\partial\mathscr{W}) \geq \varepsilon\}.$$

Consequently, $\varphi|_{\Omega\setminus D_\delta} = \bar{\phi}|_{\mathscr{U}} \circ \psi|_{\Omega\setminus D_\delta}$. It is easy to see that the map $\psi|_{\Omega\setminus D_\delta}$ is proper. Therefore, it suffices to show that $\bar{\phi}|_{\mathscr{U}} : \mathscr{U} \to \mathbb{R}$ is proper or, equivalently, that for every $b > 0$, $(\bar{\phi}|_{\mathscr{U}})^{-1}([0,b]) = \mathscr{U} \cap \bar{\phi}^{-1}([0,b])$ is compact. Let $(x,t) \in \mathscr{U} \cap \bar{\phi}^{-1}([0,b])$. Since $(x,t) \in \mathscr{U}$, by Eq. (58) we obtain that

$$|t| \leq \hat{r}(x) - \sqrt{2}\varepsilon.$$

Therefore, since $\bar{\phi}(x,t) \leq b$, we have

$$\hat{r}^2(x) - b \leq t^2 \leq (\hat{r}(x) - \sqrt{2}\varepsilon)^2.$$

That is,

$$\hat{r}(x) \leq c := \frac{2\varepsilon^2 + b}{2\sqrt{2}\varepsilon}.$$

This implies that $(\bar{\phi}|_{\mathscr{U}})^{-1}([0,b]) \subset \overline{D}_c \times [\sqrt{2}\varepsilon - c, c - \sqrt{2}\varepsilon]$ is compact, which finishes the proof. □

As a consequence of Theorem 10 and Proposition 4 we can give an alternative proof of the nonparametric version of our Calabi–Bernstein result, Theorem 8. In fact, if M^2 is complete we can apply Proposition 4 to $\Omega = M$, concluding that the function $\phi = r^2 - h^2$ is eventually positive and proper on $\Sigma(u)$. Therefore, by Corollary 1 we have that $\Sigma(u)$ is parabolic. Then, the proof follows as in the proof of Theorem 5.

8 A Local Estimate for Maximal Surfaces in a Lorentzian Product Space

In this section, we generalize the local approach given by the second author and Palmer, which is described in Sect. 2.3, to the case of maximal surfaces in a Lorentzian product space $M^2 \times \mathbb{R}_1$. Specifically, we prove the following extension of Theorem 4.

Theorem 11 ([4, Theorem 1]). *Let M^2 be an analytic Riemannian surface with nonnegative Gaussian curvature, $K_M \geq 0$, and let $f : \Sigma^2 \to M^2 \times \mathbb{R}_1$ be a maximal surface in $M^2 \times \mathbb{R}_1$. Let p be a point of Σ and $R > 0$ a positive real number such that the geodesic disc of radius R centered at p satisfies $D(p,R) \subset\subset \Sigma$. Then, for all $0 < r < R$, it holds that*

$$0 \leq \int_{D(p,r)} \|A\|^2 d\Sigma \leq c_r \frac{L(r)}{r \log(R/r)}, \tag{63}$$

where $L(r)$ denotes the length of the geodesic disc of radius r centered at p and

$$c_r = \frac{\pi^2(1+\alpha_r^2)^2}{4\alpha_r \arctan \alpha_r} > 0,$$

being $\alpha_r = -\inf_{D(p,r)} \Theta \geq 1$.

Proof. The proof is inspired by the ideas of the proof of Theorem 4. In fact, the key of the proof is to apply Lemma 1 to the smooth function on Σ given by $u = \arctan \Theta$.

Using Eqs. (28) and (31), we can compute

$$\Delta u = \frac{\Delta \Theta}{1+\Theta^2} - \frac{2\Theta \|\nabla \Theta\|^2}{(1+\Theta^2)^2} = \frac{2\Theta}{(1+\Theta^2)^2} \|A\|^2 + \frac{(\Theta^2-1)\Theta}{1+\Theta^2} \kappa_M.$$

Therefore, taking into account that $\Theta \arctan \Theta \geq 0$, $\Theta \leq -1$ and $\kappa_M \geq 0$, we obtain

$$u\Delta u = \frac{2\Theta \arctan \Theta}{(1+\Theta^2)^2} \|A\|^2 + \frac{(\Theta^2-1)\Theta \arctan \Theta}{1+\Theta^2} \kappa_M \geq \phi(\Theta) \|A\|^2,$$

where, as in Eq. (17)

$$\phi(t) = \frac{2t \arctan t}{(1+t^2)^2}.$$

Recall that the function $\phi(t)$ is strictly increasing for $t \leq -1$. Therefore, since $-\alpha_r \leq \Theta \leq -1$ on $D(p,r)$ we get

$$\phi(\Theta) \geq \phi(-\alpha_r) = \frac{2\alpha_r \arctan \alpha_r}{(1+\alpha_r^2)^2} \quad \text{on} \quad D(p,r).$$

From here the proof follows exactly as in Theorem 4. □

Remark 4. In particular, when Σ is complete, the local integral inequality (63) provides an alternative proof of the parametric version of our Calabi–Bernstein theorem, Theorem 5, in the case where M^2 is an analytic surface.

In fact, observe here that if M^2 is analytic, then Σ is also analytic since it is locally a solution of Eq. (33). On the other hand, if Σ is complete, R can approach to infinity in Eq. (63) for a fixed arbitrary $p \in \Sigma$ and a fixed r, which implies

$$\int_{D(p,r)} \|A\|^2 d\Sigma = 0.$$

Therefore, $\|A\|^2 = 0$ and Σ must be totally geodesic. From Eq. (27) we get that $\Theta = \Theta_o \leq -1$ is constant on Σ, and then Eq. (28) implies that, when $K_M > 0$ somewhere in M, it must be $\Theta_o = -1$, so Σ is a slice.

As another application of Theorem 11, at points of a maximal surface where the second fundamental form does not vanish, we are able to estimate the maximum possible geodesic radius in terms of a local positive constant.

Corollary 2 ([4, Corollary 3]). *Let M^2 be an analytic Riemannian surface with nonnegative Gaussian curvature and let $f : \Sigma^2 \to M^2 \times \mathbb{R}_1$ be a maximal surface in $M^2 \times \mathbb{R}_1$ which is not totally geodesic. Assume that $p \in \Sigma$ is a point with $\|A\|(p) \neq 0$ and let $r > 0$ be a positive real number such that $D_r = D(p,r) \subset\subset \Sigma$. Then*

$$R \leq r e^{C_r}$$

for every $R > r$ with $D(p,R) \subset\subset \Sigma$, where

$$C_r = \frac{c_r L(r)}{r \int_{D_r} \|A\|^2 d\Sigma} > 0$$

is a local positive constant depending only on the geometry of $f|_{D(p,r)}$.

A similar estimate for stable minimal surfaces in 3-dimensional Riemannian manifolds with nonnegative Ricci curvature was given by Schoen in [40]. See also [8] for another similar estimate given by the second author and Palmer for the case of non-flat space-like surfaces with nonnegative Gaussian curvature and zero mean curvature in a flat 4-dimensional Lorentzian space.

This work was partially supported by MICINN and FEDER project MTM2009-10418 and

Acknowledgments This work was partially supported by MICINN and FEDER project MTM2009-10418 and Fundación Séneca project 04540/GERM/06, Spain. This research is a result of the activity developed within the framework of the Programme in Support of Excellence Groups of the Región de Murcia, Spain, by Fundación Séneca, Regional Agency for Science and Technology (Regional Plan for Science and Technology 2007–2010).

References

1. Ahlfors, L.V.: Sur le type d'une surface de Riemann. C. R. Acad. Sci. Paris **201**, 30–32 (1935)
2. Albujer, A.L.: Geometría global de superficies espaciales en espacio producto lorentzianos, Ph. D. Thesis, Universidad de Murcia, Spain (2008)
3. Albujer, A.L.: New examples of entire maximal graphs in $\mathbb{H}^2 \times \mathbb{R}_1$. Differential Geom. Appl. **26**, 456–462 (2008)
4. Albujer, A.L., Alías, L.J.: A local estimate for maximal surfaces in Lorentzian product spaces. Mat. Contemp. **34**, 1–10 (2008)
5. Albujer, A.L., Alías, L.J.: Calabi-Bernstein results for maximal surfaces in Lorentzian product spaces. J. Geom. Phys. **59**, 620–631 (2009)

6. Albujer, A.L., Alías, L.J.: Parabolicity of maximal surfaces in Lorentzian product spaces. Math. Z. **267**, 453–464 (2011)
7. Alías, L.J., Mira, P.: On the Calabi-Bernstein theorem for maximal hypersurfaces in the Lorentz-Minkowski space. In: Proceedings on the meeting Lorentzian-Geometry-Benalmádena 2001, Benalmádena, Málaga, Spain. Publ. de la RSME, vol. 5, pp. 23–55. Madrid, Spain (2003)
8. Alías, L.J., Palmer, B.: Zero mean curvature surfaces with non-negative curvature in flat Lorentzian 4-spaces. R. Soc. Lond. Proc. Ser. A Math. Phys. Eng. Sci. **455**, 631–636 (1999)
9. Alías, L.J., Palmer, B.: A duality result between the minimal surface equation and the maximal surface equation. An. Acad. Brasil Ciênc. **73**, 161–164 (2001)
10. Alías, L.J., Palmer, B.: On the Gaussian curvature of maximal surfaces and the Calabi-Bernstein theorem. Bull. London Math. Soc. **33**, 454–458 (2001)
11. Alías, L.J., Romero, A., Sánchez, M.: Uniqueness of complete spacelike hypersurfaces of constant mean curvature in generalized Robertson-Walker spacetimes. Gen. Relat. Gravit. **27**, 71–84 (1995)
12. Brill, D., Flaherty, F.: Isolated maximal surfaces in spacetime. Comm. Math. Phys. **50**, 157–165 (1976)
13. Caballero, M., Romero, A., Rubio, R.M.: Uniqueness of maximal surfaces in Generalized Robertson-Walker spacetimes and Calabi-Bernstein type problems. J. Geom. Phys. **60**, 394–402 (2010)
14. Calabi, E.: Examples of Bernstein problems for some nonlinear equations. Proc. Symp. Pure Math. **15**, 223–230 (1970)
15. Cheng, S.Y., Yau, S.T.: Maximal space-like hypersurfaces in the Lorentz-Minkowski space. Ann. of Math. **104**, 407–419 (1976)
16. Chern, S.S.: Simple proofs of two theorems on minimal surfaces. Enseign. Math. **15**, 53–61 (1969)
17. Duc, D.M., Hieu, N.V.: Graphs with prescribed mean curvature on Poincaré disk. Bull. London Math. Soc. **27**, 353–358 (1995)
18. Estudillo, F.J.M., Romero, A.: On maximal surfaces in n-dimensional Lorentz-Minkowski space. Geom. Dedicata **38**, 167–174 (1991)
19. Estudillo, F.J.M., Romero, A.: Generalized maximal surfaces in Lorentz-Minkowski space $\mathbb{L}^3$. Math. Proc. Cambridge Philos. Soc. **111**, 515–524 (1992)
20. Estudillo, F.J.M., Romero, A.: On the Gauss curvature of maximal surfaces in the 3-dimensional Lorentz-Minkowski space. Comment. Math. Helv. **69**, 1–4 (1994)
21. Fernández, I., López, F.J.: On the uniqueness of the helicoid and Enneper's surface in the Lorentz-Minkowski space $\mathbb{R}^3_1$. Trans. Amer. Math. Soc. **363**, 4603–4650 (2011)
22. Frankel, T.: Applications of Duschek's formula to cosmology and minimal surfaces. Bull. Amer. Math. Soc. **81**, 579–582 (1975)
23. Gálvez, J.A., Rosenberg, H.: Minimal surfaces and harmonic diffeomorphisms from the complex plane onto a Hadamard surface. Amer. J. Math. **132**, 1249–1273 (2010)
24. Grigor'yan, A.: Analytic and geometric background of recurrence and non-explosion of the Brownian motion on Riemannian manifolds. Bull. Amer. Math. Soc. **36**, 135–249 (1999)
25. Huber, A.: On subharmonic functions and differential geometry in the large. Comment. Math. Helv. **32**, 13–72 (1957)
26. Kazdan, J.L.: Parabolicity and the Liouville property on complete Riemannian manifolds, Seminar on new results in nonlinear partial differential equations (Bonn, 1984), Aspects Math., E10, pp. 153–166. Vieweg, Braunschweig, 1987,
27. Kobayashi, O.: Maximal surfaces in the 3-dimensional Lorentz-Minkowski space $\mathbb{L}^3$. Tokyo J. Math. **6**, 297–309 (1983)
28. Kobayashi, S., Nomizu, K.: Foundations of Differential Geometry, vol. 2. Interscience, New York (1969)
29. Latorre, J.M., Romero, A.: New examples of Calabi-Bernstein problems for some nonlinear equations. Differ. Geom. Appl. **15**, 153–163 (2001)

30. Lee, H.: Extensions of the duality between minimal surfaces and maximal surfaces. Geom. Dedicata **151**, 373–386 (2011)
31. Lichnerowicz, A.: L'intégration des équations de la gravitation relativiste et le problème des *n* corps. J. Math. Pures Appl. **23**, 37–63 (1944)
32. McNertey, L.V.: One-parameter families of surfaces with constant curvature in Lorentz 3-space, Ph.D. thesis, Brown University, USA (1980)
33. Meeks, W., Pérez, J.: Conformal properties on classical minimal surface theory. Surveys in differential geometry, vol. 9, pp. 275–335. Int. Press, Somerville (2004)
34. Montaldo, S., Onnis, I.I.: A note on surfaces in $\mathbb{H}^2 \times \mathbb{R}$. Boll. Unione Mat. Ital. Sez. B Artic. Ric. Mat. **10**, 939–950 (2007)
35. Nelli, B., Rosenberg, H.: Minimal surfaces in $\mathbb{H}^2 \times \mathbb{R}$. Bull. Braz. Math. Soc. (N. S.) **33**, 263–292 (2002)
36. O'Neill, B.: Semi-Riemannian Geometry with Applications to Relativity. Academic, New York (1983)
37. Onnis, I.I.: Superficies em certos espaços homogêneos tridimensionais, Ph.D. thesis, Universidade Estadual de Campinas, Brazil (2005)
38. Pérez, J.: Parabolicity and minimal surfaces. Joint work with F.J. López. Clay Math. Proc. **2**, Global theory of minimal sufaces, pp. 163–174. Amer. Math. Soc., Providence (2005)
39. Romero, A.: Simple proof of Calabi-Bernstein's theorem on maximal surfaces. Proc. Amer. Math. Soc. **124**, 1315–1317 (1996)
40. Schoen, R.: Estimates for stable minimal surfaces in three dimensional manifolds, Seminar on minimal submanifolds. Ann. of Math. Stud., vol. 103, pp. 111–126. Princeton Univ. Press, Princeton (1983)

Umbilical-Type Surfaces in SpaceTime

José M.M. Senovilla

1 Introduction

Space-like surfaces play a crucial role in gravitational physics (such as in general relativity and in any other geometrical theory based on a Lorentzian manifold), especially those which are (marginally) (outer) trapped and closed—compact with no boundary—see Sect. 2.3.

A few years ago I presented a complete local classification of space-like surfaces in 4-dimensional Lorentzian manifolds [45] and discussed its generalization to arbitrary dimensions. The classification was carried out according to the *extrinsic* properties of the surface: it is an algebraic classification based, at each point, on the properties of two independent Weingarten operators. Specifically, I used two *null* Weingarten operators A_ℓ and A_k (where ℓ and k are the two independent null vectors fields orthogonal to the surface, see below.) Each Weingarten operator is a self-adjoint matrix which can be readily classified algebraically according to the signs of their (real) eigenvalues. This produces eight different types for each matrix, and therefore 64 types of points for generic space-like surfaces.

To my surprise, this was not enough for a complete classification, and I had to introduce an extra parameter to each point taking into account the *relative orientation* of the two null Weingarten operators at the chosen point. As far as I know, this extra parameter had not been considered in the literature—not even for the proper Riemannian case. The parameter can be chosen as the angle between the two orthonormal (ON) eigen-bases for A_ℓ and A_k. Therefore, it takes values on a

J.M.M. Senovilla (✉)
Física Teórica, Universidad del País Vasco, Apartado 644, 48080 Bilbao, Spain
e-mail: josemm.senovilla@ehu.es

M. Sánchez et al. (eds.), *Recent Trends in Lorentzian Geometry*, Springer Proceedings in Mathematics & Statistics 26, DOI 10.1007/978-1-4614-4897-6_3,

finite closed interval of $\mathbb{R}$. Actually, one can prove [45] that the parameter is simply related to the commutator

$$[A_k, A_\ell]$$

of the two null Weingarten operators.

The meaning and interpretation of this parameter became an important open question, and the goal of this chapter is to answer it. The main theorems to be proven are the following:

Theorem 1. *Consider a space-like surface S immersed in a 4-dimensional Lorentzian manifold $(\mathscr{V}, g)$. The necessary and sufficient condition for S to be umbilical along a normal direction is that two independent Weingarten operators—and, a fortiori, all of them—commute.*

The umbilical direction is then uniquely determined—unless the surface is totally umbilical.

This happens to be equivalent to the condition that the shape tensor be diagonalizable on S.

Theorem 2. *The necessary and sufficient condition for S to be umbilical along a normal direction is that the normal curvature of S equals the "tangent-normal" part of the Riemann tensor of $(\mathscr{V}, g)$.*

Corollary 1. *In particular, for locally conformally flat $(\mathscr{V}, g)$ (including Lorentz space forms) the necessary and sufficient condition for S to be umbilical along a normal direction is that the normal curvature vanishes.*

A precise formulation of Theorem 2 is presented in Remark 5 after the necessary notions and notations have been introduced.

There are several other interesting consequences of these theorems as well as explicit formulas for the umbilical direction. These will be presented in Sect. 5. In Sect. 2.4, I introduce a new vector field G, normal to the surface, which characterizes the umbilical property and, together with the traditional mean curvature vector H, defines the main properties of the surface.

The main results extend to *non-null* surfaces in 4-dimensional semi-Riemannian manifolds of *arbitrary* signature as I discuss succinctly at the end of this chapter. The higher dimensional case is, however, an open problem.

2 Basic Concepts and Notation

Let $(\mathscr{V},g)$ be a 4-dimensional, oriented and time-oriented, Lorentzian manifold with metric tensor g of signature $(-,+,+,+)$. At every $x \in \mathscr{V}$, the isomorphism between tangent vectors and one-forms, that is, between $T_x\mathscr{V}$ and $T_x^*\mathscr{V}$ is denoted as follows

$$\begin{aligned} \flat : T_x\mathscr{V} &\longrightarrow T_x^*\mathscr{V} \\ v &\longmapsto v^\flat \end{aligned}$$

and defined by $v^\flat(w) = g(v,w)$, $\forall w \in T_x\mathscr{V}$. Its inverse map is denoted by $\sharp$. These maps extend naturally to the tangent and cotangent bundles.

An immersed surface is given by the pair (S,Φ) where S is a 2-dimensional manifold and $\Phi : S \longrightarrow \mathscr{V}$ is an immersion. Such an S does not have to be necessarily orientable. However, as the computations herein presented will be local, I will tacitly assume—without loss of generality—that S is embedded and oriented. For instance, given that any point of an immersed S has an open neighborhood which can be identified with its image in $(\mathscr{V},g)$, to avoid confusion and unnecessary complications in the notation, S will be identified with its image $\Phi(S)$ in $\mathscr{V}$.

The *first fundamental form* of S in $\mathscr{V}$ is simply $\bar{g} \equiv \Phi^* g$, where Φ^* is the pullback of Φ. From now on, $\bar{g}$ will be assumed to be positive definite on S, which implies that every tangent vector in $T_xS, \forall x \in S$ is space-like and then S is said to be *space-like*. For such an S, at any $x \in S$, one has the orthogonal decomposition

$$T_x\mathscr{V} = T_xS \oplus T_xS^\perp .$$

Let $\mathfrak{X}(S)$ (respectively $\mathfrak{X}(S)^\perp$) denote the set of smooth vector fields tangent (resp. orthogonal) to S. In what follows, and for the sake of brevity, I will often give definitions and properties on $\mathfrak{X}(S)$, but they of course have always a previous, more fundamental, version on each T_xS. Thus, for instance, the volume element 2-form associated to $(S,\bar{g})$—denoted here by $\bar{\varepsilon}$—together with the volume element 4-form ε of $(\mathscr{V},g)$ induces a volume element 2-form on each $T_xS^\perp$, denoted by $\varepsilon^\perp$. The corresponding Hodge dual operator (see, e.g., [7]) is written and defined as

$$\star^\perp N \equiv (i_N \varepsilon^\perp)^\sharp, \; \forall N \in \mathfrak{X}(S)^\perp .$$

The vector field $\star^\perp N$ defines the unique normal direction in $\mathfrak{X}(S)^\perp$ orthogonal to the normal $N \in \mathfrak{X}(S)^\perp$.

The surface S with the first fundamental form is a Riemannian manifold $(S,\bar{g})$. As is well known, its *Levi-Civita connection* $\bar{\nabla}$, with $\overline{\nabla}\bar{g} = 0$ and no torsion, can be defined as [29, 33]

$$\overline{\nabla}_X Y \equiv (\nabla_X Y)^T , \qquad \forall X,Y \in \mathfrak{X}(S)$$

where ∇ is the canonical connection of $(\mathscr{V},g)$.[1]

The *normal connection* D acts, in turn, on $\mathfrak{X}(S)^{\perp}$

$$D_X : \mathfrak{X}(S)^{\perp} \longrightarrow \mathfrak{X}(S)^{\perp}$$

for $X \in \mathfrak{X}(S)$, and is given by the standard definition [33]

$$D_X N \equiv (\nabla_X N)^{\perp}, \quad \forall N \in \mathfrak{X}(S)^{\perp} \quad \forall X \in \mathfrak{X}(S).$$

2.1 Extrinsic Geometry: Second Fundamental Forms and Weingarten Operators

The basic extrinsic object is $\mathrm{I\!I} : \mathfrak{X}(S) \times \mathfrak{X}(S) \longrightarrow \mathfrak{X}(S)^{\perp}$, called the *shape tensor* or *second fundamental form tensor* of S in $\mathscr{V}$ and defined by [29, 33]

$$-\mathrm{I\!I}(X,Y) \equiv (\nabla_X Y)^{\perp} = \nabla_X Y - \overline{\nabla}_X Y, \quad \forall X,Y \in \mathfrak{X}(S)$$

(observe the choice of sign that may be unusual in some contexts and is actually opposite to [29, 33]). $\mathrm{I\!I}$ contains the information concerning the "shape" of S within $\mathscr{V}$ along *all* directions normal to S. Notice that

$$\mathrm{I\!I}(X,Y) = \mathrm{I\!I}(Y,X).$$

Given any normal direction $N \in \mathfrak{X}(S)^{\perp}$, the *second fundamental form* of S in $(\mathscr{V},g)$ relative to N is the 2-covariant symmetric tensor field on S defined by

$$K_N(X,Y) \equiv g\left(N, \mathrm{I\!I}(X,Y)\right), \qquad \forall X,Y \in \mathfrak{X}(S).$$

The *Weingarten operator*

$$A_N : \mathfrak{X}(S) \longrightarrow \mathfrak{X}(S)$$

associated to $N \in \mathfrak{X}(S)^{\perp}$ is defined by

$$A_N(X) \equiv (\nabla_X N)^T, \qquad \forall X \in \mathfrak{X}(S).$$

Observe that

$$\bar{g}(A_N(X),Y) = K_N(X,Y), \quad \forall X,Y \in \mathfrak{X}(S),$$

hence, at each $x \in S$, $A_N|_x$ is a self-adjoint linear transformation on T_xS. As such, it is always diagonalisable over $\mathbb{R}$.

---

[1] There is a long-standing tradition among mathematicians who study submanifolds to use the opposite convention, that is, ∇ for the inherited connection and $\bar{\nabla}$ for the background connection. I stress this point here in the hope that this will avoid any possible confusion.

2.2 *Special Bases on $\mathfrak{X}(S)^{\perp}$*

S having co-dimension 2, there are two *independent* normal vector fields on *S*. They can be appropriately chosen to form an ON basis on $\mathfrak{X}(S)^{\perp}$, in which case I will denote them by $u, n \in \mathfrak{X}(S)^{\perp}$, with

$$g(n,n) = -g(u,u) = 1, \; g(u,n) = 0.$$

Of course, any two such ON bases are related by a Boost (Lorentz transformation):

$$\begin{pmatrix} u' \\ n' \end{pmatrix} = \begin{pmatrix} \cosh\beta & \sinh\beta \\ \sinh\beta & \cosh\beta \end{pmatrix} \begin{pmatrix} u \\ n \end{pmatrix} \tag{1}$$

where β is a smooth function on *S*.

The two independent normal vector fields can also be chosen to be null (and future-pointing say), and I will denote these by $k, \ell \in \mathfrak{X}(S)^{\perp}$, so that

$$g(\ell,\ell) = g(k,k) = 0, \quad g(\ell,k) = -1$$

the last of these being a convenient normalization condition. Observe that, to any ON basis $\{u,n\}$ on $\mathfrak{X}(S)^{\perp}$, one can associate a null basis given by $\sqrt{2}\ell = u + n$ and $\sqrt{2}k = u - n$. The previous boost freedom becomes now simply

$$\ell \longrightarrow \ell' = e^{\beta}\ell, \quad k \longrightarrow k' = e^{-\beta}k \tag{2}$$

so that the two independent null *directions* are uniquely determined on *S*.

The orientations of $(\mathcal{V}, g)$ and of the imbedded surface $(S, \bar{g})$ will be chosen such that the operator $\star^{\perp}$ acts on the previous bases as follows

$$\star^{\perp}u = n, \; \star^{\perp}n = u; \quad \star^{\perp}\ell = \ell, \; \star^{\perp}k = -k.$$

2.3 *The Mean Curvature Vector Field H and Its Causal Character*

The shape tensor decomposes as

$$\mathrm{I\!I}(X,Y) = -K_k(X,Y)\,\ell - K_\ell(X,Y)\,k \tag{3}$$

in a null basis, or as

$$\mathrm{I\!I}(X,Y) = -K_u(X,Y)\,u + K_n(X,Y)\,n \tag{3'}$$

in any ON basis, $\forall X,Y \in \mathfrak{X}(S)$. Note that these formulae are invariant under the boost freedom (1) and (2). The *mean curvature vector* $H \in \mathfrak{X}(S)^{\perp}$ is defined as the trace of the shape tensor with respect to $\bar{g}$, or explicitly

$$H = -(\mathrm{tr}A_k)\,\ell - (\mathrm{tr}A_\ell)\,k \tag{4}$$

in a null basis, or

$$H = -(\mathrm{tr}A_u)\,u + (\mathrm{tr}A_n)\,n \tag{4'}$$

in ON bases. In the physics literature, each component of H along a particular normal direction $g(H,N)= \mathrm{tr}A_N$ is termed "expansion along N" of S [5, 29, 45]. In particular, $\mathrm{tr}A_\ell$ and $\mathrm{tr}A_k$ are called the *null expansions*.

Notice that H and

$$\star^\perp H = -(\mathrm{tr}A_k)\,\ell + (\mathrm{tr}A_\ell)\,k = -(\mathrm{tr}A_u)\,n + (\mathrm{tr}A_n)\,u$$

are well-defined; they are invariant under the boost gauge freedom and actually under arbitrary changes of basis. Observe also that $\star^\perp H$ is a (generically unique) direction with vanishing expansion: $\mathrm{tr}A_{\star^\perp H} = 0$. This fact is important in physics sometimes.

A very important type of surface (or submanifolds) in Riemannian geometry is the *minimal* ones. They are characterized by the vanishing of the mean curvature vector, that is, by the condition $H = 0$. Observe, however, that in Riemannian geometry any vector can only have either zero or positive norm and, hence, the only distinguished case for H is when it vanishes: the minimal surfaces. In semi-Riemannian geometry, though, vectors such as H can realize all signs for $g(H,H)$, in particular H can be time-like ($g(H,H) < 0$) or null ($g(H,H) = 0$), in the last case with $H \neq 0$. And these new cases provide new types of surfaces (and submanifolds) in the Lorentzian case.

Actually, the most important surfaces in Gravitation are defined according to such causal orientations of H. For instance, the simple condition

$$H^\flat \wedge (\star^\perp H)^\flat = 0$$

is equivalent to saying that H is null everywhere on S. These will be called *null $*$-surfaces* due to the nomenclature introduced in [45], see also [28, 47]. In the mathematical literature, surfaces with a null H were considered for instance in [39, 40] for the Minkowski spacetime under the name of "pseudo-minimal" or "quasi-minimal" surfaces, see also [12]. Among null $*$-surfaces, an important case is when H (and hence $\star^\perp H$) points along one of the null directions ℓ or k everywhere, then they are called *marginally outer trapped surfaces* (MOTS) (also called *null dual*). They have received a great deal of attention lately, in particular concerning their stability [2–4, 10] which leads to the study of an elliptic operator similar to the stability operator for minimal surfaces.

If in addition to H pointing along one of the null normal directions its causal orientation does not change on S, that is to say, it is everywhere null future or everywhere null past, then S is called a *marginally (future or past) trapped surface*, [23, 31, 45, 47, 51]. For references concerning this type of surfaces in the mathematical literature, one can consult the recent book [12].

If on the other hand H keeps its future (or past) causal orientation everywhere on S (but it can change from null to time-like from point to point), then the surface is usually called *weakly (future or past) trapped* [23, 31, 45, 47, 51]. If H is also time-like nonzero all over S, then it is said to be (future or past) trapped [5, 12, 23, 24, 35, 45, 51]. The concept of closed trapped surface—here closed means compact with no boundary—was introduced by Penrose [35] in a seminal paper where the first modern-type singularity theorem was proven. It was immediately realized that the concept of trapping is essential in many important developments concerning gravitational collapse and the formation of black holes, such as the singularity theorems [19, 23, 24, 44], the so-called "cosmic censorship conjecture" [37] together with the related subject of iso-perimetric or Penrose inequalities [20, 21, 30, 36], and the hoop conjecture [15, 32, 46].

2.4 The Extrinsic Vector Field G

One can also define another normal vector field $G \in \mathfrak{X}(S)^{\perp}$ by using a second invariant of the matrices A_N. Unfortunately, there are no other linear invariants. In spite of that, for each $N \in \mathfrak{X}(S)^{\perp}$ one can set

$$\sigma_N^2 \equiv (\mathrm{tr}A_N)^2 - 4\,\mathrm{det}A_N$$

which is called the *shear* along N [23, 51] and can also be expressed as the square of the difference of the two eigenvalues of A_N. An alternative formula is

$$\det\left(A_N - \frac{1}{2}\mathrm{tr}A_N\mathbf{1}\right) = -\frac{1}{4}\sigma_N^2.$$

It should be noted that the matrix $A_N - \frac{1}{2}\mathrm{tr}A_N\mathbf{1}$ is traceless, and therefore its two eigenvalues have opposite signs: $\sigma_N^2/4$ is the square of either of them. However, fixing the sign of σ_N so that it becomes a differentiable function on S is not free from ambiguities.[2] Whatever the signs chosen, I set by definition

$$\boxed{G \equiv \sigma_k \ell + \sigma_\ell k}$$

The two possible signs for each of σ_k, σ_ℓ provide four distinct possibilities for G which define, by ignoring overall orientations, two orthogonal directions. However, these two directions can always be described by G together with

$$\star^{\perp} G = \sigma_k \ell - \sigma_\ell k$$

independently of the chosen signs. Observe that both G and $\star^{\perp} G$ are invariant under the boost freedom (2).

---

[2]If one chooses, say, σ_N to be the positive root of $\sqrt{\sigma_N^2}$ then it may fail to be differentiable at points where the two eigenvalues of A_N coincide, that is, at points where $\sigma_N = 0$. Of course, one can always set an "initial" condition for $G|_x$ at any point on $x \in S$ and then the differentiable solution for the vector field G is fixed. Nevertheless, this initial condition is arbitrary.

It is important to remark that, due to the nonlinearity of the invariants σ_N^2,

$$G \neq \sigma_u u - \sigma_n n$$

in general. There are points where the equality holds, and they will turn out to be precisely the umbilical points.

The vector field G is intimately related to the umbilical properties of a surface S, as I am going to prove presently.

2.5 The Normal Connection One-Form s

For a fixed ON basis on $\mathfrak{X}(S)^{\perp}$, a one-form $s \in \Lambda^1(S)$ is defined by

$$s(X) \equiv -g(u, D_X n) = g(D_X u, n), \quad \forall X \in \mathfrak{X}(S).$$

For $\sqrt{2}\ell = u + n$ and $\sqrt{2}k = u - n$, one can alternatively write

$$s(X) \equiv -g(k, D_X \ell) = g(D_X k, \ell), \quad \forall X \in \mathfrak{X}(S).$$

Therefore, for all $X \in \mathfrak{X}(S)$

$$D_X u = s(X) n,\ D_X n = s(X) u; \quad D_X \ell = s(X) \ell,\ D_X k = -s(X) k.$$

Observe that s is not invariant under boost rotations (1) or (2). Actually, s is a "connection" and transforms as $s'(X) = s(X) + X(\beta)$ under those transformations, or simply

$$s' = s + d\beta.$$

It follows that $ds = ds'$ is invariant and well defined. It will be proven in the next section that this is actually related to the normal curvature on S, see formula (5), confirming the connection character of s. In the mathematical literature on Riemannian geometry, s is sometimes called the *third fundamental form* of S in $(\mathscr{V}, g)$; see, e.g., [52].

2.6 Curvatures: Gauss and Ricci Equations

The intrinsic curvature for $(S, \bar{g})$ has the usual definition

$$\overline{R}(X,Y)Z \equiv \overline{\nabla}_X \overline{\nabla}_Y Z - \overline{\nabla}_Y \overline{\nabla}_X Z - \overline{\nabla}_{[X,Y]} Z, \quad \forall X,Y,Z \in \mathfrak{X}(S).$$

Similarly, the normal curvature is defined on S by

$$R^{\perp}(X,Y)N \equiv D_X D_Y N - D_Y D_X N - D_{[X,Y]} N, \quad \forall X,Y \in \mathfrak{X}(S), \quad \forall N \in \mathfrak{X}(S)^{\perp}.$$

A simple calculation provides

$$R^{\perp}(X,Y)N = ds(X,Y) \star^{\perp} N, \qquad \forall X,Y \in \mathfrak{X}(S), \ \forall N \in \mathfrak{X}(S)^{\perp}. \tag{5}$$

This justifies that s describes the normal connection and that ds defines its curvature.

The Gauss equation relating the curvatures of $(S,\bar{g})$ and $(\mathscr{V},g)$ can be written as

$$R(W,Z,X,Y) = \overline{R}(W,Z,X,Y) + g\left(\mathbb{I}(X,Z),\mathbb{I}(Y,W)\right) - g\left(\mathbb{I}(Y,Z),\mathbb{I}(X,W)\right) \tag{6}$$

for all $X,Y,Z,W \in \mathfrak{X}(S)$, where I use the notation[3]

$$R(W,Z,X,Y) \equiv g(W,R(X,Y)Z)$$

and analogously for $\overline{R}$. However, as S is 2-dimensional, its curvature is uniquely determined by its Gaussian curvature $K(S)$. Therefore, the previous relation can be written as a single scalar equation. To that end, let me define a new extrinsic object, quadratic in the shape tensor $\mathbb{I}$, as follows. For any ON basis $\{e_1,e_2\}$ in $\mathfrak{X}(S)$, set by definition

$$\mathbb{J}(X,Y) \equiv \sum_{i=1}^{2} g\left(\mathbb{I}(e_i,X),\mathbb{I}(e_i,Y)\right), \qquad \forall X,Y \in \mathfrak{X}(S).$$

$\mathbb{J}(X,Y)$ is a 2-covariant symmetric tensor field on S. Then, define $B : \mathfrak{X}(S) \to \mathfrak{X}(S)$ by

$$g(B(X),Y) \equiv \mathbb{J}(X,Y), \qquad \forall X,Y \in \mathfrak{X}(S).$$

B is sometimes called *the Casorati operator* of S in $(\mathscr{V},g)$ [12] and has been mainly studied in the Riemannian case; see e.g., [16, 22] and references therein. In the Lorentzian case under consideration in this chapter, a straightforward calculation allows one to check that B is the anticommutator of the two null Weingarten operators:

$$B = -\{A_k, A_\ell\}. \tag{7}$$

Once more, let me remark that B is invariant under the boost freedom (2). Observe furthermore that

$$\mathrm{tr}B = g(\mathbb{I},\mathbb{I})$$

which is sometimes called the Casorati curvature [12].

With the previous notation the Gauss equation (6) becomes

$$2K(S) = \mathscr{S} - 4\mathrm{Ric}(\ell,k) + 2R(\ell,k,\ell,k) + g(H,H) - \mathrm{tr}B \tag{8}$$

where Ric and $\mathscr{S}$ are the Ricci tensor and the scalar curvature of $(\mathscr{V},g)$.

[3]Notice the sign convention, which may not coincide with the preferred one for everybody.

With regard to the Ricci equation, relating the normal curvature $R^{\perp}$ with the tangent-normal part of the spacetime curvature R on S, one can write

$$(R(X,Y)N)^{\perp} = \mathrm{I\!I}(X,A_N(Y)) - \mathrm{I\!I}(Y,A_N(X)) + R^{\perp}(X,Y)N$$
$$= \mathrm{I\!I}(X,A_N(Y)) - \mathrm{I\!I}(Y,A_N(X)) + ds(X,Y)\star^{\perp}N$$

for all $X,Y \in \mathfrak{X}(S)$ and for all $N,M \in \mathfrak{X}(S)^{\perp}$, where in the last equality I have used Eq. (5). An alternative possibility, which will reveal itself as very useful in the sequel, is

$$R(M,N,X,Y) = g\left([A_M,A_N](Y),X\right) + ds(X,Y)\, g(\star^{\perp}N,M), \tag{9}$$

for all $X,Y \in \mathfrak{X}(S)$ and all $N,M \in \mathfrak{X}(S)^{\perp}$.

3 Umbilical-Type, Pseudo-umbilical, and Related Surfaces

The concept of umbilical point is classical in semi-Riemannian geometry. When the co-dimension of a submanifold is higher than one, then there are several possible directions along which a point can be umbilic. Specifically:

Definition 1 (Umbilical Points on S). A point $x \in S$ is called *umbilical* with respect to $N|_x \in T_x^{\perp}S$ (or simply N-umbilical) if the corresponding Weingarten operator is proportional to the Identity

$$A_N|_x = \frac{1}{2}F\,\mathbf{1}.$$

Obviously, in that case $F = \mathrm{tr}A_N|_x = g(H,N|_x)$ necessarily. An equivalent characterization is

$$K_N|_x = \frac{1}{2}g(H,N|_x)\,\bar{g}|_x.$$

Definition 2 (N-Umbilical Surfaces). Thus, S is said to be *umbilical along a vector field* $N \in \mathfrak{X}(S)^{\perp}$ if

$$A_N = \frac{1}{2}g(H,N)\mathbf{1} \tag{10}$$

or equivalently, if $K_N = \frac{1}{2}g(H,N)\,\bar{g}$.

This concept was studied in the Riemannian case many years ago under some special circumstances, e.g. [13, 14], see [12] for the general semi-Riemannian case.

Observe that *minimal surfaces*, that is, those with zero mean curvature vector $H = 0$, can be considered as a limit case of N-umbilical surfaces only in the case that the whole Weingarten operator vanishes $A_N = 0$. This motivates the following definition [45].

Definition 3 (N-subgeodesic Surface). A space-like surface S is called *N-subgeodesic*, for $N \in \mathfrak{X}(S)^{\perp}$, if $A_{\star\perp N} = 0$.

This means that *any* geodesic $\gamma : I \subset \mathbb{R} \longrightarrow S$ of the surface $(S, \bar{g})$ is a sub-geodesic [43] with respect to N on the spacetime $(\mathscr{V}, g)$: its tangent vector γ' satisfies the relation

$$\nabla_{\gamma'}\gamma' = fN$$

where the function f on γ is fully determined by the relation $fN = \mathbb{II}(\gamma', \gamma')$.

Obviously, a surface is subgeodesic with respect to two independent normal vector fields N and M ($N^\flat \wedge M^\flat \neq 0$) if and only if it is *totally geodesic* ($\mathbb{II} = 0$) [33], or equivalently, if and only if $A_N = 0$, $\forall N \in \mathfrak{X}^{\perp}(S)$.

Remark 1. In traditional Riemannian geometry, there is the concept of *first normal space* $\mathscr{N}_1$ for immersed submanifolds S, defined at each $p \in S$ by

$$\mathscr{N}_1 = \text{Span}\{\mathbb{II}(X,Y); X, Y \in T_pS\}.$$

This generalizes immediately to the general semi-Riemannian case, and then N-subgeodesic surfaces have $\dim \mathscr{N}_1 \leq 1$, because $\mathbb{II}(X,Y)^\flat \wedge N^\flat = 0$ for all $X, Y \in \mathfrak{X}(S)$. Actually, N-subgeodesic surfaces are characterized by

$$\mathbb{II}(X,Y)^\flat \wedge \mathbb{II}(Z,W)^\flat = 0 \quad \forall X, Y, Z, W \in \mathfrak{X}(S)$$

and then the direction N can be determined by computing $\mathbb{II}(X,X)$ for any $X \in \mathfrak{X}(S)$ such that $\mathbb{II}(X,X) \neq 0$—and whenever S is not totally geodesic, of course. In other words, all possible second fundamental forms, or all the Weingarten operators, are proportional to each other as follows from the fact that $\mathbb{II}(X,Y) = K(X,Y)N$ for some fixed[4] rank-2 symmetric covariant tensor field K in S. Given that all submanifolds with $\dim \mathscr{N}_1 \leq 1$ are trivial A-submanifolds, a concept introduced in [11] for Riemannian manifolds—see also [17, 25, 42] and references therein for some simple Lorentzian cases—then N-subgeodesic surfaces are in particular trivial A-submanifolds.

A standard possibility for umbilical surfaces in submanifolds with co-dimension higher than one is that the umbilical direction is given by the mean curvature vector. These are called pseudo-umbilical surfaces [12].

Definition 4 (Pseudo-umbilical Surface). S is said to be pseudo-umbilical if it is umbilical with respect to $N = H$, so that

$$A_H = \frac{1}{2} g(H,H)\,\mathbf{1}.$$

[4]Up to proportionality factors, this K coincides with K_N if N is non-null. If N is null, then N can be chosen to be either ℓ or k, and K is $-K_k$ or $-K_\ell$, respectively.

In Riemannian geometry this kind of submanifolds has been studied since long ago; see e.g., [34] and specifically [52] for the co-dimension 2 situation. Probably, the first study in the semi-Riemannian case was performed in [38] and then only much later in [27, 49]. Some results concerning pseudo-umbilical submanifolds in semi-Riemannian geometry can be consulted in [1, 6, 9, 12, 26, 48], not much of it specific for Lorentzian geometry. Thus, as far as I am aware, very few things are known for pseudo-umbilical surfaces in general Lorentzian manifolds.

Less common is the idea of S being umbilical along the *unique* direction orthogonal to H in $\mathfrak{X}(S)^{\perp}$. Actually, this idea does not appear to have been considered previously, so that the following definition is new and I made the name for this type of surface up—maybe not too skillfully.

Definition 5 (Ortho-umbilical Surface). A surface S will be called *ortho-umbilical* if it is umbilical with respect to $N = \star^{\perp} H$, so that

$$A_{\star^{\perp}H} = 0.$$

As a matter of fact, for the case of co-dimension two under consideration, one can prove the following equivalence between ortho-umbilical and N-subgeodesic surfaces.

Proposition 1. *The following conditions are equivalent for a non-minimal space-like surface S in $(\mathscr{V}, g)$:*

1. *S is ortho-umbilical*
2. *S is N-subgeodesic for some $N \in \mathfrak{X}(S)^{\perp}$*
3. *S is H-subgeodesic*

Proof. $1 \Longrightarrow 3$ Assume S is ortho-umbilical and $H \neq 0$. This means that $A_{\star^{\perp}H} = 0$ which is the definition of H-subgeodesic.

$3 \Longrightarrow 2$ Trivial

$2 \Longrightarrow 1$ If S is N-subgeodesic then $A_{\star^{\perp}N} = 0$, so that $K_{\star^{\perp}N} = 0$ too. Let $M \in \mathfrak{X}(S)^{\perp}$ be any normal vector field such that $N^{\flat} \wedge M^{\flat} \neq 0$, ergo $\text{span}\{N, M\} = \mathfrak{X}(S)^{\perp}$. In the basis $\{N, M\}$ one obviously has

$$\mathrm{I\!I}(X,Y) = K_1(X,Y)N + K_2(X,Y)M, \quad \forall X, Y \in \mathfrak{X}(S)$$

for some rank-2 covariant symmetric tensor fields K_1, K_2 on S. As a matter of fact, K_1 and K_2 are determined by K_N and K_M as follows: $K_N = g(N,N)K_1 + g(N,M)K_2$ and $K_M = g(N,M)K_1 + g(M,M)K_2$. The property $K_{\star^{\perp}N} = 0$ implies, on using that by definition N and $\star^{\perp}N$ are mutually orthogonal, that

$$g(\star^{\perp}N, M)K_2(X,Y) = 0 \quad \forall X, Y \in \mathfrak{X}(S).$$

Notice, however, that $g(\star^{\perp}N, M) \neq 0$ because the unique direction in $\mathfrak{X}^{\perp}(S)$ orthogonal to $\star^{\perp}N$ is actually N, and the choice of M prevents that M and N be proportional. Thus, necessarily $K_2 = 0$ implying that $\mathbb{I}(X,Y) = K_1(X,Y)N$ for all $X,Y \in \mathfrak{X}(S)$ and, as a consequence, that

$$H = \mathrm{tr}A_1 N$$

where $A_1 : \mathfrak{X}(S) \to \mathfrak{X}(S)$ is an operator *à la* Weingarten associated to K_1, that is to say, defined by $\bar{g}(A_1(X),Y) = K_1(X,Y)$ for all $X,Y \in \mathfrak{X}(S)$. As S is non-minimal $H \neq 0$ and thus $\mathrm{tr}A_1 \neq 0$. Therefore, one finally arrives at

$$\mathbb{I}(X,Y) = \frac{1}{\mathrm{tr}A_1} K_1(X,Y)H \qquad \forall X,Y \in \mathfrak{X}(S) \tag{11}$$

and thus $K_{\star^\perp H} = 0$ from which $A_{\star^\perp H} = 0$ follows. □

Remark 2. As a consequence, and due to Remark 1, ortho-umbilical surfaces, are trivial A-surfaces, and have *all* Weingarten operators proportional to each other with $\dim \mathscr{N}_1 = 1$ (unless S is totally geodesic, in which case of course $\dim \mathscr{N}_1 = 0$).

Example 1. Taking into account that the concept of ortho-umbilical S seems to be new, I present some simple examples. Take $\mathscr{V} = \mathbb{R} \times \Sigma$ for a 3-dimensional manifold Σ and let $g = \mp dt^2 \oplus g_{\Sigma_\pm}$ where $g_{\Sigma_\pm}$ is a Riemannian (+) or Lorentzian (-) metric on Σ, so that $(\mathscr{V},g)$ is a Lorentzian manifold. Now, take an arbitrary (space-like) surface S immersed in Σ. If K is the second fundamental form of S in $(\Sigma, g_{\Sigma_\pm})$ (with respect to the unit normal m of S in $(\Sigma, g_{\Sigma_\pm})$), then the shape tensor of S in $(\mathscr{V},g)$ can be easily shown to take the form $\mathbb{I}(X,Y) = \pm K(X,Y)M$ for all $X,Y \in \mathfrak{X}(S)$, where $M \in \mathfrak{X}^\perp(S)$ is the normal that corresponds to m. The mean curvature vector H is then proportional to M and S happens to be umbilical with respect to $\star^\perp M$, that is, ortho-umbilical (and also M-subgeodesic).

A surface can be pseudo-umbilical and ortho-umbilical at the same time. This can only happen in some special cases with a null H, to be enumerated and derived rigorously later in Remark 8, or in the traditional cases of minimal surfaces or of totally umbilical surfaces, which is a particular case of the above and can be defined as:

Definition 6 (Totally Umbilical Surfaces). S is called totally umbilical if it is umbilical with respect to all possible $N \in \mathfrak{X}(S)^\perp$:

$$\forall N \in \mathfrak{X}(S)^\perp, \quad A_N = \frac{1}{2} g(H,N)\,\mathbf{1}.$$

Equivalently,

$$\mathbb{I}(X,Y) = \frac{1}{2}\bar{g}(X,Y)H, \qquad \forall X,Y \in \mathfrak{X}(S). \tag{12}$$

This provides a preliminary interpretation of the vector field G.

Proposition 2. *Totally umbilical surfaces can be characterized by*

$$G = 0.$$

Proof. If S is totally umbilical, then $A_N = (1/2)g(H,N)\mathbf{1}$ for all $N \in \mathfrak{X}(S)^\perp$, in particular $A_k = (1/2)g(H,k)\mathbf{1}$ and $A_\ell = (1/2)g(H,\ell)\mathbf{1}$ so that $\sigma_k = \sigma_\ell = 0$ and thus $G = 0$. Conversely, if $G = 0$ then $\sigma_k = \sigma_\ell = 0$ and thus

$$K_\ell = \frac{1}{2}\mathrm{tr}A_\ell\, \bar{g}\,; \qquad\qquad K_k = \frac{1}{2}\mathrm{tr}A_k\, \bar{g}.$$

Using now formulas (3) and (4) one derives Eq. (12). □

In what follows I am going to prove that, letting this case aside, the umbilical direction, if it exists, is always given by either G or $\star^\perp G$ (Corollary 2).

4 Proof of the Main Theorems

We are now ready to proof Theorems 1 and 2.

Remark 3. The results of the theorems can be stated at a point $x \in S$. For instance, "a point $x \in S$ is N-umbilical if and only if two independent Weingarten operators commute at x." However, for the sake of simplicity, I am going to omit the sub-index x, and therefore the proofs are valid for the entire surface and in accordance with their form presented in the Introduction. One should keep in mind, though, that the result may be valid only at some points of the surface in general.

Proof (of Theorem 1).
$\Longrightarrow$ Assume that $N \in \mathfrak{X}(S)^\perp$ is an umbilical direction. In (say) the null basis $N = -g(N,k)\ell - g(N,\ell)k$, and the umbilical condition (10) can be written as

$$-g(N,k)A_\ell - g(N,\ell)A_k = \frac{1}{2}g(H,N)\mathbf{1}. \tag{13}$$

By taking here the commutator with A_ℓ, or with A_k, one immediately derives (for $N \neq 0$)

$$[A_\ell, A_k] = 0.$$

Now, all possible Weingarten operators are linear combinations of any two of them, that is, for any $M \in \mathfrak{X}(S)^\perp$, there exist scalars a and b such that

$$A_M = aA_k + bA_\ell$$

and therefore

$$[A_M, A_{\tilde{M}}] = 0, \quad \forall M, \tilde{M} \in \mathfrak{X}(S)^\perp.$$

$\Longleftarrow$ Conversely, assume that $[A_\ell, A_k] = 0$. This implies that there exists a common ON eigen-basis such that both A_ℓ and A_k are diagonal. Let $\{\lambda_1, \lambda_2\}$ and $\{\nu_1, \nu_2\}$ denote the corresponding eigenvalues for A_k and A_ℓ, respectively. Then, the Eq. (13) to determine the umbilical direction N becomes in this eigen-basis

$$-g(N,\ell)\begin{pmatrix}\lambda_1 & 0\\ 0 & \lambda_2\end{pmatrix}-g(N,k)\begin{pmatrix}\nu_1 & 0\\ 0 & \nu_2\end{pmatrix}=\frac{1}{2}g(H,N)\begin{pmatrix}1 & 0\\ 0 & 1\end{pmatrix}. \tag{14}$$

Introducing here that

$$g(H,N)=-g(N,\ell)\mathrm{tr}A_k-g(N,k)\mathrm{tr}A_\ell=-g(N,\ell)(\lambda_1+\lambda_2)-g(N,k)(\nu_1+\nu_2)$$

the system of equations (14) collapses to a *single* equation

$$g(N,\ell)(\lambda_1-\lambda_2)+g(N,k)(\nu_1-\nu_2)=0.$$

Its solution is clearly unique (up to proportionality factors) and explicitly given by $g(N,k)=-\lambda_1+\lambda_2$ and $g(N,\ell)=\nu_1-\nu_2$, that is to say

$$\boxed{N_{umb}=(\lambda_1-\lambda_2)\ell-(\nu_1-\nu_2)k} \tag{15}$$

unless $\lambda_1-\lambda_2=\nu_1-\nu_2=0$, in which case the surface is totally umbilical as proven in the next Corollary 2. □

Remark 4. As a consequence, there exists a (generically unique) ON basis in which *all* possible Weingarten operators diagonalize simultaneously.

Corollary 2. *The unique umbilical direction N_{umb} is given, at each $x\in S$, either by $G|_x$ or $\star^{\perp}G|_x$.*

Proof. It is straightforward to note that

$$(\lambda_1-\lambda_2)^2=\sigma_k^2,\quad(\nu_1-\nu_2)^2=\sigma_\ell^2$$

so that the unique solution (15) for N_{umb}, at each $x\in S$, is either $\pm G|_x$ or $\pm\star^{\perp}G|_x$. The only exceptional case is defined by $N_{umb}=G=0$, but this characterizes the totally umbilical case, as follows from Proposition 2. □

Under the hypothesis of this corollary and Theorem 1 one can also use the formula $G=\sigma_u u-\sigma_n n$ which does not hold in general. This is due to the commutativity property of all Weingarten operators in this case.

The causal character of the umbilical direction can be easily sorted out due to the explicit formula (15), which allows us to compute

$$g\left(N_{umb},N_{umb}\right)=2(\lambda_1-\lambda_2)(\nu_1-\nu_2)=4\,\mathrm{tr}(A_kA_\ell)-2\,\mathrm{tr}A_k\,\mathrm{tr}A_\ell. \tag{16}$$

Using here the expression (7) for B, this can be invariantly rewritten as

$$g\left(N_{umb},N_{umb}\right)=g(H,H)-2\,\mathrm{tr}B.$$

Thus, the following criteria provide the causal character of the umbilical direction if it exists:

$$g(H,H) - 2\,\mathrm{tr}B \begin{cases} < 0 \Rightarrow N_{umb} \text{ is time-like} \\ > 0 \Rightarrow N_{umb} \text{ is space-like} \\ = 0 \Rightarrow N_{umb} \quad \text{is null} \end{cases}$$

An alternative way of expressing the same utilizes the *ordered* eigen-bases for A_ℓ and A_k, where ordered means for instance that the first eigenvector corresponds to the larger eigenvalue. This has some relevance concerning the classification presented in [45]. Thus, from Eq. (16)

$$N_{umb} \text{ is } \begin{cases} \text{space-like if the ordered eigen-bases of } A_\ell \text{ and } A_k \text{ agree} \\ \text{time-like if the ordered eigen-bases of } A_\ell \text{ and } A_k \text{ are opposite} \\ \text{null if one of the eigen-bases of } A_\ell \text{ or } A_k \text{ cannot be ordered} \end{cases}$$

Let us now prove the second main theorem that, with the introduced notation, can be reformulated as:

Remark 5 (Reformulation of Theorem 2). The necessary and sufficient condition for S to be umbilical along a normal direction is

$$R^\perp(X,Y)N = (R(X,Y)N)^\perp, \qquad \forall X,Y \in \mathfrak{X}(S), \ \ \forall N \in \mathfrak{X}(S)^\perp$$

This is yet equivalent to

$$R(M,N,X,Y) = ds(X,Y)\, g(\star^\perp N, M), \ \forall X,Y \in \mathfrak{X}(S), \ \ \forall N,M \in \mathfrak{X}(S)^\perp \tag{17}$$

Proof (of Theorem 2). Using the Ricci equation (9) and noting that $[A_M, A_N] = 0$ due to Theorem 1, one obtains Eq. (17). Eliminating M in this expression, one can also write

$$(R(X,Y)N)^\perp = ds(X,Y) \star^\perp N, \ \forall X,Y \in \mathfrak{X}(S), \ \ \forall N \in \mathfrak{X}(S)^\perp$$

which together with Eq. (5) proves the result. □

Finally, I give the proof of Corollary 1.

Proof (of Corollary 1). We must prove that the necessary and sufficient condition for S to be umbilical along a normal direction is that

$$R^\perp = 0$$

for locally conformally flat spacetimes. It is well known [18, 50] that locally conformally flat semi-Riemannian manifolds are characterized by the vanishing of the Weyl conformal curvature tensor C, defined by [18]

$$C(v,w,y,z) \equiv R(v,w,y,z) + \frac{\mathscr{S}}{6}\left(g(v,y)g(w,z) - g(v,z)g(w,y)\right)$$
$$-\frac{1}{2}\left[Ric(v,y)g(w,z) - Ric(v,z)g(w,y) - Ric(w,y)g(v,z) + Ric(w,z)g(v,y)\right]$$

for all $v,w,y,z \in T\mathscr{V}$. It is then easily verified that in general

$$(R(X,Y)N)^{\perp} = (C(X,Y)N)^{\perp}, \qquad \forall X,Y \in \mathfrak{X}(S), \quad \forall N \in \mathfrak{X}(S)^{\perp}$$

and consequently, if $(\mathscr{V},g)$ is locally conformally flat, then

$$(R(X,Y)N)^{\perp} = 0, \quad \forall X,Y \in \mathfrak{X}(S), \ \forall N \in \mathfrak{X}(S)^{\perp}$$

so that from Theorem 2 one gets $ds = 0$, or equivalently

$$R^{\perp} = 0. \qquad \square$$

5 Some Important Corollaries and Consequences

In this section, I present several consequences of Theorems 1 and 2 for the special cases of pseudo-umbilical and ortho-umbilical surfaces.

Corollary 3 (Pseudo-umbilical *S*). *A non-minimal space-like surface S is pseudo-umbilical if and only if B is proportional to the Identity.*

Remark 6. A more precise corollary will present the same statement at a point $x \in S$. I recall here Remark 3 where this was carefully explained. For the sake of clarity, however, let me re-state now the previous corollary in its more precise version:

At a non-minimal point $x \in S$, a space-like surface S is pseudo-umbilical if and only if $B|_x$ is proportional to the Identity.

The same happens with all results in this chapter. As the proof is always essentially the same as the one given, I will simply omit any further mention of this in what follows.

Proof. Assume that S is pseudo-umbilical and $H \neq 0$. This means that $N^{\flat}_{umb} \wedge H^{\flat} = 0$ which, on using expressions (4) for H and (15) for N_{umb}, becomes

$$(\lambda_1 - \lambda_2)(v_1 + v_2) + (\lambda_1 + \lambda_2)(v_1 - v_2) = 0$$

that is to say

$$\lambda_1 v_1 - \lambda_2 v_2 = 0.$$

But then, on the common eigen-basis for A_k and A_ℓ—this eigen-basis does exist due to Theorem 1—formula (7) implies

$$B = 2\begin{pmatrix} -\lambda_1 \nu_1 & 0 \\ 0 & -\lambda_2 \nu_2 \end{pmatrix} = -2\lambda_1 \nu_1 \begin{pmatrix} 1 & 0 \\ 0 & 1 \end{pmatrix}$$

or in other words

$$B = \frac{1}{2}\mathrm{tr}B\,\mathbf{1}. \tag{18}$$

Conversely, if Eq. (18) holds then from Eq. (7)

$$A_k A_\ell + A_\ell A_k = -\frac{1}{2}\mathrm{tr}B\,\mathbf{1}$$

and commuting here with A_k and with A_ℓ one derives, respectively,

$$\left[A_\ell, A_k^2\right] = 0, \quad \left[A_k, A_\ell^2\right] = 0.$$

Using now the Cayley–Hamilton theorem ($A^2 - \mathrm{tr}A\,A + \det A\,\mathbf{1} = 0$ for every 2×2-matrix A), they become, respectively,

$$\mathrm{tr}A_k\,[A_\ell, A_k] = 0, \quad \mathrm{tr}A_\ell\,[A_\ell, A_k] = 0$$

so that $[A_\ell, A_k] = 0$ follows unless $\mathrm{tr}A_k = \mathrm{tr}A_\ell = 0$, that is, unless $H = 0$. Theorem 1 then implies that (for $H \neq 0$) S is umbilical along the direction (15), and the calculation above (18) can be reversed to check that this umbilical direction N_{umb} is parallel to H. □

Note that the condition (18) of the previous corollary can be invariantly characterized by

$$(\mathrm{tr}B)^2 - 4\det B = 0.$$

Corollary 4 (Ortho-umbilical S). *A non-minimal space-like surface S is ortho-umbilical if and only if*

$$\mathrm{I\!I}(X,Y)^\flat \wedge H^\flat = 0, \quad \forall X, Y \in \mathfrak{X}(S). \tag{19}$$

Proof. If S is ortho-umbilical, from Remark 2 one knows that all Weingarten operators are proportional to each other so that, on using expression (11), one immediately derives (19). Conversely, assume that Eq. (19) holds (and $H \neq 0$). Then, there must exist a rank-2 symmetric covariant tensor field κ on S such that

$$\mathrm{I\!I}(X,Y) = \kappa(X,Y)H, \quad \forall X, Y \in \mathfrak{X}(S) \tag{20}$$

and therefore $K_{\star^\perp H}(X,Y) = g(\star^\perp H, \mathrm{I\!I}(X,Y)) = 0$ for arbitrary $X, Y \in \mathfrak{X}(S)$, that is to say, $K_{\star^\perp H} = 0$, which leads to $A_{\star^\perp H} = 0$. □

Observe that comparing Eq. (11) with Eq. (20), one has $\kappa = K_1/\mathrm{tr}A_1$. Defining $\tilde{\kappa} : \mathfrak{X}(S) \to \mathfrak{X}(S)$ by $g(\tilde{\kappa}(X),Y) = \kappa(X,Y)$ for all $X,Y \in \mathfrak{X}(S)$ it follows that

$$\mathrm{tr}\tilde{\kappa} = 1\,. \tag{21}$$

It is interesting to compare the totally umbilical condition (12) with the more general ortho-umbilical one given by Eq. (20) together with Eq. (21).

The computation of B for ortho-umbilical surfaces provides, by means of Eq. (20), the expression

$$B = g(H,H)\,\tilde{\kappa}^2$$

so that one has

$$\mathrm{tr}B = g(H,H)\mathrm{tr}\tilde{\kappa}^2 = g(H,H)(1 - 2\det\tilde{\kappa}) \tag{22}$$

where in the last step I have used, once more, the Caley–Hamilton theorem for $\tilde{\kappa}$ together with Eq. (21). Introducing the last formula in the Gauss equation (8), the following corollary follows.

Corollary 5. *Ortho-umbilical surfaces satisfy the following relation between their Gaussian curvature, the curvature of the spacetime, and its normalized Lipschitz-Killing curvature* $\det\tilde{\kappa}$*:*

$$2K(S) = \mathscr{S} - 4Ric(\ell,k) + 2R(\ell,k,\ell,k) + 2g(H,H)\det\tilde{\kappa} \tag{23}$$

Remark 7. Recall that the Lipschitz-Killing curvature relative to $N \in \mathfrak{X}(S)$ is simply defined as $\det A_N$; see, e.g., [41]. Given that, for ortho-umbilical surfaces, all Weingarten operators are essentially the same and can be described up to proportionality factors by the unit-trace matrix $\tilde{\kappa}$, the concept of normalized Lipschitz-Killing curvature, represented by $\det\tilde{\kappa}$, makes sense and is well defined.

Proof. From the Gauss equation (8) and Eq. (22) one gets Eq. (23) at once. □

Corollary 6. *Ortho-umbilical surfaces in Lorentz space forms have vanishing normal curvature* $R^\perp = 0$*, and they also satisfy the following relation between the constant curvature* $\mathscr{K}$ *of* $(\mathscr{V},g)$*, the Gaussian curvature of S, and its normalized Lipschitz-Killing curvature:*

$$K(S) = \mathscr{K} + g(H,H)\det\tilde{\kappa}\,.$$

Proof. If $(\mathscr{V},g)$ has constant curvature $\mathscr{K}$ it is in particular locally conformally flat so that Corollary 1 implies $R^\perp = 0$. Then a trivial calculation using the constant-curvature hypothesis provides

$$\mathscr{S} - 4\,\mathrm{Ric}(\ell,k) + 2R(\ell,k,\ell,k) = 2\mathscr{K}$$

so that Eq. (23) proves the result. □

Remark 8 (The case when S is pseudo- and ortho-umbilical). If a non-minimal S is pseudo-umbilical as well as ortho-umbilical then, from the previous corollaries, clearly

$$B = g(H,H)\,\tilde{\kappa}^2 \quad \text{and} \quad B = \frac{1}{2}\mathrm{tr}B\,\mathbf{1}$$

This actually implies that either

1. $2\tilde{\kappa} = \mathbf{1}$, that is $2\kappa = \bar{g}$, so that from Eq. (20) S is actually totally umbilical, or
2. $$g(H,H) = 0, \quad B = 0$$

   so that they have a shape tensor of the form

   $$\mathrm{II}(X,Y) = -K_k(X,Y)\,\ell \;\text{ or }\; -K_\ell(X,Y)\,k \qquad \forall X,Y \in \mathfrak{X}(S)$$

   Thus, they are H-subgeodesic MOTS and also 0-isotropic in the sense of [8].

Let me finally state an instance where there always exist umbilical-type surfaces. Consider a spacetime with an integrable conformal Killing vector ξ (no causal character for ξ is required nor necessary here)[18, 50], that is, such that

$$\forall v,w \in T\mathscr{V}, \quad 2\phi\, g(v,w) = g(\nabla_v \xi, w) + g(\nabla_w \xi, v) \tag{24}$$

and also

$$\xi^\flat \wedge d\xi^\flat = 0.$$

This last condition implies that ξ is orthogonal to an integrable distribution; in other words, locally there exist functions F and τ such that $\xi^\flat = F d\tau$; hence, τ =const. is a family of hypersurfaces orthogonal to ξ. Consider any space-like surface S imbedded in any of these orthogonal hypersurfaces (such that $\xi|_S \neq 0$). Then, $\xi \in \mathfrak{X}(S)^\perp$ and one can define its Weingarten operator A_ξ. From Eq. (24) one has

$$\forall X,Y \in \mathfrak{X}(S), \quad 2\phi|_S\, g(X,Y) = g(\nabla_X \xi, Y) + g(\nabla_Y \xi, X) =$$
$$= -g(A_\xi(X),Y) - g(A_\xi(Y),X) = -2g(A_\xi(X),Y)$$

ergo

$$A_\xi = -\phi|_S\, 1.$$

Thus, any such S is ξ-umbilical and it satisfies all the properties shown above for them.

6 Final Considerations

Even though this chapter focuses on space-like surfaces in Lorentzian 4-dimensional manifolds, the concepts and ideas can also be considered in other dimensions and signatures and for other types of surfaces. As a matter of fact, the main result of this chapter, the commutativity of the Weingarten operators for umbilical-type surfaces, holds true, *mutatis mutandis*, for space-like surfaces in 4-dimensional semi-Riemannian manifolds of *arbitrary signature*. The theorems are also valid for *time-like* surfaces. In both generalizations, one only has to rewrite the proofs in ON bases (leaving the appropriate signs free to cover all possibilities).

Unfortunately, the result is exclusive, however, of dimension four and co-dimension two. A simple analysis shows that:

1. Co-dimension two space-like submanifolds in semi-Riemannian manifolds of *higher dimensions* will also have *two* independent Weingarten operators, and their commutativity at a point can be seen to be a necessary condition for the point to be umbilic. However, it cannot be sufficient in general. For n-dimensional manifolds, the problem resides in the fact that any Weingarten operator is an $(n-2)\times(n-2)$ matrix, so that in diagonal form the number of equations to determine a relation between the two independent components of the would-be umbilical direction N_{umb} is too large, and has no solution in general.
2. If the co-dimension is greater than two, then there are *more than two* independent Weingarten operators, and their commutativity is not even a necessary condition, as can be easily checked. There can be a linear combination of three or more matrices which is proportional to the identity while the matrices do not commute.

It will be interesting to know if there are any generalizations of the results in this chapter to arbitrary dimension.

Acknowledgments I thank Miguel Sánchez and the referees for some comments. Supported by grants FIS2010-15492 (MICINN), UFI 11/55 and GIU06/37 (UPV/EHU) and P09-FQM-4496 (J. Andalucía—FEDER).

References

1. Alías, L.J., Estudillo, F.J.M., Romero, A.: Spacelike submanifolds with parallel mean curvature in pseudo-Riemannian space forms. Tsukuba J. Math. **21**, 169–179 (1997)
2. Andersson, L., Metzger, J.: The area of horizons and the trapped region. Commun. Math. Phys. **290**, 941–972 (2009)
3. Andersson, L., Mars, M., Simon, W.: Local existence of dynamical and trapping horizons. Phys. Rev. Lett. **95**, 111102 (2005)
4. Andersson, L., Mars, M., Simon, W.: Stability of marginally outer trapped surfaces and existence of marginally outer trapped tubes. Adv. Theor. Math. Phys. **12**, 853–888 (2008)

5. Beem, J.K., Ehrlich, P.E., Easley, K.L.: Global Lorentzian geometry. In: Pure and Applied Math. vol. 202. Marcel Dekker, New York (1996)
6. Bektaş, M., Ergüt, M.: Compact space-like submanifolds with parallel mean curvature vector of a pseudo-Riemannian space. Acta Univ. Palack. Olomuc. Fac. Rerum Natur. Math. **38**, 17–24 (1999)
7. Bray, H., Hayward, S., Mars, M., Simon, W.: Generalized inverse mean curvature flows in spacetime. Commun. Math. Phys. **272**, 119–138 (2007)
8. Cabrerizo, J.L., Fernández, M., Gómez, J.S.: Isotropy and marginally trapped surfaces in a spacetime. Class. Quantum Grav. **27**, 135005 (2010)
9. Cao, Xi-F.: Pseudo-umbilical spacelike submanifolds in the indefinite space form. Balkan J. Geom. Appl. **6**, 117–121 (2001)
10. Carrasco, A., Mars, M.: Stability of marginally outer trapped surfaces and symmetries. Class. Quantum Grav. **26**, 175002 (2009)
11. Chen, B.Y.: Geometry of Submanifolds. Marcel Dekker, New York (1973)
12. Chen, B.Y.: Pseudo-Riemannian Geometries, δ-Invariants, and Applications. World Scientific, Singapore (2011)
13. Chen, B.Y., Yano, K.: Submanifolds umbilical with respect to a non-parallel normal subbundle. Kōdai Math. Sem. Rep. **25**, 289–296 (1973)
14. Chen, B.Y., Yano, K.: Umbilical submanifolds with respect to a nonparallel normal direction. J. Diff. Geom. **8**, 589–597 (1973)
15. Cvetic, M., Gibbons, G.W., Pope, C.N.: More about Birkhoff's Invariant and Thorne's Hoop Conjecture for Horizons. Class. Quantum Grav. **28**, 195001 (2011)
16. Decu, S., Haesen, S., Verstraelen, L.: Optimal inequalities characterising quasi-umbilical submanifolds. J. Inequal. Pure Appl. Math. **9**(3), 79 (2008)
17. Dursun, U.: On Chen immersions into Lorentzian space forms with nonflat normal space. Publ. Math. Debrecen **57**, 375–387 (2000)
18. Eisenhart, L.P., Riemannian Geometry. Princeton University Press, Princeton (1949)
19. Galloway, G.J., Senovilla, J.M.M.: Singularity theorems based on trapped submanifolds of arbitrary co-dimension. Class. Quantum Grav. **27**, 152002 (2010)
20. Gibbons, G.W.: The isoperimetric and Bogomolny inequalities for black holes. In: Willmore, Y., Hitchin, H. (eds.) Global Riemannian Geometry, pp. 194–202. Ellis Horwood, Chichester (1984)
21. Gibbons, G.W.: Collapsing shells and the isoperimetric inequality for black holes. Class. Quantum Grav. **14**, 2905–2915 (1997)
22. Haesen, S., Kowalczyk, D., Verstraelen, L.: On the extrinsic principal directions of Riemannian submanifolds. Note Mat. **29**, 41–53 (2009)
23. Hawking, S.W., Ellis, G.F.R.: The large scale structure of spacetime. Cambridge Univ. Press, Cambridge (1973)
24. Hawking, S.W., Penrose, R.: The singularities of gravitational collapse and cosmology. Proc. Roy. Soc. London A **314**, 529–548 (1970)
25. Houh, C.S.: On Chen surfaces in a Minkowski spacetime. J. Geom. **32**, 40–50 (1988)
26. Hu, Y.J., Ji, Y.Q., Niu, D.Q.: Space-like pseudo-umbilical submanifolds with parallel mean curvature in de Sitter spaces. J. Ningxia Univ. Nat. Sci. Ed. **26**, 121–124 (2005)
27. Kim, Y.H., Kim, Y.W.: Pseudo-umbilical surfaces in a pseudo-Riemannian sphere or a pseudo-hyperbolic space. J. Korean Math. Soc. **32**, 151–160 (1995)
28. Khuri, M.: A note on the non-existence of generalized apparent horizons in Minkowski space. Class. Quantum Grav. **26**, 078001 (2009)
29. Kriele, M.: Spacetime. Springer, Berlin (1999)
30. Mars, M.: Present status of the Penrose inequality. Class. Quantum Grav. **26**, 193001 (2009)
31. Mars, M., Senovilla, J.M.M.: Trapped surfaces and symmetries. Class. Quantum Grav. **20**, L293–L300 (2003)
32. Misner, C.W., Thorne, K.S., Wheeler, J.A.: Gravitation. W.H. Freeman and Co., New York (1973)

33. O'Neill, B.: Semi-Riemannian Geometry: With Applications to Relativity. Academic, New York (1983)
34. Ôtsuki, T.: Pseudo-umbilical submanifolds with M-index ≤ 1 in Euclidean spaces. Kōdai Math. Sem. Rep. **20**, 296–304 (1968)
35. Penrose, R.: Gravitational collapse and spacetime singularities. Phys. Rev. Lett. **14**, 57–59 (1965)
36. Penrose, R.: Techniques of Differential Topology in Relativity, Regional Conference Series in Applied Math. vol. 7. SIAM, Philadelphia (1972)
37. Penrose, R.: Naked singularities. Ann. N.Y. Acad. Sci. **224**, 125 (1973)
38. Roşca, R.: Sur les variétés lorentziennes 2-dimensionnelles immergées pseudo-ombilicalement dans une variét relativiste. C. R. Acad. Sci. Paris Sr. A-B **274**, A561–A564 (1972)
39. Roşca, R.: Varietatile bidimensionale dei spatiul Minkowski pentru care curburite lui Otsuki sint nule. St. cerc. Mat. **24**, 133–141 (1972)
40. Roşca, R.: Sous-varietés pseudo-minimales et minimales d'une varieté pseudo-Riemannienne structure per une connexion spin-euclidienn. C. R. Ac. Sci. Paris (Serie A-B) **290**, 331–333 (1980)
41. Rouxel, B.: A-submanifolds in Euclidean space. Kōdai Math. J. **4**, 181–188 (1981)
42. Rouxel, B.: Sur les A-surfaces d'un espace-temps de Minkowski M^4. Riv. Mat. Univ. Parma **8**, 309–315 (1982)
43. Schouten, J.A.: Ricci Calculus. Springer, Berlin (1954)
44. Senovilla, J.M.M.: Singularity theorems and their consequences. Gen. Rel. Grav. **30**, 701–848 (1998)
45. Senovilla, J.M.M.: Classification of spacelike surfaces in spacetime. Class. Quantum Grav. **24**, 3091–3124 (2007)
46. Senovilla, J.M.M.: A reformulation of the hoop conjecture. Europhys. Lett. **81**, 20004 (2008)
47. Senovilla, J.M.M.: Trapped surfaces. Int. J. Mod. Phys. D **20**, 2139–2168 (2011)
48. Song, W.D., Pan, X.Y.: Pseudo-umbilical spacelike submanifolds in de Sitter spaces. J. Math. Res. Exposition **26**, 825–830 (2006)
49. Sun, H.: On spacelike submanifolds of a pseudo-Riemannian space form. Note Mat. **15**, 215–224(1995)
50. Stephani, H., Kramer, D., MacCallum, M.A.H., Hoenselaers, C., Herlt, E.: Exact Solutions to Einstein's Field Equations, 2nd edn. Cambridge University Press, Cambridge (2003)
51. Wald, R.M.: General Relativity. The University of Chicago Press, Chicago (1984)
52. Yano, K., Ishihara, S.: Pseudo-umbilical submanifolds of co-dimension 2. Kōdai. Math. Sem. Rep. **21**, 365–382 (1969)

Stability of Marginally Outer Trapped Surfaces and Applications

Marc Mars

1 Introduction

In geometric theories of gravity, the gravitational field is a manifestation of the curvature of a spacetime, namely, an n-dimensional smooth manifold with a metric of Lorentzian signature $\{-,+\cdots,+\}$. An immediate consequence is that freely falling observers feel no gravitational field at sufficiently small scales. This is similar to the fact that any Riemannian manifold is infinitesimally flat, and hence any sufficiently local geometric measurement can be approximated by a Euclidean measurement. As a consequence, the notion of strong gravitational field becomes necessarily a subtle one in any geometric theory of gravity. This implies, in particular, that no useful notion of intense gravitational field can be attached to one single spacetime point and that a less local notion becomes necessary. A reasonable possibility that has proved very successful is the use of space like, codimension-two embedded surfaces. Such surfaces have the distinctive feature that may serve as initial events for sending pulses of light. Since the codimension is two (and assuming orientability of various objects), two independent future-directed surface-orthogonal pulses of light can be emitted, one towards one side of the surface (say inwards) and another towards the other side (say outwards). These pulses of light (i.e. null geodesics starting on the surface with tangent vector orthogonal to the surface) will generate two null hypersurfaces which are smooth sufficiently near the starting surface. In this situation, one may analyse the dependence in time of the area of those light fronts. In a "normal" situation, i.e. when the gravitational field is weak, the pulse of light sent outwards will increase its area, while the pulse of light sent inwards will have decreasing area. However, light is bent by the

M. Mars (✉)
Departamento de Física Fundamental and Instituto de Física Fundamentaly Matemáticas (IUFFyM), Universidad de Salamanca, Plaza de la Merced s/n, 37008 Salamanca, Spain
e-mail: marc@usal.es

M. Sánchez et al. (eds.), *Recent Trends in Lorentzian Geometry*, Springer Proceedings in Mathematics & Statistics 26, DOI 10.1007/978-1-4614-4897-6_4,

gravitational field (which in geometric terms translates into the obvious statement that geodesics depend on the geometry). Consequently, if the gravitational field near the surface is intense and directed, say inwards, it is possible that the outward light geodesics may bend inwards sufficiently so that the area of the light fronts *decreases*. This geometric fact may be taken as a convincing indication that the gravitational field is intense. Surfaces where this behaviour (or a variant thereof) occurs are typically called *trapped surfaces*. Surfaces exhibiting a behaviour which is borderline between the "normal" situation and the strong gravitational field situation are typically called "marginally trapped". In the literature, several specific definitions have been used, each one with their peculiarities (see, e.g. [41] for a classification). However, they all share the general pattern described above. The marginal-type surfaces are of particular interest because they may, in principle, locate transition zones from strong to weak gravitational fields. It should be emphasized that, for the discussion above to be physically sound, it is necessary to restrict the surfaces to being compact (in the non-compact case, the area element may increase due to physical reasons different from those exposed above, in fact non-compact spacelike surfaces with decreasing area element in *both* orthogonal null directions exist even in the Minkowski spacetime, see Example 4.1 in [40]).

Another completely different approach to define, and locate, strong gravitational fields is to consider causally disconnected regions. The motion of particles, and more generally of any signals, is restricted to lie within the null cones of the spacetime metric at any point. This is one of the ways in which the gravitational field affects to propagation of particles and fields. Intuitively (although difficult to put in precise terms), one may think that the gravitational field bends the null cones towards the source of the gravitational field. If the field is sufficiently strong, the null cones will bend so much that no particle or signal will be able to escape from a predetermined spacetime region. This general idea leads to the concept of black hole spacetime and of black hole region (i.e. the non-escape region in a black hole spacetime). The precise definition of black hole is, however, not simple because one needs to define the "non-escape" region in a sensible way (for instance, the causal future of an event is by construction a non-escape region, but this does not capture the idea of non-escaping a predetermined region). The most natural framework admitting such definition is the class of spacetimes which contain a large region that qualifies as "infinity". In deliberately vague terms, a black hole region is then a region that cannot be observed from infinity. As a consequence, the notion of black hole requires imposing strong global causal assumptions on a spacetime. Black holes are certainly among the most interesting spacetime objects (or rather spacetimes). They have very interesting physical and geometrical properties. However, they also have a fundamental drawback. With our present understanding of the gravitational field equations, it is not possible to know if a black hole will form during the evolution of a given initial configuration. This is because the very concept of black hole needs complete and detailed knowledge of the causal properties of the maximal Cauchy development of the given initial configuration.

It should be clear so far that the two notions of strong gravitational fields outlined above are very different from each other. However, they share the basic fact that "bending" of light is involved in one way or another. It is reasonable to ask whether, despite all appearances, there is a relationship between those two concepts. This relationship indeed exists. It is well-known (see Propositions 12.2.3 and 12.2.4 in [42] and Theorem 6.1 in [13], and also [15]) that trapped surfaces lie necessarily inside the black hole region in a black hole spacetime satisfying the so-called dominant energy condition (DEC). This last notion (defined below) states, physically, that any observer measures energy fluxes which propagate at most at the speed of light. This confinement result is very satisfactory because, in some sense, it shows that the global notion of strong gravitational field (black holes) captures the quasi-local notion (trapped surfaces). It is most natural to ask whether the converse is also true, namely: are spacetimes containing trapped surfaces (and satisfying suitable reasonable conditions) always black hole spacetimes? In view of what has been said above on our present impossibility of determining whether a given initial data evolves to form a black hole or not, it is clear that this question cannot be answered so far. There are, however, indications that it might be true. The most important one comes from the so-called *singularity theorems*. These are fundamental results in gravity that predict geodesic causal incompleteness of all spacetimes satisfying suitable properties. The detailed statements of the various singularity theorems do not concern us here (see, e.g. [40, 42]). However, they all share three basic conditions, namely an energy condition, a causality condition and a quasi-local condition that guarantees that the gravitational field is sufficiently strong in some spacetime region. One of this quasi-local conditions is the existence of a trapped surface. Thus, spacetimes containing trapped surfaces (and satisfying a number of additional physically reasonable conditions) are singular in the sense that they are causally geodesically incomplete. The previous question can therefore be rephrased as follows:

Are singularities that form during the evolution of a regular initial data set visible from infinity?

In this form, the question is even more fundamental and important than before because it addresses the basic issue of whether the gravity theory involved (say, e.g. general relativity) is predictable, at least, in the asymptotic region at infinity. It is widely believed that the question above has a negative answer (for generic initial data). This is the content of the *weak cosmic censorship conjecture* put forward by Penrose [38]. This problem is very difficult indeed, and little is known at present on its validity (see [1, 43] for reviews and [12, 18] for rigorous results in spherical symmetry). It is clear, however, that there is a close connection between the validity of the weak cosmic censorship conjecture and the fact that trapped surfaces may signal the presence of a black hole spacetime.

In fact, in evolutionary approaches to spacetimes (most notably in approaches where the gravitational field equations are solved via numerical methods) marginally outer trapped surfaces (MOTS) (defined below) are routinely taken as good indicators of the location of the black hole boundary (the so-called event horizon). In numerical relativity, it is even customary to abuse notation and call these surfaces "black holes" even though no global information is available to make sure that a black hole spacetime will indeed form. This is just a manifestation of the widespread opinion among general relativists that MOTSs are good indicators of the eventual formation of a black hole spacetime. Even more, it is expected that MOTSs should approximate the event horizon at late times after all dynamical processes have taken place and an equilibrium configuration is eventually approached.

In order to see whether these expectations are confirmed, it becomes necessary to study in detail the geometry of MOTSs. Although many issues remain open, a number of results have been obtained in the last years. In particular, the notion of stability of MOTSs has proved to be important and useful. The aim of this chapter is to describe this notion and to review various places where it has turned out to be relevant.

2 Definition of Marginally Outer Trapped Surface

In this work, $(M, g^{(n)})$ will denote an n-dimensional ($n \geq 4$) oriented manifold M together with a smooth metric $g^{(n)}$ of Lorentzian signature $\{-,+,\cdots,+\}$. $(M, g^{(n)})$ is always assumed to be time-oriented. We take all manifolds to be smooth and connected. Manifolds are without boundary unless otherwise stated, in which case ∂M will be used for its boundary. Scalar product with $g^{(n)}$ is denoted by $< , >$. If necessary, Greek indices will be used for spacetime tensors. The Lie derivative on M will be denoted by $\mathcal{L}$. For any metric h, we use $\mathrm{Ric}(h)$, $\mathrm{Ein}(h)$ and $\mathrm{Scal}(h)$ respectively for the Ricci, Einstein and scalar curvatures of h. For the spacetime metric $g^{(n)}$, we simply write Ric, Ein, Scal for the corresponding curvature tensors and ∇ for the Levi–Civita covariant derivative.

The notation for submanifolds is as follows: If Σ is a smooth manifold of dimension $s \leq n$ and $\Phi : \Sigma \to M$ is an immersion, we say that $\Phi(\Sigma)$ is an **immersed submanifold**. If Φ is injective, we say that $\Phi(\Sigma)$ is a **submanifold**. If moreover, the two topologies inherited by $\Phi(\Sigma)$ from Φ and from the inclusion map onto M agree, then $\Phi(\Sigma)$ is an **embedded submanifold**. In this latter case, we often identify Σ and $\Phi(\Sigma)$ whenever appropriate. An (immersed/embedded) submanifold is space like if the first fundamental from $\Phi^{\star}(g^{(n)})$ is positive definite. Throughout this work, we use the following definition for **surface**.

Definition 1. A **surface** S is a smooth, orientable, closed (i.e. compact and without boundary), codimension-two, space like embedded submanifold of $(M, g^{(n)})$ (with embedding Φ_S). Moreover, a surface will be connected unless otherwise stated.

2.1 Geometry of Spacelike Surfaces

It is straightforward to show that the normal bundle NS of a surface admits a basis of future null normals $\{\ell^+, \ell^-\}$. More specifically, denoting by $\mathfrak{X}(S)^{\perp}$ the set of sections of the normal bundle, there exist $\ell^{\pm} \in \mathfrak{X}(S)^{\perp}$ such that $\{\ell^+|_p, \ell^-|_p\}$ are

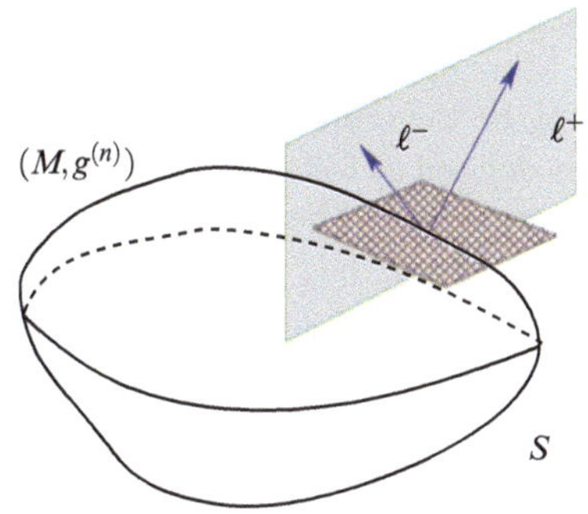

Fig. 1 Schematic representation of a surface in a four-dimensional spacetime. The vertical plane represents the normal space N_pS orthogonal to S at p. The null basis $\ell^\pm$ of the normal space is also depicted. The representation is obviously not accurate because the tangent space T_pS and the normal space N_pS intersect only at the zero vector

future directed, null and linearly independent at each point $p \in S$. This null basis is obviously not unique. We will partially fix the basis by demanding $< \ell^+, \ell^- > = -2$ everywhere on S. The remaining freedom are the "boosts", namely, transformations $\ell^\pm \longrightarrow F^{\pm 1}\ell^\pm$ where F is a smooth positive function $F \in C^\infty(S, \mathbb{R}^+)$.

S being space like, the tangent space $T_pM, p \in S$, admits a direct sum decomposition $T_pM = T_pS \oplus N_pS$, where T_pS and N_pS are, respectively, the tangent and normal spaces to S at p, see Fig. 1. According to this direct sum, a vector $V \in T_pM$ decomposes as $V = V^\parallel + V^\perp$. The first fundamental form on S, called h in the following, is a positive definite Riemannian metric. The corresponding Levi–Civita covariant derivative is denoted by ∇_h.

As usual, we define the second fundamental form vector as the symmetric, bilinear map:

$$\chi(X,Y) = -(\nabla_X Y)^\perp, \qquad X,Y \in \mathfrak{X}(S).$$

The **null extrinsic curvatures** are the projections of $\chi(X,Y)$ along the null normals, $\chi_\pm(X,Y) \stackrel{\text{def}}{=} < \chi(X,Y), \ell^\pm >$. Taking traces on S we have the **mean curvature vector**, $H = \text{tr}_S\chi$, and the **null expansions** $\theta_\pm = < H, \ell^\pm >$. These definitions imply the decomposition

$$H = -\frac{1}{2}\left(\theta_+\ell^- + \theta_-\ell^+\right).$$

The remaining extrinsic information of the embedded surface S is encoded in the normal connection. Given $\zeta \in \mathfrak{X}(S)^\perp$, its covariant derivative along a tangent vector $X \in \mathfrak{X}(S)$ is defined as $\nabla^\perp_X \zeta \stackrel{\text{def}}{=} (\nabla_X \zeta)^\perp$. Since the scalar products of basis vectors $\ell^\pm$ are constant, the normal covariant derivative is fully determined by the connection one-form $\mathbf{s}(X) \stackrel{\text{def}}{=} -\frac{1}{2} < \ell^-, \nabla_X \ell^+ >$.

Regarding orientability issues, a choice of orientation on S fixes the orientation of the normal bundle NS. More specifically, assume that a choice of orientation on S has been made and let $\eta_{\mathbf{S}}$ be the corresponding volume form. The volume form $\eta_{\mathbf{S}^\perp}$ on the normal space is defined as follows: Let η be the spacetime volume form (recall that all spacetimes are oriented here). For any pair of normal vector fields $\zeta_1, \zeta_2 \in \mathfrak{X}(S)^\perp$ define $\eta_{\mathbf{S}^\perp}(\zeta_1,\zeta_2) \stackrel{\text{def}}{=} \eta\left(\zeta_1,\zeta_2,X_1,X_2\right)/\eta_{\mathbf{S}}(X_1,X_2)$, where $X_1, X_2 \in \mathfrak{X}(S)$ are arbitrary except for the condition of being linearly independent at each point. Alternatively, a choice of orientation on the normal space fixes uniquely an orientation on S satisfying the previous relationship.

Assume that an orientation has been chosen on S (or on its normal space) and let $\xi \in \mathfrak{X}(S)^\perp$. The dual vector $\xi^\star \in \mathfrak{X}(S)^\perp$ is defined (c.f. [8]) as $<\xi^\star,\zeta> \stackrel{\text{def}}{=} \eta_{\mathbf{S}^\perp}(\xi,\zeta)$ for any vector $\zeta \in \mathfrak{X}(S)^\perp$. The following properties are immediate:

$$<\xi^\star,\xi>=0, \qquad <\xi^\star,\xi^\star>=-<\xi,\xi>. \tag{1}$$

Note that, given a non-null $\xi|_p$, Eqs. (1) admit a unique solution $\xi^\star|_p$ up to sign. This sign reflects the freedom in choosing the orientation on NS (or equivalently on TS).

2.2 *Marginally Outer Trapped Surfaces*

As discussed in the introduction, a possible way of determining the strength of a gravitational field at a quasi-local level is by studying the behaviour of light fronts of pulses originating on a surface S. The collection of null geodesics starting on S along the orthogonal null direction ℓ^+ (ℓ^-) define a null hypersurface $\mathscr{N}^+$ ($\mathscr{N}^-$) which is smooth near S. Light fronts are sections of $\mathscr{N}^\pm$ (i.e. smooth embedded surfaces in $\mathscr{N}^\pm$ intersecting each null geodesic orthogonal to S precisely once). Taking the light fronts sufficiently close to S, we can evaluate their area by using the first variation of area of submanifolds.

Let ν be a normal variation vector on S, i.e. a vector field defined in a spacetime neighbourhood of S which is orthogonal to S on the surface. Choose ν to be compactly supported (this entails no restriction since S itself is compact). This vector field generates a one-parameter local group of transformations $\{\varphi_\tau\}_{\tau\in I}$, where $I \subset \mathbb{R}$ is an interval containing $\tau = 0$ and τ is the canonical parameter of the group. The variation of S along ν is the one-parameter family of surfaces $S_\tau \equiv \varphi_\tau(S)$. Obviously we have $S_{\tau=0} = S$. For any covariant tensor Γ defined on S and depending on its geometry (intrinsic or extrinsic), let us denote by Γ_τ the corresponding (formally analogous) tensor defined on the surface $S_\tau = \varphi_\tau(S)$. The variation of Γ along ν is defined as $\delta_\nu \Gamma \equiv \frac{d}{d\tau}\left[\varphi_\tau^*(\Gamma_\tau)\right]\big|_{\tau=0}$, where φ_τ^* denotes the pull-back of φ_τ. It is immediate to check that this variation only depends on the values of ν on S and not on its extension off S.

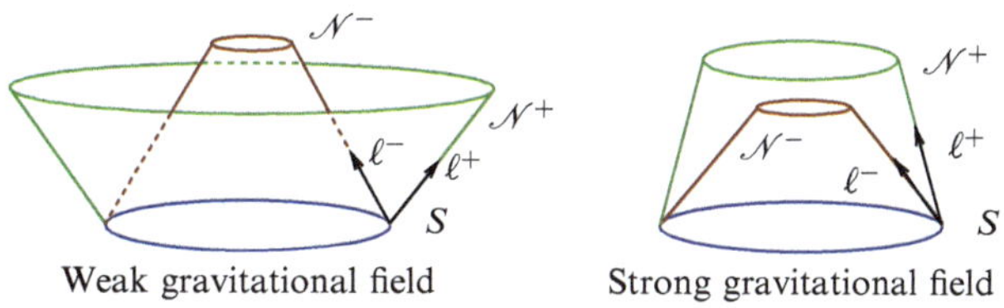

Fig. 2 Schematic figure showing the behaviour of light pulses emitted from S along its two independent orthogonal directions. In the weak gravitational field case, the area of one of the pulses increases and the other decreases. In the strong gravitational field case, light rays are bent "inwards" and the area of the light fronts decreases in both directions

For the first variation of area, let $|S_\tau|$ denote the area of the surface S_τ. The formula of the first variation of area states (see e.g. [11])

$$\delta_v |S| \stackrel{\text{def}}{=} \left. \frac{d|S_\tau|}{d\tau} \right|_{\tau=0} = \int_S < H, v > \eta_S. \tag{2}$$

Variations of S along $\mathcal{N}^\pm$ are defined by vector fields v satisfying $v|_S = \psi \ell^\pm$ where $\psi \in C^\infty(S, \mathbb{R}^+ \cup 0)$. A necessary condition for the area of the light fronts to decrease along $\mathcal{N}^\pm$ is that $\delta_{\psi\ell^\pm}|S| \leq 0$ for any such ψ. By the first variation of area, this is equivalent to $< H, \ell^\pm > \leq 0$, i.e. that the mean curvature vector H is future causal on S. This, in turn, is equivalent to the null expansion $\theta_\pm$ being non-positive, $\theta_\pm \leq 0$. If these inequalities are strict (i.e. if H is future time like) then the area of S is strictly larger than that of S_τ, for $\tau > 0$ sufficiently small. This is the condition defining a future trapped surface and which signals the presence of a strong gravitational field as discussed in the introduction, see Fig. 2. The marginal case corresponds to the situation when the area is stationary along one of the null directions and non-increasing along the other null direction, i.e. $\{\theta_+ = 0, \theta_- \leq 0\}$ or $\{\theta_- = 0, \theta_+ \leq 0\}$.

The fact that the definition of marginally trapped surfaces involves both equalities *and* inequalities makes them somewhat cumbersome to study. It is natural to ask whether relaxing the inequality part still defines a class of surfaces where the gravitational field is strong. More specifically, the question is whether surfaces satisfying $H \propto \ell^+$ (or $H \propto \ell^-$) are still useful to define strong gravitational fields. In physical terms, one expects this to be true provided the surface S admits a well-defined notion of "exterior", and that the null expansion which vanishes is the one along the exterior direction. This is because light pulses emitted to the exterior are bent inwards by the gravitational attraction and even more so are the light pulses emitted in the inward direction. If the gravitational field is sufficiently intense, the corresponding outer null expansion will be negative (i.e. the outward directed light pulses will focus) or zero in the borderline case. However, this physical expectation is tied to the existence of a sensible notion of "exterior", which may easily be a non-trivial problem in itself. A convenient approach is to study *all* surfaces for which one of the null expansions vanishes and, later on, restrict the class further in order to ensure that an adequate notion of exterior exists. These considerations lead to the notion of **MOTS**.

Definition 2. A **marginally outer trapped surface** (MOTS, for short) is a surface with stationary area with respect to variations along $\mathscr{N}^+$ or along $\mathscr{N}^-$. Equivalently, a surface is a MOTS if and only if $H \propto \ell^+$ or $H \propto \ell^-$ everywhere.

Remark. After renaming ℓ^+ and ℓ^- if necessary, we can always assume that a MOTS satisfies $H \propto \ell^+$, or equivalently $\theta_+ = 0$. For MOTS, we will always choose the orientation of the normal bundle so that $\eta_{S^\perp}(\ell^+, \ell^-) < 0$. As discussed above, this also fixes the orientation on S.

3 Stability of MOTS

MOTS are by definition critical points of the area functional for a certain class of variations. Minimal surfaces, on the other hand, are stationary points of area with respect to *all* possible variations. Thus, although both concepts involve properties of the area functional, there are fundamental differences between them. In the Lorentzian case, a codimension-two spacelike surface admits a normal bundle with two very special directions, namely the null normal directions. For codimension-two embedded submanifolds in a Riemannian ambient space, all normal directions are, in principle, equivalent and there is no intrinsic way of selecting specific directions along which to perform particular variations. It is clear that minimal codimension-two embedded surfaces (which must satisfy two scalar conditions, namely the vanishing of the whole mean curvature vector) are much more restrictive than MOTS. In view of this fundamental difference, one should expect very few similarities between MOTS and minimal codimension-two surfaces. Consider, on the other hand, minimal hypersurfaces in a Riemannian ambient manifold. They are also critical points of the area functional and now the space of variations has the same dimensionality as for MOTS. Since MOTS are $(n-2)$-dimensional objects in n-dimensional spacetimes, one might hope that MOTS could share important similarities with minimal hypersurfaces in $(n-1)$-dimensional Riemannian ambient spaces. This, in principle, rather vague relationship has turned out to be much deeper than originally expected. One instance where such close similarities arise involves the notion of **stability**.

We define "minimal hypersurface" as a codimension-one closed, orientable, embedded submanifold with vanishing mean curvature. We take the ambient manifold to be Riemannian although everything below would also hold for non-degenerate codimension-one submanifolds in an ambient space of arbitrary signature. The area functional is stationary on a minimal hypersurface. The second variation of area gives information about the extremal (minimum or saddle point) properties of the area functional on a minimal hypersurface S_m. Let m be a unit normal to S_m and $\psi \in C^\infty(S_m, \mathbb{R})$, see Fig. 3.

The second variation of area defines the stability operator L_0 via

$$\delta^2_{\psi m}|S_m| = \int_{S_m} \psi \delta_{\psi m} H \eta_{\mathbf{S_m}} = \int_{S_m} \psi L_0(\psi)\, \eta_{\mathbf{S_m}}. \tag{3}$$

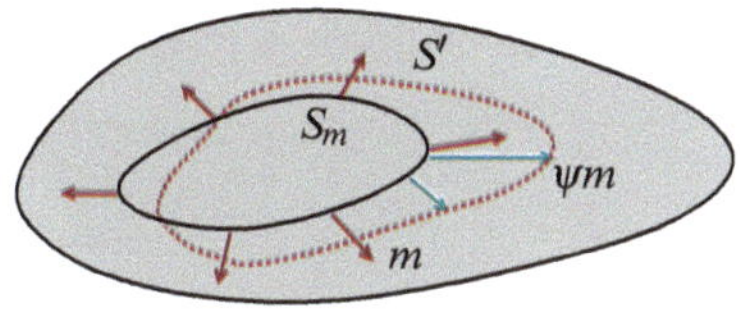

Fig. 3 Variation along ψm of a minimal hypersurface S_m. The space of variation vectors contains one arbitrary smooth function ψ on S_m

Its explicit form reads (see, e.g. [11])

$$L_0(\psi) = -\Delta_h \psi - \left(\mathrm{Ric}(h)(m,m) + ||\kappa||_h^2\right)\psi, \tag{4}$$

where Δ_h is the Laplacian on (S_m, h) and κ is the second fundamental form of S_m along m. The operator L_0 is linear, elliptic and formally self-adjoint with respect to the L^2 product with the metric volume form η_{S_m}. This implies that the principal eigenvalue λ_0 can be represented by the Rayleigh–Ritz formula

$$\lambda_0 = \inf_{\substack{\psi \in C^\infty(S_m,\mathbb{R}) \\ \psi \neq 0}} \frac{<\psi, L\psi>_{L^2}}{<\psi,\psi>_{L^2}}.$$

This immediately gives a lower bound for any second variation of area

$$\delta^2_{\psi m}|S_m| \geq \lambda_0 \int_{S_m} \psi^2 \eta_{S_m}.$$

Equality occurs if and only if ψ is a principal eigenfunction i.e $L_0\psi = \lambda_0\psi$. Thus, $\delta^2_{\psi m}|S_m| \geq 0$ is equivalent to $\lambda_0 \geq 0$. A minimal surface satisfying $\lambda_0 \geq 0$ is called **stable minimal surface**.

MOTS are codimension-two surfaces, and the area functional is critical only for variations along ℓ^+. It is natural to evaluate the second variation of area along ℓ^+. The result is then given by the well-known Raychaudhuri equation (see e.g. [42]) and reads

$$\delta^2_{\psi\ell^+}|S| = -\int_S \psi^2 \left(\mathrm{Ein}(\ell^+,\ell^+) + \mathrm{tr}_h(\chi_+^2)\right)\eta_S.$$

The result is algebraic in ψ, so no differential operator similar to the stability operator above can be defined in this way. Moreover, if the spacetime $(M, g^{(n)})$ satisfies the so-called **null energy condition (NEC)**, namely, $\mathrm{Ein}(\ell,\ell) \geq 0$ for all null vector fields ℓ, then the second variation is always non-positive and MOTS are always local maxima of area. No useful notion of stability can be obtained from this second variation and any connection between stability and area is lost. However, as discussed in [3, 4], the second variation of area for minimal surfaces is essentially equivalent to the *first* variation of its mean curvature H (this is already apparent from the first equality in (3)). MOTS are defined by the vanishing of the scalar quantity θ_+, so its first variation may give useful information for MOTS related to

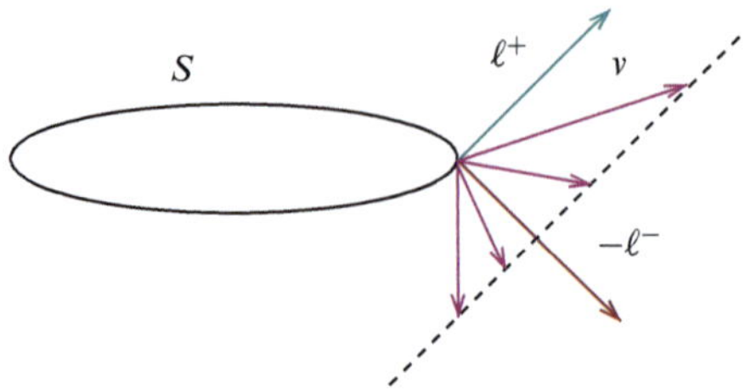

Fig. 4 The vector field v is a (unique) representative of a section of the bundle of normal directions to S provided this section is nowhere tangent to ℓ^+. Note that if a section is somewhere tangent to ℓ^+, the corresponding vector v would diverge to infinity. This restriction on the section is dictated by the fact that the first variation of θ_+ along ℓ^+ gives an algebraic operator instead of a differential operator; see [4] (in particular Lemma 3.1) for further details

the stability properties for minimal surfaces. As already mentioned, it turns out to be useful to restrict the class of variations to dimension one. This amounts to choosing a section of the bundle of normal directions to S. Since the variation of θ_+ along ℓ^+ is algebraic (instead of differential) , it is convenient to restrict to sections which are nowhere tangent to ℓ^+. Given a fixed ℓ^+, such a section is uniquely represented by a vector field $v \in \mathfrak{X}(S)^\perp$ satisfying (see Fig. 4)

$$v = -\frac{1}{2}\ell^- + V\ell^+, \quad V \in C^\infty(S,\mathbb{R}). \tag{5}$$

Thus, these sections may be labelled by smooth functions V on S. Note that the variation direction v is not restricted to being of any specific causal character. For instance, the choice $V = 0$ gives $v = -\ell^-/2$, i.e. variations along the null direction ℓ^-.

For any function ψ on S, we can calculate the first-order variation $\delta_{\psi v}\theta_+$. This defines [4] the so-called stability operator L_v via $L_v(\psi) \stackrel{\text{def}}{=} \delta_{\psi v}\theta_+$. A formula for the first variation of θ_+ was derived by Newman in [37] for arbitrary immersed spacelike submanifolds. The derivation was simplified in [4] (see also [28] and [9]). The explicit form of the stability operator reads

$$\begin{aligned} L_v(\psi) &= -\Delta_h\psi + 2\mathbf{s}(\nabla_h\psi) \\ &+ \left(\frac{\mathrm{Scal}(h)}{2} - \mathrm{Ein}(\ell^+, v^\star) - V\mathrm{tr}_h(\chi_+^2) - ||\mathbf{s}||_h^2 + \mathrm{div}_h\mathbf{s}\right)\psi, \end{aligned} \tag{6}$$

where $v^\star = \frac{1}{2}\ell^- + V\ell^+$ in accordance with the general definition of the $\star$ operation and our choice of orientation for the normal space of MOTS. Since $< v^\star, v >= 0$ and $< v^\star, v^\star >= - < v, v >$, it follows that $v^\star$ is time like if and only if v is space like, and vice versa (and hence that $v^\star$ is null if and only if v is null). Recall the following definition of **dominant energy condition**.

Definition 3. A spacetime $(M, g^{(n)})$ satisfies the DEC if $-\text{Ein}(u, \cdot)$ is future causal for any future causal vector field u.

It follows that the term $\text{Ein}(\ell^+, v^\star)$ is non-negative in spacetimes satisfying the DEC provided v is space like or null. Obviously, this term is also non-negative in Ricci-flat spacetimes irrespectively of the choice of direction v.

3.1 Principal Eigenvalue of the Stability Operator

The operator L_v plays a similar role for MOTS as the stability operator L_0 for minimal surfaces. Like L_0, L_v is an elliptic, second-order operator. However, L_v is *not* self-adjoint in general (not even allowing for different measures in the definition of the L^2 product). Nevertheless, as a consequence of the Krein–Rutman theorem [34] there exists a unique **principal eigenvalue** defined as follows. L_v being not self-adjoint, its eigenvalues lie on the complex plane. It turns out however, that the infimum λ_v of the real parts of all eigenvalues is finite. Moreover, there exists one single eigenvalue whose real part is λ_v. This eigenvalue is called principal eigenvalue [22] and it has the property that it is always real (and hence coincides with λ_v). The eigenspace of λ_v is one-dimensional [7] and all eigenfunctions in this space have constant sign [21, 22]. All these properties were originally obtained for bounded domains in $\mathbb{R}^n$ and for the Dirichlet problem. However, they extend easily to the compact manifold case (see Appendix B in [4] for more details).

Let us show next in which sense can the sign of the principal eigenvalue of the stability operator for MOTS be related to stability properties of such surfaces. Returning to minimal surfaces, stability means that the area functional does not decrease to second-order. If the second variation is positive (as opposed to non-negative) for non-zero variations, then the minimal surface is said to be strictly stable. In terms of the principal eigenvalue, strict stability corresponds to $\lambda_0 > 0$. It is a well-known fact that minimal hypersurfaces are stable (strictly stable) if and only if there exists a positive variation along the normal m which, to first order, does not decrease (does increase) the mean curvature.

For MOTS the second variation of area along directions tangent to v gives no useful information because the first variation is nonzero along those directions and this term is dominant. However, by comparison with the previous discussion, one may ask under which conditions there exist nearby surfaces exterior to S ("exterior" being defined as the direction to which v points) which have positive null expansion. The answer is closely tied to the sign of the principal eigenvalue [4], as follows:

Lemma 1. *The principal eigenvalue λ_v of the stability operator L_v of a MOTS S is non-negative if and only if there exists an outward variation ψv ($\psi \geq 0$, $\psi \not\equiv 0$) such that $\delta_{\psi v} \theta_+ \geq 0$.*

The principal eigenvalue is positive if and only if there exists an outward variation such that $\delta_{\psi v} \theta_+ > 0$.

This lemma leads to the following definition [4] of **stable** and **strictly stable** MOTS.

Definition 4. A MOTS S is called:

- Strictly stable along v if $\lambda_v > 0$.
- Stable along v if $\lambda_v \geq 0$.
- Marginally stable along v if it is stable but not strictly stable along v (i.e. $\lambda_v = 0$).
- Unstable along v if $\lambda_v < 0$.

In order to study stability of MOTS, it becomes necessary to understand under which conditions the principal eigenvalue is non-negative (or positive) and analyse its consequences. Despite the similarities between the properties of the principal eigenvalue of general second-order elliptic operators and standard properties of the lowest eigenvalue of self-adjoint operators, there are also fundamental differences. One of them comes from the possible ways of characterizing the principal eigenvalues. For self-adjoint operators, the Rayleigh–Ritz quotient characterization is of paramount importance. For MOTS, this characterization is no longer true. However, Donsker and Varadhan [21] were able to obtain two alternative min-max characterizations of the principal eigenvalue as follows

$$\lambda_v = \inf_{\mu_S \in \mathscr{P}(S)} \sup_{\psi \in C^\infty(S,\mathbb{R}^+)} \int_S \frac{L_v(\psi)}{\psi} \mu_S,$$

$$\lambda_v = \sup_{\psi \in C^\infty(S,\mathbb{R}^+)} \inf_{x \in S} \frac{L_v(\psi)(x)}{\psi(x)}.$$

Here $\mathscr{P}(S)$ denotes the space of probability measures on S.

Despite its interest, these characterizations are in general difficult to work with (because they involve a minimization of suprema, or vice versa). On the other hand, the Rayleigh–Ritz characterization is of the main technical tools for studying properties of stable minimal surfaces (see, e.g. [16]). It is natural to ask whether some type of characterization that resembles the Rayleigh–Ritz characterization can also be obtained for MOTS. This was done in [4] based on the Hodge decomposition of one-forms on compact manifolds. Indeed, the obstruction to being self-adjoint for the stability operator for MOTS is encoded in the connection one-form s of the normal bundle. For any (smooth) one-form s on a compact Riemannian manifold, the Hodge decomposition asserts that there exists a smooth function f and a smooth, divergence-free one-form $\mathbf{z}$ (i.e. a one-form satisfying $\mathrm{div}_h \mathbf{z} = 0$), such that

$$\mathbf{s} = df + \mathbf{z},$$

(note that this decomposition can be refined further because $\mathbf{z}$ can be written as a sum of co-exact one-form and a harmonic form, but the decomposition above is more useful for our purposes here). This decomposition is unique up to an arbitrary additive constant in f. In [4], the following lemma was proved.

Lemma 2. *Let* (S,h) *be a compact Riemannian manifold and* $L_v(\psi) \stackrel{def}{=} -\Delta_h \psi + 2\mathbf{s}(\nabla_h \psi) + c\psi$ *where* **s** *and c are, respectively, a smooth one-form and a smooth scalar on S. Then, the principal eigenvalue* λ_v *of* L_v *admits the Rayleigh–Ritz type characterization*

$$\lambda_v = \inf_u \int_S \left(||\nabla_h u||_h^2 + Qu^2 - ||d\omega_u + \mathbf{z}||_h^2 u^2 \right) \eta_S, \quad u \in C^\infty(S, \mathbb{R}^+), \quad ||u||_{L^2} = 1,$$

where $Q \stackrel{def}{=} c + ||\mathbf{s}||_h^2 - div_h \mathbf{s}$ *and* ω_u *is the unique solution of*

$$-\Delta_h \omega_u - 2u^{-1} < \nabla_h \omega_u, \nabla_h u >= 2u^{-2} 1 \mathbf{z}(\nabla_h u), \quad \int_S u^{-1} \omega_u \eta_S = 0. \tag{7}$$

When **s** is a gradient, then we have $\mathbf{z} \equiv 0$ and the elliptic problem (7) has unique solution $\omega_u = 0$ for all functions u and we recover the standard Rayleigh–Ritz quotient for self-adjoint operators.

A simple consequence of this lemma is that the principal eigenvalue of a non-self-adjoint operator L_v can be bounded above and below by the principal eigenvalues of two canonically defined self-adjoint operators. Given L_v as in Lemma 2, let $L_s(\psi) \stackrel{def}{=} -\Delta_h \psi + Q\psi$ and $L_z(\psi) = -\Delta_h \psi + \left(Q - ||\mathbf{z}||_h^2\right) \psi$ and denote by λ_s and λ_z the corresponding principal eigenvalues. It follows [4].

Lemma 3. *Let* λ_s, λ_v *and* λ_s *be defined as before. Then* $\lambda_s \geq \lambda_v \geq \lambda_z$.

It is clear that the "symmetrized stability operator" L_s is non-negative (i.e. has a non-negative principal eigenvalue) on any stable MOTS. The converse is, however, not true, so it makes sense to define the notion of a MOTS being *symmetrized stable along a direction v* whenever the symmetrized stability operator L_s has non-negative principal eigenvalue. It is well-known that the topology of stable minimal hypersurfaces is restricted when the ambient Riemannian manifold satisfies curvature inequalities. It may be expected that similar topological restrictions should exist for stable MOTS provided the ambient spacetime satisfies also suitable curvature inequalities. Galloway and Schoen have shown [27] that this is the case and that, in fact, the notion of symmetrized stability suffices.

Theorem 1 (Galloway and Schoen [27]). *Let* $(M, g^{(n)})$ *be a spacetime satisfying the DEC and S is a MOTS which is symmetrized stable with respect to a direction v satisfying* $||v||_{g^{(n)}} \geq 0$ *everywhere. Then S is of non-negative Yamabe type (i.e. admits a metric of non-negative constant curvature). Moreover, the Yamabe type is positive unless* (S,h) *is Ricci flat,* $Ein(\ell^+, v^\star) = 0$ *and* $\chi^+ = 0$.

Remark. This theorem was stated in [27] only for stable MOTS along spacelike directions. However, stability of the MOTS was only used to show (by direct estimates) that the Rayleigh–Ritz quotient is non-negative:

$$\inf_{\substack{\psi \in C^\infty(S,\mathbb{R}) \\ \psi \neq 0}} \frac{\int_S \psi L_s(\psi) \eta_S}{\int_S \psi^2 \eta_S} \geq 0.$$

This inequality states precisely that the principal eigenvalue of the symmetrized stability operator L_s is non-negative, so the theorem covers this case as well. It can also be checked that the proof extends to the more general case of "achronal" directions (i.e. $||v||_{g^{(n)}} \geq 0$), as stated above.

Remark. In the case of four dimensional spacetimes, Theorem 1 states that stable MOTS are of spherical topology, or else are a flat torus with vanishing null second fundamental form χ^+. This result extends a previous theorem by Hawking [29].

3.2 Dependence of the Stability Properties on the Direction

In the definition of the stability operator, the direction v along which the variations are performed is fixed. It is of interest to study how the stability operator depends on the direction v. A natural possibility is to fix one direction and to compare the stability operator along any direction v with respect to this fixed direction. The choice of reference direction v_0 will depend on the geometric structure available. For instance, imagine that S is known to lie within a codimension-one embedded submanifold Σ_0 and that ℓ^+ is nowhere tangent to Σ_0. Then there is a unique vector field v_0 along S which is normal to S, tangent to Σ_0 and satisfying (5) (note that we are not making any causality assumption on Σ_0 besides the fact that it is nowhere tangent to ℓ^+). In this context, it is natural to choose v_0 as the reference vector and compare all other stability operators with L_{v_0}. If no additional structure of this sort exists, or if one wishes to work exclusively with spacetime information, then the only privileged direction v is the unique null direction orthogonal to S which is linearly independent to ℓ^+, namely, $v_0 = -\frac{1}{2}\ell^-$. This direction being geometrically distinguished, we simplify its notation and write L_- instead of $L_{-\ell^-/2}$. We also write λ_- instead of $\lambda_{-\ell^-/2}$. Since the function V vanishes for this vector, it follows from (6) that

$$L_-(\psi) = -\Delta_h \psi + 2\mathbf{s}(\nabla_h \psi) + \frac{1}{2}\left(\mathrm{Scal}(h) - \mathrm{Ein}(\ell^+, \ell^-) - 2||\mathbf{s}||_h^2 + 2\mathrm{div}_h \mathbf{s}\right)\psi,$$
$$L_v(\psi) = L_-(\psi) - VW\psi. \tag{8}$$

where we have defined the function

$$W \stackrel{\text{def}}{=} \mathrm{tr}_h(\chi_+^2) + \mathrm{Ein}(\ell^+, \ell^+).$$

If $(M, g^{(n)})$ satisfies the NEC, then $W \geq 0$. Thus, making V larger (more positive) will tend to decrease the principal eigenvalue. Increasing V makes the direction of v closer to the direction of ℓ^+, so the decreasing of the principal eigenvalue is consistent with the fact that the second variation of area along ℓ^+ is always non-positive (provided NEC holds). In fact, if $W \not\equiv 0$, then there necessarily exists a direction sufficiently close to ℓ^+ for which the principal eigenvalue is negative.

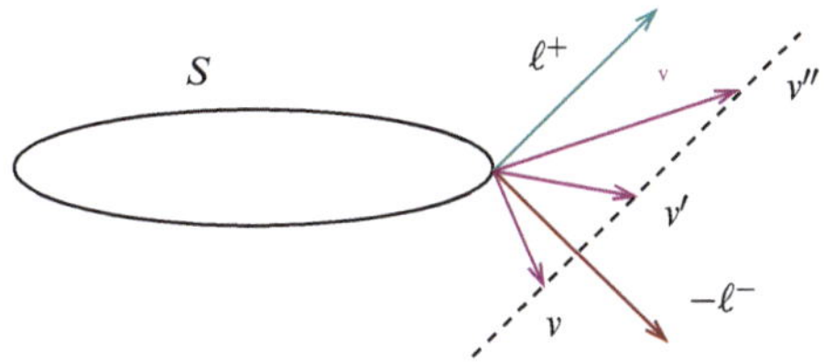

Fig. 5 The direction v'' is more tilted towards ℓ^+ than v' and the same holds for v' with respect to v. Note that one dimension has been suppressed in S and two have been suppressed when plotting normal directions. It is clear that, in general, given two normal directions, neither is tilted towards ℓ^+ with respect to the other

Conversely, for directions sufficiently close to $-\ell^+$ (in the sense that $V \to -\infty$), then the zero-order term in L_v can be made as large and positive as desired. This implies that, as long as $W \not\equiv 0$, there always exists directions (close to $-\ell^+$) for which the principal eigenvalue is positive. More quantitatively, the following estimate is easily derived [4].

Lemma 4. *Let v, v' define directions nowhere tangent to ℓ^+ and let V, V' be the corresponding functions according to (5). Then, the eigenvalues λ_v, $\lambda_{v'}$ satisfy the estimates*

$$\lambda_{v'} + \inf_S [(V' - V)W] \leq \lambda_v \leq \lambda_{v'} + \sup_S [(V' - V)W].$$

The following facts can be easily deduced from this lemma (see Fig. 5 for a schematic representation of the various vectors involved):

(a) If $W \equiv 0$, then λ_v is independent of direction.
(b) Assume that v' is tilted towards ℓ^+ with respect to v, i.e. the function $V' \geq V$ everywhere (see Fig. 5) and that $(M, g^{(n)})$ satisfies the NEC, then $\lambda_{v'} \leq \lambda_v$.
(c) Under NEC, a MOTS which is stable with respect to a spacelike direction, is also stable with respect to the null direction $-\ell^-/2$.
(d) Under NEC, a marginally stable MOTS along $-\ell^-/2$ (i.e. $\lambda_- = 0$) is unstable along any spacelike direction v unless $\mathrm{Ein}(\ell^+, \ell^+) = 0$ and $\chi_+ = 0$ on S.
(e) If the NEC is satisfied and $W \not\equiv 0$ then $\lambda_{v''} < 0$ for some direction v'' sufficiently close to ℓ^+.

The following corollary follows directly from Theorem 1 and Lemma 4 and appears to have been unnoticed before.

Corollary 1. *Let S be a MOTS in a spacetime $(M, g^{(n)})$ satisfying the DEC. Assume that S is stable with respect to some spacelike direction v and that its Yamabe type is zero. Then, in addition to Ricci flatness, $\chi^+ = 0$, and $Ein(\ell^+, v^\star) = 0$ on S (as stated in Theorem 1), we also have $Ein(\ell^+, u)|_S = 0$ for any vector u orthogonal to S.*

Proof. The DEC implies the NEC. So S is also stable with respect to the null direction $-\ell^-/2$. Since its Yamabe type is zero and $(\ell^-)^\star = -\ell^-$ it follows from Theorem 1 that $\mathrm{Ein}(\ell^+,\ell^-) = 0$. Since v is spacelike, $v^\star$ is time like and hence linearly independent to ℓ^- at every point, from which the claim follows. □

Remark. In fact this corollary also holds for MOTS which are merely symmetrized stable with respect to some spacelike direction v. This follows from the fact that Lemma 4 also holds if we replace λ_v by λ_s and $\lambda_{v'}$ by $\lambda_{s'}$, where λ_s and $\lambda_{s'}$ are, respectively, the principal eigenvalue of the symmetrized stability operator corresponding to v and to v'. This, in turn, is a direct consequence of the fact that the transformation from the stability operator into its symmetrized counterpart only involves the connection one-form $\mathbf{s}$, which is independent of v.

Remark. A simple consequence of this corollary is that a stable (or even symmetrized stable) MOTS with respect to an achronal direction can be of zero Yamabe type and satisfy at the same time $\mathrm{Ein}(\ell^+,u)|_S \not\equiv 0$ for some normal field $u \in \mathfrak{X}(S)^\perp$ only if S is marginally stable along $-\ell^-/2$ and $\mathrm{Ein}(\ell^+,\ell^+)|_S \not\equiv 0$.

4 Barrier Properties of MOTS

An important property of minimal hypersurfaces is that they act as barriers for other hypersurfaces. In its simplest form, this is a consequence of the maximum principle for minimal surfaces, which states that two distinct minimal surfaces cannot touch (recall that two surfaces S_1 and S_2 "touch" each other if their intersection is non-empty, and there exists a neighbourhood $U_1 \subset S_1$ of $S_1 \cap S_2$ and a tubular neighbourhood of U_1 such that S_2 intersects U_1 only on one side).

MOTSs are codimension-two objects and the notion of "touching" is necessarily subtler because one needs to define the meaning of "being on one side". As mentioned above, given a spacelike surface S, one can generate a null hypersurface $\mathscr{N}^+$ by sending pulses of light orthogonally to S along the direction ℓ^+. Null hypersurfaces, like any other hypersurface in a manifold with a metric, admit a well-defined notion of second fundamental form. Despite the fact that the first fundamental form on a null hypersurface is degenerate (and hence cannot be inverted), it is nevertheless possible to define the trace of the second fundamental form by passing to a suitable quotient space (see e.g. [25] for details). This defines a scalar function θ on the null hypersurface $\mathscr{N}^+$ called "null expansion" of $\mathscr{N}^+$. Let $p \in \mathscr{N}^+$ be arbitrary and $\theta|_p$ be the null expansion with respect to the normal vector ℓ^+. It is a remarkable consequence of the degeneracy of the induced metric that *any* spacelike surface $\hat{S}$ embedded in $\mathscr{N}^+$ and passing through p has null expansion along ℓ^+ (note that this vector is necessarily orthogonal to $\hat{S}$) with exactly the same value as the hypersurface null expansion $\theta|_p$. The maximum principle for null hypersurfaces [24][1] implies a maximum principle for MOTS. The appropriate notion of "touching" is that the surfaces intersect and, locally near the intersection set, one of the surfaces lies to the future of the null hypersurface $\mathscr{N}^+$ generated by the other surface. The precise statement is as follows (cf. Proposition 3.1 in [6] and Proposition 2.4 in [5] when the surfaces are restricted to lying on a *spacelike* hypersurface).

Theorem 2 (Maximum principle for null expansions). *Let S_1, S_2 be codimension-2, spacelike surfaces intersecting at a point p and assume that their corresponding tangent planes at p coincide. Let $\ell_+(S_1)$, $\ell_+(S_2)$ be future-directed null normals which agree at p.*

Assume that there exists a spacetime neighbourhood U_p of p. Which is causal (i.e. free of closed causal curves) and satisfying $\sup_{S_1 \cap U_p} \theta_+(S_1) \leq 0 \leq \inf_{S_2 \cap U_p} \theta_+(S_2)$. Suppose that $S_1 \cap U_p$ lies in the causal future of $\mathscr{N}^+(S_2)$ (in U_p) and S_1 and S_2 are contained in a hypersurface Σ transverse to $\mathscr{N}^+(S_2)$. Then S_1 and S_2 coincide on U_p.

This theorem states, in particular, that MOTS cannot touch each other (in the sense of this theorem) if their null normals agree at the touching point.

Let us, for the remainder of this section, restrict the discussion to MOTS lying on a fixed hypersurface Σ. More precisely, consider an embedded hypersurface Σ with no a priori restriction on its causal character. We say that a surface S is a **MOTS lying on Σ** if there is an embedding $\Phi : S \to \Sigma$ and a future null normal ℓ^+ to S such that ℓ^+ is nowhere tangent to Σ and the null expansion θ_+ of S along ℓ^+ vanishes. Obviously, when Σ is space like, the condition that ℓ^+ is nowhere tangent to Σ is automatically satisfied. Let S be a MOTS lying on Σ with corresponding null normal ℓ^+. Let v be the unique normal to S which is tangent to Σ and satisfies $< v, \ell^+ >= 1$. A rescaling of ℓ^+ changes v but without affecting its orientation. Given a two sided neighbourhood D of S in Σ, define the **exterior part** D^+ as the one one-sided part of D which lies in the positive direction of v. Similarly, we define the **interior part** D^- as the one-sided part of D which lies in the negative direction of v. Thus, a MOTS with a selected ℓ^+ defines locally near S a unique notion of exterior and interior within Σ. Note that if the MOTS has non-zero mean curvature somewhere, this notion of exterior/interior is unambiguous (because then ℓ^+, and hence v, are uniquely fixed up to positive rescaling). However, if $H \equiv 0$ on S then a choice of null normal ℓ^+ is necessary in order to define exterior/interior in Σ. Conversely, let S be an embedded surface in Σ with a selected normal v tangent to Σ. We define $\theta_{+,v}$ to be the null expansion along the unique future directed null normal ℓ^+ to S satisfying $< \ell^+, v >= 1$. The following definition is natural in this context [4].

[1] This result states, roughly speaking, that if two null hypersurfaces touch each other at p and satisfy the property that, locally near p, the hypersurface lying to the past has non-negative null expansion while the one lying to the future has non-positive null expansion, then the two null hypersurfaces must coincide in a neighbourhood of p. The precise statement can be found in Theorems 2.1 and 3.4 in [24].

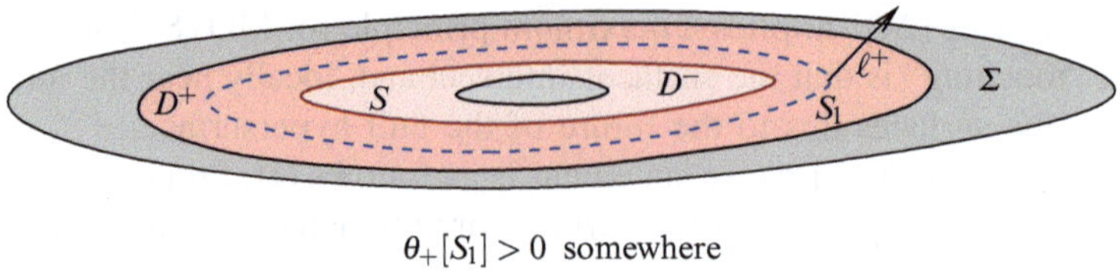

Fig. 6 This figure shows a MOTS S lying on a hypersurface Σ. If S is locally outermost, then there exists a sufficiently small two-sided neighbourhood D of S such that any surface S_1 fully contained in the exterior part D^+ and homologous to S must have positive null expansion somewhere with respect to the outer direction ℓ^+

Definition 5. Let Σ be an embedded hypersurface in $(M, g^{(n)})$ and S a MOTS lying on Σ. Assume that a future null normal ℓ^+ satisfying $\theta_+[S] = 0$ has been chosen. S is called **locally outermost in Σ** if there exists a two-sided neighbourhood $D^+ \cup D^- \subset \Sigma$ of S such that the exterior D^+ contains no surface S_1 satisfying:

(a) $S \cup S_1$ bounds a domain $\mathscr{D}$ in D^+.
(b) $\theta_{+,v}[S_1] \leq 0$ where v is a normal to S_1 in Σ pointing outside the domain $\mathscr{D}$.

This notion of locally outermost (see Fig. 6) captures the idea of a surface being (locally) an outer barrier for weakly trapped surfaces (i.e. surfaces with non-positive null expansion with respect to the outer null direction). In the case of minimal surfaces, a similar notion of barrier turns out to be intimately related to stability properties of the surface. For MOTS, an analogous result holds [4].

Proposition 1. *Let Σ be an embedded hypersurface in $(M, g^{(n)})$ and S a MOTS lying on Σ with a selected null normal ℓ^+ for which $\theta_+ = 0$. Let v be the unique normal to S which is tangent to Σ and satisfies $< v, \ell^+ >= 1$.*

(a) If S is strictly stable along v, then S is locally outermost in Σ.
(b) If S is locally outermost in Σ, then S is stable along v.

For locally outermost MOTS lying on a spacelike hypersurface, the topological result [27] (Theorem 1 above) can be strengthened [26].

Theorem 3 (Galloway [26]). *Let $(M, g^{(n)})$ be a spacetime satisfying the DEC and Σ a space like embedded hypersurface. If S is a locally outermost MOTS in Σ then S is of positive Yamabe type.*

From a spacetime perspective, it is natural to ask whether this theorem extends to the case of hypersurfaces which are either space like or null at every point. As far as I know, neither a proof nor a counterexample to such a statement is known. It would be of interest to settle this point.

A most natural question regarding barrier properties of MOTS is whether there exists a globally outermost MOTS in a given hypersurface. A fundamental theorem along these lines has been proved by Andersson and Metzger [5] in spacetime dimension four and extended to dimension up to and including eight by Eichmair using different methods [23]. We need some notation first.

Consider a compact space like embedded hypersurface Σ with smooth boundary $\partial\Sigma$. Assume that this boundary can be split in two non-empty disjoint component $\partial\Sigma = \partial^-\Sigma \cup \partial^+\Sigma$. A (not necessarily connected) surface S embedded in $\Sigma \setminus \partial^+\Sigma$ is called **bounding** if $S \cup \partial^+\Sigma$ is the boundary of a domain $\Omega^+(S)$ in Σ. This domain is called "exterior" of S in Σ. For bounding surfaces (in particular for $\partial^-\Sigma$), we always denote by θ_+ the null expansion with respect to the (unique) outer null normal ℓ^+ satisfying that its tangential projection on Σ has unit norm and points towards the exterior of S in Σ. For $\partial^+\Sigma$, the null normal is chosen so that its tangential projection is unit and points outside Σ. Combining the results of Andersson and Metzger [5] and Eichmair [23], the next existence and uniqueness result of an outermost MOTS in Σ follows (Theorem 3.3 in [2]).

Theorem 4 (Andersson et al. [2]). *Let $(M, g^{(n)})$ be a spacetime of dimension $4 \le n \le 8$. Let Σ be a compact, space like embedded hypersurface with smooth boundary $\partial\Sigma = \partial^+\Sigma \cup \partial^-\Sigma$, as before. Assume that $\theta_+[\partial^-\Sigma] \le 0$ and $\theta_+[\partial^+\Sigma] > 0$. Then there exists a unique (not necessarily connected) bounding MOTS S in Σ satisfying the property that no bounding surface in Σ with $\theta_+ \le 0$ intersects the exterior $\Omega^+(S)$.*

Being globally outermost, S is also locally outermost and hence stable and of positive Yamabe type.

4.1 MOTS and Symmetries

Spacetimes with symmetries play an important role in gravitational theories. For instance, the gravitational collapse that necessarily occurs when an initial data set contains a trapped surface is believed to settle down to a stationary black hole once all relevant dynamical processes (like, e.g. accretion of matter into the black hole or emission of gravitational waves) have already taken place. It is clear that a proper analysis of MOTS in stationary spacetimes is important in order to understand whether this type of surfaces are useful as quasi-local replacements for black holes. A first natural step is to study the restrictions on MOTS arising from the presence of spacetime symmetries. Since symmetries of various sorts are typically defined by imposing restrictions on its vector field generators, it is convenient to study MOTS, and in particular their stability properties, in spacetimes containing special types of vector fields. Let us start with the following simple observation [10]:

Assume that a spacetime admits a Killing vector ξ and let S be a MOTS. It is clear that the motion along the local one-parameter group of transformations φ_τ generated by ξ changes no geometric properties of S. Hence, the surface $S_\tau = \varphi_\tau(S)$ is also a MOTS, which implies that the first variation of θ_+ along ξ vanishes. Assume that ξ is nowhere tangent to S and linearly independent to ℓ^+ on S. Then, there exists a unique vector v orthogonal to S and satisfying $< v, \ell^+ >= 1$ such that $\xi = Fv + \xi^{\parallel}$, where $\xi^{\parallel}$ is tangent to S. It follows

$$0 = \delta_\xi \theta_+ = \delta_{Fv+\xi^{\parallel}} \theta_+ = \delta_{Fv}\theta_+ = L_v(F),$$

where the third equality follows because tangential variations of a vanishing quantity are obviously zero. Thus, by this symmetry argument, it follows that the stability operator L_v admits 0 as an eigenvalue with eigenfunction $F = <\xi, \ell^+>$. However, the stability operator (6) is written in terms of the geometry of S, and it is far from obvious by looking at that expression that $L_v(F)$ should vanish. The lesson is that the form (6) of the stability operator may not be the most convenient one for studying the interplay of MOTS and symmetries. One needs an alternative expression of L_v adapted to spacetime information. Since generators of symmetries are typically defined in terms of conditions on the Lie derivative of the metric along the generator (as, for instance, for Killing vector, homotheties, conformal Killing vectors, Kerr–Schild vectors [17], etc.), the following definition turns out to be useful.

Definition 6. Let ξ be a vector field in a spacetime $(M, g^{(n)})$. The **metric deformation tensor** a^{ξ} is the symmetric, two-covariant tensor defined as

$$a^{\xi} \stackrel{\text{def}}{=} \mathscr{L}_{\xi}\, g^{(n)}.$$

The following lemma [10] gives an explicit form for the variation of θ_+ on a MOTS along an arbitrary vector ξ in terms of its deformation tensor. A more general version valid for any normal expansion (not necessarily along $\ell^{\pm}$) and for arbitrary surfaces (not necessarily MOTS) can be found in [10].

Lemma 5. *Let ξ be a vector field with deformation tensor a^{ξ} in a spacetime $(M, g^{(n)})$. Let S be a MOTS with embedding Φ_S, then*

$$\delta_{\xi}\, \theta_+ = -\frac{1}{4}\theta_- a^{\xi}(\ell^+, \ell^+) - tr_h\left(a^{\xi,S} \cdot \chi_+\right) + h^{\alpha\beta}\ell^{+\gamma}\left[\frac{1}{2}\nabla_{\gamma} a^{\xi}_{\alpha\beta} - \nabla_{\alpha} a^{\xi}_{\gamma\beta}\right]\bigg|_S .$$

where $a^{\xi,S} \stackrel{\text{def}}{=} \Phi_S^{\star}(a^{\xi})$, $tr_h(a^{\xi,S} \cdot \chi_+)$ is the double trace (with the metric h) of $a^{\xi,S} \otimes \chi_+$ in the first-third indices and in the second-fourth indices, and $h^{\alpha\beta}$ is the projector tangent to S.

When the vector field ξ is Killing, then its deformation tensor vanishes identically and the expression above gives $\delta_{\xi}\, \theta_+ = 0$. We already knew this, but the difference is that the conclusion now follows from an explicit calculation rather than from a symmetry argument. For vector fields generating symmetries other than isometries, the expression above is necessary because no direct symmetry invariance argument is available.

Several results can be obtained by studying the interplay between stable MOTS and special types of vector fields by means of Lemma 5. Here only two examples of the type of results that can be derived are included. I refer to [10] for details and further results (see also [6, 35] for previous works on the interplay between different types of trapped surfaces and symmetries).

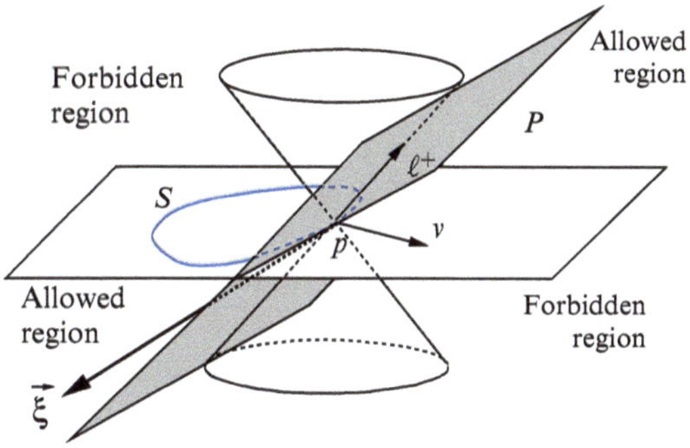

Fig. 7 The planes $T_p\Sigma$ and $P \equiv T_pS \oplus \text{span}\{\ell^+|_p\}$ divide the tangent space T_pM in four regions. Theorem 5 implies that if S is strictly stable and ξ is a Killing vector or a homothety in a spacetime satisfying the NEC which points above Σ everywhere, then ξ cannot enter into the forbidden region at any point (and, similarly, if ξ points below Σ everywhere). The allowed region includes the plane P. However, if there is a point with $W \neq 0$ where ξ is not tangent to Σ, then the result is also valid for stable MOTS with P belonging to the forbidden region

Theorem 5. *Let S be a stable MOTS with respect to a direction v. Suppose that $(M, g^{(n)})$ admits a conformal Killing vector ξ, i.e. $\mathscr{L}_\xi\, g^{(n)} = 2\phi g^{(n)}$ (including homotheties $\phi = C$ and isometries $\phi = 0$):*

(a) If $0 \not\equiv 2\ell^+(\phi) - <\xi, v^\star> W|_S \leq 0$, then $<\xi, \ell^+>|_S < 0$.
(b) If S is strictly stable and $2\ell^+(\phi) - <\xi, v^\star> W|_S \leq 0$, then $<\xi, \ell^+>|_S \leq 0$ and vanishes at one point only if it vanishes everywhere.

This result is true also if all inequalities are reversed (this follows immediately by applying the theorem to $-\xi$). This result can be particularized for Killing vector fields or homotheties in spacetimes satisfying the NEC. Then $\ell^+(\phi) = 0$ and $W \geq 0$, so the theorem gives restrictions directly on the scalar product $<\xi, v^\star>$, i.e. on the relative position of ξ with respect to the normal stability direction v. Figure 7 describes the restrictions in this case.

If the Killing vector is future or past, the result can be strengthened [10].

Theorem 6. *Let S be a MOTS stable with respect to $-\ell^+/2$. Assume that $(M, g^{(n)})$ satisfies the NEC and admits a causal Killing vector or homothety ξ which is future directed everywhere on S:*

(a) Then $\chi_+ = 0$ and $Ein(\ell^+, \ell^+) = 0$ at every point p where $\xi(p) \neq 0$.
(b) If S is strictly stable, then $\xi \propto \ell^+$ everywhere.

5 MOTS and Killing Horizons

Stationary black holes have the property that the stationary Killing vector is tangent to the event horizon. If the Killing vector is hypersurface orthogonal, then the event horizon is in fact a Killing horizon, i.e. a null hypersurface where the Killing vector is tangent, null and nonzero. More generally, event horizons are also Killing horizons in stationary and axially symmetric four-dimensional black holes. As already mentioned several times, in the context of cosmic censorship, one expects collapsing, asymptotically flat spacetimes to settle down to a stationary black hole. One also expects that tubes of MOTS eventually approach the event horizon (see however [44]), which by the uniqueness theorems of stationary black holes should in turn be an approximate Killing horizon (at least in four spacetime dimensions, and possibly also in higher dimensions).

A first step towards this aim would be understanding the interplay between MOTS and Killing horizons. This is the aim of a recent investigation [36]. Here I only summarize some of the results. The reader is referred to [36] for the proofs and for further details.

5.1 *Killing Horizons*

Let us assume that the spacetime $(M, g^{(n)})$ admits a Killing vector ξ which is nonzero and null along a smooth null hypersurface $\mathscr{H}_\xi$, a so-called **Killing horizon**. To be more precise, we adopt the following definition of Killing horizon, which includes a topological restriction (c.f. Condition 2.1 in [39]).

Definition 7 (Killing horizon). A Killing horizon $\mathscr{H}_\xi$ of a Killing vector ξ is a smooth hypersurface embedded in $(M, g^{(n)})$ with embedding $\Phi_\xi : \mathscr{H}_\xi \to M$, satisfying the following properties:

(a) $\mathscr{H}_\xi$ is diffeomorphic to $S \times \mathbb{R}$ where S is a closed $(n-2)$-manifold.
(b) The first fundamental form $g_{\mathscr{H}_\xi} \stackrel{\mathrm{def}}{=} \Phi_\xi^\star(g^{(n)})$ is degenerate along the $\mathbb{R}$ factor at every point $p \in \mathscr{H}_\xi$.
(c) For each $p \in \mathscr{H}_\xi$, the Killing vector $\xi|_{\Phi_\xi(p)}$ is non-zero, null, future directed and tangent to $\Phi_\xi(\mathscr{H}_\xi)$.

Being null and tangent to $\mathscr{H}_\xi$ it follows that ξ points along the $\mathbb{R}$ factor at every point $p \in \mathscr{H}_\xi$. The condition of ξ being future directed is adopted here only for notational simplicity later (in particular when we choose $\ell^+ = \xi$ on any section of a Killing horizon). Obviously, if $\mathscr{H}_\xi$ satisfies all properties above except for the condition of ξ being future directed, then $\mathscr{H}_\xi$ is a Killing horizon with respect to $-\xi$.

The degeneracy of the first fundamental form of $\mathscr{H}_\xi$ implies that ξ is a nowhere zero normal vector to $\mathscr{H}_\xi$. The square norm of ξ vanishes on $\mathscr{H}_\xi$, which implies that its gradient is normal to $\mathscr{H}_\xi$. The proportionality factor between this gradient and ξ defines the surface gravity. More precisely (see e.g. [42]).

Definition 8. The **surface gravity** κ_ξ of a Killing horizon $\mathscr{H}_\xi$ is a function $\mathscr{H}_\xi \to \mathbb{R}$ defined by

$$\nabla\lambda|_p = 2\kappa_\xi \xi|_p, \quad \forall p \in \mathscr{H}_\xi, \qquad \text{where} \quad \lambda \stackrel{\mathrm{def}}{=} -<\xi,\xi>.$$

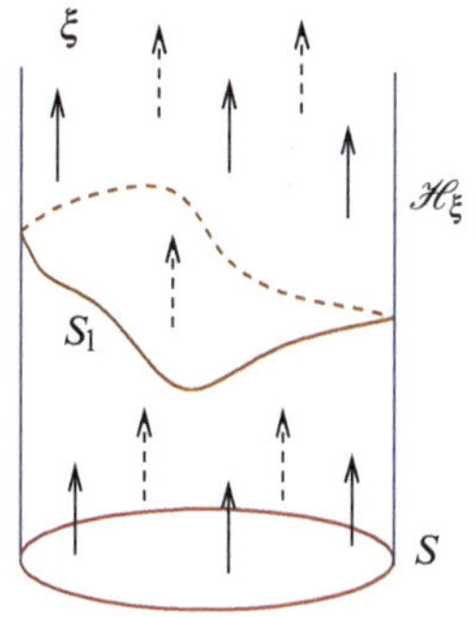

Fig. 8 Schematic figure for a Killing horizon. The vector field ξ is null, tangent to $\mathscr{H}_\xi$ and nowhere zero. Furthermore, it is the restriction on $\mathscr{H}_\xi$ of a Killing vector of the spacetime. A section of the Killing horizon is a smooth spacelike surface embedded in $\mathscr{H}_\xi$

The surface gravity can be equivalently defined by

$$\nabla_\xi \xi \stackrel{\mathscr{H}_\xi}{=} \kappa_\xi \xi, \tag{9}$$

i.e. κ_ξ measures the departure of the Killing parametrization of the orbits of ξ on $\mathscr{H}_\xi$ (which, from general properties of null hypersurfaces, are necessarily null geodesics) from being an affine parametrization. The following definition is standard.

Definition 9. A Killing horizon $\mathscr{H}_\xi$ is degenerate if $\kappa_\xi \equiv 0$ and non-degenerate if $\kappa_\xi|_p \neq 0$ for some point $p \in \mathscr{H}_\xi$.

Killing horizons are well-studied objects. I summarize some well-known properties (see e.g. [31, 42] for details).

Since $\mathscr{H}_\xi$ is a (trivial) bundle over S, we can define $\pi : \mathscr{H}_\xi \simeq S \times \mathbb{R} \rightarrow S$ as the projection onto S. The fact that ξ is a Killing vector implies that there exists a Riemannian metric on S, denoted by h, such that $g_{\mathscr{H}_\xi} = \pi^\star(h)$. Moreover, for any section S_1 of $(\mathscr{H}_\xi, S, \pi)$ (see Fig. 8) the spacetime metric (or equivalently the first fundamental form $g_{\mathscr{H}_\xi}$) induces a Riemannian metric h_1 on S_1. The Riemannian spaces (S_1, h_1) and (S, h) are isometric with isometry $\pi|_{S_1}$. As a simple consequence of this fact (see e.g. [25]), the second fundamental form vector $\chi(S_1)$ of any section S_1 (as a spacetime submanifold) satisfies $\chi_\xi(S_1) \stackrel{\text{def}}{=} < \chi(S_1), \xi >= 0$. In particular, the null expansion along ξ vanishes:

$$\theta_\xi(S_1) \stackrel{\text{def}}{=} \text{tr}_{S_1} \chi_\xi(S_1) = 0.$$

The Raychaudhuri equation for the null expansion $\theta_\xi(S_1)$ combined with $\chi_\xi(S_1) = 0$ immediately implies $\text{Ein}(\xi, \xi) = 0$ everywhere on a Killing horizon $\mathscr{H}_\xi$.

All sections of a Killing horizon are therefore MOTS with respect to the null normal ξ. On a given section S_0, let us define k as the unique future directed null normal satisfying $< k, \xi >= -2$.

Regarding the surface gravity κ_ξ, it is well-known (see, e.g. [42]) that κ_ξ is constant on any Killing horizon in a spacetime satisfying the DEC. The surface gravity is also constant on any Killing horizon of an integrable Killing vector (i.e. a Killing vector satisfying $\xi \wedge d\xi = 0$), irrespectively of any energy condition. In fact, both statements are a simple consequence of the identity (see [42] p. 334 for a derivation in four dimensions which holds true in any dimension):

$$\xi \wedge d\kappa_\xi \stackrel{\mathscr{H}_\xi}{=} -\xi \wedge \text{Ric}_\xi,$$

where $\text{Ric}_\xi \stackrel{\text{def}}{=} \text{Ric}(\xi, \cdot)$ and κ_ξ stands for any smooth extension of the surface gravity to a neighbourhood of $\mathscr{H}_\xi$.

5.2 Stability Operator of MOTS in Killing Horizons

The aim of this subsection is to study the stability properties of MOTS embedded in Killing horizons. The first immediate observation is that the stability operator L_v is independent of the direction v. This follows from (8) and the fact that $W = \text{tr}_h(\chi_+^2) + \text{Ein}(\xi, \xi)$ vanishes identically on a Killing horizon. So, the stability operator is a property of the section itself, not of the normal direction along which the variations are performed. Let us write L_{S_1} for this stability operator. The following lemma follows from combining properties of Killing horizons with the general expression for the stability operator in terms of deformation tensors (see [36] for details).

Proposition 2 (Stability operator of a section of a Killing horizon). *Let $\mathscr{H}_\xi$ a Killing horizon and S_0 a section of $\mathscr{H}_\xi$. Let s be the connection one-form of the normal bundle of S_0 in the basis $\{\xi(=\ell^+), k = (\ell^-)\}$, namely $s(X) \equiv -\frac{1}{2} < k, \nabla_X \xi >$. Let $s = dF + z$ be the Hodge decomposition of s and define $u \stackrel{\text{def}}{=} e^{2F}$. Then, the stability operator of S_0 is*

$$L_{S_0}(\psi) = -div_h\left(u\nabla_h\left(\frac{\psi}{u}\right)\right) + 2\mathbf{z}(\nabla_h\psi) - \frac{1}{2}\kappa_\xi\theta_k\psi. \tag{10}$$

A natural question is to ask how does the stability operator (or, more generally, the stability properties) depend on the section S_0. The case of degenerate horizons is an easy consequence of Proposition 2.

Lemma 6. *Let S_0 be a section of a degenerate horizon $\mathscr{H}_\xi$. Then S_0 is marginally stable.*

In order to determine the dependence of the stability on the section for non-degenerate horizons, we need to compare two arbitrary sections of $\mathscr{H}_\xi$. Let us therefore fix a section S_0 of $\mathscr{H}_\xi$. Any other section $S[f]$ is defined by a graph over S_0, $f : S_0 \to \mathbb{R}$, see Fig. 9. Let $\pi_f : S[f] \to S_0$ be the natural projection along orbits of the Killing vector. π_f is a diffeomorphism between these two spaces (in fact an isometry with their respective induced metrics).

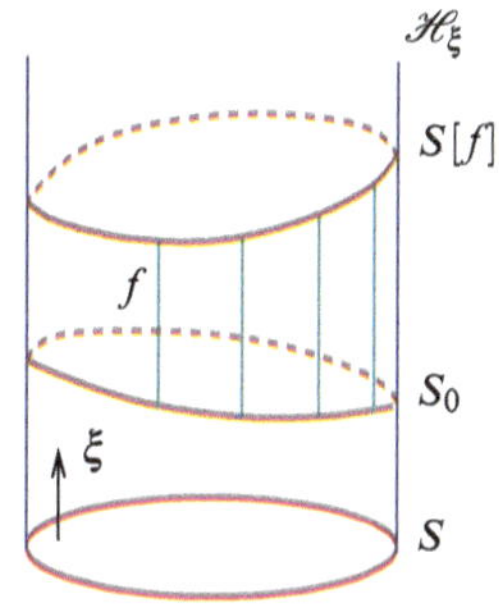

Fig. 9 An arbitrary section $S[f]$ of a Killing horizon is defined uniquely by a graph function $f : S_0 \to \mathbb{R}$ on a fixed section S_0. Lemma 7 establishes how the geometric properties of $S[f]$ relate to the geometric properties of S_0

In order to determine the behaviour of the stability operator, we need to relate the one-forms $\mathbf{s}$ of S_0 and $\mathbf{s[f]}$ of $S[f]$ and the null expansions θ_k of S_0 and $\theta_k[f]$ of $S[f]$. The following lemma [36] establishes these relationships.

Lemma 7. *Let S_0, $S[f]$ and π_f as before. Let k_f be the null normal to $S[f]$ satisfying $< \xi, k_f > \stackrel{S[f]}{=} -2$. Denote by $\mathbf{s[f]}$ the connection of $S[f]$ with respect to k_f, by $\chi_f^{k_f}$ the second fundamental form of $S[f]$ along k_f and by $\theta_k[f]$ its trace. Then*

$$\begin{aligned} \mathbf{s[f]} &= \pi_f^\star \left(\mathbf{s} + \kappa_\xi df\right), && (11)\\ \chi_f^{k_f} &= \pi_f^\star \left(\chi^k + 2Hess_h f + 2\kappa_\xi df \otimes df + 2df \otimes \mathbf{s} + 2\mathbf{s} \otimes df\right), && (12)\\ \theta_k[f] &= \left(\theta_k + 2\Delta_h f + 2\kappa_\xi ||\nabla_h f||_h^2 + +4\mathbf{s}(\nabla_h f)\right) \circ \pi_f, && (13) \end{aligned}$$

where $Hess_h$ denotes the Hessian with the h-metric connection and $\mathbf{s}$, χ^k and θ_k are, respectively, the connection one-form, the second fundamental form and the null expansion of S_0 along the null normal k satisfying $< \xi, k > \stackrel{S_0}{=} -2$.

This transformation lemma allows us to obtain the relationship between the stability operator of S_0 and $S[f]$ in the case of constant surface gravity.

Proposition 3 (Dependence of stability operator on the section). *Assume that κ_ξ is constant. The stability operator of $S[f]$ is related to the stability operator of S_0 by*

$$L_{S[f]}\left(\psi \circ \pi_f\right) = e^{-\kappa_\xi f} L_{S_0}\left(e^{\kappa_\xi f}\psi\right) \circ \pi_f, \qquad \forall \psi \in C^2(S_0, \mathbb{R}).$$

The following is an immediate consequence.

Corollary 2. *Let $\mathscr{H}_\xi$ be a Killing horizon with constant surface gravity. Then the principal eigenvalue is independent of the section.*

Although the case of constant surface gravity is the most interesting one (because κ_ξ is constant if the DEC is satisfied), it is natural to ask whether the result above extends to the case of general κ_ξ. It turns out that the constancy of the surface gravity is a necessary condition for the validity of the conclusion of Corollary 2 [36].

The analysis of stability of MOTS lying on Killing horizons can be used to clarify the relationship between area–angular momentum inequalities obtained in different contexts. The first one was obtained by Hennig, Ansorg and Cederbaum [30] in the context of stationary and axially symmetric black holes. The second was obtained by Jaramillo, Reiris and Dain [33] for stable MOTS along the null direction $-\ell^-/2$. A similar inequality for minimal surfaces lying on maximal spacelike hypersurfaces was also obtained by Dain and Reiris [19]. The relationship between the result for black holes and the result for minimal surfaces has been clarified recently by Chruściel, Eckstein, Nguyen and Szybka [14]. I refer to [36] for a discussion on how the analysis of the stability properties of Killing horizons (in four dimensions) allows for a clarification of the relationship between the results in [30] and [33]. Here I only summarize the statement of the inequality as given by Jaramillo, Reiris and Dain. For more details on this beautiful inequality, see [32] and the recent review article [20].

6 Axially Symmetric MOTS and Angular Momentum

A remarkable consequence of stability of MOTS is that, in the axially symmetric case and in spacetimes satisfying the DEC, the area of the MOTS is always bounded below by a universal expression involving the angular momentum of the surface. We start with the definition of "axial symmetry" in this context. This definition captures the minimal ingredients needed for the proof of the inequality as given in [33].

Definition 10. A MOTS S is axially symmetric if there exists a vector field $\eta \in \mathfrak{X}(S)$ with closed orbits satisfying

(a) $\mathscr{L}_\eta h = 0$.
(b) $\mathscr{L}_\eta \mathbf{s} = 0$, for some choice of basis $\{\ell^+, \ell^-\}$.
(c) η commutes with the stability operator L_v for some choice of normal vector v.

The following standard definition of "angular momentum" is motivated by the Komar expression of conserved quantities associated to Killing vectors.

Definition 11. The **angular momentum** of an axially symmetric, two-dimensional MOTS is the integral

$$J = \frac{1}{8\pi} \int_S \mathbf{s}(\eta) \eta_S.$$

The following theorem establishes a remarkable inequality between area and angular momentum for stable, axially symmetric MOTS in four-dimensional spacetimes.

Theorem 7 (Jaramillo et al. [33]). *Let $(M, g^{(4)})$ be a spacetime satisfying the DEC. Let S be a MOTS in $(M, g^{(4)})$ and assume S to be axially symmetric and stable with respect to the null direction $-\ell^-/2$. Then*

$$|S| \geq 8\pi |J|.$$

Acknowledgements Financial support under the projects FIS2009- 07238 (Spanish MEC) and P09-FQM-4496 (Junta de Andalucía and FEDER funds) are acknowledged.

References

1. Andersson, L.: The global existence problem in general relativity. In: Chruściel, P.T., Friedrich, H. (eds.) The Einstein Equations and the Large Scale Behaviour of Gravitational Fields. Birkhäuser, Basel (2004)
2. Andersson, L., Eichmair, M., Metzger, J.: Jang's equation and its applications to marginally trapped surfaces. Proceedings of the Complex Analysis & Dynamical Systems IV Conference, Nahariya, Israel, May 2009 (arXiv:1006.4601)
3. Andersson, L., Mars, M., Simon, W.: Local existence of dynamical and trapping horizons. Phys. Rev. Lett. **95**, 111102 (4 pp.) (2005)
4. Andersson, L., Mars, M., Simon, W.: Stability of marginally outer trapped surfaces and existence of marginally outer trapped tubes. Adv. Theor. Math. Phys. **12**, 853–888 (2008)
5. Andersson, L., Metzger, J.: The area of horizons and the trapped region. Commun. Math. Phys. **290**, 941–972 (2009)
6. Ashtekar, A., Galloway, G.J.: Some uniqueness results for dynamical horizons. Adv. Theor. Math. Phys. **9**, 1–30 (2005)
7. Berestycki, H., Nirenberg L., Varadhan S.R.S.: The ground state and maximum principle for 2nd order elliptic operators in general domains. C.R. Acad. Sci. Paris, **317**, Série I, 51–56 (1993)
8. Bray, H., Hayward, S., Mars, M., Simon. W.: Generalized inverse mean curvature flows in spacetime. Commun. Math. Phys. **272**, 119–138 (2007)
9. Cai, M., Galloway, G.J.: On the topology and area of higher dimensional black holes. Class. Quantum Grav. **18**, 2707–2718 (2001)
10. Carrasco, A., Mars, M.: Stability of marginally outer trapped surfaces and symmetries. Class. Quantum Grav. **26**, 175002 (19 pp.) (2009)
11. Chavel, I.: Riemannian geometry, a modern introduction. Cambridge Studies in Advanced Mathematics, vol. 98, Cambridge University Press, Cambridge (2006)
12. Christodoulou, D.: The instability of naked singularities in the gravitational collapse of a scalar field. Ann. Math. **149**, 183–217 (1999)
13. Chruściel, P.T., Galloway, G.J., Solis, D.: Topological censorship for Kaluza-Klein spacetimes. Ann. Henri Poincaré **10**, 893–912 (2009)
14. Chruściel, P.T., Eckstein, M., Nguyen, L., Szybka, S.J.: Existence of singularities in two-Kerr black holes. Class. Quantum Grav. **28**, 245017 (2011)
15. Claudel, C.-M.: Black holes and closed trapped surfaces: a revision of a classic theorem. arXiv:gr-qc/0005031
16. Colding, T.H., Minicozzi W.P.: Minimal surfaces. Courant Lecture Notes in Mathematics, vol. 4. Courant Institute of Mathematical Sciences, New York (1999)
17. Coll, B., Hildebrandt, S., Senovilla, J.M.M.: Kerr–Schild symmetries. Gen. Rel. Grav. **33**, 649–670 (2001)
18. Dafermos, M.: Spherically symmetric spacetimes with a trapped surface. Class. Quantum Grav. **22**, 2221–2232 (2005)
19. Dain, S., Reiris, M.: Area-angular momentum inequality for axisymmetric black holes. Phys. Rev. Lett. **107**, 051101 (2011)
20. Dain, S.: Geometric inequalities for axially symmetric black holes. Classical and Quantum Gravity **29**, 073001 (2012) (arXiv:1111.3615)
21. Donsker, M.D., Varadhan S.R.S.: On a variational formula for the principal eigenvalue for operators with maximum principle. Proc. Nat. Acad. Sci. USA **72**, 780–783 (1975)

22. Donsker, M.D., Varadhan, S.R.S.: On the principal eigenvalue of second-order elliptic differential operators. Commun. Pure Appl. Math. **29**, 591–621 (1976)
23. Eichmair, M.: The plateau problem for marginally trapped surfaces. J. Diff. Geom. **83**, 551–584 (2009)
24. Galloway, G.J.: Maximum principles for null hypersurfaces and null splitting theorems. Ann. Poincaré Phys. Theor. **1**, 543–567 (2000)
25. Galloway, G.J.: Null geometry and the Einstein equations. In: Chrściel, P.T., Friedrich, H. (eds.) The Einstein Equations and the Large Scale Behaviour of Gravitational fields. Birkhäuser, Basel (2004)
26. Galloway, G.J.: Rigidity of marginally trapped surfaces and the topology of black holes. Comm. Anal. Geom. **16**, 217–229 (2008)
27. Galloway, G.J., Schoen, R.: A generalization of Hawking's black hole topology theorem to higher dimensions. Commun. Math. Phys. **266**, 571–576 (2006)
28. Hayward, S.A.: General laws of black-hole dynamics. Phys. Rev. D **49**, 6467–6474 (1994)
29. Hawking, S.W.: The event horizon, in black holes. In: DeWitt, C., DeWitt, B.S. (eds.) Les Houches lectures. North Holland, Amsterdam (1972)
30. Hennig, J., Ansorg, M., Cederbaum, C.: A universal inequality between the angular momentum and the horizon area for axisymmetric and stationary black holes with surrounding matter. Class. Quantum Grav. **25**, 162002 (2008)
31. Heusler, M.: Black hole uniqueness theorems. Cambridge Lecture Notes in Physics, vol. 6. Cambridge University Press, Cambridge (2006)
32. Jaramillo, J.L., Area-angular momentum inequality in stable marginally trapped surfaces, in this volume.
33. Jaramillo, J.L., Reiris, M., Dain, S.: Black hole Area-Angular momentum inequality in non-vacuum spacetimes. Phys. Rev D **84**, 121503 (2011) arXiv:1106.3743 (gr-qc)
34. Krein. M., Rutman M.A.: Linear operators leaving invariant a cone in a Banach space. Usp. Mat. Nauk. (N.S.) **3**, 59–118 (1948); English translation in Amer. Math. Soc. Trans. Ser. (1), **10** 199–325 (1962)
35. Mars, M., Senovilla, J.M.M.: Trapped surfaces and symmetries. Class. Quantum Grav. **20**, L293–L300 (2003)
36. Mars, M.: Stability of MOTS in totally geodesic null horizons, Class. Quantum Grav. **29**, 145019 (2012)
37. Newman, R.P.A.C.: Topology and stability of marginal 2-surfaces. Class. Quantum Grav. **4**, 277–290 (1987)
38. Penrose, R.: Gravitational collapse—the role of general relativity. Nuovo Cimiento **1**, 252–276 (1965)
39. Rácz, I., Wald, R.M.: Extensions of spacetimes with Killing horizons. Class. Quantum Grav. **9**, 2643–2656 (1992)
40. Senovilla, J.M.M.: Singularity theorems and their consequences. Gen. Rel. Grav. **30**, 701–848 (1998)
41. Senovilla, J.M.M.: Classification of spacelike surfaces in spacetime. Class. Quantum Grav. **24**, 3091–3124 (2007)
42. Wald, R.M.: General Relativity. University of Chicago Press, Chicago (1984)
43. Wald, R.M.: Gravitational collapse and cosmic censorship, In: Iyer, B.R., Bhawal, B. (eds.) Black Holes, Gravitational Radiation and the Universe. Fundamental Theories of Physics, vol. 100, pp. 69–85. Kluwer Academic, Dordrecht (1999)
44. Williams, C.: A black hole with no marginally trapped tube asymptotic to its event horizon. Proceedings of the Complex Analysis & Dynamical Systems IV Conference, Nahariya, Israel, May 2009 (arXiv:1005.5401)

Area Inequalities for Stable Marginally Trapped Surfaces

José Luis Jaramillo

1 Introduction

The Lorentzian nature of spacetime geometry, with its inherent notion of null cone, controls the rich features of light bending in general relativity. This includes in particular the possibility of causal disconnection between spacetime regions as well as the convergence behavior of (trapped) light rays. Both aspects, related by the notion of (weak) cosmic censorship [46], lay at the basis of the concept of black hole in general relativity. In spite of the complexity of the generic situation, it is remarkable that stationary and vacuum black hole spacetimes are completely characterized by a few parameters with physical interpretation, namely the total mass M, the angular momentum J, and certain charges Q_i. This is referred to as the *no hair* property of stationary vacuum black holes, leading to the Kerr–Newman family of spacetimes. The parameters M, J, and Q_i in this family of solutions fulfill a class of geometric inequalities that bound the mass from below. Such inequalities can also be expressed in terms of the area A of the horizon (instead of M) presenting the general form

$$(A/(4\pi))^2 \geq (2J)^2 + \left(\sum_i Q_i^2\right)^2. \quad (1)$$

In this form, the inequalities are written completely in terms of the quasi-local geometry of the stationary black hole horizon. As a second remarkable point, these quasi-local geometric inequalities extend to fully generic dynamical and non-vacuum situations, providing general lower bounds for the black hole horizon area.

In the stationary axisymmetric case with matter surrounding the horizon, these quasi-local inequalities have been proved to hold for the Killing horizon in

J.L. Jaramillo (✉)
Max-Planck-Institut für Gravitationsphysik, Albert Einstein Institut, Am Mühlenberg 1, D-14476 Potsdam, Germany
e-mail: Jose-Luis.Jaramillo@aei.mpg.de

M. Sánchez et al. (eds.), *Recent Trends in Lorentzian Geometry*, Springer Proceedings in Mathematics & Statistics 26, DOI 10.1007/978-1-4614-4897-6_5,

[5, 6, 40, 41]. Regarding dynamical situations, and after the study in [21–27] of the global vacuum axisymmetric inequalities involving M, the quasi-local vacuum case has been studied in [1, 28, 31, 33] using axisymmetric initial data. Finally, in [30, 34, 35, 43, 51] a purely spacetime (Lorentzian) perspective has been adopted, permitting to identify and refine the key assumptions, this leading to the extension of the inequalities to generic dynamical scenarios with matter. The study of these inequalities in higher dimensions has been started in [42]. A general review on geometric inequalities in axially symmetric black holes is presented in [29].

Here, we discuss these quasi-local inequalities, placing the emphasis on the involved Lorentzian aspects, namely, the notion of stability of marginally outer trapped surfaces. The latter provides a Lorentzian analogue to the notion of stable minimal surfaces in Riemannian geometry. New results are presented regarding the incorporation in the inequalities of Yang–Mills charges and a geometric charge for certain divergence-free vectors on closed surfaces. We also comment on the interpretation of the integral characterization of the stability condition as an *energy flux* inequality.

2 Geometric and Physical Elements

Let (M, g_{ab}) be a four-dimensional Lorentzian manifold satisfying Einstein's equations

$$G_{ab} + \Lambda g_{ab} = 8\pi T_{ab}, \tag{2}$$

where $G_{ab} = R_{ab} - \frac{1}{2}R g_{ab}$ is the Einstein tensor associated with the Levi–Civita connection ∇_a, Λ is the cosmological constant, and T_{ab} the stress-energy tensor. Unless otherwise stated, in the following, the stress-energy tensor is assumed to satisfy a dominant energy condition (namely given a future-oriented causal vector v^a, then $-T^a{}_b v^b$ is a future-oriented causal vector) and the cosmological constant to be nonnegative $\Lambda \geq 0$. We use throughout Penrose's abstract index notation (e.g., [53]).

2.1 Geometry of 2-Surfaces

Let us consider a closed space-like orientable 2-surface $\mathscr{S}$ embedded in (M, g_{ab}) (in the following, we shall assume that surfaces $\mathscr{S}$ are closed, space-like, and orientable, unless otherwise stated). Let us denote the induced metric on $\mathscr{S}$ by q_{ab}, with Levi–Civita connection D_a, Ricci scalar ${}^2R$ and volume element ε_{ab} (we will denote by dS the area measure on $\mathscr{S}$). Decomposing the tangent plane T_pM at each point $p \in \mathscr{S}$ as $T_pM = T_p\mathscr{S} \oplus T_p^{\perp}\mathscr{S}$, let us consider future-oriented null vectors ℓ^a

and k^a spanning the normal plane $T_p^{\perp}\mathscr{S}$ and normalized as $\ell^a k_a = -1$. This leaves a (boost) rescaling freedom $\ell'^a = f\ell^a$, $k'^a = f^{-1}k^a$. We can write

$$q_{ab} = g_{ab} + k_a \ell_b + \ell_a k_b. \tag{3}$$

Regarding the extrinsic curvature elements that we need in our analysis, let us consider the deformation tensors $\Theta^{(\ell)}_{ab}$ and $\Theta^{(k)}_{ab}$ along ℓ^a and k^a, respectively,

$$\Theta^{(\ell)}_{ab} \equiv q^c{}_a q^d{}_b \nabla_c \ell_d, \;\; \Theta^{(k)}_{ab} \equiv q^c{}_a q^d{}_b \nabla_c k_d. \tag{4}$$

They determine the second fundamental form $\mathscr{K}^c_{ab}$ of $(\mathscr{S}, q_{ab})$ into (M, g_{ab}), namely, $\mathscr{K}^c_{ab} \equiv q^d{}_a q^e{}_b \nabla_d q^c{}_e = k^c \Theta^{(\ell)}_{ab} + \ell^c \Theta^{(k)}_{ab}$ (cf. J.M.M. Senovilla's contribution in this volume). In particular, the expansion $\theta^{(\ell)}$ and the shear $\sigma^{(\ell)}_{ab}$ associated with the null normal ℓ^a, are given respectively by the trace and traceless parts of $\Theta^{(\ell)}_{ab}$

$$\theta^{(\ell)} \equiv q^{ab}\Theta^{(\ell)}_{ab} = q^{ab}\nabla_a \ell_b, \;\; \sigma^{(\ell)}_{ab} \equiv \Theta^{(\ell)}_{ab} - \frac{1}{2}\theta^{(\ell)} q_{ab} \,. \tag{5}$$

In addition, we consider the normal fundamental form $\Omega^{(\ell)}_a$

$$\Omega^{(\ell)}_a = -k^c q^d{}_a \nabla_d \ell_c, \tag{6}$$

that provides a connection on the normal bundle $T^*_{\perp}\mathscr{S}$. More specifically considering a form $v_a \in T^*_{\perp}\mathscr{S}$, expressed as $v_a = \alpha \ell_a + \beta k_b$, we can write $q^c{}_a \nabla_c v_b = \Theta^{(v)}_{ab} + D^{\perp}_a v_b$, where $\Theta^{(v)}_{ab} = q^c{}_a q^d{}_b \nabla_c v_d = \alpha \Theta^{(\ell)}_{ab} + \beta \Theta^{(k)}_{ab}$ and

$$D^{\perp}_a v_b = D^{\perp}_a(\alpha \ell_b + \beta k_b) = (D_a \alpha + \Omega^{(\ell)}_a \alpha)\ell_b + (D_a \beta - \Omega^{(\ell)}_a \beta) k_b. \tag{7}$$

Transformation rules under a null normal rescaling $\ell'^a = f\ell^a$, $k'^a = f^{-1}k^a$ are

$$\theta^{(\ell')} = f\theta^{(\ell)}, \;\; \sigma^{(\ell')}_{ab} = f\sigma^{(\ell)}_{ab}, \;\; \Omega^{(\ell')}_a = \Omega^{(\ell)}_a + D_a(\ln f). \tag{8}$$

2.1.1 Axisymmetry

The introduction of a canonical angular momentum J on $\mathscr{S}$ requires imposing axisymmetry. In this context, we require the geometry of $\mathscr{S}$ to be axisymmetric with axial Killing vector η^a on $\mathscr{S}$. More specifically we require

$$\mathscr{L}_\eta q_{ab} = 0, \;\; \mathscr{L}_\eta \Omega^{(\ell)}_a = 0, \;\; \mathscr{L}_\eta \ell^a = \mathscr{L}_\eta k^a = 0, \tag{9}$$

where η^a has closed integral curves, vanishes (in the spherical case) exactly at two points on $\mathscr{S}$, and is normalized so that its integral curves have an affine length of 2π. We adopt a tetrad $(\xi^a, \eta^a, \ell^a, k^a)$ on $\mathscr{S}$, where the unit vector ξ^a tangent to $\mathscr{S}$

satisfies $\xi^a\eta_a = \xi^a\ell_a = \xi^a k_a = 0$, $\xi^a\xi_a = 1$. We can then write $q_{ab} = \frac{1}{\eta}\eta_a\eta_b + \xi_a\xi_b$, with $\eta = \eta^a\eta_a$, and

$$\Omega_a^{(\ell)} = \Omega_a^{(\eta)} + \Omega_a^{(\xi)}, \quad \Omega_a^{(\ell)}\Omega^{(\ell)a} = \Omega_a^{(\eta)}\Omega^{(\eta)a} + \Omega_a^{(\xi)}\Omega^{(\xi)a}, \tag{10}$$

with $\Omega_a^{(\eta)} = \eta^b\Omega_b^{(\ell)}\eta_a/\eta$ and $\Omega_a^{(\xi)} = \xi^b\Omega_b^{(\ell)}\xi_a$. We can introduce now a canonical (gravitational) angular momentum as

$$J_{\mathrm{K}} = \frac{1}{8\pi}\int_{\mathscr{S}} \Omega_a^{(\ell)}\eta^a dS, \tag{11}$$

where the divergence-free character of η^a together with the transformations properties of $\Omega_a^{(\ell)}$ in Eq. (8) guarantee the invariance of J under a rescaling of the null normals. This angular momentum on $\mathscr{S}$ coincides with the Komar one, namely $J_{\mathrm{Komar}} = \frac{1}{8\pi}\int_{\mathscr{S}} \nabla_a\eta_b dS^{ab}$ with $dS^{ab} = \frac{1}{2}(k^a\ell^b - \ell^a k^b)dS$, if η^a can be extended as a Killing vector to a spacetime neighborhood of $\mathscr{S}$.

2.2 *Electromagnetic Field*

Let us consider an electromagnetic field on (M, g_{ab}) with field strength (Faraday) tensor F_{ab}. On a local chart we can express F_{ab} in terms of a vector potential A_a as $F_{ab} = \nabla_a A_b - \nabla_b A_a$ (note that A_a is not necessarily globally defined; see the next section). The electromagnetic stress-energy tensor is given by

$$T_{ab}^{\mathrm{EM}} = \frac{1}{4\pi}\left(F_{ac}F_b{}^c - \frac{1}{4}g_{ab}F_{cd}F^{cd}\right). \tag{12}$$

Given $\mathscr{S}$, we denote the electric and magnetic field components normal to $\mathscr{S}$ as

$$E_\perp = F_{ab}\ell^a k^b, \quad B_\perp = {}^*F_{ab}\ell^a k^b, \tag{13}$$

where ${}^*F_{ab}$ is the dual of F_{ab}, namely, ${}^*F_{ab} = \frac{1}{2}\varepsilon_{abcd}F^{cd}$ with ε_{abcd} the volume element of g_{ab}. Electric and magnetic charges can be expressed as (e.g., [10, 17])

$$Q_{\mathrm{E}} = \frac{1}{4\pi}\int_{\mathscr{S}} E_\perp dS, \; Q_{\mathrm{M}} = \frac{1}{4\pi}\int_{\mathscr{S}} B_\perp dS. \tag{14}$$

When discussing the angular momentum in the presence of an electromagnetic field, we add $\mathscr{L}_\eta A_a = 0$ to the axisymmetry requirements Eq. (9). Then, the following canonical notion of total angular momentum can be introduced on $\mathscr{S}$ [8, 19, 29, 50]

$$J = J_{\mathrm{K}} + J_{\mathrm{EM}} = \frac{1}{8\pi}\int_{\mathscr{S}} \Omega_a^{(\ell)}\eta^a dS + \frac{1}{4\pi}\int_S (A_a\eta^a)E_\perp dS. \tag{15}$$

2.2.1 Yang–Mills Fields

Given a Yang–Mills theory with gauge Lie group G, the dynamical fields are given in terms of a 1-form A_a evaluated on the Lie algebra $\mathscr{G}$ of G. More properly, A_a is a connection on a principal $G-$bundle P over the spacetime M. Denoting the generators in $\mathscr{G}$ as T_i and writing the Lie-algebra commutation relations as

$$[T_i, T_j] = C^k_{ij} T_k, \tag{16}$$

the Cartan–Killing quadratic form on $\mathscr{G}$ is given by

$$\mathrm{k}_{ij} = C^k_{il} C^l_{jk}, \tag{17}$$

which is nondegenerate for semisimple Lie algebras. For real compact Lie algebras, k_{ij} is non-degenerate and positive-definite. More generally, for gauge groups G given by the direct product of compact real Lie groups and $U(1)$ factors, a real symmetric positive-definite matrix g_{ij} is constructed with diagonal blocks determined by (see, e.g., [54]): (1) the Cartan–Killing form k_{ij} for the compact semisimple factors, and (2) by a positive constant for the $U(1)$ factors (a basis $\{T_i\}$ of $\mathscr{G}$ such that $\mathrm{g}_{ij} = \delta_{ij}$ is usually employed). Writing the Yang–Mills connection as $\mathrm{A}_a = {A_a}^i T_i$, the Yang–Mills tensor $\mathrm{F}_{ab} = {F_{ab}}^i T_i$ is given by the curvature of A_a, that is $\mathrm{F}_{ab} = (d\mathrm{A})_{ab} + \mathrm{A}_a \wedge \mathrm{A}_b = \left(\nabla_a {A_b}^k - \nabla_b {A_a}^k + C^k_{ij} A^i_a A^j_b\right) T_k$. The Yang–Mills stress-energy tensor can be written as

$$T^{\mathrm{YM}}_{ab} = \frac{1}{4\pi} \mathrm{g}_{ij} \left({F_{ac}}^i {F_b}^{cj} - \frac{1}{4} g_{ab} {F_{cd}}^i F^{cdj} \right). \tag{18}$$

We can define Yang–Mills electric and magnetic charges [11, 20, 52] as

$$Q^{\mathrm{YM}}_{\mathrm{E}} = \frac{1}{4\pi} \int |\mathrm{E}^{\mathrm{YM}}_{\perp}|\, dS, \quad Q^{\mathrm{YM}}_{\mathrm{M}} = \frac{1}{4\pi} \int |\mathrm{B}^{\mathrm{YM}}_{\perp}|\, dS, \tag{19}$$

where

$$|\mathrm{E}^{\mathrm{YM}}_{\perp}| = \left[\left({F_{ab}}^i k^a \ell^b\right) \mathrm{g}_{ij} \left({F_{cd}}^j k^c \ell^d\right)\right]^{\frac{1}{2}}, \; |\mathrm{B}^{\mathrm{YM}}_{\perp}| = \left[\left({{}^*F_{ab}}^i k^a \ell^b\right) \mathrm{g}_{ij} \left({{}^*F_{cd}}^j k^c \ell^d\right)\right]^{\frac{1}{2}}. \tag{20}$$

Electromagnetic theory corresponds to the commutative case $G = U(1)$ [note that the matrix g_{ij} in expressions (18) and (20) reduces then to a factor 1]. In particular, the Yang–Mills principal fiber-bundle perspective sheds light on the topological nature of the magnetic charge Q_{M}, offering an understanding of magnetic monopoles as associated with the nontriviality of the $U(1)$-bundle over M (see, e.g., [49, 55]). Although the potential A_a does not exist globally, it is well defined as a connection on the $U(1)$-bundle, and, in particular, the magnetic charge Q_{M} is controlled by the first Chern class of such nontrivial $U(1)$-bundle.

3 Stability of Marginally Outer Trapped Surfaces

The stability for marginally trapped surfaces is the crucial element in the present discussion of the area inequalities. This notion is extensively reviewed in the contribution by M. Mars in this volume. We discuss the basic elements here needed.

3.1 Basic Definitions

First, we choose conventionally ℓ^a as the *outgoing* null vector at $\mathscr{S}$ and refer to $\mathscr{S}$ as a *marginally outer trapped surface* (MOTS) if $\theta^{(\ell)} = 0$. Note that no condition is required on the ingoing expansion $\theta^{(k)}$. The stability of MOTS is introduced in terms of the deformation operator δ_v on $\mathscr{S}$ that controls the infinitesimal variations of geometric objects defined on $\mathscr{S}$ under an infinitesimal deformation of the surface along a vector v^a on $\mathscr{S}$ (here, v^a will be always normal to $\mathscr{S}$). This operator δ_v, discussed in detail in [3, 4] (see also M. Mars' contribution and [16, 18]), is the analogue in the Lorentzian setting to the deformation operator employed in the discussion of minimal surfaces in Riemannian geometry. We require $\mathscr{S}$ to be *stably outermost* in the sense introduced in [3,4] (see also [37,48]). More specifically, we require $\mathscr{S}$ to be *spacetime stably outermost* [30, 43]:

Definition 1. A closed orientable marginally outer trapped surface $\mathscr{S}$ is referred to as *spacetime stably outermost* if there exists an outgoing ($-k^a$-oriented) vector $X^a = \gamma\ell^a - \psi k^a$, with $\gamma \geq 0$ and $\psi > 0$, with respect to which $\mathscr{S}$ is stably outermost:

$$\delta_X \theta^{(\ell)} \geq 0. \tag{21}$$

If, in addition, X^a (i.e., γ, ψ) and $\Omega_a^{(\ell)}$ are axisymmetric, we will refer to $\delta_X \theta^{(\ell)} \geq 0$ as an (axisymmetry-compatible) spacetime stably outermost condition.

Alternatively, one could introduce the notion of stability for MOTS in terms of the non-negativity of the principal eigenvalue λ_v of the stability operator L_v associated with δ_v, namely, $L_v \psi = \delta_{\psi v}\theta^{(\ell)}$. Although L_v is not self-adjoint in the generic case, its principal eigenvalue (i.e., the eigenvalue with smallest real part) is indeed real. Then, the characterization Eq. (21) of a stably outermost MOTS is proved as a lemma [3, 4]. This is the strategy followed in the contribution by M. Mars in this volume.

Finally, note that the characterization of MOTSs as spacetime stably outermost is independent of the choice of future-oriented null normals ℓ^a and k^a. Indeed, given $f > 0$, for $\ell'^a = f\ell^a$ and $k'^a = f^{-1}k^a$ we can write $X^a = \gamma\ell^a - \psi k^a = \gamma'\ell'^a - \psi' k'^a$ (with $\gamma' = f^{-1}\gamma \geq 0$ and $\psi' = f\psi > 0$), and it holds $\delta_X \theta^{(\ell')} = f \cdot \delta_X \theta^{(\ell)} \geq 0$.

3.2 Integral-Inequality Characterizations of MOTS Stability

The first step in the proofs of area inequalities (1) consists in casting condition (21) as an integral geometric inequality over $\mathcal{S}$. Condition (21) plays, for MOTS in a Lorentzian context, a role analogous to that of the stability condition for minimal surfaces in Riemannian geometry. In the Riemannian case, this refers to the minimization of the area of $\mathcal{S}$ with respect to arbitrary deformations along αs^a, where s^a is the normal to $\mathcal{S}$ in a given 3-slice and α is an arbitrary function on $\mathcal{S}$. In contrast, the stability condition in Definition 1 only states the existence of a positive function ψ (and, secondarily, of $\gamma \geq 0$). The proof of area inequalities involving the angular momentum requires writing Eq. (21) as an integral inequality in terms of arbitrary (axisymmetric) functions α. The following lemma [43] provides this.[1]

Lemma 1. *Given a closed orientable marginally outer trapped surface $\mathcal{S}$ satisfying the spacetime stably outermost condition for an axisymmetric X^a, then for all axisymmetric functions α on $\mathcal{S}$*

$$\int_{\mathcal{S}} \left[D_a\alpha D^a\alpha + \frac{1}{2}\alpha^2\, {}^2\!R \right] dS \geq$$

$$\int_{\mathcal{S}} \left[\alpha^2 \Omega_a^{(\eta)} \Omega^{(\eta)a} + \alpha\beta\sigma_{ab}^{(\ell)}\sigma^{(\ell)ab} + G_{ab}\alpha\ell^a(\alpha k^b + \beta\ell^b) \right] dS, \qquad (22)$$

where $\beta = \alpha\gamma/\psi$. If in addition we assume that the right hand side in the inequality (22) is not identically zero, then $\mathcal{S}$ has a S^2 topology, under Einstein's equations.

Proof. We basically follow the discussion in Sect. 3.3. of [2] (cf. Theorem 2.1 in [36] for a similar reasoning, essentially reducing a non time-symmetric case to a time-symmetric one). First, we evaluate $\delta_X\theta^{(\ell)}/\psi$ for $X^a = \gamma\ell^a - \psi k^a$ in Definition 1, with axisymmetric γ and ψ. For this we use (e.g., Eqs. (2.23) and (2.24) in [16])

$$\begin{aligned}
\delta_{\alpha\ell}\theta^{(\ell)} &= \kappa^{(\alpha\ell)}\theta^{(\ell)} - \alpha\left[\sigma_{ab}^{(\ell)}\sigma^{(\ell)ab} + G_{ab}\ell^a k^b + \frac{1}{2}\left(\theta^{(\ell)}\right)^2\right], \\
\delta_{\beta k}\theta^{(\ell)} &= \kappa^{(\beta k)}\theta^{(\ell)} + {}^2\!\Delta\beta - 2\Omega_a^{(\ell)}D^a\beta \\
&\quad + \beta\left[\Omega_a^{(\ell)}\Omega^{(\ell)a} - D^a\Omega_a^{(\ell)} - \frac{1}{2}{}^2\!R + G_{ab}k^a\ell^b - \theta^{(\ell)}\theta^{(k)}\right],
\end{aligned}$$

[1] Alternatively, one could start characterizing MOTS stability in terms of the principal eigenvalue λ_X. Then, the expression of λ_X in a Rayleigh–Ritz type characterization [4] leads essentially to the integral inequality. See M. Mars' contribution, where the role of α is played by a function u.

where $\kappa^{(v)} = -v^a k^b \nabla_a \ell_b$. Imposing $\theta^{(\ell)} = 0$, we can write for $X^a = \gamma \ell^a - \psi k^a$

$$\begin{aligned}\frac{1}{\psi}\delta_X \theta^{(\ell)} &= -\frac{\gamma}{\psi}\left[\sigma^{(\ell)}_{ab}\sigma^{(\ell)ab} + G_{ab}\ell^a\ell^b\right]\\ &\quad -{}^2\Delta \ln\psi - D_a \ln\psi D^a \ln\psi + 2\Omega^{(\ell)}_a D^a \ln\psi\\ &\quad -\left[-D^a\Omega^{(\ell)}_a + \Omega^{(\ell)}_c \Omega^{(\ell)c} - \frac{1}{2}{}^2R + G_{ab}k^a\ell^b\right]. \end{aligned} \tag{23}$$

We multiply by α^2 for arbitrary (axisymmetric) α and integrate on $\mathscr{S}$. Using $\int_{\mathscr{S}} \frac{\alpha^2}{\psi}\delta_X \theta^{(\ell)} dS \geq 0$ and integrating by parts, we can write

$$\begin{aligned} 0 \leq & \int_{\mathscr{S}} \alpha\beta\left[-\sigma^{(\ell)}_{ab}\sigma^{(\ell)ab} - G_{ab}\ell^a\ell^b\right] dS\\ &+ \int_{\mathscr{S}} \alpha^2\left[-\Omega^{(\ell)}_a\Omega^{(\ell)a} + \frac{1}{2}{}^2R - G_{ab}k^a\ell^b\right] dS\\ &+ \int_{\mathscr{S}} \left[2\alpha D_a\alpha D^a\ln\psi - \alpha^2 D_a\ln\psi D^a\ln\psi\right] dS\\ &+ \int_{\mathscr{S}} \left[2\alpha^2\Omega^{(\ell)}_a D^a\ln\psi - 2\alpha\Omega^{(\ell)}_a D^a\alpha\right] dS. \end{aligned} \tag{24}$$

From the axisymmetry of α and ψ, $\Omega^{(\eta)^a} D_a\alpha = \Omega^{(\eta)^a} D_a\psi = 0$, and using Eq. (10)

$$\begin{aligned} 0 \leq & \int_{\mathscr{S}} \alpha\beta\left[-\sigma^{(\ell)}_{ab}\sigma^{(\ell)ab} - G_{ab}\ell^a\ell^b\right] dS\\ &+ \int_{\mathscr{S}} \alpha^2\left[-\Omega^{(\eta)}_a\Omega^{(\eta)^a} + \frac{1}{2}{}^2R - G_{ab}k^a\ell^b\right] dS\\ &+ \int_{\mathscr{S}} \Big[2(D^a\alpha)(\alpha D_a\ln\psi - \alpha\Omega^{(\xi)}_a)\\ &-(\alpha D_a\ln\psi - \alpha\Omega^{(\xi)}_a)(\alpha D^a\ln\psi - \alpha\Omega^{(\xi)^a})\Big] dS. \end{aligned} \tag{25}$$

Making use of the following Cauchy–Schwarz inequality in the last integral

$$D^a\alpha D_a\alpha \geq 2D^a\alpha(\alpha D_a\ln\psi - \alpha\Omega^{(\xi)}_a) - |\alpha D\ln\psi - \alpha\Omega^{(\xi)}|^2, \tag{26}$$

inequality (22) follows for all axisymmetric α. Finally, if the right-hand side of Eq. (22) does not vanish, then it is positive-definite. In particular, for the third term this follows by imposing Einstein's equations and from the assumption of non-negative cosmological constant $\Lambda \geq 0$ and the dominant energy condition (note that $\alpha k^b + \beta \ell^b$ is a non-space-like vector). The sphericity of $\mathscr{S}$ follows by considering a constant α in Eq. (22): it implies a positive value for the Euler characteristic of $\mathscr{S}$.

□

The proof of the area-charge inequality, resulting from dropping the angular momentum J in Eq. (1), requires neither a symmetry assumption nor casting Eq. (21) in terms of an arbitrary α. We use the following lemma (slightly generalizing that in [30]).

Lemma 2. *Given a closed orientable marginally outer trapped surface $\mathscr{S}$ satisfying the spacetime stably outermost condition, then the following inequality holds*

$$\int_{\mathscr{S}} \left[G_{ab}\ell^a k^b + N\left(\sigma^{(\ell)}_{ab}\sigma^{(\ell)ab} + G_{ab}\ell^a\ell^b\right)\right] dS \leq 4\pi(1-g), \tag{27}$$

where g is the genus of $\mathscr{S}$ and $N = \frac{\chi}{\psi} \geq 0$. If in addition we assume that the left hand side in the inequality (27) is non-negative and not identically zero, then it follows that $g = 0$, and hence, $\mathscr{S}$ has the S^2 topology (under Einstein's equations).

Proof. The proof is slightly simpler than that of Lemma 1. We integrate directly expression (23) over $\mathscr{S}$. On the left-hand side, we use the stability condition (21). Divergence terms in the right-hand side integrate to zero and we rearrange terms as

$$-(D_a \ln\psi - \Omega^{(\ell)}_a)(D^a \ln\psi - \Omega^{(\ell)a}) = -D_a \ln\psi D^a \ln\psi + 2\Omega^{(\ell)}_a D^a \ln\psi - \Omega^{(\ell)}_c \Omega^{(\ell)c}, \tag{28}$$

so that the integral is nonpositive. From the Gauss–Bonnet theorem, we write

$$\int_{\mathscr{S}} \frac{1}{2}{}^2R dS = 4\pi(1-g). \tag{29}$$

Collecting these observations, the inequality (22) follows. If the left-hand side of the inequality (22) is nonnegative, it follows that g can be 0 or 1. If it is not identically zero, then $g = 0$, and hence, $\mathscr{S}$ has the S^2 topology (see [36] for a generalization, making use of similar techniques to those employed in Lemma 1, of the topological restrictions holding in higher dimensions for outer apparent horizons: the latter must be of Yamabe type, namely, they admit metrics of positive scalar curvature; this generalizes Hawking's result on the topology of black hole horizons). □

3.3 Variants to the Stably Outermost Condition

3.3.1 On an *Averaged* Outermost Stably Conditions for MOTS

Inequalities (22) and (27) do not require a *point-like* stability condition. We could consider an (in principle weaker) *averaged* stability condition for MOTS.

Definition 2. Given a closed orientable MOTS $\mathscr{S}$ we will refer to it as (dipole) *weight-averaged stably outermost* if there exists an outgoing ($-k^a$-oriented) vector

$x^a = \gamma\ell^a - k^a$, with $\gamma \geq 0$ such that, for all functions α on $\mathscr{S}$, the variations of $\theta^{(\ell)}$ with respect to $X^a = \alpha x^a$ fulfill the integral condition

$$\int_{\mathscr{S}} (X^a \ell_a) \delta_X \theta^{(\ell)} dS \geq 0. \tag{30}$$

Proofs of inequalities involving the angular momentum (cf. Sects. 4 and 6) could start from this averaged condition. Note that $(X^a \ell_a) = \alpha = \text{const}$ provides an *averaged stably outermost* condition, the element needed in proving area-charge inequalities (cf. Sect. 5). Finally, a (2n-moment) *weight-averaged stably outermost* condition could be introduced as $\int_{\mathscr{S}} (X^a \ell_a)^n \, \delta_X \theta^{(\ell)} dS \geq 0$, for integers n.

3.3.2 Towards Axisymmetry Relaxation

Let η^a be a divergence-free vector on $\mathscr{S}$, with squared-norm $\eta = \eta^a \eta_a$ constant along itself, i.e. $\eta^a D_a \eta = 0$ (fulfilled, in particular, by Killing vectors). As in Sect. 2.1.1, we write $q_{ab} = \frac{1}{\eta} \eta_a \eta_b + \xi_a \xi_b$, with $\xi^a \eta_a = \xi^a \ell_a = \xi^a k_a = 0$, $\xi^a \xi_a = 1$, and $\Omega_a^{(\eta)} = \eta^b \Omega_b^{(\ell)} \eta_a / \eta$ and $\Omega_a^{(\xi)} = \xi^b \Omega_b^{(\ell)} \xi_a$, so relations (10) hold. The geometric quantity

$$Q[\eta] = \frac{1}{4\pi} \int_{\mathscr{S}} \frac{1}{\sqrt{\eta}} \Omega_a^{(\ell)} \eta^a dS, \tag{31}$$

is well defined on $\mathscr{S}$ in the sense that: (1) it does not depend on the normalization of the null normal ℓ^a and (2) there is no normalization ambiguity related to η^a. The first point follows from the transformation properties (8) of $\Omega_a^{(\ell)}$ under $\ell'^a = f\ell^a$, together with the divergence-free character of $\hat{\eta}^a = \eta^a / \sqrt{\eta}$, i.e. $D_a(\hat{\eta}^a) = \frac{1}{\sqrt{\eta}} D_a \eta^a - \frac{1}{2\eta\sqrt{\eta}} \eta^a D_a \eta = 0$. Regarding the second point, $Q[\eta]$ is defined in terms of $\hat{\eta}^a$, with $\hat{\eta}^a \hat{\eta}_a = 1$. This is the analogue of the 2π-orbit normalization for axial Killing vectors in expressions (11) and (15) for the angular momentum (note that η^a need not be axial). We can then adapt the MOTS stability condition.

Definition 3. Given a closed orientable marginally outer trapped surface $\mathscr{S}$ and a divergence-free vector η^a on it, $\mathscr{S}$ is said to be η^a-compatible spacetime stably outermost if there exists an outgoing ($-k^a$-oriented) vector $X^a = -\psi k^a$, with $\psi > 0$ and $\eta^a D_a \psi = 0$, such that the variation of $\theta^{(\ell)}$ with respect to X^a fulfills the condition $\delta_X \theta^{(\ell)} \geq 0$.

The following lemma holds.

Lemma 3. *Given a closed orientable MOTS $\mathscr{S}$ satisfying the η^a-compatible spacetime stably outermost condition for X^a, then for all α such that $\eta^a D_a \alpha = 0$, it holds*

$$\int_{\mathscr{S}} \left[D_a \alpha D^a \alpha + \frac{1}{2} \alpha^2 \, {}^2\!R \right] dS \geq \int_{\mathscr{S}} \left[\alpha^2 \left(\Omega_a^{(\eta)} \Omega^{(\eta)a} + G_{ab} \ell^a k^b \right) \right] dS, \tag{32}$$

Proof. The proof proceeds exactly as in Lemma 1. It is straightforward to generalize it for $X^a = \gamma\ell^a - \psi k^a$, $\gamma \geq 0$, so that the shear and the $G_{ab}\ell^a\ell^b$ terms are incorporated. □

4 The Area-Angular Momentum Inequality

We first state the main result in this section (see [43] for further details).

Theorem 1 (cf. [43]). *Given an axisymmetric closed orientable marginally outer trapped surface $\mathscr{S}$ satisfying the (axisymmetry-compatible) spacetime stably outermost condition, in a spacetime with nonnegative cosmological constant and fulfilling the dominant energy condition, it holds the inequality*

$$A \geq 8\pi|J|, \tag{33}$$

where A and J are the area and gravitational (Komar) angular momentum of $\mathscr{S}$. If equality holds, then $\mathscr{S}$ has the geometry of an extreme Kerr throat sphere and, in addition, if the vector X^a in the stability condition can be found to be space-like then $\mathscr{S}$ is a section of a non-expanding horizon.

The proof of the area-angular momentum inequality (33) has two parts. The first one is purely geometric and provides the lower bound on the area A

$$A \geq 4\pi e^{\frac{\mathscr{M}-8}{8}}, \tag{34}$$

where $\mathscr{M}$ is a functional on the sphere geometry. The second part solves a variational problem, subject to the constraint of keeping constant the a priori given angular momentum J. In particular, it is shown [1,31,33] the existence of a minimum

$$\mathscr{M} \geq \mathscr{M}_0 \ , \qquad \text{under the constraint: } J \text{ fixed}, \tag{35}$$

such that the evaluation of Eq. (34) on $\mathscr{M}_0$ leads to inequality (33). Moreover, the minimizer is unique, this leading to a rigidity result. We focus here on the first geometric part and refer the reader to the proper references on the variational part [1,31,33].

Proof. First, we consider an axisymmetric stably outermost MOTS and apply the result in Lemma 1. We consider inequality (22) where we disregard the positive-definite gravitational radiation shear squared term and, under Einstein's equations, also the cosmological constant and matter terms (again assuming a nonnegative cosmological constant $\Lambda \geq 0$ and the dominant energy condition). Therefore

$$\int_{\mathscr{S}} \left[D_a\alpha D^a\alpha + \frac{1}{2}\alpha^2\, {}^2\!R \right] dS \geq \int_{\mathscr{S}} \alpha^2 \Omega_a^{(\eta)} \Omega^{(\eta)a} dS. \tag{36}$$

Second, we express this inequality in terms of certain potentials for the geometry of $\mathscr{S}$. Assuming a nonvanishing right-hand side in Eq. (36) (otherwise Eq. (34) is trivial), $\mathscr{S}$ has a spherical topology. On an axisymmetric sphere we can always write [9]

$$ds^2 = q_{ab}dx^a dx^b = e^{\sigma}\left(e^{2q}d\theta^2 + \sin^2\theta d\varphi^2\right), \tag{37}$$

with σ and q functions on θ satisfying $\sigma + q = c$, where c is a constant. Then, $dS = e^c dS_0$, with $dS_0 = \sin\theta d\theta d\varphi$. In addition, the squared norm η of the axial Killing vector $\eta^a = (\partial_\varphi)^a$ is given by

$$\eta = e^{\sigma}\sin^2\theta. \tag{38}$$

Choosing $\alpha = e^{c-\sigma/2}$ [31], the evaluation of the left-hand side in Eq. (36) results in

$$\int_{\mathscr{S}}\left[D_a\alpha D^a\alpha + \frac{1}{2}\alpha^2\,{}^2\!R\right]dS = e^c\left[4\pi(c+1) - \int_{\mathscr{S}}\left(\sigma + \frac{1}{4}\left(\frac{d\sigma}{d\theta}\right)^2\right)dS_0\right]. \tag{39}$$

To evaluate the right-hand side in Eq. (36) we note that, due to the S^2 topology of $\mathscr{S}$, we can always express $\Omega^{(\ell)}_a$ in terms of a divergence-free and an exact form

$$\Omega^{(\ell)}_a = \varepsilon_{ab}D^b\tilde{\omega} + D_a\lambda, \tag{40}$$

with $\tilde{\omega}$ and λ fixed up to a constant. From the axisymmetry of q_{ab} and $\Omega^{(\ell)}_a$ (functions $\tilde{\omega}$ and λ are then axially symmetric), it follows $\Omega^{(\eta)}_a = \varepsilon_{ab}D^b\tilde{\omega}$ and $\Omega^{(\xi)}_a = D_a\lambda$. Before proceeding further, we evaluate the angular momentum J. Writing $\eta^a\Omega^{(\ell)}_a = \eta^a\Omega^{(\eta)}_a = \varepsilon_{ab}\eta^a D^b\tilde{\omega}$ and expressing ξ^a as $\xi_b = \eta^{-1/2}\varepsilon_{ab}\eta^a$, we have

$$\Omega^{(\ell)}_a\eta^a = \eta^{1/2}\xi^a D_a\tilde{\omega}. \tag{41}$$

Plugging this into Eq. (11) [or Eq. (15), since $J_{\mathrm{EM}} = 0$] and using Eq. (37) we find

$$J = \frac{1}{8}\int_0^{\pi} 2\eta\frac{d\tilde{\omega}}{d\theta} = \frac{1}{8}\int_0^{\pi}\frac{d\bar{\omega}}{d\theta} = \frac{1}{8}[\bar{\omega}(\pi) - \bar{\omega}(0)], \tag{42}$$

where we have introduced the potential $\bar{\omega}$ as $d\bar{\omega}/d\theta \equiv (2\eta)d\tilde{\omega}/d\theta$. The use of $\bar{\omega}$, rather than $\tilde{\omega}$, permits to control directly the angular momentum in terms of the values of $\bar{\omega}$ at the axis. This is crucial to implement the constraint $J = \mathrm{const}$ in the variational problem. We use $\bar{\omega}$ in the following, rather than $\tilde{\omega}$. Further geometric intuition is gained by noting that, if the axial vector η^a on $\mathscr{S}$ extends to a spacetime neighborhood of $\mathscr{S}$ (something not needed in the present discussion), we can define the *twist* vector of η^a as $\omega_a = \varepsilon_{abcd}\eta^b\nabla^c\eta^d$ and the relation $\xi^a\omega_a = \xi^a D_a\bar{\omega}$ holds on $\mathscr{S}$. In the vacuum case, a spacetime twist potential ω satisfying $\omega_a = \nabla_a\omega$ can be defined, so that $\bar{\omega}$ and ω coincide on $\mathscr{S}$ up to a constant. Note however that $\bar{\omega}$ can be defined always on $\mathscr{S}$, even in the presence of matter.

From Eqs. (40) and (37) and the adopted choice for α, we have

$$\alpha^2 \Omega_a^{(\eta)} \Omega^{(\eta)a} = \frac{\alpha^2}{4\eta^2} D_a \bar{\omega} D^a \bar{\omega} = \frac{1}{4\eta^2} \left(\frac{d\bar{\omega}}{d\theta} \right)^2. \tag{43}$$

Plugging this into Eq. (36) and using Eq. (39) we get

$$8(c+1) \geq \mathscr{M}[\sigma, \bar{\omega}], \tag{44}$$

with

$$\mathscr{M}[\sigma, \bar{\omega}] = \frac{1}{2\pi} \int_{\mathscr{S}} \left[\left(\frac{d\sigma}{d\theta} \right)^2 + 4\sigma + \frac{1}{\eta^2} \left(\frac{d\bar{\omega}}{d\theta} \right)^2 \right] dS_0. \tag{45}$$

Using these expressions and $A = 4\pi e^c$ leads to inequality (34). This completes the first stage in the proof. In a second stage, by solving the variational problem defined by $\mathscr{M}[\sigma, \bar{\omega}]$ with J constant as a constraint, one can prove [1, 33]

$$\mathscr{M} \geq \mathscr{M}_0 = 8\ln(2|J|) + 8. \tag{46}$$

This, namely $e^{(\mathscr{M}-8)/8} \geq 2|J|$, together with Eq. (34) leads to the area-angular momentum inequality (33). Actually, the only minimizer for $\mathscr{M}_0$ in Eq. (46) is a section of the horizon of extremal Kerr, this leading to a rigidity result [31, 43]: if equality in Eq. (33) holds, first, the intrinsic geometry of $\mathscr{S}$ is that of an extreme Kerr throat sphere [28] and, second, the vanishing of the positive-definite terms in Eq. (22) implies, for space-like X^a in Eq. (21), the vanishing of the shear $\sigma_{ab}^{(\ell)}$ so that $\mathscr{S}$ is an *instantaneous* (non-expanding) isolated horizon [15]. □

5 The Area-Charge Inequality Generalizations

Remarkably, if we drop the angular momentum from the inequality (1), the resulting area-charge inequality requires neither the use of a variational principle nor the assumption of any symmetry.

Theorem 2 (cf. [30]). *Given a closed orientable marginally outer trapped surface $\mathscr{S}$ satisfying the spacetime stably outermost condition, in a spacetime satisfying Einstein's equations with non-negative cosmological constant Λ and non-electromagnetic matter fields T_{ab} satisfying the dominant energy condition, then*

$$A \geq 4\pi \left(Q_{\mathrm{E}}^2 + Q_{\mathrm{M}}^2 \right), \tag{47}$$

where A, Q_{E} and Q_{M} are the area, electric and magnetic charges of $\mathscr{S}$.

Proof. We start from Lemma 2 and use inequality (27) and Einstein's equation (2). Since the vector $k^a + \gamma/\psi \ell^a$ is time-like or null, using that the tensor T_{ab} satisfies the dominant energy condition (and in particular the null energy condition), that

Λ is non-negative and the term proportional to N is positive-definite, we get from Eq. (27)

$$8\pi \int_{\mathscr{S}} T^{\rm EM}_{ab} \ell^a k^b dS \leq 4\pi(1-g). \tag{48}$$

The term $T^{\rm EM}_{ab} \ell^a k^b$ can be written as

$$T^{\rm EM}_{ab} \ell^a k^b = \frac{1}{8\pi} \left[\left(\ell^a k^b F_{ab} \right)^2 + \left(\ell^a k^{b*} F_{ab} \right)^2 \right]. \tag{49}$$

This result is purely algebraic, something crucial for the later generalization to Yang–Mills fields. To derive Eq. (49), we use the decomposition Eq. (3) for g_{ab} and calculate

$$F_{ab}F^{ab} = -2 \left(\ell^a k^b F_{ab} \right)^2 - 4q^{ab} k^c F_{ac} \ell^d F_{bd} + F_{ab} F_{cd} q^{ac} q^{bd}, \tag{50}$$

and

$$\ell^a k^c F_{ab} F_c{}^b = \left(\ell^a k^b F_{ab} \right)^2 + q^{ab} k^c F_{ac} \ell^d F_{bd}. \tag{51}$$

Noting that the pullback of F_{ab} on the surface $\mathscr{S}$ is proportional to the volume element ε_{ab} of the surface $\mathscr{S}$, we can evaluate $F_{ab}F_{cd}q^{ac}q^{bd}$ and $(\varepsilon^{ab}F_{ab})^2$ to obtain

$$F_{ab}F_{cd}q^{ac}q^{bd} = \frac{1}{2} \left(\varepsilon^{ab} F_{ab} \right)^2 = 2 \left({}^*F_{ab} \ell^a k^b \right)^2, \tag{52}$$

where the identity ${}^*F_{ab}\ell^a k^b = \frac{1}{2} F_{ab}\varepsilon^{ab}$ follows from the relation $\varepsilon_{ab} = \varepsilon_{abcd}\ell^c k^d$. Inserting these expressions into Eq. (12), we obtain Eq. (49). Then, using relation (49) into inequality (48) we get

$$\int_{\mathscr{S}} \left[\left(\ell^a k^b F_{ab} \right)^2 + \left(\ell^a k^{b*} F_{ab} \right)^2 \right] dS \leq 4\pi(1-g). \tag{53}$$

If the left-hand side is identically zero then the charges are zero and the inequality (47) is trivial. We assume that it is not zero at some point, and hence, we have $g = 0$. To bound the left-hand side of inequality (53), we use Hölder inequality on $\mathscr{S}$. For integrable functions f and h, Hölder inequality is given by

$$\int_{\mathscr{S}} fh dS \leq \left(\int_{\mathscr{S}} f^2 dS \right)^{1/2} \left(\int_{\mathscr{S}} h^2 dS \right)^{1/2}. \tag{54}$$

If we take $h = 1$, then we obtain

$$\int_{\mathscr{S}} f dS \leq \left(\int_{\mathscr{S}} f^2 dS \right)^{1/2} A^{1/2}, \tag{55}$$

where A is the area of $\mathscr{S}$. Using this inequality in Eq. (53), we obtain

$$A^{-1}\left[\left(\int_{\mathscr{S}}\ell^a k^b F_{ab} dS\right)^2+\left(\int_{\mathscr{S}}\ell^a k^{b*}F_{ab} dS\right)^2\right]\leq 4\pi. \tag{56}$$

Finally, we use the expression of the charges (14) to express the left-hand-side of Eq. (56) in terms of $Q_{\rm E}$ and $Q_{\rm M}$. Hence, the inequality (47) follows. □

5.1 Yang–Mills Charges

The derivation of the area-charge inequality (47) does not involve Maxwell equations; only the algebraic form of the electromagnetic stress-energy tensor Eq. (12) is used. Given the similar structure of the Yang–Mills stress-energy tensor Eq. (18), the result generalizes to include Yang–Mills charges Eq. (19), for compact Lie groups.

Corollary 1. *Under the conditions of Theorem 2, for a Yang–Mills theory with compact simple Lie group G (more generally with G given by a direct product of compact simple Lie groups and $U(1)$ factors), it holds the inequality*

$$A\geq 4\pi\left[\left(Q_{\rm E}^{\rm YM}\right)^2+\left(Q_{\rm M}^{\rm YM}\right)^2\right]. \tag{57}$$

Proof. Proceeding exactly as in Theorem 2 and writing

$$T_{ab}^{\rm YM}\ell^a k^b=\frac{1}{8\pi}g_{ij}\left[\left(\ell^a k^b F_{ab}{}^i\right)\left(\ell^c k^d F_{cd}{}^j\right)+\left(\ell^a k^{b*}F_{ab}{}^i\right)\left(\ell^c k^{c*}F_{cd}{}^j\right)\right], \tag{58}$$

we derive the analogue of inequality (53)

$$4\pi\geq\int_{\mathscr{S}}\left[\left(\ell^a k^b F_{ab}{}^i\right)g_{ij}\left(\ell^c k^d F_{cd}{}^j\right)+\left(\ell^a k^{b*}F_{ab}{}^i\right)g_{ij}\left(\ell^c k^{c*}F_{cd}{}^j\right)\right]dS. \tag{59}$$

In this case, we can write the form (55) of Hölder inequality as

$$\int_{\mathscr{S}}g_{ij}V^iV^j\geq\frac{1}{A}\left(\int_{\mathscr{S}}\left(g_{ij}V^iV^j\right)^{\frac{1}{2}}dS\right)^2, \tag{60}$$

for compact Lie algebras, for which g_{ij} is positive-definite [just take $f^2=g_{ij}V^iV^j\geq 0$ in Eq. (55)]. Using inequality (60) in Eq. (59) leads to inequality (57). □

5.2 Further Generalizations

The area–charge inequality can be extended to incorporate the quantity $Q[\eta]$ in Eq. (31).

Corollary 2. *Under the conditions of Lemma 3, the following inequality holds*

$$A \geq 4\pi Q[\eta]^2. \tag{61}$$

Proof. Starting from inequality (32), choose $\alpha = 1$ and drop the electromagnetic or Yang–Mills components (it is straightforward to include them). Then we can write

$$4\pi \geq \int_{\mathscr{S}} \Omega_a^{(\eta)} \Omega^{(\eta)a} dS = \int_{\mathscr{S}} \frac{1}{\eta} \left(\Omega_a^{(\ell)} \eta^a\right)^2 dS = \int_{\mathscr{S}} \left(\frac{1}{\sqrt{\eta}} \Omega_a^{(\ell)} \eta^a\right)^2 dS. \tag{62}$$

Using again inequality (55), now with $f = \frac{1}{\sqrt{\eta}} \Omega_a^{(\ell)} \eta^a$, we obtain

$$\int_{\mathscr{S}} \left(\frac{1}{\sqrt{\eta}} \Omega_a^{(\ell)} \eta^a\right)^2 dS \geq \frac{1}{A} \left(\int_{\mathscr{S}} \frac{1}{\sqrt{\eta}} \Omega_a^{(\ell)} \eta^a dS\right)^2, \tag{63}$$

from which inequality (61) follows when using expression (31) for $Q[\eta]$. □

Two remarks are in order. First, inequality (61) does not reduce to the area-angular momentum inequality (33), even if η^a is an axial Killing vector. Even in this case, the quantity $Q[\eta]$ is not an angular momentum due to the $1/\sqrt{\eta}$ factor (this is easily seen on dimensional grounds). However, whenever existing, $Q[\eta]$ is a geometric quantity providing a non-trivial lower bound for the area. Second, the area-charge geometric inequalities (47), (57) and (61) can be collected in the more general form

$$A \geq 4\pi \left[Q_{\mathrm{E}}^2 + Q_{\mathrm{M}}^2 + \left(Q_{\mathrm{E}}^{\mathrm{YM}}\right)^2 + \left(Q_{\mathrm{M}}^{\mathrm{YM}}\right)^2 + Q[\eta]^2\right], \tag{64}$$

assuming that the individual terms make sense.

5.2.1 The Cosmological Constant and Stability Operator Eigenvalue

The area–charge inequality has been extended in [51] to include the cosmological constant Λ and the principal eigenvalue λ of the stability operator associated with the deformation operator δ (λ is a real number [3,4]). The inequality reads

$$\Lambda^* A^2 - 4\pi(1-g)A + (4\pi)^2 \sum_i Q_i^2 \leq 0, \tag{65}$$

where $\Lambda^* \equiv \Lambda + \lambda$ and Q_i correspond to Q_{E}, Q_{M}, $Q_{\mathrm{E}}^{\mathrm{YM}}$, $Q_{\mathrm{M}}^{\mathrm{YM}}$ and $Q[\eta]$. The previous inequality (64) follows from the stability condition $\Lambda^* > 0$ and $g = 0$. We highlight the remarkable fact that the cosmological constant and the principal eigenvalue enter formally in exactly the same manner. This suggests the possibility of linking global and quasi-local notions of stability in black hole spacetime geometries.

5.2.2 Energy Flux Terms

From a physical perspective, it is suggestive to rewrite the previous inequality (65) without dropping neither the matter terms nor the piece proportional to N in Eq. (27). Following [13, 14] we define $\mathscr{F}_{\text{grav}} \equiv \frac{1}{16\pi}\int_{\mathscr{S}} N\sigma^{(\ell)}_{ab}\sigma^{(\ell)ab} dS$ as the instantaneous flux of (transverse [38, 39]) gravitational radiation measured by an (Eulerian) observer associated with a foliation with lapse function N. Expressing the flux of matter energy as $\mathscr{F}_{\text{matter}} \equiv \int_{\mathscr{S}} T^{\text{M}}_{ab}\ell^a t^b dS$ (with $t^a = k^b + N\ell^b$ a time-like vector) and the electromagnetic Poynting flux as $\mathscr{F}_{\text{EM}} = \int_{\mathscr{S}} NT^{\text{EM}}_{ab}\ell^a\ell^b dS$, we write (with $g = 0$)

$$\frac{1}{2} \geq \frac{\Lambda^*}{2}\left(\frac{A}{4\pi}\right) + \frac{1}{2}\left(\frac{4\pi}{A}\right)\sum_i Q_i^2 + \mathscr{F}_{\text{EM}} + \mathscr{F}_{\text{matter}} + 2\mathscr{F}_{\text{grav}}. \tag{66}$$

This emphasizes the role of the integral inequalities (22), (27) and (32) in Lemmas 1, 2 and 3 as *energy flux* inequalities. In particular, the flux inequality (66) indicates that the instantaneous flux of energy into a stable black hole horizon is bounded from above so that it cannot be arbitrarily large.

6 The Area-Angular Momentum-Charge Inequality

After discussing the area–angular momentum and area–charge inequalities, we address now the inequality with all the relevant quantities in Einstein–Maxwell theory.

Theorem 3 (cf. [34, 35]). *Given an axisymmetric closed orientable marginally outer trapped surface $\mathscr{S}$ satisfying the (axisymmetry-compatible) spacetime stably outermost condition, in a spacetime with nonnegative cosmological constant and matter content fulfilling the dominant energy condition, it holds the inequality*

$$(A/(4\pi))^2 \geq (2J)^2 + (Q_{\text{E}}^2 + Q_{\text{M}}^2)^2, \tag{67}$$

where A is the area of $\mathscr{S}$ and J, Q_{E} and Q_{M} are, respectively, the total (gravitational and electromagnetic) angular momentum, the electric and the magnetic charges associated with $\mathscr{S}$. If equality holds, then $\mathscr{S}$ has the geometry of an extreme Kerr–Newman throat sphere and, in addition, if vector X^a in the stability condition can be found to be space-like then $\mathscr{S}$ is a section of a non-expanding horizon.

Proof. The proof [34, 35] follows the steps in Theorem 1, namely, with a first stage in which a lower bound Eq. (34) on the area is derived, followed by the resolution of a variational problem under the constraints of keeping J, Q_{E}, and Q_{M} fixed.

First, starting from inequality (22), proceeding then as in the derivation of Eq. (36) and using relations (49) and (13), we obtain

$$\int_{\mathscr{S}} \left[|D\alpha|^2 + \frac{1}{2}\alpha^2\, {}^2\!R \right] dS \geq \int_{\mathscr{S}} \alpha^2 \left[|\Omega^{(\eta)}|^2 + (E_\perp^2 + B_\perp^2) \right] dS. \tag{68}$$

From this expression, contact can be made [34] with the proof in [41] to establish inequality (67) for vanishing Q_{M}. Here, we rather follow [34, 35] the strategy in Sect. 4. In order to identify the relevant action functional $\mathscr{M}$ for the variational problem, in particular its dependence on appropriate potentials permitting to control the constraints on J, Q_{E}, and Q_{M}, we adopt again a coordinate system (37) on the axisymmetric sphere and use the decomposition Eq. (40) introducing the potential $\tilde{\omega}$. From expressions (14) for Q_{E} and Q_{M} and (15) for J, we write (see details in [35])

$$\begin{aligned}
Q_{\mathrm{E}} &= \frac{1}{2}\int_0^\pi E_\perp e^c \sin\theta d\theta = \frac{1}{2}[\psi(\pi) - \psi(0)] \\
Q_{\mathrm{M}} &= \frac{1}{4\pi}\int_0^\pi \frac{dA_\varphi}{d\theta} d\theta = \frac{1}{2}[\chi(\pi) - \chi(0)] \\
J &= \frac{1}{8}\int_0^\pi \left(2\eta\frac{d\tilde{\omega}}{d\theta} + 2\chi\frac{d\psi}{d\theta} - 2\psi\frac{d\chi}{d\theta} \right) = \frac{1}{8}[\omega(\pi) - \omega(0)],
\end{aligned} \tag{69}$$

where we have introduced the new potentials ω, χ and ψ on $\mathscr{S}$

$$\begin{aligned}
\frac{d\psi}{d\theta} &= E_\perp e^c \sin\theta \ , \quad \chi = A_\varphi, \\
\frac{d\omega}{d\theta} &= 2\eta\frac{d\tilde{\omega}}{d\theta} + 2\chi\frac{d\psi}{d\theta} - 2\psi\frac{d\chi}{d\theta} = \frac{d\bar{\omega}}{d\theta} + 2\chi\frac{d\psi}{d\theta} - 2\psi\frac{d\chi}{d\theta}.
\end{aligned} \tag{70}$$

Therefore fixing ω, χ and ψ on the axis does control the values of Q_{E} and Q_{M} and J in the variational problem. Using these potentials in Eq. (68), with $\alpha = e^{c-\sigma/2}$, we get

$$8(c+1) \geq \mathscr{M}[\sigma, \omega, \psi, \chi], \tag{71}$$

where (here the squares are calculated with the round sphere)

$$\begin{aligned}
\mathscr{M}[\sigma, \omega, \psi, \chi] = \frac{1}{2\pi}\int_{\mathscr{S}} \Big[& 4\sigma + |D\sigma|^2 \\
& + \frac{|D\omega - 2\chi D\psi + 2\psi D\chi|^2}{\eta^2} + \frac{4}{\eta}(|D\psi|^2 + |D\chi|^2) \Big] dS_0,
\end{aligned} \tag{72}$$

from which an inequality (34) is recovered by using $A = 4\pi e^c = 4\pi e^{\sigma(0)}$. The proof of the area-charge-angular momentum inequality (67) is completed by showing that

$$\mathscr{M} \geq \mathscr{M}_0 = 8\ln\sqrt{(2J)^2 + (Q_{\mathrm{E}}^2 + Q_{\mathrm{M}}^2)^2} + 8, \tag{73}$$

under the constraint of keeping J, Q_E and Q_M fixed. Here $\mathscr{M}_0$ corresponds to the evaluation of $\mathscr{M}$ on extremal Kerr-Newman with J, with Q_E and Q_M given. The details of this variational problem are discussed in [35], where rigidity is also proved. □

7 Discussion

We have reviewed a set of geometric inequalities holding for stably outermost marginally outer trapped surfaces embedded in generic dynamical, non-necessarily axisymmetric spacetimes with ordinary matter that can extend to and cross the black hole horizon. These inequalities provide lower bounds for the area A, in terms of expressions involving (linearly) the angular momentum J and (quadratically) the electric and magnetic charges, Q_E and Q_M. Extensions including Yang–Mills charges, Q_E^{YM} and Q_M^{YM}, as well as a charge $Q[\eta]$ for certain divergence-free vectors, have also been discussed. If J is present, axisymmetry is required on the surface (and only on the surface). Otherwise the inequalities involve no symmetry requirements.

We have adopted a purely quasi-local spacetime Lorentzian approach. However, it is worthwhile to note that these inequalities were initially discussed on initial data in spatial 3-slices by using Riemannian techniques, in particular minimal surfaces. Although more stringent in their spacetime requirements, whenever applicable, such versions also hold on more general surfaces than marginally outer trapped surfaces. We have however focused here on the specific context of black hole horizons. In this setting, the adoption of a spacetime perspective based entirely on purely Lorentzian concepts has offered crucial geometric insights into the problem: all geometric elements in the proof acquire a clear spacetime meaning. This has led to a refinement in the required conditions permitting, in particular, the generic incorporation of matter in the discussion. The crucial ingredient enabling the shift to a purely Lorentzian discussion has been the identification of the stably outermost condition for marginally outer trapped surfaces as the elementary notion involved. In essence, this is the only required ingredient. In this sense, the fulfillment of inequalities (1) is just a fundamental and direct (irreducible) consequence of the Lorentzian structure of spacetime. This is the main conclusion that we want to stress in these notes.

Strictly speaking, the inclusion of the angular momentum in the inequalities requires two further (related) elements: axisymmetry on the surface and an analytic variational principle. This is in contrast with inequalities in which J is absent, that are straightforward geometric consequences of the stability condition. Certainly, the identification of potentials σ, ω, χ and ψ for the functional $\mathscr{M}$ is related to the spherical topology of $\mathscr{S}$, ultimately controlled by the stability condition for MOTS. However, it is indeed of relevance to assess the role of the axisymmetry and variational treatment requirements in this problem. First, relaxing local axisymmetry in the angular momentum characterization is of interest in astrophysical contexts. Second, the success of the variational problem is intrinsically related to the existence of particular spacetimes, namely extremal stationary (axisymmetric) black holes, that saturate the inequality and simultaneously provide a (unique) minimum for the functionals $\mathcal{M}$. A better understanding of the structural role played by the variational principle in the proofs could offer insight into the properties of the space of solutions of the theory. In particular, the observation that spacetimes admitting symmetries are singular points in the space of solutions of (vacuum) Einstein's equations [7,32] (namely conical singularities) could shed some light on the relation between the presence of symmetries and the need of a variational principle.

From a physical perspective, stable marginally trapped surfaces are sections of quasi-local models for black hole horizons. More precisely, the spacetime stably outermost condition is essentially the *outer condition* introduced in [37] for trapping horizons, namely worldtubes of apparent horizons. From an initial data perspective, the (strictly) stably outermost condition is precisely the condition that guarantees the evolution of an initial apparent horizon into a dynamical horizon [3,4] with a unique foliation by marginally outer trapped surfaces [12]. The inequalities here studied provide a characterization of the notion of black hole horizon (sub)extremality [17]. Moreover, the rigidity results imply that the saturation of the inequalities characterizes the extremality of the horizon geometry (see also [44,45]). These considerations endorse the discussion of the first law of thermodynamics in dynamical horizons [13, 14] where, in particular, the non-negativity of the surface gravity is equivalent to the fulfillment of the discussed inequalities. Equivalently, support is given for the physical validity of the Christodoulou mass, as a function growing with the area (for fixed J and Q_i). Beyond the inequalities among A, J and the charges Q_i, but still in the context of energy balance equations, we have noted in Sect. 5.2.2 that the integral characterization of the stability condition can be interpreted as an *energy flux* inequality.

In the general context of the standard picture of gravitational collapse [47], the inequalities here studied provide a set of quasi-local geometric probes into black hole dynamics in generic situations. In this sense, it is of interest to explore a possible connection between these inequalities and aspects of the cosmic censorship conjecture (e.g. through their link to related global inequalities [35]), or possible implications in the understanding of partial problems in black hole stability.

We would like to conclude by emphasizing that these inequalities represent a particular example of the extension to a Lorentzian setting of tools and concepts employed in the discussion of minimal surfaces in a Riemannian context. In this sense, this family of problems provides a concrete bridge between research in Riemanian and Lorentzian geometries.

Acknowledgments This work is fully indebted to the close scientific collaboration with S. Dain, M.E. Gabach Clément, M. Reiris and W. Simon. I would like to express here my gratitude to them. I would also like to thank A. Aceña, M. Ansorg, C. Barceló, M. Mars and J.M.M. Senovilla for useful discussions. I thank M.E. Gabach Clément for her careful reading of this chapter. I acknowledge the support of the Spanish MICINN (FIS2008-06078-C03-01) and the Junta de Andalucía (FQM2288/219).

References

1. Acena, A., Dain, S., Clement, M.E.G.: Horizon area–angular momentum inequality for a class of axially symmetric black holes. Class. Quant. Grav. **28**, 105,014 (2011). DOI 10.1088/0264-9381/28/10/105014
2. Andersson, L., Eichmair, M., Metzger, J.: Jang's equation and its applications to marginally trapped surfaces (2010)
3. Andersson, L., Mars, M., Simon, W.: Local existence of dynamical and trapping horizons. Phys. Rev. Lett. **95**, 111,102 (2005)
4. Andersson, L., Mars, M., Simon, W.: Stability of marginally outer trapped surfaces and existence of marginally outer trapped tubes. Adv. Theor. Math. Phys. **12**, 853–888 (2008)
5. Ansorg, M., Hennig, J., Cederbaum, C.: Universal properties of distorted Kerr-Newman black holes. Gen. Rel. Grav. **43**, 1205–1210 (2011). DOI 10.1007/s10714-010-1136-8
6. Ansorg, M., Pfister, H.: A universal constraint between charge and rotation rate for degenerate black holes surrounded by matter. Class. Quant. Grav. **25**, 035,009 (2008). DOI 10.1088/0264-9381/25/3/035009
7. Arms, J., Marsden, J., Moncrief, V.: The structure of the space of solutions of Einstein's equations. II. Several Killing fields and the Einstein Yang-Mills equations. Annals Phys. **144**, 81–106 (1982). DOI 10.1016/0003-4916(82)90105-1
8. Ashtekar, A., Beetle, C., Lewandowski, J.: Mechanics of rotating isolated horizons. Phys.Rev. **D64**, 044,016 (2001). DOI 10.1103/PhysRevD.64.044016
9. Ashtekar, A., Engle, J., Pawlowski, T., Van Den Broeck, C.: Multipole moments of isolated horizons. Class. Quant. Grav. **21**, 2549 (2004)
10. Ashtekar, A., Fairhurst, S., Krishnan, B.: Isolated horizons: Hamiltonian evolution and the first law. Phys. Rev. **D62**, 104,025 (2000)
11. Ashtekar, A., Fairhurst, S., Krishnan, B.: Isolated horizons: Hamiltonian evolution and the first law. Phys.Rev. **D62**, 104,025 (2000). DOI 10.1103/PhysRevD.62.104025
12. Ashtekar, A., Galloway, G.J.: Some uniqueness results for dynamical horizons. Adv. Theor. Math. Phys. **9**, 1 (2005)
13. Ashtekar, A., Krishnan, B.: Dynamical horizons: Energy, angular momentum, fluxes and balance laws. Phys. Rev. Lett. **89**, 261,101 (2002)
14. Ashtekar, A., Krishnan, B.: Dynamical horizons and their properties. Phys. Rev. D **68**, 104,030 (2003)
15. Ashtekar, A., Krishnan, B.: Isolated and dynamical horizons and their applications. Liv. Rev. Relat. **7**, 10 (2004). URL http://www.livingreviews.org/lrr-2004-10. URL (cited on 9 March 2012) http://www.livingreviews.org/lrr-2004-10
16. Booth, I., Fairhurst, S.: Isolated, slowly evolving, and dynamical trapping horizons: geometry and mechanics from surface deformations. Phys. Rev. **D75**, 084,019 (2007)
17. Booth, I., Fairhurst, S.: Extremality conditions for isolated and dynamical horizons. Phys. Rev. **D77**, 084,005 (2008). DOI 10.1103/PhysRevD.77.084005
18. Cao, L.M.: Deformation of Codimension-2 Surface and Horizon Thermodynamics. JHEP **1103**, 112 (2011). DOI 10.1007/JHEP03(2011)112
19. Carter, B.: Republication of: Black hole equilibrium states part ii. general theory of stationary black hole states. Gen. Relat. Gravit. **42**, 653–744 (2010). URL http://dx.doi.org/10.1007/s10714-009-0920-9. 10.1007/s10714-009-0920-9
20. Chrusciel, P., Kondracki, W.: Some global charges in classical Yang-Mills theory. Phys.Rev. **D36**, 1874–1881 (1987). DOI 10.1103/PhysRevD.36.1874
21. Chruściel, P.T.: Mass and angular-momentum inequalities for axi-symmetric initial data sets I. Positivity of mass. Annals Phys. **323**, 2566–2590 (2008). DOI 10.1016/j.aop.2007.12.010
22. Chruściel, P.T., Li, Y., Weinstein, G.: Mass and angular-momentum inequalities for axi-symmetric initial data sets. II. Angular-momentum. Annals Phys. **323**, 2591–2613 (2008). DOI 10.1016/j.aop.2007.12.011

23. Chrusciel, P.T., Lopes Costa, J.: Mass, angular-momentum, and charge inequalities for axisymmetric initial data. Class. Quant. Grav. **26**, 235,013 (2009). DOI 10.1088/0264-9381/26/23/235013
24. Costa, J.L.: A Dain Inequality with charge. arXiv:0912.0838 (2009)
25. Dain, S.: Angular-momentum-mass inequality for axisymmetric black holes. Phys. Rev. Lett. **96**, 101,101 (2006)
26. Dain, S.: Proof of the (local) angular momemtum-mass inequality for axisymmetric black holes. Class. Quant. Grav. **23**, 6845–6856 (2006). DOI 10.1088/0264-9381/23/23/015
27. Dain, S.: Proof of the angular momentum-mass inequality for axisymmetric black holes. J. Diff. Geom. **79**, 33–67 (2008)
28. Dain, S.: Extreme throat initial data set and horizon area-angular momentum inequality for axisymmetric black holes. Phys. Rev. **D82**, 104,010 (2010). DOI 10.1103/PhysRevD.82.104010
29. Dain, S.: Geometric inequalities for axially symmetric black holes. Class. Quant. Grav. **29**, 073,001 (2012). DOI 10.1088/0264-9381/29/7/073001
30. Dain, S., Jaramillo, J.L., Reiris, M.: Area-charge inequality for black holes. Class.Quant.Grav. **29**, 035,013 (2012). DOI 10.1088/0264-9381/29/3/035013
31. Dain, S., Reiris, M.: Area - Angular momentum inequality for axisymmetric black holes. Phys.Rev.Lett. **107**, 051,101 (2011). DOI 10.1103/PhysRevLett.107.051101
32. Fischer, A., Marsden, J., Moncrief, V.: The structure of the space of solutions of Einsteins equations I. One Killing field. Ann. Inst. H. Poincaré **33**, 147–194 (1980)
33. Gabach Clement, M.E.: Comment on Horizon area-Angular momentum inequality for a class of axially symmetric black holes. arXiv:1102.3834 (2011)
34. Gabach Clement, M.E., Jaramillo, J.L.: Black hole Area-Angular momentum-Charge inequality in dynamical non-vacuum spacetimes. Accepted in Phys. Rev. D. arXiv:1111.6248 (2012)
35. Gabach Clement, M.E., Jaramillo, J.L., Reiris, M.: Proof of the area-angular momentum-charge inequality for axisymmetric black holes. arXiv:1207.6761 (2012)
36. Galloway, G.J., Schoen, R.: A Generalization of Hawking's black hole topology theorem to higher dimensions. Commun.Math.Phys. **266**, 571–576 (2006). DOI 10.1007/s00220-006-0019-z
37. Hayward, S.: General laws of black-hole dynamics. Phys. Rev. D **49**, 6467 (1994)
38. Hayward, S.: Energy conservation for dynamical black holes. Phys. Rev. Lett. **93**, 251,101 (2004)
39. Hayward, S.A.: Energy and entropy conservation for dynamical black holes. Phys. Rev. D **70**, 104,027 (2004)
40. Hennig, J., Ansorg, M., Cederbaum, C.: A universal inequality between the angular momentum and horizon area for axisymmetric and stationary black holes with surrounding matter. Class. Quant. Grav. **25**, 162,002 (2008)
41. Hennig, J., Cederbaum, C., Ansorg, M.: A universal inequality for axisymmetric and stationary black holes with surrounding matter in the Einstein-Maxwell theory. Commun. Math. Phys. **293**, 449–467 (2010). DOI 10.1007/s00220-009-0889-y
42. Hollands, S.: Horizon area-angular momentum inequality in higher dimensional spacetimes. Class.Quant.Grav. **29**, 065,006 (2012). DOI 10.1088/0264-9381/29/6/065006
43. Jaramillo, J.L., Reiris, M., Dain, S.: Black hole Area-Angular momentum inequality in non-vacuum spacetimes. Phys.Rev. **D84**, 121,503 (2011). DOI 10.1103/PhysRevD.84.121503
44. Jaramillo, J.L.: A note on degeneracy, marginal stability and extremality of black hole horizons. Class.Quant.Grav. **29** 177,001 (2012). DOI 10.1088/0264-9381/29/17/177001
45. Mars, M.: Stability of MOTS in totally geodesic null horizons. Class.Quant.Grav. **29** 145,019 (2012). DOI 10.1088/0264-9381/29/14/145019
46. Penrose, R.: Gravitational collapse: The role of general relativity. Riv. Nuovo Cim. **1**, 252 (1969)
47. Penrose, R.: Naked singularities. Annals N. Y. Acad. Sci. **224**, 125 (1973)
48. Racz, I.: A simple proof of the recent generalisations of Hawking's black hole topology theorem. Class. Quant. Grav. **25**, 162,001 (2008). DOI 10.1088/0264-9381/25/16/162001

49. Ryder, L.: Dirac monopoles and the Hopf map S(3) to S(2). J. Phys. A **A13**, 437–447 (1980). DOI 10.1088/0305-4470/13/2/012
50. Simon, W.: Gravitational field strength and generalized Komar integral. Gen. Rel. Grav. **17**, 439 (1985). DOI 10.1007/BF00761903
51. Simon, W.: Bounds on area and charge for marginally trapped surfaces with a cosmological constant. Class. Quant. Grav. **29**, 062,001 (2012). DOI 10.1088/0264-9381/29/6/062001
52. Sudarsky, D., Wald, R.M.: Extrema of mass, stationarity, and staticity, and solutions to the Einstein Yang-Mills equations. Phys. Rev. **D46**, 1453–1474 (1992). DOI 10.1103/PhysRevD.46.1453
53. Wald, R.M.: General Relativity. Chicago University Press, Chicago (1984)
54. Weinberg, S.: The quantum theory of fields. Modern applications, vol. 2. Cambridge University Press, New York (1996)
55. Wu, T.T., Yang, C.N.: Concept of nonintegrable phase factors and global formulation of gauge fields. Phys. Rev. D **12**, 3845–3857 (1975). DOI 10.1103/PhysRevD.12.3845. URL http://link.aps.org/doi/10.1103/PhysRevD.12.3845

Infinitesimal and Local Convexity of a Hypersurface in a Semi-Riemannian Manifold

Erasmo Caponio

1 Introduction

Let Ω be an open of subset of $\mathbb{R}^2$ and $f\colon \Omega \to \mathbb{R}$ be a differentiable function in Ω. We say that f is locally convex at a point $(x_0, y_0) \in \Omega$ if there exists a neighborhood $U \subset \Omega$ of (x_0, y_0) such that the graph of the restriction of f to U is nowhere below the tangent plane at $\big(x_0, y_0, f(x_0, y_0)\big)$ or, equivalently, all the straight lines through the point $\big(x_0, y_0, f(x_0, y_0)\big)$ are nowhere above the graph of $f_{|U}$. If f is twice differentiable, then the above condition is satisfied if the Hessian of f is positive semidefinite in a neighborhood of (x_0, y_0), while the positive semi-definiteness of $\mathrm{d}^2 f$ at the single point (x_0, y_0) is necessary but not sufficient, as the function $f(p) = f(x,y) = -(x^4 + y^4)$ at $(0,0)$ shows.

Let H be the hypersurface in $\mathbb{R}^3$ which is the graph of the function f; the following quadratic form associated to the Hessian of f at a point (x,y), i.e.,

$$\Pi_{(x,y)}\big((u,v),(u,v)\big) = \frac{f_{xx}(x,y)u^2 + 2f_{xy}(x,y)uv + f_{yy}(x,y)v^2}{\sqrt{1 + f_x^2(x,y) + f_y^2(x,y)}},$$

is the *second fundamental form* of H at $\big(x, y, f(x,y)\big)$. Thus, the convexity of a twice differentiable function f in a neighborhood of $(x_0, y_0) \in \Omega$ is equivalent to the fact that the Π is positive semidefinite in the same neighborhood.

Now assume that H is a smooth embedded hypersurface in a Riemannian manifold (M, g). The natural generalizations of the above notions are the following ones:

E. Caponio (✉)
Dipartimento di Meccanica, Matematica e Managment, Politecnico di Bari, via Orabona 4, 70125, Bari, Italy
e-mail: caponio@poliba.it

M. Sánchez et al. (eds.), *Recent Trends in Lorentzian Geometry*, Springer Proceedings in Mathematics & Statistics 26, DOI 10.1007/978-1-4614-4897-6_6,

1. The first one becomes the requirement that there exists a neighborhood $\tilde{U}$ in M of $p_0 \in H$ such that the intersections of the images of the geodesics through p_0 with velocity vector at p_0 tangent to H with $\tilde{U}$ are contained on the closure of one of the two connected component $\tilde{U} \setminus H$ or, equivalently, there exists a neighborhood O of $0 \in T_{p_0}H$ and a neighborhood $\tilde{U}$ of p_0 such that the $\exp_{p_0}(O)$ is contained $H \cup C$, where C is one of the two connected component of $\tilde{U} \setminus H$; this condition is called *local convexity* at p_0;
2. The second one becomes the requirement that there exists an open neighborhood $U \subset H$ of p_0 such that the second fundamental form of U, with respect to a smooth unit normal vector field n on U, is positive semidefinite[1] for each $p \in U$, i.e.

$$\Pi_p(v,v) = g_p(n, \nabla_V V) \geq 0, \tag{1}$$

where ∇ is the Levi-Civita connection of (M,g) and V is a vector field on M extending $v \in T_pH$; this condition is called *infinitesimal convexity* in U.

Remark 1. The global versions of these definitions require that the embedded hypersurface H is *orientable*, i.e., H must admit a smooth unit normal vector field n. In such a case, H is said infinitesimally convex if $\Pi(v,v) \geq 0$, for each $v \in TH$, where Π is the second fundamental form of H with respect to n.

As in the case of the graph of a function, infinitesimal convexity at a neighborhood of $p_0 \in H$ implies local convexity. Surprisingly enough, this property was proved only relatively recently for a hypersurface in a Riemannian manifold having constant sectional curvature in [15] and in the general case in [9]. Besides, the proof in [9] is rather involved and it seems to depend drastically on the Riemannian setting. In Sect. 2, we will describe which are the obstacles one encounters trying to extend Bishop's proof when the ambient manifold is a Finsler or a semi-Riemannian one. The implication in the Finsler setting was obtained recently in [2]. In Sect. 3, we will show that the techniques in [2] also give the implication in a semi-Riemannian manifold, and in Sect. 4, we will consider the same problem in relation to the causal character of the geodesics and to another notion of convexity called *geometric convexity*. Finally, we will see some applications to geodesic connectedness of an open subset of a Lorentzian manifold by means of causal geodesics.

[1] According to the case when H is the graph of a function and n is the smooth unit vector which lies on the same side of the graph as the canonic unit vector defining the axis of the values of the function, from [9], we know that the geodesics issuing from p_0 and tangent to H are locally contained in the closure of the component of a tubular neighborhood of H individuated by $-n$. As the sign of Π changes if one changes n with $-n$, it is clear that what is really important in this definition is the fact that Π is semidefinite in U, either positive or negative. Clearly in the case when Π is negative semidefinite the geodesics tangent at p_0 to H are locally contained in the component individuated by n itself.

2 A Review of Bishop's Proof

The proof in [9] that infinitesimal convexity in a neighborhood U of a point $p_0 \in H$ implies local convexity at p_0 is based on a reduction of the general problem to the case of a Riemannian manifold (M,g) of dimension two. Such a reduction is quite natural, and it is based on the following idea. For any geodesic γ starting from p_0 with initial velocity tangent to $H \cap U$ at p_0, consider the surface $S(\gamma)$ ruled by the geodesics orthogonal to $H \cap U$ and passing through γ. Two such surfaces $S(\gamma)$, $S(\bar{\gamma})$, $\gamma \neq \bar{\gamma}$, intersect along the geodesic σ which is orthogonal to H at p_0 (see Fig. 1). Then, for each γ, define the unit vector field $E \in TS(\gamma)$ which is the geodesic field (with respect to the Riemannian metric induced on $S(\gamma)$ by g) such that E_p is orthogonal to σ for all $p \in \sigma$. The surface $S(\gamma)$ can be parametrized by the coordinates u and v which are, respectively, the affine parameters of σ and γ (thus γ is represented on $S(\gamma)$ by the equation $u = 0$). The unit vector field X in $TS(\gamma)$ which is tangent to $H \cap S(\gamma)$ can be written as $X_{(u,v)} = c(u,v)E_{(u,v)} + s(u,v)F_{(u,v)}$, where F is the smooth unit vector field in $TS(\gamma)$ orthogonal to E (oriented in such way that $F_{p_0} = -n_{p_0}$). Using the infinitesimal convexity assumption, it is possible to obtain a differential inequality, satisfied by the function s, implying that s is non positive on $H \cap S(\gamma)$. Then, local convexity at p follows from the observation that $X(u) = cE(u) + sF(u) = sF(u) \geq 0$.

The delicate point in this argument is that E could not be a well defined vector field in $\tilde{U} \setminus \{\sigma\}$, for any convex neighborhood $\tilde{U}$ of p_0 in M.

By *convex neighborhood* of a point p in a Riemannian or semi-Riemannian manifold, here and hereafter, we mean an open neighborhood $\tilde{U}$ which has the property that any two point in $\tilde{U}$ are connected by a unique geodesic whose

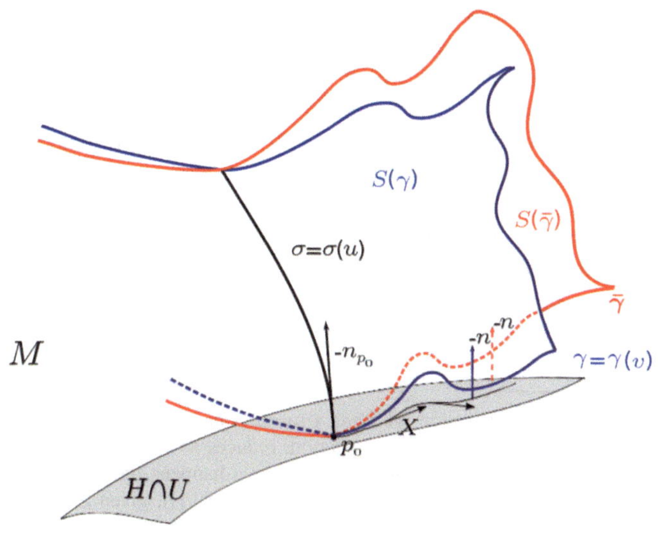

Fig. 1 The reduction to a surface $S(r)$

support is contained in $\tilde{U}$ (it is well known than any point on a Riemannian or semi-Riemannian manifold has a convex neighborhood, see, e.g., [27, Chap. 5, Proposition 7]).

Bishop rules out this pathological case by using an estimate on the distance in M of the focal points of σ in $S(\gamma)$. Such estimate is based, on the one hand, on the fact that the Gauss curvatures of the surfaces $S(\gamma)$ are uniformly bounded from above (in a neighborhood of p_0) by the sectional curvature of the ambient manifold M (provided that the metric is at least C^4) and, on the other one, on the fact that the geodesics in $S(\gamma)$ having E as their velocity vector fields, leave σ orthogonally and then the distance from σ (in M) of their points can be evenly controlled.

Such type of arguments have no straightforward extension when the ambient manifold (M,g) is semi-Riemannian. Indeed, as shown in [26], if the sectional curvature is bounded for all time-like or space-like planes at a point p, then it is constant at p. Moreover, the Morse index of space-like geodesics is always $+\infty$. Thus, Rauch comparison theorems (see [8] for the case of geodesics starting orthogonally to a given geodesic) do not generally hold in semi-Riemannian geometry. Some results are available in literature, but they concern either causal geodesics (cf. [5, Sect. 11.2] for geodesics connecting two given points and [16,24] for more general boundary conditions) or Jacobi fields along a given single geodesic of any causal character satisfying two different types of boundary conditions [23].

In [2], Bishop's result was extended to a hypersurface in a Finsler manifold (M,F) by using a different approach from that in [9]. In the Finslerian setting, the problem of directly extending the proof in [9] is not related to the lack of satisfactory comparison results (cf. [1, Sect. 9.1]) but rather on the fact that the differential inequality satisfied by the function s above (that comes from the evaluation of the second fundamental form of the curve $S(\gamma)\cap H$, in the two-dimensional manifold $S(\gamma)$, which assumes by construction the same values of that of the hypersurface H) is obtained by using, in an essential way, the metric compatibility of the Levi-Civita connection. For a Finsler manifold, the role of the Levi-Civita connection can be played by the Chern connection which is metric compatible (and torsion-free) if and only the Finsler structure is Riemannian (see, e.g. [1, Exercise 2.4.2]).[2]

The approach followed in [2] will be briefly described in the next section. Here, we would like to emphasize that when the Finsler metric comes from a Riemannian one (i.e., $F(v)=\sqrt{g(v,v)}$) the result in [2] improves the one in [9] with respect to the smoothness assumption on the metric and on the hypersurface. Indeed from [2], it is enough that the metric is differentiable with locally Lipschitz differential (and we will assume the same on the semi-Riemannian metric g), while in [9], it assumed that it is C^4.

[2]In some cases, the Chern connection which is a linear connection on the vector bundle π^*TM over $TM\setminus\{0\}$, $\pi\colon TM\to M$ the natural projection, reduces to a linear connection on M, even if the Finsler metric is not a Riemannian one. When this happens the Finsler metric is said of *Berwald* type. Since from a theorem of Szabó (cf. [1, Sect. 10.1]), given a Berwald metric F on M, there exists a Riemannian metric such that its Levi-Civita connection coincides with the Chern connection of F, we can state, as already observed in [10], that Bishop's proof is also valid in any Berwald space.

Remark 2. In regard to the smoothness of the hypersurface, we slightly improve the requirement in [2] that the hypersurface is an embedded submanifold locally described as the regular level of a twice differentiable function having Lipschitz second differential. Since at each point of an embedded submanifold H, there is a coordinate system adapted to H (see e.g., [27, Chap. 1, Proposition 28]), it will be clear in the proof of Lemma 2 that it is enough to assume that the hypersurface H is a C^2 embedded submanifold of M (of course, also (M,g) must have a differentiable structure of class at least C^2).

3 Infinitesimal Convexity Implies Local Convexity

Let (M,g) be a semi-Riemannian manifold of dimension m and H be an embedded non-degenerate hypersurface in M. Let ϕ be a function defined in a neighborhood $\tilde{U} \subset M$ of $p_0 \in H$ such that $\phi^{-1}(0) = H \cap \tilde{U}$ and its gradient $\nabla\phi$ at any $p \in \phi^{-1}(0)$ has the same orientation of the smooth unit normal unit vector field n in $H \cap \tilde{U}$ with respect to which the second fundamental form of $H \cap \tilde{U}$ is defined. Let $H_\phi := \nabla(\mathrm{d}\phi)$ be the Hessian of ϕ, where ∇ is the Levi-Civita connection of (M,g). In a coordinate system $(W,(x^1,\dots,x^m))$ of M, using the Einstein summation convention, we get

$$(H_\phi)_p(u,v) = \frac{\partial^2\phi}{\partial x^i \partial x^j}(p)u^i v^j - \frac{\partial\phi}{\partial x^l}(p)\Gamma^l_{ij}(p)u^i v^j, \tag{2}$$

for any $p \in V$ and any $u = (u^1,\dots,u^m)$, $v = (v^1,\dots,v^m) \in T_pM$, where Γ^l_{ij} are the Christoffel symbols of the metric g. From Eq. (2), it's easy to see that if $\gamma\colon (a,b) \to M$ is a geodesic then

$$(\phi\circ\gamma)''(s) = (H_\phi)_{\gamma(s)}\big(\dot\gamma(s),\dot\gamma(s)\big). \tag{3}$$

The following lemma, concerning the equivalence between the second fundamental form of $H \cap \tilde{U}$ and the Hessian of a function ϕ as above, is well known. We give here a proof for the sake of completeness.

Lemma 1. *For all $V \in T(H\cap\tilde{U})$, $H_\phi(V,V) = -|\nabla\phi|\Pi(V,V)$*

Proof. By definition, $H_\phi(V,V) = \tilde{V}(\tilde{V}(\phi)) - \nabla_{\tilde{V}}\tilde{V}(\phi)$, where $\tilde{V}$ is any vector field on M extending V and $\nabla\phi$ is the gradient of ϕ. Using the metric compatibility of the Levi-Civita connection, we get $\tilde{V}(\tilde{V}(\phi)) - \nabla_{\tilde{V}}\tilde{V}(\phi) = g(\nabla_{\tilde{V}}\nabla\phi,\tilde{V})$. As V and $\nabla\phi$ are orthogonal at any point of $H\cap\tilde{U}$, we get on H

$$0 = V(g(\nabla\phi,V)) = g(\nabla_{\tilde{V}}\nabla\phi,\tilde{V}) + g(\nabla\phi,\nabla_{\tilde{V}}\tilde{V}),$$

hence $H_\phi(V,V) = -g(\nabla\phi,\nabla_{\tilde{V}}\tilde{V})$ on H. As $n = \frac{\nabla\phi}{|\nabla\phi|}|_H$ and $\Pi(V,V) = g(n,\nabla_{\tilde{V}}\tilde{V})$ we get the thesis. □

Remark 3. As already observed in [4, Sect. 3], locally defining the hypersurface H as a regular level set of a function allows to extend the notion of infinitesimal convexity to hypersurfaces having points where the tensor induced by the ambient metric is degenerate. At these points, the notion of second fundamental form is meaningless. In what follows, unless differently specified, we will assume that H is an embedded hypersurface in (M,g), not necessarily non-degenerate.

Definition 1. Let (M,g) be a semi-Riemannian manifold and H be a C^2 embedded hypersurface in M. We say that H is infinitesimally convex in a neighborhood $U \subset H$ of a point $p_0 \in H$ if there exists a neighborhood $\tilde{U}$ of p_0 in M and a C^2 function $\phi\colon \tilde{U} \to \mathbb{R}$ such that $U = \tilde{U} \cap H = \phi^{-1}(0)$, 0 is a regular value of ϕ, and $H_\phi(V,V)$ is semidefinite (either negative or positive), for all $V \in TU$.

Remark 4. From Lemma 1, it is clear that this definition is independent of the function ϕ if H is a nondegenerate hypersurface. A posteiori, from Theorem 3, the same is true also in the degenerate case. Indeed if there exists a function ϕ with respect to which H is infinitesimally convex in U, by Theorem 3, H is locally convex in U and then it is infinitesimally convex with respect to any other function. Moreover, the "local convex side" of H (i.e., the closure of the connected component of $\tilde{U} \setminus H$ where the geodesics starting at the points of U with velocity vector tangent to H are locally contained) is $\{x \in \tilde{U} : \phi(x) \leq 0\}$, if $H_\phi(V,V) \leq 0$, and $\{x \in \tilde{U} : \phi(x) \geq 0\}$, if $H_\phi(V,V) \geq 0$.

A very natural approach in trying to prove that infinitesimal convexity in a neighborhood of a point p_0 of a hypersurface implies local convexity is to evaluate ϕ along any geodesic arc γ starting at p_0 with initial velocity vector tangent to H. The infinitesimal convexity assumption can be used to get a differential inequality satisfied by the function $\rho = \phi \circ \gamma$, at least when the image of γ is contained on the side of H which is the candidate convex one (recall Remark 4). The differential inequality and the initial conditions satisfied by ρ imply that ρ is actually constant and equal to 0, that is the geodesic γ is contained in H. This is an intermediate fundamental step to prove local convexity at p_0. The remaining part of the proof is a quite trivial consequence of the fact that any point in a semi-Riemannian manifold has a convex neighborhood.

Lemma 2. *Assume that H is infinitesimally convex in a neighborhood U of $p_0 \in H$. Let $\tilde{U}$ be a neighborhood of p_0 in M and $\phi\colon \tilde{U} \to \mathbb{R}$ such that 0 is a regular value of ϕ and $\phi^{-1}(0) = H \cap \tilde{U} = U$, and $H_\phi(V,V) \leq 0$ for each $V \in TU$. Let $\gamma\colon [0,b] \to M$ be a geodesic satisfying the initial conditions $\gamma(0) = p_0$, $\dot{\gamma}(0) \in T_{p_0}H$ and such that $\gamma([0,b]) \subset \{x \in \tilde{U}\colon \phi(x) \geq 0\}$, then $\gamma([0,b]) \subset H$.*

Proof. Since H is an embedded submanifold, we can assume, without loss of generality, that $\tilde{U}$ is the domain of a coordinate system, $\varphi\colon \tilde{U} \to \mathbb{R}^m$, $\varphi = (x^1,\dots,x^m)$ centered at p_0 and that $x^m(p) = 0$, for each $p \in H \cap \tilde{U}$, and the function ϕ is equal to x^m. Then the map $\mathscr{P}\colon \tilde{U} \to H$, $\mathscr{P}(p) = \varphi^{-1}\big((x^1(p),\dots x^{m-1}(p),0\big)$, is well defined. Let γ_H be the projection on H of the geodesic γ, $\gamma_H(s) = \mathscr{P}(\gamma(s)) = \varphi^{-1}(\gamma^1(s),\dots,\gamma^{m-1}(s),0)$, where γ^i, $i = 1,\dots,m$ are the components

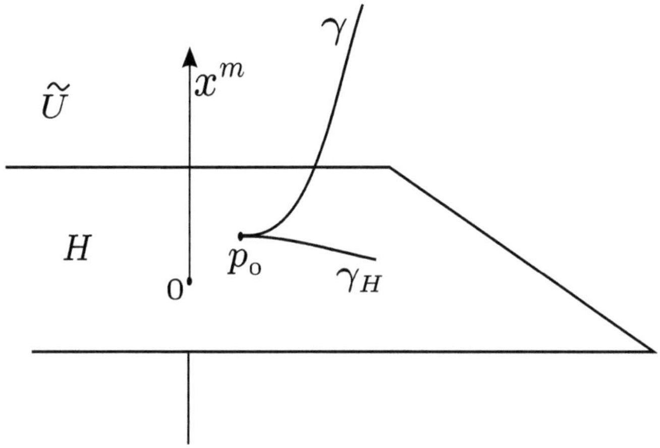

Fig. 2 The projection on H in the coordinate system $(\tilde{U},\varphi)$

of the geodesic γ in the coordinate system $(\tilde{U},\varphi)$ (see Fig. 2). In our notations, the function $\rho = x^m(\gamma)$ is the m-th component γ^m of γ. Thus, from Eqs. (3) and (2), the second derivative of ρ is given by

$$\ddot{\rho}(s) = (H_{x^m})_{\gamma(s)}(\dot{\gamma}(s),\dot{\gamma}(s)) = -\Gamma^m_{ij}(\gamma(s))\dot{\gamma}^i(s)\dot{\gamma}^j(s).$$

Since H is infinitesimally convex in U,

$$(H_{x^m})_{\gamma_H(s)}(\dot{\gamma}_H(s),\dot{\gamma}_H(s)) = -\Gamma^m_{ij}(\gamma_H(s))\dot{\gamma}^i_H(s)\dot{\gamma}^j_H(s) \le 0.$$

Thus, we can estimate $\ddot{\rho}$ as follows:

$$\begin{aligned}
\ddot{\rho}(s) &\le -\Gamma^m_{ij}(\gamma(s))\dot{\gamma}^i(s)\dot{\gamma}^j(s) + \Gamma^m_{ij}(\gamma_H(s))\dot{\gamma}^i_H(s)\dot{\gamma}^j_H(s) \\
&= -\Gamma^m_{ij}(\gamma(s))\dot{\gamma}^i(s)\dot{\gamma}^j(s) + \Gamma^m_{ij}(\gamma_H(s))\dot{\gamma}^i(s)\dot{\gamma}^j(s) \\
&\quad + \Gamma^m_{ij}(\gamma_H(s))(\dot{\gamma}^i(s)+\dot{\gamma}^i_H(s))(\dot{\gamma}^j_H(s)-\dot{\gamma}^j(s)) \\
&= \Big(\Gamma^m_{ij}(\gamma_H(s)) - \Gamma^m_{ij}(\gamma(s))\Big)\dot{\gamma}^i(s)\dot{\gamma}^j(s) \\
&\quad + \Gamma^m_{ij}(\gamma_H(s))(\dot{\gamma}^i(s)+\dot{\gamma}^i_H(s))(\dot{\gamma}^j_H(s)-\dot{\gamma}^j(s))
\end{aligned} \tag{4}$$

For each $p \in U$, let us call $\Gamma^m(p)$ the symmetric bilinear operator on $\mathbb{R}^m$ defined by the Christoffel symbols Γ^m_{ij} at p. The first term in the summand in Eq. (4) is bounded above by $\|\Gamma^m(\gamma_H(s)) - \Gamma^m(\gamma(s))\|\,|\dot{\gamma}(s)|^2$, where $\|\cdot\|$ is the norm of the bilinear operators on $\mathbb{R}^m$ and $|\gamma(s)|$ is the Euclidean norm of the vector $(\dot{\gamma}^i(s))_i \in \mathbb{R}^m$. Since γ is smooth on $[0,b]$ and the Christoffel symbols are Lipschitz on U, this last quantity is bounded above by $a_1|\gamma_H(s)-\gamma(s)|$, where a_1 is a positive constant depending on γ

and $|\gamma_H(s)-\gamma(s)|$ is the Euclidean norm of $(\gamma_H^i(s)-\gamma^i(s))_i \in \mathbb{R}^m$; hence, it is equal to $|\gamma^m(s)| = |\rho(s)| = \rho(s)$. By continuity, the second term in Eq. (4) is bounded above by $a_2|\dot\gamma_H^j(s)-\dot\gamma^j(s)| = a_2|\dot\gamma^m(s)| = a_2|\dot\rho(s)|$, where a_2 is a positive constant depending on γ. Therefore, ρ satisfies the differential inequality

$$\ddot\rho \le c(\rho + |\dot\rho|) \tag{5}$$

where c is a positive constant, and the initial condition $\rho(0) = \dot\rho(0) = 0$, and from [2, Lemma 3.1], it must be equal to 0 on $[0,b]$ (i.e., $x^m \circ \gamma = \phi \circ \gamma \equiv 0$). □

Theorem 1. *Let (M,g) be a semi-Riemannian manifold and H be an embedded C^2 hypersurface in M. Let $p \in H$ and U be a neighborhood of p in H; then H is infinitesimally convex in U if and only if it is locally convex in U.*

Proof. Let $\tilde U$ be an open subset of M and $\phi\colon \tilde U \to \mathbb{R}$ be a function such that 0 is a regular value, $\phi^{-1}(0) = H \cap \tilde U$ and $U = H \cap \tilde U$.

Assume that H is locally convex at any point $p \in U$, i.e., there exists a neighborhood O of $0 \in T_pM$ such that $\exp_p(O \cap T_pH)$ is contained in the closure one of the connected component of $\tilde U \setminus H$, say $\{x \in \tilde U\colon \phi(x) \le 0\}$. Let $v \in O \cap T_pH$ and consider the affinely parametrized geodesic γ such that $\gamma(0) = p$, $\dot\gamma(0) = v$. Thus there exists $a \in (0,+\infty)$ such that the function $\phi\colon \gamma\colon [-a,1] \to \mathbb{R}$ is well defined and has a maximum point at 0. Therefore $(H_\phi)_p(v,v) = (\phi \circ \gamma)''(0) \le 0$.

Assume now that H is infinitesimally convex in U, with $H_\phi(V,V) \le 0$ for all $V \in TU$. Let $p \in U$; we are going to show that for any open convex neighborhood B of p in M the set $C\colon = B \cap \{x \in \tilde U\colon \phi(x) > 0\}$ is also convex, i.e., any two points in C are joined by a unique geodesic whose support is contained in C. Let A be the subset of $C \times C$ given by the couple of points that can be connected by a unique geodesic with support in C. As C is a connected subset of M, it is enough to show that A is nonempty and it is an open and closed subset in $C \times C$. Clearly, each couple $(p_1, p_1) \in C \times C$ can be connected by a constant geodesic, i.e., $A \neq \emptyset$. If $(p_1, p_2) \in A$ and α is the unique geodesic connecting them and whose inner points are in C, by smooth dependence of geodesics by boundary conditions in a convex neighborhood (cf. [27, Chap. 5, Lemma 9]), we can consider two small enough neighborhoods U_1 and U_2 of p_1 and, respectively, p_2 in C, such that the unique geodesic in B connecting $\bar p_1 \in U_1$ to $\bar p_2 \in U_2$ lies in a small neighborhood of α; hence, its points are contained in C and A is open. Now let $(p_1, p_2) \in \overline{A} \subset C \times C$ and consider a sequence $(p_n^1, p_n^2)_n \subset A$ converging to (p_1, p_2). The sequence of geodesics α_n, parametrized on $[0,1]$ and connecting p_n^1 to p_n^2 in B, converges in the C^2 topology to the geodesic $\alpha\colon [0,1] \to B$ connecting p_1 to p_2. Thus, α lies in $\{x \in \tilde U\colon \phi(x) \ge 0\}$. As the points p_1 and p_2 are in $\{x \in \tilde U\colon \phi(x) > 0\}$, from Lemma 2, α cannot be tangent to H at any of its inner points.

Having proved that C is a convex neighborhood in M, the rest of the proof follows by contradiction. Indeed, assume that there exists $p \in U$ and a sequence of vectors $(v_n) \subset T_pH$ such that $v_n \to 0$ and $p_n\colon = \exp_p(v_n) \in C$. Let γ_n be the geodesic in C connecting the first point of the sequence p_1 to p_n. As p_n converges to p, the sequence γ_n converges uniformly (actually in the C^2 topology) to the geodesic $\gamma\colon [0,1] \to B$ connecting p to p_1 in B. By uniform convergence, the image of γ must be contained in $\{x \in \tilde{U}\colon \phi(x) \geq 0\}$. Since $\gamma(0) = p \in H$ and $\dot{\gamma}(0) = v_1 \in T_pH$, from Lemma 2, $\gamma([0,1]) \subset H$. But $\gamma(1) = p_1 \notin H$. □

4 Some Remarks and Applications

4.1 *Convexity with Respect to the Geodesics Having the Same Causal Character*

In a semi-Riemannian manifold, the set of geodesics through a point p can be divided into three disjoint subsets according to the causal character of the initial velocity vector at p. It is natural to ask if the equivalence in Theorem 1 holds restricting the geodesics, or equivalently the tangent vectors to H at any $p \in U \subset H$, involved in the definitions of local, or infinitesimal convexity, to one of these subsets.

To be more precise, let us give the definition of local and infinitesimal convexity, taking into account causality. Let H be a nondegenerate embedded hypersurface of (M,g), $p \in H$ and W_p be the set of the time-like (resp. light-like, space-like) vectors in T_pM. Let $U \subset H$ be a neighborhood of p in H and n be a unit normal smooth vector field on U. We say that H is *time-* (resp. *null-*, *space-*) *locally convex* at $p_0 \in H$ if there exists a neighborhood O of $0 \in T_{p_0}H$ such that $\exp_{p_0}(W_{p_0} \cap O)$ is contained in the closure of the connected component of a tubular neighborhood of $H \cap U$ individuated by $-n$. Analogously, we say that a nondegenerate embedded hypersurface H is *time-* (resp. *null-*, *space-*) *infinitesimally convex* in a neighborhood $U \subset H$ of p_0 if its second fundamental form with respect to n is positive semidefinite on time-like (resp. null, space-like) vector field on TU, i.e., $\Pi_p(v,v) \geq 0$, for each $p \in U$ and for all vectors $v \in W_p \cap T_pH$. By Lemma 1, this last condition is equivalent to $(H_\phi)_p(v,v) \leq 0$, for all $v \in W_p \cap T_pH$, where ϕ is a function as in Definition 1. As in Remark 3, defining time- (resp. null-, space-) infinitesimal convexity in this last way allows one to consider also degenerate hypersurface.

Remark 5. Since the null cone at a point $p \in M$ is the boundary of both the subsets of space-like and time-like vectors at p, if H is time- or space-infinitesimally convex at a point $p_0 \in H$ then, by continuity, it is also null-infinitesimally convex.

We observe that Lemma 2 continues to hold for time- and space-infinitesimal convexities, up to taking a smaller $b > 0$ as the upper limit of the interval of definition of the geodesic γ. Indeed, the projection map $\mathscr{P}$ is smooth, and therefore, since $\dot{\gamma}_H(0) = \dot{\gamma}(0)$ is time-like (resp. space-like), the curve γ_H remains time-like (resp. space-like) on a right neighborhood of $0 \in \mathbb{R}$; thus, inequality (4) is true for s in such neighborhood (clearly, for null-infinitesimal convexity, this argument is invalid).

Remark 6. Under the weaker hypothesis that H is only time- (resp. space-) infinitesimally convex in U, the proof of the "only if" part of Theorem 1 becomes wrong because the set A, now defined as the set of the couples of points in C that can be joined by a time-like (resp. space-like) geodesic contained in C, is no longer closed; in fact, the limit of a sequence of time-like or space-like geodesics might be a null geodesic, Remark 5. Moreover, allowing the limit be a null geodesic leads to change the definition of A including points that can be connected by a null geodesic but then A becomes non-open. Thus, we leave the following as an open problem:

Problem 1. Prove or disprove that for any smooth embedded hypersurface H in a semi-Riemannian manifold (or, at least, in a Lorentzian manifold) time-, null- or space- infinitesimal convexity in a neighborhood $U \subset H$ of a point $p \in H$ implies the same type (time-, null- or space-) of local convexity at p.

Remark 7. A positive answer to this problem has been given in [12, Corollary 3.5 and Remark 3.8] for null/time-infinitesimal and local convexities of a hypersurface of the type $H_0 \times \mathbb{R}$ in a standard stationary Lorentzian manifold $M_0 \times \mathbb{R}$, by reducing the problem to a Finslerian one (cf. [13, 14] or [22]) and using the above-mentioned result [2].

4.2 Geometric Convexity

Let Ω be an open subset of the semi-Riemannian manifold (M,g) with C^2 differentiable boundary $\partial\Omega$. Let $\phi : M \to \mathbb{R}$ be a C^2 function such that 0 is a regular value, $\phi^{-1}(0) = \partial\Omega$, and $\phi(x) > 0$, for each $x \in \Omega$.

In what follows, infinitesimal convexity of $\partial\Omega$ will be always considered as defined in terms of the function ϕ globally defining $\partial\Omega$ as a regular level set.

Observe that under these assumptions $\partial\Omega$ is C^2 embedded and oriented hypersurface in (M,g) (the orientation is given by the transversal vector field $\nabla^R\phi$, where ∇^R is the gradient operator with respect to any auxiliary Riemannian metric on M).

From Lemma 2, it immediately follows that if $\partial\Omega$ is infinitesimally convex at any of its points, then the following condition, called (e.g. in [2–4, 29]) *geometric convexity* of $\partial\Omega$ or, equivalently, of Ω, holds:

1. For any two points $p,q \in \Omega$ and for any geodesic arc γ: $[0,1] \to M$ from p to q, if $\gamma([0,1])$ is contained in $\Omega \cup \partial\Omega$ then, actually, it is contained in Ω.

(observe that the same also happens for respectively time/space-infinitesimal convexity of $\partial\Omega$ and time-like and space-like geodesics).

This fact was already observed in [18, Theorem 6] for a C^3 domain Ω in a complete Riemannian manifold. We emphasize that, from a technical point of view, local convexity at each point of $\partial\Omega$ (which clearly implies geometric convexity) is harder to prove than the other convexity notions except in the case when *strongly* (i.e., (1) is satisfied with the strict inequality) infinitesimal convexity holds (cf. [29, Sect. 1], where strongly infinitesimal convexity is called *strictly* infinitesimal convexity); for example, in trying to prove that infinitesimal convexity in a neighborhood of a point $p \in \partial\Omega$ implies local convexity at p, we must encompass the possibility of a geodesic $\gamma = \gamma(t)$ oscillating between Ω and $M \setminus \bar{\Omega}$, as $t \to t_0$, with $\gamma(t_0) = p$. In other words, the fact that ρ is nonnegative is essential in differential inequality (5). Anyway, if geometric convexity holds, we immediately get, as in the first part of the proof of Theorem 1, that $\partial\Omega$ is infinitesimal convex at any of its points. Thus, we can state:

Corollary 1. *For any C^2 open subset Ω of a semi-Riemannian manifold local, infinitesimal and geometric convexity of $\partial\Omega$ are equivalent; moreover, time/space-infinitesimal convexity are respectively equivalent to time/space-geometric convexity.*

We recall that the first chain of equivalence holds also for a C^2 open subset of a Finsler manifold (see [2, Corollary 1.2] and recall Remark 2), while both the equivalences between full infinitesimal and geometric convexity and time/space-infinitesimal convexity and time/space-geometric convexity were obtained in [4, Theorems 4.3, 4.4 and Appendix A] for a C^3 open subset of the type $\Omega = \Omega_0 \times \mathbb{R}$ in a standard stationary Lorentzian manifold $M = M_0 \times \mathbb{R}$. The equivalence in the null case, already obtained in [4] for a static standard region, was proved in [3, Theorem 2.5]. We observe that this last equivalence also follows by the results in [12] as in Remark 7.

Space-, null- and time-geometric convexity, in a Lorentzian setting, were introduced in [6] and were used there, together with variational methods, to prove existence and multiplicity results about the number of space-like and time-like geodesics connecting a couple of points (for time-like ones, only certain chronologically related points are to be considered) in an open subset $\Omega = \Omega_0 \times \mathbb{R}$ having space or time-geometrically convex boundary of a standard static Lorentzian manifold $M = M_0 \times \mathbb{R}$. These results were extended to standard stationary Lorentzian manifolds in [19]. Remarkably, null- and time-geometric convexity of some open subsets of this type, contained in the outer Schwarzschild, Reissner-Nordström and Kerr spacetimes, have been proved in [25, Sect. 7] (see also [7, 12, 17]). We stress that these results were obtained by showing that the boundaries of such open subsets are strongly infinitesimally convex and indeed the advantage of the notion of infinitesimal convexity with respect to the other ones relies, of course, in its direct computability.

4.3 Some Applications to Geodesic Connectedness

It is worth to observe that geometric convexity extends the classical notion of convexity of a subset of $\mathbb{R}^m$. For example, assume that (M, g) is a smooth complete Riemannian manifold and Ω is a smooth, connected open subset of M having geometrically convex boundary $\partial\Omega$; then there exists a (non necessarily unique) geodesic connecting p to q, contained in Ω and having length equal to the distance in Ω between p and q. A way to prove that is to apply both a minimization and a penalization argument to the energy functional of the Riemannian manifold. We refer to [25, Corollary 4.4.7] for a proof, using these variational methods, of the existence of a geodesic connecting p to q; the minimizing property of such a geodesic can be proved as in [21, Remark at p.448] while the case when $\partial\Omega$ is not differentiable and/or (M,g) is not complete has been studied in [4].

This result also holds in a forward or backward complete Finsler manifold [2, Theorem 1.3]. A more general result is obtained replacing the completeness of the Finsler manifold (M,F) with the assumption that the closure in Ω of any ball, with respect to the symmetric distance associated to the pseudo-distance induced by F on Ω, is compact [2, Theorem 1.3] and [12, Remark 4.2].

Another way to prove such results is by using a shortening argument to the length functional. In an open subset of a Lorentzian manifold satisfying good causality properties, similar and in some aspects dual (cf. [27, p. 409]) arguments are applicable. We refer to [27, Chap. 14] for more details on the definitions and notations about causality that we are going to use.

Let (M,g) be a time-oriented Lorentzian manifold and γ be a causal curve on M (i.e., assuming for simplicity that γ is piecewise smooth, $g(\dot\gamma,\dot\gamma)\leq 0$)). The Lorentzian length of γ is defined as $L(\gamma)=\int_\gamma\sqrt{-g(\dot\gamma,\dot\gamma)}$. Let Ω be a subset of M and p, q two points in Ω. We will denote by $\mathscr{C}^\Omega_{pq}$ the set of the future pointing causal curves $\gamma\colon [0,1]\to\Omega$ such that $\gamma(0)=p$ and $\gamma(1)=q$. We say that q *is causally related to* p *in* Ω, and we write $p<^\Omega q$, if $\mathscr{C}^\Omega_{pq}\neq\emptyset$. Let us define the *time separation in* Ω, $\tau^\Omega(p,q)\in[0,+\infty]$, as $\tau^\Omega(p,q)=\sup_{\gamma\in\mathscr{C}^\Omega_{pq}}L(\gamma)$, if $\mathscr{C}^\Omega_{pq}\neq\emptyset$ otherwise, $\tau^\Omega(p,q)=0$.

Proposition 3. *Let (M,g) be a time-oriented Lorentzian manifold and Ω be a C^2, open subset such that $\bar\Omega$ is compact and strongly causal. Assume that $\partial\Omega$ is convex (recall the first part of Corollary 1), then for any $p,q\in\Omega$ with $p<^\Omega q$ there is a future pointing causal geodesic $\gamma\in\mathscr{C}^\Omega_{pq}$ such that $L(\gamma)=\tau^\Omega(p,q)$ (hence $\tau^\Omega(p,q)$ is finite).*

Proof. Let $\{\gamma_n\}$ be a sequence of curves, $\gamma_n:[0,1]\to\Omega$, such that $\gamma_n(0)=p$, $\gamma_n(1)=q$ and $L(\gamma_n)\to\tau^\Omega(p,q)$. From [27, Chap. 14, Proposition 8] there is a limit sequence $p_0=p<^M p_1<^M\ldots<p_j<^M\ldots$ (see [27, Chap. 14, Def. 7]) for $\{\gamma_n\}$. Arguing as in the proof of [27, Chap. 14, Lemma 14] the limit sequence must be finite, and the last point p_m has to be equal to q. From the definition of a limit sequence, there exist a subsequence $\{\gamma_k\}$ and sequences $\{s_{kj}\}$, $j=0,\ldots,m$, such that $\gamma_k(s_{kj})\to p_j$ and then $p_j\in\bar\Omega$; moreover, there are convex neighborhoods $\{C_j\}_{j=0,\ldots,m-1}$ associated to the limit sequence $\{p_j\}_{j=0,\ldots,m}$ such that $p_j,p_{j+1}\in C_j$, for each $j=0,\ldots,m$. From the second part of the proof of Theorem 1, we know that the sets $C_j\cap\Omega$ are convex neighborhoods. Consider the future pointing causal broken geodesic λ from p to q, with vertices p_j, having one segment λ_j in each convex set C_j (such broken geodesic is a *quasi-limit* of $\{\gamma_n\}$, [27, p. 406]). Consider also the future pointing causal geodesics λ_{kj} connecting $\gamma_k(s_{kj})$ with $\gamma_k\big(s_{k(j+1)}\big)$ in[3] C_j. By the smooth dependence of geodesics by boundary conditions in a convex neighborhood ([27, Chap. 5, Lemma 9]) the geodesics λ_{kj} converge in the C^2 topology to λ_j and $L(\lambda_{kj}) \to L(\lambda_j)$. Hence λ is contained in $\bar{\Omega}$ and, by the length maximizing property of each causal geodesic segment in C_j (see [27, Chap. 5, Proposition 34]), we get $\tau^{\Omega}(p,q) = \lim_k L(\gamma_k) \leq \lim_k L(\lambda_k) = L(\lambda)$, where λ_k is the broken geodesics in Ω with vertices $\gamma_k(s_{kj})$ and segments λ_{kj}. Since p belongs to Ω, the portion of λ given by the first segment λ_0 plus a small part $\lambda_1^s \subset \lambda_1$ starting at p_1 and contained in C_0 cannot be entirely contained in $\partial\Omega$, and then, by Lemma 2, it cannot be a pregeodesic. Therefore, there exists a causal future pointing geodesic α_0 having length strictly greater than $L(\lambda_0)+L(\lambda_1^s)$, connecting p to the end point p_1^s of λ_1^s. By choosing a sequence of points $\{p_{h1}^s\} \subset C_0 \cap \Omega$, such that $p_{h1}^s \to p_1^s$, the future pointing causal geodesics defined by the points p_0 and p_{h1}^s and having support in $C_0 \cap \Omega$ converge, in the C^2 topology, to α_0. Hence by Lemma 2, α_0 does not intersect $\partial\Omega$ in any of its inner points. As the end point p_1^s belongs to $C_0 \cap C_1$, in an analogous way, replacing C_0 with C_1, we can construct a broken geodesic segment α_2 between p and p_2, longer than α_0 plus the segment of λ from p_1^s and p_2, which does not intersect $\partial\Omega$ in any of its inner point. Thus in m steps, we construct a future pointing causal broken geodesic α connecting p to q, entirely contained in Ω and such that $\tau^{\Omega}(p,q) \leq L(\lambda) < L(\alpha)$. This contradiction comes from the fact that we have assumed that λ is not a geodesic. Using again Lemma 2, we conclude that λ must be contained in Ω and $\tau^{\Omega}(p,q) = L(\lambda)$. □

Since a globally hyperbolic Lorentzian manifold is strongly causal and the intersection of the causal future $J^+(p)$ of p with the causal past $J^-(q)$ of q is compact, as in Proposition 3 we get the following.

Corollary 2. *Let (M,g) be a globally hyperbolic Lorentzian manifold and Ω be a C^2 open subset. Assume that $\partial\Omega$ is convex, then for any $p,q \in \Omega$ with $p <^{\Omega} q$, there is a future pointing causal geodesic $\gamma \in \mathscr{C}_{pq}^{\Omega}$ such that $L(\gamma) = \tau^{\Omega}(p,q)$.*

We don't know if we can replace the assumption of (infinitesimal or, equivalently, local) convexity of $\partial\Omega$ in Proposition 3 with time-infinitesimal convexity (or even time- and null-local convexity). In fact, in that case, we cannot state that the sets $C_i \cap \Omega$ associated to the limit sequence in the proof of Proposition 3 are convex neighborhoods (recall Remark 6).

Problem 2. Does the conclusion of Proposition 3 (or of Corollary 2) hold assuming that $\partial\Omega$ is time-infinitesimally convex?

In some globally hyperbolic stationary Lorentzian manifold we can obtain the full geodesic connectedness of an open connected subset having convex boundary. Indeed the following proposition holds.

[3]Observe that such geodesics must be causal and future pointing as the points $\gamma_k(s_{kj})$ and $\gamma_k\big(s_{k(j+1)}\big)$ are causally related, see [27, Chap. 14, Lemma 2].

Proposition 4. *Let (M,g) be a Lorentzian manifold endowed with a complete time-like Killing vector field K and admitting a smooth, space-like, complete Cauchy hypersurface. Let Ω be a C^2 open connected subset of M having convex boundary and which is invariant for the flow of K. Then for each p and q in Ω, there is a geodesic connecting p to q whose support is in Ω. Moreover, if Ω is also non contractible, a sequence $\{\gamma_n\}$ of space-like geodesics in Ω connecting p to q exists such that $g(\dot\gamma_n,\dot\gamma_n)\to+\infty$ and, if $l\colon \mathbb{R}\to M$ is a flow line of K with $l(\mathbb{R})\subset\Omega$, then the number of time-like future pointing geodesics in Ω connecting each $p\in\Omega$ to $l(s)$ diverges as $s\to+\infty$.*

The above proposition extends the results in [28], where K is also assumed to be static (i.e., its orthogonal distribution is integrable). It is based on the fact that the existence of a smooth, space-like, Cauchy hypersurface and the completeness of K imply that (M,g) is isometric by the map $\psi\colon S\times\mathbb{R}\to M$, where ψ is the flow of K and S the Cauchy hypersurface, to a standard stationary spacetime $S\times\mathbb{R}$ (cf. [11, Theorem 2.3]). As Ω is invariant for the flow of K, then $\Omega=\Omega_0\times\mathbb{R}$, with $\Omega_0=S\cap\Omega$, and since S is complete, $\Omega_0\cup\partial\Omega_0$ is complete. Then, as in the case without boundary in [11,20], one can prove that the projections on Ω_0 of the curves in a sublevel of the restriction of the energy functional of the standard stationary region $\Omega_0\times\mathbb{R}$ to the manifold of the H^1 curves γ connecting p to q and such that $g(\dot\gamma,K)$ is constant a.e., are contained in a compact subset of $\Omega_0\cup\partial\Omega_0$. Then, using a penalization argument as in [25, Sect. 4.2], the result follows as in [25, Sects. 4.4 and 4.5].

We point out that, due to the particular variational setting, last part of the statement of the above proposition follows as well assuming that $\partial\Omega$ is only time-infinitesimally convex (recall Problem 2). For a standard stationary region $\Omega_0\times\mathbb{R}$, a result of this type, without assuming global hyperbolicity of the ambient spacetime (M,g), has been obtained in [12, Theorem 4.6].

Acknowledgements I would like to thank M. Sánchez for suggesting to investigate the topic of this note and for several useful comments. Moreover, I thank the local organizing committee of the "VI International Meeting on Lorentzian Geometry, Granada 2011" for the financial support and the hospitality during the workshop.

References

1. Bao, D., Chern, S.S., Shen, Z.: An Introduction to Riemann-Finsler geometry. Graduate Texts in Mathematics. Springer, New York (2000)
2. Bartolo, R., Caponio, E., Germinario A., Sánchez, M.: Convex domains of Finsler and Riemannian manifolds. Calc. Var. Part. Differ. Equat. **40**, 335–356 (2011)
3. Bartolo, R., Germinario, A.: Convexity conditions on the boundary of a stationary spacetime and applications. Commun. Contemp. Math. **11**, 739–769 (2009)
4. Bartolo, R., Germinario, A., Sánchez, M.: A note on the boundary of a static Lorentzian manifold. Differ. Geom. Appl. **16**, 121–131 (2002)

5. Beem, J.K., Ehrlich, P.E., Easley, K.L.: Global Lorentzian Geometry, 2nd edn. Monographs and Textbooks in Pure and Applied Mathematics. Marcel Dekker, New York (1996)
6. Benci, V., Fortunato, D., Giannoni, F.: Geodesics on static Lorentzian manifolds with convex boundary. In: "Proc. Variational Methods in Hamiltonian Systems and Elliptic Equations". Pitman Res. Notes Math. Ser. vol. 243, pp. 21–41. Longman (1990)
7. Benci, V., Fortunato, D., Giannoni, F.: On the existence of geodesics in static Lorentz manifolds with singular boundary. Ann. Scuola Norm. Sup. Pisa Cl. Sci. **19**(4), 255–289 (1992)
8. Berger, M.: An extension of Rauch's metric comparison theorem and some applications. Illinois J. Math. **6**, 700–712 (1962)
9. Bishop, R.L.: Infinitesimal convexity implies local convexity. Indiana Univ. Math. J. **24**, 169–172 (1974/75)
10. Borisenko, A.A., Olin, E.A.: The global structure of locally convex hypersurfaces in Finsler-Hadamard manifolds. Mat. Zametki **87**, 163–174 (2010)
11. Candela, A.M., Flores, J.L., Sánchez, M.: Global hyperbolicity and Palais-Smale condition for action functionals in stationary spacetimes. Adv. Math. **218**, 515–536 (2008)
12. Caponio, E., Germinario, A., Sánchez, M.: Geodesics on convex regions of stationary spacetimes and Finslerian Randers spaces. Preprint URL http://arxiv.org/abs/1112.3892v1 (2011)
13. Caponio, E., Javaloyes, M.A., Masiello, A.: On the energy functional on Finsler manifolds and applications to stationary spacetimes. Math. Ann. **351**, 365–392 (2011)
14. Caponio, E., Javaloyes, M.A., Sánchez, M.: On the interplay between Lorentzian causality and Finsler metrics of Randers type. Rev. Mat. Iberoamericana **27**, 919–952 (2011)
15. do Carmo, M.P., Warner, F.W.: Rigidity and convexity of hypersurfaces in spheres. J. Differ. Geom. **4**, 133–144 (1970)
16. Ehrlich, P., Kim, S.B.: A focal comparison theorem for null geodesics. Commun. Korean Math. Soc. **6**, 73–87 (1991)
17. Flores, J.L., Sánchez, M.: Geodesics in stationary spacetimes. Application to Kerr spacetime. Int. J. Theor. Phys. Group Theory Nonlinear Opt. **8**, 319–336 (2002)
18. Germinario, A.: Homoclinics on Riemannian manifolds with convex boundary. Dynam. Syst. Appl. **4**, 549–566 (1995)
19. Giannoni, F., Masiello, A.: On the existence of geodesics on stationary Lorentz manifolds with convex boundary. J. Funct. Anal. **101**, 340–369 (1991)
20. Giannoni, F., Piccione, P.: An intrinsic approach to the geodesical connectedness of stationary Lorentzian manifolds. Comm. Anal. Geom. **1**, 157–197 (1999)
21. Gordon, W.B.: The existence of geodesics joining two given points. J. Differ. Geom. **9**, 443–450 (1974)
22. Javaloyes, M.Á.: Conformally standard stationary spacetimes and Fermat Metrics, these proceedings
23. Javaloyes, M.Á., Piccione, P.: Comparison results for conjugate and focal points in semi-Riemannian geometry via Maslov index. Pacific J. Math. **243**, 43–56 (2009)
24. Kim, S.B.: The focal Lorentzian comparison theorems for timelike geodesics. Honam Math. J. **12**, 17–25 (1990)
25. Masiello, A.: Variational Methods in Lorentzian Geometry. Pitman Research Notes in Mathematics Series. Longman Scientific and Technical, NewYork (1994)
26. Nomizu, K.: Remarks on sectional curvature of an indefinite metric. Proc. Amer. Math. Soc. **89**, 473–476 (1983)
27. O'Neill, B.: Semi-Riemannian geometry. Pure and Applied Mathematics. Academic, New York (1983)
28. Piccione, P.: Existence of geodesics in static Lorentzian manifolds with convex boundary. Proc. Roy. Soc. Edinburgh Sect. A **130**, 189–215 (2000)
29. Sánchez, M.: Geodesic connectedness of semi-Riemannian manifolds. Nonlinear Anal. **47**, 3085–3102 (2001)

Global Geodesic Properties of Gödel-type SpaceTimes

Rossella Bartolo, Anna Maria Candela, and José Luis Flores

Mathematics Subject Classification (2010): 53C50, 53C22, 58E10.

1 Introduction

Classical critical point theorems and standard Morse theory are directly applicable to functionals bounded from below which satisfy compactness assumptions, such as the Palais–Smale condition (see Sect. 2), and whose critical points have finite Morse index. Unluckily, these tools cannot be applied to many interesting problems involving functionals that are strongly indefinite. For example, geodesics joining two points z_p, z_q on an indefinite semi-Riemannian manifold $(\mathcal{M}, \langle\cdot,\cdot\rangle_L)$ are the critical points of the strongly indefinite C^1 action functional

$$f(z) = \int_0^1 \langle \dot{z}, \dot{z} \rangle_L \mathrm{d}s \tag{1}$$

defined on the Hilbert manifold Ω of all the H^1–curves joining z_p to z_q in $\mathcal{M}$ (for more details, see Sect. 2). Anyway, starting from the seminal paper [4], in

R. Bartolo (✉)
Dipartimento di Meccanica Matematica e Management, Politecnico di Bari,
Via E. Orabona 4, 70125 Bari, Italy
e-mail: r.bartolo@poliba.it

A.M. Candela
Dipartimento di Matematica, Università degli Studi di Bari "Aldo Moro",
Via E. Orabona 4, 70125 Bari, Italy
e-mail: candela@dm.uniba.it

J.L. Flores
Departamento de Álgebra, Geometría y Topología, Facultad de Ciencias,
Universidad de Málaga, Campus Teatinos, 29071 Málaga, Spain
e-mail: floresj@agt.cie.uma.es

M. Sánchez et al. (eds.), *Recent Trends in Lorentzian Geometry*, Springer Proceedings in Mathematics & Statistics 26, DOI 10.1007/978-1-4614-4897-6_7,

some particular settings, and according to the properties of the manifold $\mathcal{M}$ and its indefinite metric $\langle\cdot,\cdot\rangle_L$, the functional f in Eq. (1) has been widely studied by using variational methods, also obtaining sometimes optimal results at least in the Lorentzian case (we refer to the book [22] and to the updated survey paper [12] and references therein). A typical situation occurs when the Lorentzian metric tensor $\langle\cdot,\cdot\rangle_L$ presents symmetries (i.e., Killing vector fields): one gets rid of the negative contributions in the directions of the Killing fields and, by means of some variational principles, it is possible to handle with simpler functionals, which essentially depend only on a Riemannian metric, so that they are bounded from below and satisfy the Palais–Smale condition under reasonable assumptions. This is the case of standard stationary and Gödel-type spacetimes.

Definition 1. A Lorentzian manifold $(\mathcal{M},\langle\cdot,\cdot\rangle_L)$ is a *standard stationary spacetime* if there exist a smooth, finite–dimensional Riemannian manifold $(\mathcal{M}_0,\langle\cdot,\cdot\rangle_R)$, a vector field δ and a positive smooth function β on $\mathcal{M}_0$ such that $\mathcal{M}=\mathcal{M}_0\times\mathbb{R}$ and the Lorentz metric (under natural identifications) is

$$\langle\cdot,\cdot\rangle_L \;=\; \langle\cdot,\cdot\rangle_R+2\langle\delta(x),\cdot\rangle_R\,dt-\beta(x)\,dt^2. \tag{2}$$

When the cross term vanishes ($\delta\equiv 0$), the spacetime is called *standard static*. This is a warped product $\mathcal{M}_0\times_{\sqrt{\beta}}\mathbb{R}$ with Riemannian base and negative definite fiber.

Recall that every stationary spacetime (i.e., a spacetime admitting a timelike Killing vector field K) is locally a standard stationary one with $K=\partial_t$.

On the other hand, Gödel-type spacetimes are Lorentzian manifolds admitting a pair of commuting Killing vector fields which span a two-dimensional distribution where the metric has index 1 (the causal characters of the Killing vectors could change on the manifold, see [13, Example 5.1]). More precisely, we use the following definition (according to [10]):

Definition 2. A Lorentzian manifold $(\mathcal{M},\langle\cdot,\cdot\rangle_L)$ is a *Gödel-type spacetime*, briefly *GTS*, if a smooth, finite–dimensional Riemannian manifold $(\mathcal{M}_0,\langle\cdot,\cdot\rangle_R)$ exists such that $\mathcal{M}=\mathcal{M}_0\times\mathbb{R}^2$ and the metric $\langle\cdot,\cdot\rangle_L$ is described as

$$\langle\cdot,\cdot\rangle_L \;=\; \langle\cdot,\cdot\rangle_R+A(x)dy^2+2B(x)dydt-C(x)dt^2, \tag{3}$$

where $x\in\mathcal{M}_0$, the variables (y,t) are the natural coordinates of $\mathbb{R}^2$, and A, B, C are C^1 scalar fields on $\mathcal{M}_0$ satisfying

$$H(x) \;=\; B^2(x)+A(x)C(x)>0 \quad \text{for all } x\in\mathcal{M}_0. \tag{4}$$

Let us observe that condition (4) implies that metric (3) is Lorentzian. It is also interesting to point out that *GTS* are not necessarily time–orientable (e.g., cf. [13, Remark 1.2]).

In [18] Gödel gives an exact solution of Einstein's field equations with homogeneous perfect fluid distribution, the so-called classical Gödel universe. This spacetime, described in Example $1(e_1)$ (see also [16, 21] where its geodesic equations are explicitly integrated), has a five dimensional group of isometries, is geodesically complete, and admits closed causal curves (e.g., cf. [19]). In [28], Raychaudhuri and Thakurta start the study of homogeneity properties of GTS investigating homogeneity conditions of a class of cylindrically symmetric metrics; later on, in [29], Rebouças and Tiomno introduce a definition for Gödel metrics in four dimensions and study their homogeneity conditions (see also [6, 15]).

Example 1. The class of GTS depicted in Definition 2 is wide; indeed, this definition covers very different kinds of spacetimes, including some physically relevant examples.

(e_1) The Gödel universe (cf. [18]) is a GTS with

$$\mathscr{M}_0 = \mathbb{R}^2, \quad \langle\cdot,\cdot\rangle_R = dx_1^2 + dx_2^2$$

and with coefficients in Eq. (3) given by

$$A(x) = -\mathrm{e}^{2\sqrt{2}\omega x_1}/2, \quad B(x) = -\mathrm{e}^{\sqrt{2}\omega x_1}, \quad C(x) \equiv 1$$

($\omega > 0$ represents the magnitude of the vorticity of the flow). In [10], by a direct integration of the geodesic equations, it is constructed a geodesic joining each couple of points in $\mathscr{M}$.

(e_2) The Gödel–Synge spacetimes (cf. [31]) are GTS with $\mathscr{M}_0 = \mathbb{R}^2$ and

$$\langle\cdot,\cdot\rangle_L = dx_1^2 + dx_2^2 - g(x_1)dy^2 - 2h(x_1)dydt - dt^2,$$

where g, h are smooth functions of x_1 with $g > 0$. If $2g = h^2$ and $h = \mathrm{e}^{x_1}$, this metric reduces to the Gödel classical one.

(e_3) Some Kerr–Schild spacetimes (e.g., cf. [20]) are GTS with again $\mathscr{M}_0 = \mathbb{R}^2$ and

$$\langle\cdot,\cdot\rangle_L = dx_1^2 + dx_2^2 + dy^2 - dt^2 + V(x_1,x_2)(dy+dt)^2,$$

where V is an arbitrary function on $\mathbb{R}^2$. In this case, the coefficients in Eq. (3) are

$$A(x) = 1 + V(x), \quad B(x) = V(x), \quad C(x) = 1 - V(x),$$

and thus, $H(x) \equiv 1$.

(e_4) Some standard stationary spacetimes are GTS with $\mathscr{M} = \mathscr{M}_0 \times \mathbb{R}^2$, being $(\mathscr{M}_0 \times \mathbb{R}, \langle\cdot,\cdot\rangle_R + dy^2)$ the Riemannian part and

$$\langle\cdot,\cdot\rangle_L = \langle\cdot,\cdot\rangle_R + dy^2 + 2\delta(x)dydt - \beta(x)dt^2$$

the stationary metric with $\delta(x,y) \equiv \delta(x) \in \mathbb{R}$ and $\beta(x,y) \equiv \beta(x) > 0$ in $\mathcal{M}_0 \times \mathbb{R}$. Clearly, they are GTS with metric coefficients $A(x) \equiv 1$, $B(x) = \delta(x)$ and $C(x) = \beta(x)$.

Vice versa, some GTS are standard stationary spacetimes when $A(x)C(x) > 0$ on $\mathcal{M}_0$, being standard static if, in addition, $B \equiv 0$. For example, if $A(x) > 0$ on $\mathcal{M}_0$, the spatial part of the stationary spacetime corresponds to $\mathcal{M}_0 \times \mathbb{R}$ equipped with the Riemannian metric $\langle \cdot,\cdot \rangle_R + A(x)dy^2$ (which is complete if so is $\langle \cdot,\cdot \rangle_R$), the vector field becomes $\delta(x,y) = (0, B(x)) \in T\mathcal{M}_0 \times \mathbb{R}$, and the scalar field is $\beta(x,y) = C(x) > 0$ for each $(x,y) \in \mathcal{M}_0 \times \mathbb{R}$.

(e_5) Some examples of general plane fronted waves are also GTS. More precisely, a *general plane fronted wave* is a Lorentzian manifold $\mathcal{M}_0 \times \mathbb{R}^2$ equipped with the metric

$$\langle \cdot,\cdot \rangle_L = \langle \cdot,\cdot \rangle_R + 2dydt + H_0(x,t)dt^2,$$

where $(\mathcal{M}_0, \langle \cdot,\cdot \rangle_R)$ is a Riemannian manifold, (y,t) are the natural coordinates of $\mathbb{R}^2$, and the smooth scalar field H_0 on $\mathcal{M}_0 \times \mathbb{R}$ satisfies $H_0 \not\equiv 0$. Clearly, when $H_0(x,t)$ is autonomous (i.e., it does not depend on t), this spacetime is a GTS. Results on geodesic completeness and connectedness for these spacetimes can be found in [7].

The importance of the spacetimes above justifies the study of global properties such as geodesic connectedness and geodesic completeness. However, one cannot expect to prove general results, as these properties depend strongly on the metric coefficients (see respectively Theorems 2 and 3 and related comments). This dependence is also evident in the study of causality properties for GTS: it is well known that the classical Gödel universe is not chronological and, on the other side, stationary spacetimes can be globally hyperbolic (cf. [30, Corollary 3.4] and [14, Theorem 4.3]).

The chapter is organized as follows. In Sect. 2, we recall some variational principles for geodesics on static spacetimes and GTS. In Sect. 3, we present a new result on geodesic connectedness, and compare it with the previous ones in [2], showing its accuracy by examples. In Sect. 4, we deal with geodesic completeness and state a sufficient condition in order to obtain it. Finally, in the Appendix, we fix some widely known notations about the variational set up.

2 The Variational Principle

According to notations and statements contained in the Appendix, there is a correspondence between geodesics joining two given points z_p, z_q on a semi–Riemannian manifold $(\mathcal{M}, \langle \cdot,\cdot \rangle_L)$ and critical points of the action functional f in Eq. (1) on the Hilbert manifold $\Omega^1(z_p, z_q)$. As already remarked, if $\langle \cdot,\cdot \rangle_L$ is not Riemannian then f is strongly indefinite, but, in some Lorentzian manifolds, this difficulty can be overcome by introducing a new suitable functional.

The kernel of our approach is a variational principle stated in [5, Theorem 2.1] for static Lorentzian manifolds $\mathcal{M} = \mathcal{M}_0 \times \mathbb{R}$, with $\langle\cdot,\cdot\rangle_L$ as in Eq. (2) and $\delta \equiv 0$ (extended to stationary spacetimes in [17, Theorem 2.2], see also [8]). It is based on the fact that $\langle\partial_t,\dot{z}\rangle_L$ is constant along each geodesic z, because of the Killing character of ∂_t. Namely, $z_p = (x_p,t_p)$, $z_q = (x_q,t_q) \in \mathcal{M}$ are connected by a geodesic $\bar{z} = (\bar{x},\bar{t})$, which is a critical point of the functional f in (1) on $\Omega^1(z_p,z_q) = \Omega^1(x_p,x_q) \times W(t_p,t_q)$, if and only if $\bar{x}$ is a critical point of the functional

$$J(x) \;=\; \frac{1}{2}\int_0^1 \langle\dot{x},\dot{x}\rangle_R\,\mathrm{d}s - \frac{\Delta_t^2}{2}\left(\int_0^1 \frac{1}{\beta(x)}\,\mathrm{d}s\right)^{-1} \quad \text{on } \Omega^1(x_p,x_q), \tag{5}$$

with $\Delta_t := t_q - t_p$.

Next, let us consider the more general setting of GTS with $\mathcal{M} = \mathcal{M}_0 \times \mathbb{R}^2$ and $\langle\cdot,\cdot\rangle_L$ as in Definition 2. For each $x \in H^1(I,\mathcal{M}_0)$, let us define

$$a(x) = \int_0^1 \frac{A(x)}{H(x)}\,\mathrm{d}s,\;\; b(x) = \int_0^1 \frac{B(x)}{H(x)}\,\mathrm{d}s,\;\; c(x) = \int_0^1 \frac{C(x)}{H(x)}\,\mathrm{d}s, \tag{6}$$

$$\mathcal{L}(x) = b^2(x) + a(x)c(x). \tag{7}$$

As every GTS admits two Killing vector fields ∂_y, ∂_t (not necessarily timelike), an extension of the previous variational principle can be stated (cf. [10, Proposition 2.2]). In this setting, fixing $z_p = (x_p,y_p,t_p)$, $z_q = (x_q,y_q,t_q) \in \mathcal{M}$, with x_p, $x_q \in \mathcal{M}_0$ and (y_p,t_p), $(y_q,t_q) \in \mathbb{R}^2$, we have that $\bar{z} : I \to \mathcal{M}$ is a geodesic joining z_p to z_q in $\mathcal{M}$ if and only if it is a critical point of the action functional (1), with $\langle\cdot,\cdot\rangle_L$ as in Eq. (3), defined on the manifold $\Omega^1(z_p,z_q) = \Omega^1(x_p,x_q) \times W(y_p,y_q) \times W(t_p,t_q)$. Let $x \in \Omega^1(x_p,x_q)$ be such that $\mathcal{L}(x) \neq 0$ (cf. (7)). For all $s \in I$ we define

$$\phi_y(x)(s) := y_p + \frac{\Delta_y\, b(x) - \Delta_t\, c(x)}{\mathcal{L}(x)} \int_0^s \frac{B(x)}{H(x)}\mathrm{d}\sigma + \frac{\Delta_y a(x) + \Delta_t\, b(x)}{\mathcal{L}(x)} \int_0^s \frac{C(x)}{H(x)}\mathrm{d}\sigma,$$

$$\phi_t(x)(s) := t_p - \frac{\Delta_y\, b(x) - \Delta_t\, c(x)}{\mathcal{L}(x)} \int_0^s \frac{A(x)}{H(x)}\mathrm{d}\sigma + \frac{\Delta_y a(x) + \Delta_t\, b(x)}{\mathcal{L}(x)} \int_0^s \frac{B(x)}{H(x)}\mathrm{d}\sigma,$$

with $\Delta_y := y_q - y_p$ and $\Delta_t := t_q - t_p$. Standard arguments imply that the functions ϕ_y and ϕ_t, which go from $\Omega^1(x_p,x_q)$ to $W(y_p,y_q)$ and $W(t_p,t_q)$, respectively, are C^1.

Then, the following proposition holds (see [10, Proposition 2.2]).

Proposition 1. *Let $(\mathcal{M},\langle\cdot,\cdot\rangle_L)$ be a GTS and x_p, $x_q \in \mathcal{M}_0$ be such that $|\mathcal{L}(x)| > 0$ for all $x \in \Omega^1(x_p,x_q)$. Then, the following statements are equivalent:*

1. *$\bar{z} \in \Omega^1(z_p,z_q)$ is a critical point of the action functional f in Eq. (1);*
2. *setting $\bar{z} = (\bar{x},\bar{y},\bar{t})$, the curve $\bar{x} \in \Omega^1(x_p,x_q)$ is a critical point of the C^1 functional*

$$\mathcal{J}(x) = \frac{1}{2}\int_0^1 \langle\dot{x},\dot{x}\rangle_R\,\mathrm{d}s + \frac{\Delta_y^2 a(x) + 2\Delta_y\Delta_t b(x) - \Delta_t^2 c(x)}{2\mathcal{L}(x)} \quad \text{on } \Omega^1(x_p,x_q) \tag{8}$$

(see Eqs. (6)–(7)), while $\bar{y} = \phi_y(\bar{x})$, $\bar{t} = \phi_t(\bar{x})$, with ϕ_y, ϕ_t as above.

Furthermore,

$$\mathscr{J}(x) = f(x, \phi_y(x), \phi_t(x)) \quad \textit{for all } x \in \Omega^1(x_p, x_q).$$

Thus, the geodesic connectedness problem in the standard static and GTS cases reduces to give conditions on the functionals J in Eq. (5) and $\mathscr{J}$ in Eq. (8), respectively, which allows us to apply the classical critical point theorem below (see [27, Theorem 2.7]).

Theorem 1. *Assume that Ω is a complete Riemannian manifold and F is a C^1 functional on Ω which satisfies the Palais–Smale condition, i.e., any sequence $(x_k)_k \subset \Omega$ such that*

$$(F(x_k))_k \textit{ is bounded} \quad \textit{and} \quad \lim_{k \to +\infty} F'(x_k) = 0,$$

converges in Ω, up to subsequences. Then, if F is bounded from below, it attains its infimum.

Remark 8. In order to obtain a multiplicity result on geodesics joining two fixed points in standard static spacetimes or in GTS, the Ljusternik–Schnirelman theory can be applied to J in Eq. (5) or $\mathscr{J}$ in Eq. (8) whenever the Riemannian part has a "rich topology" (for the static case see [3] and references therein, and for GTS, see [2, 10, 11]).

In order to avoid technicalities, hereafter we assume that $\mathscr{M}_0$ is complete, so that $\Omega^1(x_p, x_q)$ is also complete for each $x_p, x_q \in \mathscr{M}_0$. Moreover, let us recall that a functional F on $\Omega^1(x_p, x_q)$ is *coercive* if

$$F(x) \to +\infty \quad \text{if} \quad \|\dot{x}\| \to +\infty,$$

where $\|\dot{x}\|^2 = \int_0^1 \langle \dot{x}, \dot{x} \rangle_R \, ds$.

The following result holds (cf. [3, Proposition 4.3] and [2, Lemma 5.3]).

Lemma 1. *Let $(\mathscr{M}_0, \langle \cdot, \cdot \rangle_R)$ be a C^3 complete Riemannian manifold and fix two points x_p, x_q in $\mathscr{M}_0$. Then, the following statements hold:*

(*a*) *if $\mathscr{M} = \mathscr{M}_0 \times \mathbb{R}$ is a static Lorentzian manifold and J in Eq. (5) is coercive on $\Omega^1(x_p, x_q)$, then J satisfies the Palais–Smale condition on $\Omega^1(x_p, x_q)$;*
(*b*) *if $\mathscr{M} = \mathscr{M}_0 \times \mathbb{R}^2$ is a GTS, $\mathscr{J}$ in Eq. (8) is coercive on $\Omega^1(x_p, x_q)$ and there exists $\nu > 0$ such that*

$$|\mathscr{L}(x)| \geq \nu \quad \textit{for all } x \in \Omega^1(x_p, x_q),$$

then $\mathscr{J}$ satisfies the Palais–Smale condition on $\Omega^1(x_p, x_q)$.

Summing up, geodesic connectedness of the mentioned spacetimes is guaranteed by conditions implying the coercivity and lower boundedness of the "Riemannian" functional associated to the problem.

For instance, in the case of J in Eq. (5), these conditions correspond to restrictions on the growth of the (positive) metric coefficient β in Eq. (2): β bounded in the pioneer paper [5] or, more in general, β subquadratic or growing at most quadratically with respect to the distance $d(\cdot,\cdot)$ induced on $\mathcal{M}_0$ by its Riemannian metric $\langle\cdot,\cdot\rangle_R$, i.e., existence of $\lambda \geq 0$, $k \in \mathbb{R}$ and a point $x_0 \in \mathcal{M}_0$ such that

$$0 < \beta(x) \leq \lambda d^2(x,x_0) + k \quad \text{for all } x \in \mathcal{M}_0 \tag{9}$$

(cf. [3, Theorem 1.1] and references therein). Remarkably, this second growth condition on β is optimal, as showed in [3, Sect. 7] by constructing a family of geodesically disconnected static spacetimes with superquadratic, but arbitrarily close to quadratic, coefficients β.

3 Geodesic Connectedness in *GTS*

At a first glance the problem in GTS can be handled in the same manner as in the static case. However, we cannot expect optimality by applying this variational approach. In fact, the classical Gödel universe cannot be studied by our tools, due to the lack of the assumption $\mathcal{L}(x) \neq 0$ on $\Omega^1(x_p,x_q)$ for each couple of points $x_p, x_q \in \mathbb{R}^2$ (cf. Example 1(e_1)). In this section, we state and prove a new theorem on geodesic connectedness for GTS (in addition to the previous ones in [2, 10]), which, even if not optimal, is accurate in the sense described below (see Corollary 1 and Example 2).

Theorem 2. *Let $(\mathcal{M} = \mathcal{M}_0 \times \mathbb{R}^2, \langle\cdot,\cdot\rangle_L)$ be a Gödel-type spacetime such that:*

(h_1) *$(\mathcal{M}_0, \langle\cdot,\cdot\rangle_R)$ is a C^3 complete Riemannian manifold;*
(h_2) *there exists $\nu > 0$ such that $\mathcal{L}(x) \geq \nu > 0$ for all $x \in H^1(I,\mathcal{M}_0)$;*
(h_3) *$m(x) \geq h(x) > 0$ for all $x \in H^1(I,\mathcal{M}_0)$, with $m(x) := \max\{a(x), -c(x)\}$ and*

$$h(x) := \int_0^1 \frac{ds}{\lambda d^2(x(s),x_0) + k} \qquad \text{for some } \lambda \geq 0,\ k \in \mathbb{R} \text{ and } x_0 \in \mathcal{M}_0.$$

Then, $(\mathcal{M}, \langle\cdot,\cdot\rangle_L)$ is geodesically connected.

Proof. Let us take any $z_p = (x_p, y_p, t_p)$, $z_q = (x_q, y_q, t_q) \in \mathcal{M}$, with x_p, $x_q \in \mathcal{M}_0$ and (y_p, t_p), $(y_q, t_q) \in \mathbb{R}^2$. From hypothesis (h_2) (in particular $\mathcal{L}(x) \neq 0$), Proposition 1 can be applied, and so the existence of geodesics joining z_p to z_q reduces to find critical points of $\mathcal{J}$ in Eq. (8) on $\Omega^1(x_p,x_q)$. Following the arguments developed in [2, Sect. 5], we have that $\mathcal{J}$ can be written as follows:

$$\mathcal{J}(x) = \frac{1}{2}\|\dot{x}\|^2 - \frac{1}{2}\frac{\Delta_+^2(x)}{\lambda_-(x)} - \frac{1}{2}\frac{\Delta_-^2(x)}{\lambda_+(x)}, \tag{10}$$

where

$$\lambda_{\pm}(x) = \frac{a(x) - c(x) \pm \sqrt{(a(x)+c(x))^2 + 4b(x)^2}}{2}$$

and $\Delta_{\pm}(x)$ are suitable real maps depending also on Δ_y, Δ_t (see Eq. (6) and [2, p. 784]). Since $\mathscr{L}(x) = -\lambda_-(x)\lambda_+(x)$, necessarily, $\lambda_+(x) > 0 > \lambda_-(x)$ for all $x \in \Omega^1(x_p, x_q)$, and thus

$$\mathscr{J}(x) \geq \frac{1}{2}\|\dot{x}\|^2 - \frac{1}{2}\frac{\Delta_-^2(x)}{\lambda_+(x)}.$$

Note also that, by the definition of $m(x)$ in (h_3), we get

$$\lambda_+(x) \geq \frac{a(x) - c(x) + |a(x) + c(x)|}{2} = m(x) > 0.$$

Hence, (h_3) implies

$$\mathscr{J}(x) \geq \frac{1}{2}\|\dot{x}\|^2 - \frac{\Delta_-^2(x)}{2m(x)} \geq \frac{1}{2}\|\dot{x}\|^2 - \frac{\Delta_-^2(x)}{2}(h(x))^{-1} \quad \text{for all } x \in \Omega^1(x_p, x_q).$$

So, from [3, Theorem 1.1], it follows that $\mathscr{J}$ is bounded from below and coercive (cf. Eqs. (5) and (9)). Furthermore, by (h_2) and Lemma 1(b), the functional $\mathscr{J}$ satisfies the Palais–Smale condition. Hence, Theorem 1 can be applied, and a geodesic connecting z_p with z_q is obtained. As z_p, z_q are arbitrary, the thesis follows. □

An immediate consequence of Theorem 2 is the following result concerning some standard stationary spacetimes (cf. Example 1(e_4)).

Corollary 1. *Let* $(\mathscr{M} = \mathscr{M}_0 \times \mathbb{R}^2, \langle\cdot,\cdot\rangle_L)$ *be a standard stationary spacetime with* $\langle\cdot,\cdot\rangle_L = \langle\cdot,\cdot\rangle_R + dy^2 + 2\delta(x)dydt - \beta(x)dt^2$, *where* $\delta,\ \beta : \mathscr{M}_0 \to \mathbb{R}$, $\beta(x) > 0$ *in* $\mathscr{M}_0$. *Assume also that*

(s_1) $(\mathscr{M}_0, \langle\cdot,\cdot\rangle_R)$ *is a* C^3 *complete Riemannian manifold;*
(s_2) *there exist* $\lambda_1, \lambda_2 \geq 0$, $k_1, k_2 \in \mathbb{R}$ *and a point* $x_0 \in \mathscr{M}_0$ *such that*

$$\beta(x) \leq \lambda_1 d^2(x, x_0) + k_1, \quad \delta(x) \leq \lambda_2 d(x, x_0) + k_2 \qquad \textit{for all } x \in \mathscr{M}_0.$$

Then, $(\mathscr{M} = \mathscr{M}_0 \times \mathbb{R}^2, \langle\cdot,\cdot\rangle_L)$ *is geodesically connected.*

Proof. As the standard stationary spacetime $(\mathscr{M} = \mathscr{M}_0 \times \mathbb{R}^2, \langle\cdot,\cdot\rangle_L)$ is a GTS with $A(x) \equiv 1$, $B(x) = \delta(x)$, and $C(x) = \beta(x)$, the thesis follows from Theorem 2. □

Notice that Corollary 1 is a particular case of [1, Theorem 1.2] for general standard stationary manifolds $\mathscr{M} = \mathscr{M}_0 \times \mathbb{R}$ with $\langle\cdot,\cdot\rangle_L$ as in Eq. (2), which proof is based on fine estimates involving the metric coefficients. The following example shows the accurate character of this result.

Example 2. Let us consider $\mathbb{R}^3$ endowed with the following family of metrics:

$$\langle\cdot,\cdot\rangle_{L,\varepsilon} = dx^2 + dy^2 - \beta_\varepsilon(x)dt^2, \quad \varepsilon \geq 0,$$

where $(x,y,t) \in \mathbb{R}^3$ and β_ε is a (positive) smooth function on $\mathbb{R}$ such that

$$\begin{cases} \beta_\varepsilon(x) = 1 + |x|^{2+\varepsilon} & \text{if } x \in \mathbb{R}\setminus(-1,1) \\ \beta_\varepsilon([-1,1]) \subset [1,2]. \end{cases}$$

By Corollary 1 (with $\delta \equiv 0$), the spacetime is geodesically connected if ε=0. However, the spacetime is geodesically disconnected for any (and thus, for arbitrarily close to zero) strictly positive ε (see [3, Sect. 7]).

In order to give a more precise idea of the known results on geodesic connectedness in GTS by applying variational tools, let us review the corresponding results in [2]. In [2, Theorem 4.3], by using the expression Eq. (10) of $\mathscr{J}$, the geodesic connectedness of GTS is proven under assumptions (h_1) and (h_2) in Theorem 2, in addition to the following one:

(h_3') $A(x) - C(x) > 0$ for all $x \in \mathscr{M}_0$ and the (positive) map $\frac{H(x)}{A(x)-C(x)}$ is at most quadratic.

Indeed, these conditions imply that $\mathscr{J}$ is bounded from below and coercive, which allows us to apply Theorem 1 in view of Lemma $1(b)$.

As an immediate application of this result to Kerr–Schild spacetime (Example $1(e_3)$), observe that here $A(x) - C(x) = 2V(x)$, $H(x) \equiv 1$, and $\mathscr{L}(x) \neq 0$ on $H^1(I,\mathscr{M}_0)$. Thus, the geodesic connectedness is ensured if V is strictly positive and $(2V(x))^{-1}$ is at most quadratic.

On the other hand, in [2, Theorem 4.4], we consider the simpler case, where $\mathscr{L}(x) \leq -\nu < 0$ for all $x \in H^1(I,\mathscr{M}_0)$ and $A(x) - C(x) < 0$ for all $x \in \mathscr{M}_0$.

Finally, notice that in [2] the growth assumption involves only the metric coefficients, and not the integrals in Eq. (6). This contrasts with [10, 11], where, in order to get the coercivity of $\mathscr{J}$, it is required that

$$\left|\frac{a(x)}{\mathscr{L}(x)}\right|, \quad \left|\frac{b(x)}{\mathscr{L}(x)}\right|, \quad \left|\frac{c(x)}{\mathscr{L}(x)}\right| \quad \text{are uniformly bounded on } H^1(I,\mathscr{M}_0).$$

Remark 2. Regarding to the case $A \equiv C$ left over in [2], if A (hence C) is always different from zero, then we are in the stationary case (Example $1(e_4)$) with $\beta(x) = |A(x)|$.

In general, if $B \equiv 0$ and $H(x) = A(x)C(x) > 0$ with $A(x) > 0$ and $\beta(x) = C(x)$, then we have Example $1(e_4)$ in the static case. So, $\mathscr{J}(x) \geq J(x)$ on each $\Omega^1(x_p,x_q)$ and the optimal result in [3, Theorem 1.1] can be used. Let us point out that a direct use of (h_3') for the particular case $A \equiv 1$ would give the desired result only for $\beta(x) < 1$.

If $A \equiv C \equiv 0$, then $\mathcal{L}(x) = b^2(x)$ and GTS becomes the more general type of warped product spacetimes, with fiber the two dimensional Lorentz–Minkowski spacetime (see also [10,13] and references therein). In this case, we deal again with a functional as in Eq. (5), and we get global geodesic connectedness for the class of metrics $\langle\cdot,\cdot\rangle_R - 2\delta(x)\,dy\,dt$, where δ is a positive function with at most a quadratic growth (compare with [10, Appendix B]).

Moreover, if $a \equiv c$ on $H^1(I, \mathcal{M}_0)$, then

$$\mathcal{J}(x) \geq \frac{1}{2}\|\dot{x}\|^2 - \frac{\Delta_-^2}{|a(x)|}.$$

Hence, if $A(x) > 0$ in $\mathcal{M}_0$, we obtain geodesic connectedness by assuming that $H(x)/A(x)$ grows at most quadratically in $\mathcal{M}_0$ (cf. Eqs. (5) and (9)).

Remark 3. In [26] Piccione and Tausk generalize the Morse index theorem to semi–Riemannian manifolds admitting a smooth distribution spanned by commuting Killing vector fields and containing a maximal negative distribution for the given metric. So, they obtain Morse relations for standard stationary spacetimes and, when the nondegeneracy condition $|\mathcal{L}(x)| > 0$ holds, for GTS (cf. [26, Theorems 4.6 and 4.8]). In particular, also in our setting, Morse relations hold. In fact, under the assumptions of Theorem 2 or of [2, Theorem 4.3] (where (h_1) and (h_2) hold, while (h_3) is replaced by (h_3')), a formal power series involving the Maslov index and the reduced Maslov index can be stated as in [26, Theorem 4.8] for each pair of non–conjugate points (for this definition, cf., e.g., [25]).

4 Geodesic Completeness

In this section, we establish and prove a result on geodesic completeness for GTS.

Theorem 3. *Let $(\mathcal{M} = \mathcal{M}_0 \times \mathbb{R}^2, \langle\cdot,\cdot\rangle_L)$ be a Gödel-type spacetime such that:*

(c_1) *$(\mathcal{M}_0, \langle\cdot,\cdot\rangle_R)$ is a complete Riemannian manifold;*
(c_2) *there exist $\lambda \geq 0$, $k \in \mathbb{R}$ and a point $x_0 \in \mathcal{M}_0$ such that the map*

$$\mu : x \in \mathcal{M}_0 \mapsto C(x) - A(x) + \sqrt{(A(x)+C(x))^2 + 4B^2(x)} \in \mathbb{R}$$

(which is strictly positive by Eq. (4)) satisfies

$$1/\mu(x) \leq \lambda d^2(x, x_0) + k \quad \text{for all } x \in \mathcal{M}_0. \tag{11}$$

Then, $(\mathcal{M}, \langle\cdot,\cdot\rangle_L)$ is geodesically complete.

Proof. Let $z : [0,T) \to \mathcal{M}$, $z(s) = (x(s), y(s), t(s))$, be an inextendible geodesic. Arguing by contradiction, it is enough to prove that if $T < +\infty$, then the $\langle\cdot,\cdot\rangle_R$–length of $x(s)$ is bounded, and so, z can be extended to T against the maximality assumption (see [25, Lemma 5.8]).

As ∂_y and ∂_t are Killing vector fields, there exist constants $c_1, c_2 \in \mathbb{R}$ such that

$$\begin{cases} A(x)\dot{y} + B(x)\dot{t} \equiv c_1 \\ B(x)\dot{y} - C(x)\dot{t} \equiv c_2 \end{cases} \quad \text{for all } s \in [0,T), \tag{12}$$

with

$$\mathscr{S}(x) = \begin{pmatrix} A(x) & B(x) \\ B(x) & -C(x) \end{pmatrix} \tag{13}$$

symmetric matrix with $\det \mathscr{S}(x) = -H(x) < 0$.

Furthermore, as z is a geodesic, there exists a constant $E_z \in \mathbb{R}$ such that

$$\langle \dot{z}, \dot{z} \rangle_L = \langle \dot{x}, \dot{x} \rangle_R + A(x)\dot{y}^2 + 2B(x)\dot{y}\dot{t} - C(x)\dot{t}^2 \equiv E_z \quad \text{for all } s \in [0,T). \tag{14}$$

Thus, by Eqs. (12) and (14) we get

$$\langle \dot{x}, \dot{x} \rangle_R + c_1\dot{y} + c_2\dot{t} = E_z \quad \text{for all } s \in [0,T). \tag{15}$$

On the other hand, by Eqs. (12) and (4) we have

$$\dot{y} = \frac{c_1 C(x) + c_2 B(x)}{H(x)}, \qquad \dot{t} = \frac{c_1 B(x) - c_2 A(x)}{H(x)}.$$

Whence, by Eq. (15) and using the notation $\|\dot{x}\|_R^2 := \langle \dot{x}, \dot{x} \rangle_R$, we get

$$\|\dot{x}\|_R^2 = E_z + \frac{c_2^2 A(x) - c_1^2 C(x) - 2c_1 c_2 B(x)}{H(x)}. \tag{16}$$

Note that the symmetric matrix $\mathscr{S}(x)$ in Eq. (13) admits two (non–null) real eigenvalues

$$\Lambda_\pm(x) = \frac{A(x) - C(x) \pm \sqrt{(A(x) + C(x))^2 + 4B^2(x)}}{2}, \qquad \text{with } \Lambda_+(x) > 0 > \Lambda_-(x).$$

Recall that by standard arguments there exists an orthogonal matrix $Q(x)$ such that

$$Q(x)^T \begin{pmatrix} A(x) & B(x) \\ B(x) & -C(x) \end{pmatrix} Q(x) = \begin{pmatrix} \Lambda_+(x) & 0 \\ 0 & \Lambda_-(x) \end{pmatrix}.$$

Let us denote $(\tilde{c}_1, \tilde{c}_2) = (c_1^2 + c_2^2)^{-1/2}(c_1, c_2)$ and $(\tilde{c}_1(x), \tilde{c}_2(x)) = (\tilde{c}_1, \tilde{c}_2)Q(x)$. By definition we have $\mu(x) = -2\Lambda_-(x)$, and, by the orthogonality of $Q(x)$, we have $[\tilde{c}_i(x)]^2 \le 1$ for $i \in \{1,2\}$. So, we can rewrite Eq. (16) as:

$$
\begin{aligned}
\|\dot{x}\|_R^2 &= E_z + \frac{\begin{pmatrix} c_1 c_2 \end{pmatrix} \begin{pmatrix} A(x) & B(x) \\ B(x) & -C(x) \end{pmatrix} \begin{pmatrix} c_1 \\ c_2 \end{pmatrix}}{H(x)} \\
&= E_z + \frac{\begin{pmatrix} c_1 \; c_2 \end{pmatrix} Q(x) \begin{pmatrix} \Lambda_+(x) & 0 \\ 0 & \Lambda_-(x) \end{pmatrix} Q(x)^T \begin{pmatrix} c_1 \\ c_2 \end{pmatrix}}{H(x)} \\
&= E_z + (c_1^2 + c_2^2) \frac{\begin{pmatrix} \tilde{c}_1 \; \tilde{c}_2 \end{pmatrix} Q(x) \begin{pmatrix} \Lambda_+(x) & 0 \\ 0 & \Lambda_-(x) \end{pmatrix} Q(x)^T \begin{pmatrix} \tilde{c}_1 \\ \tilde{c}_2 \end{pmatrix}}{H(x)} \\
&= E_z + (c_1^2 + c_2^2) \frac{\begin{pmatrix} \tilde{c}_1(x) \; \tilde{c}_2(x) \end{pmatrix} \begin{pmatrix} \Lambda_+(x) & 0 \\ 0 & \Lambda_-(x) \end{pmatrix} \begin{pmatrix} \tilde{c}_1(x) \\ \tilde{c}_2(x) \end{pmatrix}}{-\Lambda_+(x)\Lambda_-(x)} \\
&= E_z - (c_1^2 + c_2^2) \left(\frac{[\tilde{c}_1(x)]^2}{\Lambda_-(x)} + \frac{[\tilde{c}_2(x)]^2}{\Lambda_+(x)} \right) \\
&\le E_z - \frac{c_1^2 + c_2^2}{\Lambda_-(x)} = E_z + 2\frac{c_1^2 + c_2^2}{\mu(x)}.
\end{aligned}
$$

Thus, by Eq. (11) there exist suitable constants $\bar{\lambda}, \bar{k} > 0$ such that:

$$
\|\dot{x}(s)\|_R \le \bar{\lambda} d(x(s), x(0)) + \bar{k} \le \bar{\lambda} \int_0^s \|\dot{x}(r)\|_R \, \mathrm{d}r + \bar{k} \quad \text{for all } s \in [0, T).
$$

In conclusion, we obtain

$$
\log\left(\bar{\lambda} \int_0^s \|\dot{x}(r)\|_R \, \mathrm{d}r + \bar{k} \right) - \log(\bar{k}) \le \bar{\lambda} s \le \bar{\lambda} T \quad \text{for all } s \in [0, T)
$$

and then the boundedness of the $\langle \cdot, \cdot \rangle_R$–length of $x(s)$ in $[0, T)$, as required. □

Remark 4. The at most quadratic behavior of the autonomous term $1/\mu$ required for the geodesic completeness of a *GTS* in Theorem 3 is consistent with the (optimal) growth estimates which imply the completeness of the solutions of certain second order differential equations on Riemannian manifolds (see [9]).

Appendix

Taking a connected, finite–dimensional semi–Riemannian manifold $(\mathcal{M}, g)$, let $H^1(I, \mathcal{M})$ be the associated Sobolev space for some auxiliar Riemannian metric on $\mathcal{M}$. Then, $H^1(I, \mathcal{M})$ is equipped with a structure of infinite–dimensional manifold modelled on the Hilbert space $H^1(I,\mathbb{R}^n)$. For any $z \in H^1(I,\mathcal{M})$, the tangent space of $H^1(I,\mathcal{M})$ at z can be written as follows:

$$T_z H^1(I,\mathcal{M}) = \{\zeta \in H^1(I,T\mathcal{M}) : \zeta(s) \in T_{z(s)}\mathcal{M} \text{ for all } s \in I\},$$

where $T\mathcal{M}$ is the tangent bundle of $\mathcal{M}$.

If $\mathcal{M}$ splits globally in the product of two semi–Riemannian manifolds $\mathcal{M}_1$ and $\mathcal{M}_2$, i.e. $\mathcal{M} = \mathcal{M}_1 \times \mathcal{M}_2$, then

$$H^1(I,\mathcal{M}) \equiv H^1(I,\mathcal{M}_1) \times H^1(I,\mathcal{M}_2)$$

and $T_z H^1(I,\mathcal{M}) \equiv T_{z_1} H^1(I,\mathcal{M}_1) \times T_{z_2} H^1(I,\mathcal{M}_2)$ for all $z = (z_1, z_2) \in \mathcal{M}$.

On the other hand, if $(\mathcal{M}_0, \langle\cdot,\cdot\rangle_R)$ is a C^3 complete Riemannian manifold, it can be smoothly and isometrically embedded in a Euclidean space $\mathbb{R}^N$ (see [24]); moreover such embedding can be chosen closed (see [23]) and this is used in the proof of Lemma 1. Hence, $H^1(I,\mathcal{M}_0)$ is a closed submanifold of the Hilbert space $H^1(I,\mathbb{R}^N)$. In this case, we denote by $d(\cdot,\cdot)$ the distance induced on $\mathcal{M}_0$ by its Riemannian metric $\langle\cdot,\cdot\rangle_R$, i.e.,

$$d(x_p,x_q) := \inf\left\{\int_a^b \sqrt{\langle\dot{x},\dot{x}\rangle_R}\, ds : x \in A_{x_p,x_q}\right\},$$

where $x \in A_{x_p,x_q}$ if $x : [a,b] \to \mathcal{M}_0$ is any piecewise smooth curve in $\mathcal{M}_0$ joining $x_p, x_q \in \mathcal{M}_0$.

Taking z_p, $z_q \in \mathcal{M}$, let us consider

$$\Omega^1(z_p,z_q) = \{z \in H^1(I,\mathcal{M}) : z(0) = z_p,\ z(1) = z_q\},$$

which is a submanifold of $H^1(I,\mathcal{M})$, complete if $\mathcal{M}$ is complete and having tangent space described as

$$T_z\Omega^1(z_p,z_q) = \{\zeta \in T_z H^1(I,\mathcal{M}) : \zeta(0) = 0 = \zeta(1)\} \quad \text{at any } z \in \Omega^1(z_p,z_q).$$

Moreover, for any l_p, $l_q \in \mathbb{R}$, let us denote

$$W(l_p,l_q) = \{l \in H^1(I,\mathbb{R}) : l(0) = l_p\ ,\ l(1) = l_q\}.$$

Clearly,

$$W(l_p,l_q) = H_0^1(I,\mathbb{R}) + l_{pq},$$

with $H_0^1(I,\mathbb{R}) = \{l \in H^1(I,\mathbb{R}) : l(0) = 0 = l(1)\}$, $l_{pq} : s \in I \mapsto (1-s)l_p + s l_q \in \mathbb{R}$. Hence, $W(l_p,l_q)$ is a closed affine submanifold of the Hilbert space $H^1(I,\mathbb{R})$ with tangent space

$$T_l W(l_p,l_q) = H_0^1(I,\mathbb{R}) \quad \text{for every } l \in W(l_p,l_q).$$

Acknowledgements The authors of this chapter acknowledge the partial support of the Spanish Grants with FEDER funds MTM2010-18099 (MICINN). Furthermore, R. Bartolo and A.M. Candela acknowledge also the partial support of M.I.U.R. Research Project PRIN2009 "Metodi Variazionali e Topologici nello Studio di Fenomeni Nonlineari" and of the G.N.A.M.P.A. Research Project 2011 "Analisi Geometrica sulle Varietà di Lorentz ed Applicazioni alla Relatività Generale"; J.L. Flores acknowledges also the partial support of the Regional J. Andalucía Grant P09-FQM-4496, with FEDER funds.

References

1. Bartolo, R., Candela, A.M., Flores, J.L.: Geodesic connectedness of stationary spacetimes with optimal growth. J. Geom. Phys. **56**, 2025–2038 (2006)
2. Bartolo, R., Candela, A.M., Flores, J.L.: A note on geodesic connectedness in Gödel type spacetimes. Differ. Geom. Appl. **29**, 779–786 (2011)
3. Bartolo, R., Candela, A.M., Flores, J.L., Sánchez, M.: Geodesics in static Lorentzian manifolds with critical quadratic behavior. Adv. Nonlinear Stud. **3**, 471–494 (2003)
4. Benci, V., Fortunato, D.: On the existence of infinitely many geodesics on spacetime manifolds. Adv. Math. **105**, 1–25 (1994)
5. Benci, V., Fortunato, D., Giannoni, F.: On the existence of multiple geodesics in static spacetimes. Ann. Inst. H. Poincaré Anal. Non Linéaire **8**, 79–102 (1991)
6. Calvão, M.O., Damião Soares, I., Tiomno, J.: Geodesics in Gödel–type spacetimes. Gen. Relat. Gravit. **22**, 683–705 (1990)
7. Candela, A.M., Flores, J.L., Sánchez, M.: On general Plane Fronted Waves. Geodesics. Gen. Relat. Gravit. **35**, 631–649 (2003)
8. Candela, A.M. , Flores, J.L. , Sánchez, M.: Global hyperboliticity and Palais–Smale condition for action functionals in stationary spacetimes. Adv. Math. **218**, 515–536 (2008)
9. Candela, A.M., Romero, A., Sánchez, M.: Completeness of the trajectories of particles coupled to a general force field. To appear in Arch. Rational Mech. Anal. ArXiv:1202.0523v1.
10. Candela, A.M., Sánchez, M.: Geodesic connectedness in Gödel type spacetimes. Differ. Geom. Appl. **12**, 105–120 (2000)
11. Candela, A.M., Sánchez, M.: Existence of geodesics in Gödel type spacetimes. Nonlinear Anal. TMA **47**, 1581–1592 (2001)
12. Candela, A.M., Sánchez, M.: Geodesics in semi–Riemannian manifolds: geometric properties and variational tools. In: Alekseevsky, D.V., Baum, H. (eds.) Recent Developments in Pseudo–Riemannian Geometry, Special Volume in the ESI–Series on Mathematics and Physics, pp. 359–418, EMS Publishing House (2008)
13. Candela, A.M., Salvatore, A., Sánchez, M.: Periodic trajectories in Gödel type spacetimes. Nonlinear Anal. TMA **51**, 607–631 (2002)
14. Caponio, E., Javaloyes M.A., Sánchez, M.: On the interplay between Lorentzian causality and Finsler metrics of Randers type. Rev. Mat. Iberoam. **27**, 919–952 (2011)
15. Carrion, H.L., Rebouças, M.J., Teixeira, A.F.F.: Gödel–type spacetimes in induced matter gravity theory. J. Math. Phys. **40**, 4011–4027 (1999)
16. Chandrasekhar, S., Wright, J.P.: The geodesics in Gödel's universe. Proc. Natl. Acad. Sci. USA **47**, 341–347 (1961)
17. Giannoni, F., Masiello, A.: On the existence of geodesics on stationary Lorentz manifolds with convex boundary. J. Funct. Anal. **101**, 340–369 (1991)
18. Gödel, K.: An example of a new type of cosmological solution of Einstein's field equations of gravitation. Rev. Mod. Phys. **21**, 447–450 (1949)
19. Hawking, S.W., Ellis, G.F.R.: The Large Scale Structure of SpaceTime. Cambridge University Press, London (1973)

20. Kramer, D., Stephani, H., Herlt, E., MacCallum, M.: Exact Solutions of Einstein's Field Equations. Cambridge University Press, Cambridge (1980)
21. Kundt, W.: Trägheitsbahnen in einem von Gödel angegebenen kosmologischen Modell. Z. Phys. **145**, 611–620 (1956)
22. Masiello, A.: Variational Methods in Lorentzian Geometry. Pitman Res. Notes Math. Ser. vol. 309. Longman Sci. Tech., Harlow (1994)
23. Müller, O.: A note on closed isometric embeddings. J. Math. Anal. Appl. **349**, 297–298 (2009)
24. Nash, J.: The imbedding problem for Riemannian manifold. Ann. Math. **63**, 20–63 (1956)
25. O'Neill, B.: Semi–Riemannian Geometry with Applications to Relativity. Academic, New York (1983)
26. Piccione, P., Tausk, D.V.: An index theory for paths that are solutions of a class of strongly indefinite variational problems. Calc. Var. Part. Differ. Equat. **15**, 529–551 (2002)
27. Rabinowitz, P.H.: Minimax Methods in Critical Point Theory with Applications to Differential Equations. CBMS Reg. Conf. Ser. Math. vol. 65. Amer. Math. Soc., Providence (1986)
28. Raychaudhuri, A.K., Thakurta, S.N.G.: Homogeneous spacetimes of the Gödel type. Phys. Rev. D **22**, 802–806 (1980)
29. Rebouças, M.J., Tiomno, J.: Homogeneity of Riemannian spacetimes of Gödel type. Phys. Rev. D **28**, 1251–1264 (1983)
30. Sánchez, M.: Some Remarks on Causality Theory and Variational Methods in Lorentzian Manifolds. Conf. Semin. Mat. Univ. Bari **265** (1997) ArXiv:0712.0600v2
31. Santaló, L.A.: Geodesics in Gödel–Synge spaces. Tensor N.S. **37**, 173–178 (1982)

The Geometry of Collapsing Isotropic Fluids

Roberto Giambò and Giulio Magli

1 Introduction

Spherical symmetry is a useful test-bed for open problems of astrophysical interest in General Relativity. Among them, a very relevant one is that of the final state of gravitational collapse and, therefore, of the validity of a "Cosmic Censorship" hypothesis. In particular, the study of spherically symmetric spacetimes modeling a collapsing isotropic fluid is a recurrent topic in relativistic literature. What makes it one of the most intriguing problem in gravitational collapse is that perfect fluids are a direct, physically interesting generalization of the so-called *Tolman–Bondi–Lemaitre* (TBL) solution, which is one of the few known-in-detail solutions dynamically collapsing to a singularity. The TBL solution is indeed long known to have naked singularities (the first example discovered in [2], and the complete analysis is due to [10]), while the case of isotropic fluids remains almost open. Some results are actually known from numerical relativity, in particular for barotropic perfect fluids with linear equation of state ($p = \mu\varepsilon$, $\mu \in \mathbb{R}$, see Sect. 4.1 below): for instance, Ori and Piran [13] studied the problem under the assumption of self-similarity, while Harada [7] investigated the same problem, detecting *globally* naked singularities in some cases. Choptuik, whose numerical analysis study on the gravitational collapse of a scalar field [1] remains one of the cornerstones about the cosmic censor problem, worked with Neilsen [11] to the limit case $\mu \to 1^-$, which was also the aim of Snajdr [16]. Unfortunately, outside the realm

R. Giambò (✉)
School of Science and Technology, Mathematics Division, University of Camerino,
Via Madonna delle Carceri, I-62032 Camerino, Italy
e-mail: roberto.giambo@unicam.it

G. Magli
Department of Mathematics, Polytechnic of Milan, Pzza Leonardo da Vinci 32,
I-20133 Milan, Italy
e-mail: giulio.magli@polimi.it

M. Sánchez et al. (eds.), *Recent Trends in Lorentzian Geometry*, Springer Proceedings in Mathematics & Statistics 26, DOI 10.1007/978-1-4614-4897-6_8,

of numerical relativity, little is known about the geometry of these spacetimes: whether a singularity is developed, and if that is the case, what is the causal structure of the solution. There are many studies which do not rely on numerical techniques but either they refer to general situations without specifying the matter properties (see, e.g., [4]) or they deal with anisotropic spacetimes (see, e.g., [5] and references therein).

Some results will be sketched here which shed new light on this problem. In particular, we report here on a quite general analysis of this problem which can be carried out for general equations of state provided that certain regularity assumptions are satisfied. These assumptions essentially require Taylor-expandability of the solution in a special system of coordinates and allow for a quite general picture of barotropic perfect fluids (with pressure proportional to energy density) as well as for some other cases of interest. The problem of how recovering the final state of dust (TBL) collapse from perfect fluids remains, however, open although some hints can be derived. Indeed, it appears that the barotropic solutions found do not converge to TBL solution as the ratio pressure over density goes to zero. Moreover, the qualitative picture emerging from these models is quite different from the TBL case. On the other side, for models where the equation of state is perturbed in a nonlinear way, a qualitative behavior of the singularity similar to the background model is recovered. All these facts likely represent an evidence of the crucial role of pressure in the neighborhood of the singular boundary to determine the causal structure of the spacetime. These results seem to confirm the analysis carried out in [9], where homogeneous dust collapse (Oppenheimer–Snyder model, see Example 31 in Sect. 3) is perturbed adding a small amount of pressure.

The chapter is organized as follows: we review general spherical models in Sect. 2, specializing to the isotropic case in Sect. 3. Qualitative results are presented and discussed in Sect. 4.

2 Relativistic Stars in Spherical Symmetry

Let us consider a generic, non-static, spherically symmetric, 4–dimensional spacetime. Using a comoving coordinate system (t,r,θ,ϕ), the metric is written in the form

$$g = -e^{2\nu}\,\mathrm{d}t^2 + e^{2\lambda}\,\mathrm{d}r^2 + R^2\,\mathrm{d}\Omega^2, \tag{1}$$

where $\mathrm{d}\Omega^2 = \mathrm{d}\theta^2 + \sin^2\theta\,\mathrm{d}\phi^2$ is the first fundamental form of $\mathbb{S}^2 \subset \mathbb{R}^3$, and ν, λ and R are functions of (t,r) only.

In the following, we will be interested in those matter models admitting a well-defined description in terms of the standard relativistic mechanics of continua. This implies that g must satisfy Einstein field equations (hereafter, Greek indices like μ,ν run from 0 to 3)

$$G^{\mu}_{\nu} = 8\pi T^{\mu}_{\nu} \tag{2}$$

where G is the Einstein tensor of g ($G = Ric - \frac{1}{2}Sg$, S being the scalar curvature) and T is the stress–energy tensor embodying the matter properties, that in the comoving system takes the form

$$8\pi T = -\varepsilon\, \mathrm{d}t \otimes \frac{\partial}{\partial t} + p_r \mathrm{d}r \otimes \frac{\partial}{\partial r} + p_t \left(\mathrm{d}\theta \otimes \frac{\partial}{\partial \theta} + \mathrm{d}\phi \otimes \frac{\partial}{\partial \phi} \right) \tag{3}$$

where ε, p_r and p_t are the *energy density*, the *radial* and the *tangential pressure*, respectively, and again are all functions of (t,r) only.

Using Eqs. (1) and (3), and introducing *Misner–Sharp mass* function

$$m = \frac{R}{2}(1 - g(\nabla R, \nabla R)) = \frac{R}{2}\left(1 - (R'e^{-\lambda})^2 + (\dot{R}e^{-\nu})^2\right) \tag{4}$$

(where a prime and a dot denote differentiation with respect to r and t, respectively, and ∇ is the gradient operator indicted by g) a complete set for Einstein field equation (2) is given by

$$m' = 4\pi\varepsilon R^2 R', \tag{5}$$

$$\dot{m} = -4\pi p_r R^2 \dot{R}, \tag{6}$$

$$\dot{R}' = \dot{\lambda} R' + \nu' \dot{R}, \tag{7}$$

$$p_r' = -(\varepsilon + p_r)\nu' - \tfrac{2R'}{R}(p_r - p_t). \tag{8}$$

In particular, Eq. (7) is equivalent to Eq. (2) for $(\mu,\nu) = (0,1)$, and using it in Eq. (2) for $(\mu,\nu) = (0,0)$ and $(\mu,\nu) = (1,1)$ we get Eqs. (5) and (6) respectively. Finally, Bianchi identity div $T = 0$ implies relation (8).

We are interested in spacetimes modeling collapsing (spherical) objects, with the aim to investigate the behavior especially near the center of symmetry of the system ($r = 0$). We stress that the "central shell" $r = 0$ is regular at initial time of observation and may possibly develop a singularity after some amount of comoving time. In view of that we will consider r as defined in a right neighborhood of $r = 0$, say $[0, r_b]$, and we will need

1. To assume some geometrical and physical reasonableness hypotheses on the metric.
2. To smoothly match the solution with an external spacetime.

Let us briefly review these assumptions.

2.1 Geometrical and Physical Assumptions

- We first demand that the internal solution satisfies the *dominant energy condition (dec)*, which means that

$$\varepsilon \geq 0, \quad |p_r| \leq \varepsilon, \quad |p_t| \leq \varepsilon, \tag{9}$$

  throughout the evolution.

- We require that there exists an initial time (say, $t = 0$) such that the solution is regular: in particular, the energy density measured on the internal solution must be finite and outward decreasing:

$$\lim_{r\to 0^+} \varepsilon(0,r) \in \mathbb{R}, \quad \frac{\mathrm{d}}{\mathrm{d}r}\varepsilon(0,r) \leq 0 \,\text{on}\, [0,r_b]. \tag{10}$$

Without loss of generality, on the initial time, the following initial condition can be imposed

$$R(0,r) = r. \tag{11}$$

- We will impose some local cartesianity conditions on the metric, to prevent bad behavior on the center of symmetry due to polar coordinate choice and to impose isotropy at the center of symmetry. This is equivalent to state

$$R(t,0) = 0, \quad R'(t,0) = e^{\lambda}(t,0), \quad p_r(t,0) = p_t(t,0) \tag{12}$$

for all $t \geq 0$ up to (possibly) singularity formation.
- We also ask for the solution to be free from *shell–crossing singularities*, that are caused by the vanishing of $R'(t,r)$. Shell-crossing singularities usually correspond to Tipler-weak divergences of the curvature, and for this reason are considered as "less important" although extendibility proofs of a spacetime beyond a shell-crossing singularity is available in literature only in some particular cases (see [12] and reference therein as [14]). We will require that no shell-crossing singularity appears prior to (possible) singularity formation due to the vanishing of $R(t,r)$. The latter are usually called *shell–focusing* singularities and are those we will be interested in. Then we assume that

$$R(t,r) > 0 \Rightarrow R'(t,r) > 0. \tag{13}$$

- Finally, in order to obtain a global model, a matching with an external space will be performed at $\Sigma = \{r = r_b\}$, requiring that the first and the second fundamental forms of the two metrics at Σ coincide (Israel–Darmois junction conditions). From Eq. (6) we observe that the radial pressure p_r in general does not vanish along Σ, which as well known is a necessary and sufficient condition to match the solution with a Schwarzschild exterior. In this more general case, a natural choice for the exterior metric is the generalized Vaidya spacetime written in radiative coordinates (V is a null coordinate)

$$g_{\text{ext}} = -\left(1 - \frac{2\mu(V,S)}{S}\right)\mathrm{d}V^2 + 2\,\mathrm{d}V\,\mathrm{d}S + S^2\mathrm{d}\Omega^2, \tag{14}$$

where $\mu(V,S)$ is an arbitrary (non negative) C^2 function. The immersion of Σ in the two spacetimes can be parameterized respectively by $(\tau,\theta,\phi) \hookrightarrow (\tau,r_b,\theta,\phi)$ and $(\tau,\theta,\phi) \hookrightarrow (V(\tau),S(\tau),\theta,\phi)$, and junction conditions are found to be

$$S(\tau) = R(\tau, r_b), \quad \dot{V}(\tau) = \frac{e^{\nu}}{R' e^{-\lambda} + \dot{R} e^{-\nu}}|_{(\tau, r_b)} \tag{15}$$

$$\mu(V(\tau), S(\tau)) = m(\tau, r_b), \qquad \frac{\partial \mu}{\partial V}(V(\tau), S(\tau)) = 0. \tag{16}$$

Equation (15) can be solved for $S(\tau)$ and $V(\tau)$ to give the parameterization of Σ into the outer spacetime, once that the inner spacetime is known on Σ. Instead, Eq. (16) impose a constraint on the mass function μ on Σ – for example, if one requires the outer solution to be a Schwarzschild exterior and then μ constant, then $\dot{m}(\tau, r_b) = 0$ which is equivalent, in view of Eq. (6), that $p_r = 0$ on Σ as stated before.

2.2 Singularity Formation and Cosmic Censor

In addition to the assumptions above reviewed, we require the interior spacetime to model a collapsing spherical object, which implies that we will be interested in those solutions such that $\dot{R}(t, r) \leq 0$ during the evolution. The collapse may either produce an asymptotically regular solution existing for all times $t \geq 0$ or a (shell-focusing) singularity, due to the vanishing of $R(t, r)$. Actually, recalling Eq. (12), singularities will be detected by the relation

$$a(t, r) := \frac{R(t, r)}{r} = 0, \tag{17}$$

thus defining a singularity curve $t_s(r)$ such that $a(t_s(r), r) = 0$. In view of Eq. (10), $t_s(t)$ (if it exists) is strictly positive $\forall r \in [0, r_b]$. Let us consider the case when $t_s(r)$ is well defined $\forall r \in [0, r_b]$. To get information about the causal structure of the spacetime, we will perform a study of radial null geodesics which can be extended (in the comoving past) up to the singular curve $t_s(r)$. For this aim a crucial role is played by the *apparent horizon*, implicitly defined by the relation $R = 2m$, which is the boundary of the *trapped region* $R < 2m$.

We will review in the final section some particular cases occurring during the collapse of isotropic models. For the moment, as an example, we sketch a situation happening, for instance, in the collapse of a spherical dust cloud (i.e., $p_r = p_t = 0$) as well as for a more general class of spacetimes [5]. It can be seen that the relation $R = 2m$ implicitly defines a curve $t_h(r)$ such that $t_h(0) = t_s(0)$ and $t_h(r) < t_s(r)$, $\forall r > 0$. Then all geodesics that can be extended in the past up to the points $t_s(r)$ for r strictly positive, are entirely confined inside the trapped region. On the other side, there is the possibility for a radial null geodesic – an infinite number of geodesics, actually – with support *outside* the trapped region, to be extended up to $t_s(0)$. This feature can be interpreted as a violation of the so-called cosmic censorship conjecture, originally stated by Roger Penrose [15].

It is possible to rephrase the existence problem for such geodesics as an existence problem for the following ODE:

$$\frac{\mathrm{d}t}{\mathrm{d}r} = e^{(\lambda-\nu)(t,r)}, \quad t(0) = t_s(0), \tag{18}$$

but this in principle is not a well-defined Cauchy problem because of the lack of regularity on the initial condition. Therefore, a remarkable property of the apparent horizon may be of help – indeed, $t_h(r)$ is a subsolution of the ODE (18). This fact implies that if there exists a supersolution of Eq. (18), say $t_*(r)$, such that $t_*(r) \le t_h(r)$ (and $=$ iff $r = 0$), then usual comparison results in ODE theory gives the existence of a family of radial null geodesics emitted from $t_s(0)$ outside the trapped region. Existence problems to Eq. (18) are then linked to solutions to differential inequalities which in principle are easier to check.

Remark 21. One can in principle think at this stage that the existence problem for *nonradial* null geodesics is still left open. However, it can be proved – using again comparison arguments in ODE theory – that if no radial geodesics exist, then nonradial geodesics too cannot be emitted from the singularity.

3 Regular Isotropic Models

It is well known that the system of Einstein field equations (5)–(8) is underdetermined, and then more relations are needed in order to "close" the system. These further relations are usually provided by an equation of state, expressing one (or more) conditions on the stress–energy tensor components. To begin, we will partially fix one of these conditions, specializing to the case of an isotropic fluid, i.e.,

$$p_r(t,r) = p_t(t,r). \tag{19}$$

Moreover, we are going to perform this study in a new coordinate setting. Indeed, we will consider coordinates (a, r, θ, ϕ) where a is given by Eq. (17). In view of Eq. (11) the internal solution will be studied on the set $[0,1] \times [0,r_b] \times \mathbb{S}^2$, and the singularity corresponds to the boundary $a = 0$. It is convenient to introduce the functions $M(a,r) = \frac{2m}{r^3}$,

$$\gamma(a,r) = \frac{p}{\varepsilon}, \tag{20}$$

and $Y(a,r) = R' e^{-\lambda}$. Moreover, we also make the positions

$$w(a,r) = a', \quad z(a,r) = \dot{a}, \tag{21}$$

that will be used to link the new coordinate system to the old one.

Example 31. In this chapter we wish to give some insight into gravitational collapse of isotropic pressure spherical models under some regularity assumption on the solution. As is well known, and as said before, the only inhomogeneous solution with isotropic pressure which is known in full generality is the TBL dust cloud, the one obtained when the pressure vanishes ($p_r = p_t = 0$). The so-called *marginally bound* case of this solution is obtained when $Y \equiv 1$ and is described by $R(t,r) = r(1-k(r)t)^{2/3}$, with $k(r) = 1 - \alpha r^n + o(r^n)$ (and $\alpha > 0$) depending on the mass profile of the solution. The special case $k(r) \equiv$ constant correspond to the homogeneous (in the sense that the singularity curve $t_s(r)$ is constant) Oppenheimer–Snyder model. In the coordinate system (a,r) the solution takes the form

$$w = -\frac{2k'(r)}{3k(r)}\left(\frac{1}{\sqrt{a}} - a\right), \quad z = -\frac{2k(r)}{3\sqrt{a}} \tag{22}$$

$$e^{\lambda} = R' = a + rw = a\left(1 - \frac{2rk'(r)}{3k(r)}\left(\frac{1}{a\sqrt{a}} - 1\right)\right), \quad e^{\nu} = 1. \tag{23}$$

We can see that this solution can be developed in power series with respect to r around $r = 0$, and this will be basically the assumption we will make on the general solution.

Before going on and write the system, we observe from Eq. (23) that the power series in r of λ badly behaves with respect to a, since every coefficient – that is actually a function of a – contains increasing power, diverging terms in a. This suggests the choice of e^{λ} as an unknown, rather than λ itself. For sake of uniformity the same choice will be made for ν. Hereafter, therefore, we introduce two new unknown functions

$$B(a,r) = e^{\lambda}, \quad F(a,r) = e^{\nu}, \tag{24}$$

in place of old variables λ and ν.

With the above positions, Einstein field equations in (a,r) coordinates become (subscript denotes partial derivative)

$$3M + M_r r + wrM_a - \varepsilon a^2(wr + a) = 0, \tag{25}$$

$$M_a + \gamma\varepsilon a^2 = 0, \tag{26}$$

$$(\gamma+1)\varepsilon(wr+a)Y_a + Y\left[(\gamma\varepsilon)_r + w(\gamma\varepsilon)_a\right]r = 0, \tag{27}$$

$$r(F_r + wF_a)Y - (wr+a)Y_a F = 0, \tag{28}$$

Moreover, since $wr + a = R' = YB$, and using Eq. (4), we have the following two functional dependencies:

$$z = -F\left(\frac{M}{a} + \frac{Y^2-1}{r^2}\right)^{1/2},$$

$$w = \frac{YB - a}{r}. \tag{29}$$

Finally, an equation is needed to express a compatibility property between w and z introduced in Eq. (21), i.e., $\dot{w} = z'$, that in the (a,r) coordinate system reads as

$$z_r + wz_a - zw_a = 0. \tag{30}$$

As outlined above we now make our last assumption: that B, F, Y, ε and M are solutions of Eqs. (25), (27), (28), and (30) that are C^n with respect to r in $[0, r_b]$, $\forall a \in]0,1]$. Then

$$M(a,r) = \sum_{i=0}^{n} M_i(a) r^i + o(r^n),$$

for n is sufficiently large, and analogously for A, B, F, and ε. The $o(r)$ above and hereafter must be intended as a function of both (a,r), of course.

We stress that the remaining unknown functions γ, p, w, and z can be expressed using the functional dependencies (20), (26), (29) and (29). With the above ansatz, the equations can be expanded in order to recover relations for the Taylor coefficients of the unknown functions. The higher degree in regularity – with respect to r – the more accurate information we can get, and an iterative scheme can be found out that determines the whole solution up to $M_0(a)$, $Y_2(a)$ and $B_i(a)$, $i \geq 1$. In particular, for example, since $\varepsilon_0(a) = \frac{3M_0(a)}{a^3}$ then from Eq. (26), we see that the choice of $M_0(a)$ determines the leading power of γ:

$$\gamma_0(a) = -\frac{aM_0'(a)}{3M_0(a)}.$$

A further specialization – in addition to Eq. (19) – can be used to prescribe each of the remaining functions up to a constant, that corresponds to the freedom to choose the initial data for the evolution.

The information about leading order terms of the unknown functions can be used to check compatibility with the assumptions stated in Sect. 2.1 in a right neighborhood of $r = 0$, and to study the behavior of the apparent horizon and the radial null geodesic possibly emitted by the singularity $a = 0$, using methods already sketched in Sect. 2.2 opportunely adapted to this new coordinate setting.

4 Qualitative Results and Discussion

In this section, we review the qualitative results that can be inferred from the form of the coefficients of the solution for different choices of the free functions corresponding to significant isotropic fluid models.

4.1 Barotropic Fluids with a Linear Equation of State

This models corresponds to the case $\gamma = \frac{p}{\varepsilon} = \mu \in \mathbb{R}$. We restrict $\mu \in [-1,1]$ to satisfy the dec (9).

This equation of state fixes all free functions except $Y_2(a)$ – for this reason, we choose to test these models with the family $Y_2(a) = y_2 a^{\alpha}$, with y_2, α real parameters. The structure of the singularity for these fluids is qualitatively much different from the TBL case, and the main reason for that is that the apparent horizon drastically changes behavior depending on the values of α and μ [8]. In some case we have a horizon "of a first type" which has a behavior similar to the TBL models leading to a black hole, and indeed here, the singularity is completely hidden inside the trapped region and is invisible to faraway observers. But there are also situations "of a second type" emerging for these fluids – occurring for negative pressure only, unless one does not violate some of the conditions expressed in Sect. 2.1 as no shell-crossing singularity formation (13) – where the horizon lets the central singularity be naked; this kind of horizon appearing here does *not* have a corresponding limit $\mu \to 0$, as the previous one.

These features make the singular boundary for these fluids qualitatively different from the geometry of dust collapse. Although central naked singularities are potentially occurring in both cases, perturbing the horizon of a TBL solution – which there gives rise to either a naked singularity or to a black hole, depending on the data of the problem – produces here a horizon of the first type, which always leads to a black hole. The horizon of the second type, possibly leading to a central naked singularity, is a distinctive feature of the $\mu \neq 0$ models.

This bifurcating behavior may be explained by observing that when $\mu \neq 0$, some of the functions determined by the procedure discussed in the previous section are not well defined in the limit $\mu \to 0$. The naked singularity formation in dust collapse can thus be seen as an unstable phenomenon with respect to (regular) linear perturbation of the equation of state and in addition, under the regularity hypotheses made in this chapter, these fluids cannot be seen at all as a proper perturbation of the case $\mu = 0$.

Finally, to make the whole picture even more complex, when $\mu \leq -\frac{1}{3}$, there are choices – always generic, except the case $\mu = \frac{1}{3}$ – of the free functions that lead to absence of horizon. This means that the whole singularity curve – not only the central ones, then – can be globally naked, which is consistent with the analysis by Cooperstock et al. [3]. Let us remark that, if we imposed the strong energy condition – that in case of isotropic fluids correspond to assume $\varepsilon + 3p \geq 0$ and $\varepsilon + p \geq 0$ – instead of the dominant, this case would have been excluded, except the special case $\mu = -\frac{1}{3}$ that, as said before, produces a globally naked singularity for a nongeneric choice of the free functions.

4.2 *Generalized Chaplygin Gas*

The above discussion of the γ-constant case suggests that the role of pressure is crucial in the singularity trapping process. Then one can think about models with a softer equation of state, and see whether naked singularity appears as it did in dust case. An example can be provided by a generalization of Chaplygin fluids, where the equation of state is given by

$$p = \mu\varepsilon^{-\alpha}, \quad \alpha \geq 0, \mu \in \mathbb{R}. \tag{31}$$

In these models, when the collapse ends into a singularity, the divergence of the energy density provides an upper bound for the (absolute value of the) pressure p. For the sake of simplicity here we report on cases $\alpha = 0$ and $\alpha = 1$ only, stressing the fact that for this class of models, the dust solution is recovered in the limit $\mu \to 0$.

In the first case, corresponding to nonzero (yet bounded) pressures in the approach to the singularity, it is seen that the horizon forms and completely covers central and noncentral singularities. Then, getting bounded pressures seems not enough to retain naked singularities, and indeed the situation dramatically changes when $\alpha = 1$. Here, the horizon still covers noncentral singularities, but the central singularity can be naked in a generic way, regardless of the pressure sign.

It clearly appears from the above reviewed cases that the pressure highly influences the qualitative behavior of the singularity and therefore, the causal structure of the collapsing model. In the linear case $p = \mu\varepsilon$ the pressure diverges with the energy density in the approach to the singularity, which results to be hidden by the apparent horizon – when the latter forms. These models also contain some interesting cases where the horizon does not even form and then the singular boundary is globally naked.

This intricate picture becomes simpler when the equation of state is perturbed in such a way that the pressure goes to zero as the energy diverges – here, these models *are* proper dust perturbations, since TBL solutions are recovered in the limit $\mu \to 0$, and in fact a central naked singularity may take place. The case when pressure remain bounded but also bounded away from zero near the singular boundary yet presents some uncertainty – preliminary studies applied to the parametric equation of state $p = e^{-1/\rho}$, $\varepsilon = \rho\, e^{-1/\rho}$ show that naked singularity may appear in the center of symmetry of the system as the energy diverges ($\rho \to +\infty$), unlike the $\alpha = 0$ Chaplygin model.

The boundedness of pressure near the singular boundary then seems a key ingredient to produce counterexamples to cosmic censorship. As is well known (see, e.g., [6] and references therein), many examples are known in literature of anisotropic models with naked singularities, where both tangential and radial pressures diverge – although in different manners – and then isotropy places a further constraint to naked singularities. But this constraint is far from simplifying the geometry of the spacetime, at least when the pressure diverges near the singularity – as seen for the linear equation of state. Of course, to get a complete picture one should be able to prove convergence theorems for the series of the unknown functions of the system, which is basically related to a global existence–uniqueness theorem for the Einstein field equations, and this would in principle cut out some of the examples above reviewed.

References

1. Choptuik, M.W.: Phys. Rev. Lett. **70**, 9 (1993)
2. Christodoulou, D.: Commun. Math. Phys. **93**, 171 (1984)
3. Cooperstock, F.I., Jhingan, S., Joshi, P.S., Singh, T.P.: Class. Quant. Grav. **14**, 2195 (1997)
4. Fayos, F., Torres, R.: Class. Quantum Grav. **28**, 215023 (2011)
5. Giambò, R., Giannoni, F., Magli, G., Piccione, P.: Commun. Math. Phys. **235**(3), 545 (2003)
6. Giambò, R., Giannoni, F., Magli, G., Piccione, P.: Class. Quantum Grav. **20**, L75 (2003)
7. Harada, T.: Phys. Rev. D **58**, 104015 (1998)
8. Iguchi, H., Harada T., Mena, F.C.: Class. Quantum Grav. **22**, 841 (2005)
9. Joshi, P.S., Malafarina, D., Saraykar, R.V.: *Int. J. Mod. Phys. D*, arxiv:1107.3749 (2011)
10. Joshi, P.S., Singh, T.P.: Class. Quant. Grav. **13**, 559 (1996)
11. Neilsen D.W., Choptuik, M.W.: Class. Quantum Grav. **17**, 761 (2000)
12. Nolan, B.C.: Class. Quantum Grav. **20**, 575 (2003)
13. Ori, A., Piran, T.: Phys. Rev. D **42**(4), 1068 (1990)
14. Papapetrou A., Hamoui, A.: Ann. Inst. H Poincaré **6**, 343 (1967)
15. Penrose, R.: R. Nuovo Cim. **1**, 252 (1969); reprinted in Gen. Rel. Grav. **34**, 1141 (2002)
16. Snajdr, M.: Class. Quantum Grav. **23**, 3333 (2006)

Conformally Standard Stationary SpaceTimes and Fermat Metrics

Miguel Angel Javaloyes

1 Introduction

Fermat's principle, say, that light rays minimize the arrival time, is linked to general relativity from its very beginning. As early as 1917, Weyl established a version for static spacetimes in [56], and several other authors, as Levi–Civita and Synge [34, 54], gave some attention to the principle. Not much later, in 1927, Levi–Civita stated the stationary version in [35] (see also [50]) that later was included in the book [33]. The general version was formulated by Kovner in 1990 [32] and rigorously established by Perlick in [47] (see also [49] for a version in Finsler spacetimes).

Independently from Fermat's principle, Randers metrics appeared as an attempt of Randers to geometrize electromagnetism in general relativity [51], but it seems that it was Ingarden the first one that thought in Randers metrics as Finsler ones in his PhD thesis [27]. By the way, Miron [45] suggested to name the Randers metrics endowed with a nonlinear Lorentz connection (associated to the Lorentz equation in electrodynamics) as *Ingarden spaces*. Afterwards, they were recovered by M. Matsumoto with the aim of giving examples of the so-called C-reducible Finsler metrics. In order to obtain these examples, he introduced the class of (α,β)-metrics in a manifold M, that is, Finsler metrics that are obtained as a homogeneous combination of the square root of a Riemannian metric h and a one-form β on M (with the notation $\alpha(v)=\sqrt{h(v,v)}$ for $v\in TM$) [40]. In particular, Randers metrics are defined as $\alpha+\beta$. This function is positively homogeneous but not reversible. Moreover, it is positive whenever the h-norm of β is less than 1 in every point. Subsequently, the Japanese school of Finsler geometry spent some time studying Randers metrics, mostly problems related with curvature [41, 53, 57].

M.A. Javaloyes (✉)
Departamento de Matemáticas, Universidad de Murcia, Campus de Espinardo,
30100 Espinardo, Murcia, Spain
e-mail: majava@um.es; ma.javaloyes@gmail.com

M. Sánchez et al. (eds.), *Recent Trends in Lorentzian Geometry*, Springer Proceedings in Mathematics & Statistics 26, DOI 10.1007/978-1-4614-4897-6_9,

Let us point out that the approach of G. Randers himself was somewhat different, since he constructed his metric from a Lorentzian metric and a one-form in the spacetime. It is also remarkable that Lichnerowicz and Thiry obtained a Randers metric when studying Jacobi–Maupertuis principle in general relativity (see [37] and [36, p. 155]).

In this chapter, we will describe some recent results that use techniques of Finsler geometry to study conformally standard stationary spacetimes and vice versa.

2 Finsler and Randers Metrics

There are several definitions of Finsler metrics [30]. But the most general case where you can extend most of the classic Riemannian results is the following. Let $\pi : TM \to M$ be the natural projection from the tangent bundle to the manifold. A *Finsler metric* is a continuous function $F : TM \to [0, +\infty)$ satisfying the following properties:

1. F is C^∞ in $TM \setminus 0$, that is, it is smooth away from the zero section,
2. F is fiberwise positively homogeneous of degree one, that is , $F(\lambda v) = \lambda F(v)$ for every $v \in TM$ and $\lambda > 0$,
3. F^2 is fiberwise strongly convex, that is, the fundamental tensor g_u defined as

$$g_u(v,w) = \frac{\partial^2}{\partial s \partial t} F^2(u + tv + sw)|_{t,s=0}, \tag{1}$$

where $u \in TM \setminus 0$ and $v, w \in T_{\pi(u)}M$, is positively defined for every $u \in TM \setminus 0$.

These conditions imply that F is positive away from the zero section, the triangle inequality holds for F in the fibers (see [3, Sect. 1.2B]) and F^2 is C^1 [55]. Property (3) above is essential to guarantee minimization properties of geodesics. The first geometers that worked with Randers metrics seemed very concerned with computation of curvatures and invariants related with connections, and, apparently, they overlooked the question of strong convexity. Let us recall that a Randers metric on a manifold M is constructed using a Riemannian metric h and a one-form β on M as

$$R(v) = \sqrt{h(v,v)} + \beta(v) \tag{2}$$

for every $v \in TM$. It turns out that it is fiberwise strongly convex if and only if it is positive for every $v \in TM$. This can be easily seen computing the fundamental tensor (see [30, Corollary 4.17]):

$$g_v(w,w) = \frac{R(v)}{\sqrt{h(v,v)}} \left(h(w,w) - \frac{1}{h(v,v)} h(v,w)^2 \right) + \left(\frac{h(v,w)}{\sqrt{h(v,v)}} + \beta(w) \right)^2,$$

with $v \in TM \setminus 0$ and $w \in T_{\pi(v)}M$. Up to our knowledge, the first time that a proof of this fact appeared was in [3, Sect. 11.1] published in 2000.

Positive homogeneity of Finsler metrics implies that the length of a piecewise smooth curve $\gamma\colon [a,b]\subseteq \mathbb{R}\to M$ given by

$$\ell_F(\gamma)=\int_a^b F(\dot\gamma)ds$$

does not depend on the orientation preserving parametrization of the curve. Then you can define the distance between two points $p,q\in M$ as

$$d(p,q)=\inf_{\gamma\in C_{p,q}}\ell_F(\gamma),$$

where $C_{p,q}$ is the space of piecewise smooth curves from p to q. This gives a generalized distance (see [58, p. 5] and also [20, 28]), but not necessarily reversible as the length of a curve depends on the orientation of the parametrization (observe that in general $F(-v)\neq F(v)$). Then, you can define two kind of balls, that is, forward and backward balls, respectively, as

$$B_F^+(p,r)=\{q\in M: d_F(p,q)<r\},\quad B_F^-(p,r)=\{q\in M: d_F(q,p)<r\},$$

for every $p\in M$ and $r>0$. Moreover, there exist several definitions for Cauchy sequences.

Definition 1. A sequence $\{x_n\}_{n\in\mathbb{N}}$ is called a *forward (resp. backward) Cauchy sequence* if for any $\varepsilon>0$, there exists $N\in\mathbb{N}$ such that $d_F(x_i,x_j)<\varepsilon$ for any $i,j\in\mathbb{N}$ satisfying $N<i<j$ (resp. $N<j<i$).

Moreover, you can also define the energy functional as

$$E_F(\gamma)=\frac{1}{2}\int_a^b F(\dot\gamma)^2ds$$

for every piecewise smooth curve $\gamma\colon [a,b]\subseteq\mathbb{R}\to M$, and geodesics as critical points of this functional. In particular, geodesics must have constant speed (see, e.g., [15, Proposition 2.1]). Let us point out that in some references as [3] geodesics are defined as critical points of the length functional and as a consequence they are not assumed to have constant speed.

3 Fermat's Principle in Conformally Standard Stationary SpaceTimes

Let us recall that a conformally stationary spacetime is a Lorentz manifold (M,g) that admits a timelike conformal vector field K. We refer to the classical books [6, 46] for the basic definitions on Lorentzian geometry and causality. Observe that K determines a time orientation in (M,g) and thus, a spacetime that, with an abuse of notation, we will denote also by (M,g). It can be shown that when K is complete and the spacetime is distinguishing (see [29]), then (M,g) splits as a *conformally standard stationary spacetime*, that is, $M = S\times\mathbb{R}$ and the metric g can be expressed as

$$g((v,\tau),(v,\tau)) = \varphi(g_0(v,v)+2\omega(v)\tau-\tau^2), \tag{3}$$

in $(x,t)\in S\times\mathbb{R}$, where $(v,\tau)\in T_xS\times\mathbb{R}$, φ is a smooth positive function on $S\times\mathbb{R}$ and ω and g_0 are respectively a one-form and a Riemannian metric on the manifold S. In this setting, the vector field K is induced from the natural lifting to M of the canonical vector field d/dt on $\mathbb{R}$, which we will denote in the following by ∂_t. Let us remark that sometimes in literature the one-form ω is expressed using the metrically equivalent vector field δ, that is, $\omega(v) = g_0(v,\delta)$ for every $v\in TS$.

We must observe that, in a Lorentzian manifold, light-like geodesics and their conjugate points are preserved by conformal changes up to parametrization (see, e.g., [44, Theorem 2.36]). This implies that studying light-like geodesics of $(S\times\mathbb{R},g)$ is equivalent to studying light-like geodesics of $(S\times\mathbb{R},\frac{1}{\varphi}g)$. As a consequence, we can assume that the spacetime is a *normalized standard stationary spacetime*, that is, a standard stationary spacetime with a unit Killing vector field and

$$g((v,\tau),(v,\tau)) = g_0(v,v)+2\omega(v)\tau-\tau^2, \tag{4}$$

in $(x,t)\in S\times\mathbb{R}$ for any $(v,\tau)\in T_xS\times\mathbb{R}$. In this case, ∂_t rather than a conformal vector field is a unit Killing vector field.

The advantage of formulating Fermat's principle in (conformally) standard stationary spacetimes is that it is possible to define a global time function given by the second coordinate in $S\times\mathbb{R}$ and it also makes sense to speak about the spatial position, that is, the first coordinate. Now fix two spatial positions x_0 and x_1 in S. Then, Fermat's principle says that the paths of light rays are critical points of the global time function between all the possible trajectories for light rays from (x_0,t_0) to the integral curve of ∂_t through (x_1,t_1), with $t_0,t_1\in\mathbb{R}$. According to general relativity, as photons are massless, the trajectories of light rays must be described by light-like curves. Therefore, the space of curves for the Fermat's principle must be composed of smooth future-pointing light-like curves. Let us observe that as the time orientation is assumed to be given by the Killing vector field ∂_t, a future-pointing causal curve is a curve $\gamma=(x,t):[a,b]\subseteq\mathbb{R}\to S\times\mathbb{R}$ satisfying $g(\dot\gamma,\dot\gamma)\leq 0$ and $\dot t>0$. If $\gamma=(x,t):[0,1]\to S\times\mathbb{R}$ is a smooth light-like curve from (x_0,t_0) to (x_1,t_1), we need to compute t_1, which is the value of the global time in $\gamma(1)$. As γ is light-like, we have that

$$g_0(\dot x,\dot x)+2\omega(\dot x)\dot t-\dot t^2=0,$$

and hence, as γ is assumed to be future-pointing (i.e., $\dot t>0$),

$$\dot t=\sqrt{g_0(\dot x,\dot x)+\omega(\dot x)^2}+\omega(\dot x).$$

Integrating the last equation, we get

$$t(s) = t_0 + \int_0^s \left(\sqrt{g_0(\dot{x},\dot{x}) + \omega(\dot{x})^2} + \omega(\dot{x}) \right) dv. \tag{5}$$

As a consequence, light-like geodesics must be critical points of the functional

$$T(\gamma) = t_1 = t_0 + \int_0^1 \left(\sqrt{g_0(\dot{x},\dot{x}) + \omega(\dot{x})^2} + \omega(\dot{x}) \right) dv.$$

This functional is, up to a constant, the length functional of the Finsler metric in S given by

$$F(v) = \sqrt{g_0(v,v) + \omega(v)^2} + \omega(v), \tag{6}$$

for every $v \in TS$. This metric is of Randers type, that is, the addition of the square of a Riemannian metric and a one-form of norm less than one in every point. We will call this metric the *Fermat metric* associated to the splitting (4) (or in general to the splitting (3)).

Remark 1. With a similar reasoning, we get that past-pointing light-like geodesics are controlled by the reverse metric of (6), that is,

$$\tilde{F}(v) = F(-v)$$

for every $v \in TS$. It is easy to see that:

(i) $d_{\tilde{F}}(p,q) = d_F(q,p)$ for every $p,q \in S$.
(ii) $\gamma\colon [0,1] \subseteq \mathbb{R} \to S$ is a geodesic from p to q of (S,F) if and only if the reverse curve $\tilde{\gamma}\colon [0,1] \to S$, $t \to \tilde{\gamma}(t) = \gamma(1-t)$ is a geodesic from q to p of $(S,\tilde{F})$.

Then, all the properties of past-pointing light-like geodesics can be also written in terms of the Fermat metric (6).

Remark 2. Let us consider the class of standard static spacetimes $(S \times \mathbb{R}, g_{st})$, with

$$g_{st}((v,\tau),(v,\tau)) = g_0(v,v) - \beta(x)\tau^2,$$

in $(x,t) \in S \times \mathbb{R}$, where $(v,\tau) \in T_xS \times \mathbb{R}$, g_0 is a Riemannian metric on S and β a positive smooth function on S. In particular, they are standard stationary, and the Fermat metric associated to them is Riemannian. Indeed, it is conformal to the metric induced by g_{st} in S, namely, g_0/β. This fact was already pointed out in [35, p. 343]. Up to the name of Fermat metric, other authors have used another name for the same concept, for example, in [18, 25], it is named as *optical metric* and in [2], *optical reference geometry*.

Remark 3. In the stationary case, it must be clarified that our terminology is different from the one introduced by Perlick in [48], where the name of *Fermat metric* is used for the Riemannian metric in S given by

$$h(v,v) = g_0(v,v) + \omega(v)^2, \tag{7}$$

for $v \in TS$. In [48], it is also used the name of *Fermat one-form* for ω. Observe that then our Fermat metric is the addition of the Fermat one-form and the square root of Perlick's Fermat metric. But our Fermat metric contains all the information, and in fact, it allows one to recover the Fermat one-form and Perlick's Fermat metric as

$$h(v,v) = \frac{1}{4}\left(F(v)+F(-v)\right)^2, \qquad \omega(v) = \frac{1}{2}\left(F(v)-F(-v)\right),$$

for any $v \in TS$, where F is given in Eq. (6).

The above computations show that in (conformally) standard stationary spacetimes, Fermat's principle relates future-pointing light-like geodesics of $(S \times \mathbb{R}, g)$ as in Eq. (4) with geodesics of the Finsler manifold (S,F) with F given in Eq. (6) up to reparameterizations. Let us state the relation including parameterizations.

Theorem 1 (Fermat's principle). *Let $(S \times \mathbb{R}, g)$ be a standard stationary spacetime as in Eq. (4). A curve $\gamma = (x,t) : [a,b] \subseteq \mathbb{R} \to S \times \mathbb{R}$ is a light-like geodesic of $(S \times \mathbb{R}, g)$ if and only if x is a geodesic for the Fermat metric F in Eq. (6) parameterized to have constant h-Riemannian speed (h as in Eq. (7)) and*

$$t(s) = t(a) + \int_a^s F(\dot{x})\mathrm{d}v,$$

for every $s \in [a,b]$.

Proof. The equivalence can be easily obtained computing the critical points of the length functional for F with h-constant Riemannian speed using the Levi–Civita connection ∇ of g_0 and then the light-like critical points of the energy functional of g using again ∇ (see, e.g., [15, Theorem 4.1]). □

Remark 4. Let us point out that Perlick [48] considers a more general case than conformally standard stationary spacetimes. Basically, he considers a conformally stationary spacetime (M,g) where the flow lines of the conformal vector field K of (M,g) have a structure of Hausdorff manifold $\hat{M}$ and the natural projection $\pi : M \to \hat{M}$ is a principal fiber bundle with structure group $\mathbb{R}$, with the action given by the flow of K. Observe that as the fiber is $\mathbb{R}$, there always exists a section of the bundle (see, e.g., [31, page 58]). But the existence of a space-like section is not guaranteed. In fact, assuming that K is complete, this happens if and only if the spacetime is distinguishing (see [29]). Given a section S of the fiber bundle, we can express the metric of (M,g) as in Eq. (4), but with g_0 not necessarily positive definite. In this case, the global function given by the second coordinate is not necessarily a time function, that is, it does not have to be strictly increasing in causal curves. As a consequence, the Fermat metric obtained in Eq. (6) can be non-positive along some directions of the tangent space. In fact, it is not difficult to see that the Fermat metric Eq. (6) is a Finsler metric (with the definition given in Sect. 2) if and only if the section S is space-like.

It can be helpful to restate the Fermat's principle as follows (see [16, Proposition 4.1]).

Proposition 1. *Let $z_0 = (x_0, t_0)$, $L_{x_1} = \{(x_1, t) : t \in \mathbb{R}\}$ be, respectively, a point and a vertical line in a standard stationary spacetime. Then, z_0 can be joined with L_{x_1} by means of a future-pointing light-like pregeodesic $t \mapsto \gamma(t) = (x_\gamma(t), t)$ starting at z_0 if and only if x_γ is a unit speed geodesic of the Fermat metric F which joins x_0 with x_1. In this case,*

$$t_1 - t_0 = \ell_F(x_\gamma|_{[t_0,t_1]}).$$

Let us observe that the Fermat metric depends on the space-like section you choose to obtain the standard splitting (which in some references as [26, 47] is called the gauge choice). The above proposition can be used to obtain the relation between two Fermat metrics associated to different splittings of the same stationary spacetime (with a fixed time-like Killing vector field K). If $(S \times \mathbb{R}, g)$ is one of the splittings (with g as in Eq. (4)), the other one is determined by a section given by a smooth function $f : S \to \mathbb{R}$ as $S_f = \{(x, f(x)) \in S \times \mathbb{R} : x \in S\}$. Then you can define the map $\psi_f : S \times \mathbb{R} \to S \times \mathbb{R}$ given as $\psi(x,t) = (x, t + f(x))$ for every $(x,t) \in S \times \mathbb{R}$. Therefore the other splitting is expressed as $(S \times \mathbb{R}, g_f)$, where $g_f = \psi_f^*(g)$ (here $*$ denotes the pullback operation).

Proposition 2. *With the above notation, the Fermat metric associated to the splitting $(S \times \mathbb{R}, g_f)$ is $F_f = F - df$, where F is the Fermat metric associated to $(S \times \mathbb{R}, g)$ and df is the differential of the smooth function f.*

Proof. Observe that given a curve $\gamma : [-\varepsilon, \varepsilon] \to S$, with $\varepsilon > 0$,

$$\hat{F}(\dot{\gamma}(0)) = \left.\frac{d}{ds}\right|_{s=0} \ell_{\hat{F}}(\gamma|_{[0,s]}),$$

for any Finsler metric $\hat{F}$. Moreover, as a consequence of Proposition 1,

$$\ell_{F_f}(\gamma) = \ell_F(\gamma) + f(\gamma(-\varepsilon)) - f(\gamma(\varepsilon)) = \int_{-\varepsilon}^{\varepsilon} (F(\dot{\gamma}) - df(\dot{\gamma}))ds. \quad (8)$$

Given $v \in TS$, consider $\gamma : [-\varepsilon, \varepsilon] \to S$ such that $\dot{\gamma}(0) = v$. Then,

$$F_f(v) = \left.\frac{d}{ds}\right|_{s=0} \ell_{F_f}(\gamma|_{[0,s]})) = F(v) - df(v)$$

for any $v \in TS$. □

Proposition 3. *Given an arbitrary function $f : S \to \mathbb{R}$, the section S_f of $S \times \mathbb{R}$ is space-like if and only if $F(v) > df(v)$ for every $v \in TS$.*

Proof. See also [16, Proposition 5.8]. □

Now, we can establish the so-called Stationary to Randers correspondence [16]. Let us call $\mathrm{Stat}(S\times\mathbb{R})$ the space of standard stationary spacetimes with normalized Killing vector field ∂_t and $\mathrm{Rand}(S)$ the space of Randers metrics on S. Then one has the bijective map

$$\mathrm{Stat}(S\times\mathbb{R})\to\mathrm{Rand}(S),\qquad g\mapsto F_g, \tag{9}$$

where F_g is determined as in Eq. (6) by the same stationary data pair (g_0,ω) which determines g in Eq. (4). Moreover, we can define in both sets equivalence relations as

$$\begin{aligned} R\sim R' &\iff R-R'=df \text{ for some smooth function f on S,}\\ g\sim g' &\iff g'=\psi_f^* g \quad \text{for some change of the initial section } \psi_f, \end{aligned}$$

and consider the corresponding quotient sets $\mathrm{Rand}(S)/\sim$, $\mathrm{Stat}(S\times\mathbb{R})/\sim$. Proposition 2 says that the bijection (9) induces a well-defined bijective map between the quotients

$$(\mathrm{Stat}(S\times\mathbb{R})/\sim)\to(\mathrm{Rand}(S)/\sim).$$

This relation constitutes a very important issue for Randers metrics, because the global invariants in the spacetime must be translated in invariants for the entire class of Randers metrics that differ in the differential of a function.

4 Causality and Fermat Metrics

As, by Proposition 1, geodesics of Fermat metrics contain all the information of light-like geodesics up to reparameterization, it turns out that Fermat metrics can be used to describe the chronological future and past of a given point. As a consequence, we can characterize the causal conditions of a standard stationary spacetime in terms of the Fermat metric. This relation was established in [16] with some previous partial results in [15]. Recall that we say that two events p and q in a spacetime are chronologically related, and write $p\ll q$ (resp. strictly causally related $p<q$) if there exists a future-pointing time-like (resp. causal) curve γ from p to q; p is causally related to q if either $p<q$ or $p=q$, denoted $p\leq q$. Then the *chronological future* (resp. *causal future*) of $p\in M$ is defined as $I^+(p)=\{q\in M: p\ll q\}$ (resp. $J^+(p)=\{q\in M: p\leq q\}$). Analogous notions appear substituting the word "future" by "past" and denoting $I^-(p),J^-(p)$.

Proposition 4. *Let $(S\times\mathbb{R},g)$ be a standard stationary spacetime as in Eq. (4) and $(x_0,t_0)\in S\times\mathbb{R}$. Then*

$$\begin{aligned} I^+(x_0,t_0)&=\cup_{s>0}\{t_0+s\}\times B_F^+(x_0,s),\\ I^-(x_0,t_0)&=\cup_{s<0}\{t_0-s\}\times B_F^-(x_0,s). \end{aligned}$$

Proof. See [16, Proposition 4.2]. □

Using the expression of the chronological future and past in terms of the forward and backward balls of the Fermat metric, we can easily obtain the characterization of the causality conditions in terms of the Fermat metric. For definitions and properties of the different levels of causality, we refer to [44].

Theorem 2. *Let* $(S \times \mathbb{R}, g)$ *be a standard stationary spacetime as in Eq. (4). Then* $(S \times \mathbb{R}, g)$ *is causally continuous. Furthermore:*

(a) It is causally simple if and only if one of the following equivalent conditions holds:

(i) $J^+(p)$ *is closed for all p,*
(ii) $J^-(p)$ *is closed for all p.*
(iii) The associated Finsler manifold (S, F) *is convex.*

(b) It is globally hyperbolic if and only if the subsets $B_F^+(x,r) \cap B_F^-(x,r)$ *are relatively compact for every* $x \in S$ *and* $r > 0$.

Moreover, a slice $S \times \{t_0\}, t_0 \in \mathbb{R}$, *is a Cauchy hypersurface if and only if the Fermat metric F on S is forward and backward complete.*

Proof. See [16, Theorems 4.3 and 4.4]. For part (b) see also [16, Proposition 2.2]. □

The static version of the last proposition can be found in [52, Proposition 3.5] (see also [6, Theorem 3.66]). Furthermore, an extension of last theorem characterizing the stationary regions that are causally simple in terms of convex regions for the Fermat metric has been achieved in [11]. Theorem 2 implies some consequences for Randers metrics. In particular we can establish a generalization of the classical Hopf–Rinow theorem.

Theorem 3. *Given a Randers manifold* (M, R), *the following conditions are equivalent:*

(i) The subsets $B_R^+(x,r) \cap B_R^-(x,r)$ *are relatively compact for every* $x \in M$ *and* $r > 0$.
(ii) The subsets that are forward and backward bounded are relatively compact.
(iii) There exists $f : M \to \mathbb{R}$ *such that* $R + df$ *is a forward and backward complete Randers metric.*

Moreover, these conditions imply the convexity of (M, R).

Proof. The equivalence between the two first conditions is standard, and it holds for any Finsler metric. For $(i) \Rightarrow (iii)$, first observe that any Randers metric can be obtained as the Fermat metric of a standard stationary spacetime (see [8, Proposition 3.1]). Now let $(S \times \mathbb{R}, g)$ be the standard stationary spacetime having as a Fermat metric R. By Proposition 2, this spacetime is globally hyperbolic, but then using [7], we obtain that there exists a smooth space-like Cauchy hypersurface S_f. Consider the splitting associated to the Cauchy hypersurface. By Proposition 2, the Fermat metric associated to the new splitting is of the form $R - df$ for a certain smooth function $f : S \to \mathbb{R}$. Moreover, by Proposition 2, $R - df$ must be forward and backward complete. For $(iii) \Rightarrow (i)$, observe that part (b) of Theorem 2 implies that the stationary spacetime associated to $R + df$ is globally hyperbolic. By Proposition 2, the stationary spacetime associated to R is the same as the one associated to $R + df$, but considering another splitting. Therefore it is globally hyperbolic and (i) follows from part (b) of Theorem 2. The convexity can be obtained from the Avez–Seifert Theorem applied to $(S \times \mathbb{R}, g)$ (see e.g., [6, Theorem 6.1]). □

It turns out that the condition of forward or backward completeness can be substituted by one of the two first equivalent conditions in Theorem 3 in some classical results of Finsler geometry as, for example, the theorems of Bonnet–Myers and Synge or the sphere theorem in its nonreversible version by Rademacher (see [16, Remark 5.3]).

Theorem 2 has been used in [17] to obtain some conditions that ensure global hyperbolicity. Recall that h is defined in Eq. (7). Given any Riemannian metric g in S, we will denote by d_g the distance in S associated to g. We say that a positive function f in S grows at most linearly with respect to d_g if given a point x_0, there exist positive constants A, B such that $f(x) \leq A \cdot d_g(x_0, x) + B$ for every $x \in S$. This condition does not depend on x_0. We also will denote

$$\|\omega\|_g = \sup_{v \in T_x S} \frac{|\omega(v)|}{\sqrt{g(v,v)}}$$

the g-norm of a one-form ω in $x \in S$ for any Riemannian metric g on S.

Theorem 4. *Let $(S \times \mathbb{R}, g)$ be a conformally standard stationary spacetime with g as in Eq. (3). Then the slices $S \times \{t\}$, $t \in \mathbb{R}$, are Cauchy hypersurfaces if one of the following conditions holds:*

(i) The metric $\frac{1}{(1+\|\omega\|_{g_0}^2)^2} h$ is complete.
(ii) The metric g_0 is complete and $\|\omega\|_{g_0}$ grows at most linearly in d_{g_0}.
(iii) There exists a proper function $f : S \to \mathbb{R}$ such that the product $\|df\|_{g_0} \cdot \|\omega\|_{g_0}$ grows at most linearly in $d_{g_0 + df \otimes df}$

Moreover, if $(S \times \mathbb{R}, g)$ is globally hyperbolic:

(iv) The slices $S \times \{t\}$, $t \in \mathbb{R}$, are Cauchy hypersurfaces if $\|\omega\|_{g_0}^2$ grows at most linearly in d_h.
(v) For any proper function $f : S \to \mathbb{R}$, $\|\omega\|_{g_0}$ grows at most linearly in $d_{g_0 + df \otimes df}$.

Proof. For (i) and (iv), see [17, Theorem 2]. For (ii), see part (1) of Proposition 2 in [17] and for (iii) and (v), [17, Theorem 4]. □

Indeed, in [17], the authors obtain several interesting pinching inequalities as

$$\frac{\sqrt{h(v,v)}}{2(1 + \|\omega\|_{g_0}^2)} \leq F(v) \leq 2\sqrt{h(v,v)}$$

and

$$\frac{\sqrt{g_0(v,v)}}{\sqrt{1+\|\omega\|^2_{g_0}}+\|\omega\|_{g_0}} \leq F(v) \leq (\sqrt{1+\|\omega\|^2_{g_0}}+\|\omega\|_{g_0})\sqrt{g_0(v,v)}$$

for every $v \in TS$ (see [17, Propositions 1 and 2]). We point out that in [52] (especially in Corollary 3.5) there are some results in the same direction as the last theorem.

As a further relation between causality of a standard stationary spacetime and Randers metrics, Cauchy developments will be constructed in terms of the Fermat metric. A subset A of a spacetime M is *achronal* if no $x,y \in A$ satisfies $x \ll y$; in this case, the *future (resp. past) Cauchy development* of A, denoted by $D^+(A)$ (resp. $D^-(A)$), is the subset of points $p \in M$ such that every past- (resp. future)-inextendible causal curve through p meets A. The union $D(A) = D^+(A) \cup D^-(A)$ is the *Cauchy development* of A. The *future (resp. past) Cauchy horizon $H^+(A)$ (resp. $H^-(A)$)* is defined as

$$H^{\pm}(A) = \{p \in \bar{D}^{\pm}(A) : I^{\pm}(p) \text{ does not meet } D^{\pm}(A)\}.$$

Intuitively, $D(A)$ is the region of M a priori predictable from data in A, and its *horizon* $H(A) = H^+(A) \cup H^-(A)$, the boundary of this region.

Proposition 5. *Let $(S \times \mathbb{R}, g)$ be a standard stationary spacetime as in Eq. (4) such that $S \times \{t_0\}$ is a Cauchy hypersurface, $A \subset S$, and $A_{t_0} = A \times \{t_0\}$ the corresponding (necessarily achronal) subset of $S \times \{t_0\}$. Then*

$$D^+(A_{t_0}) = \{(y,t) : d_F(x,y) > t - t_0 \textit{ for every } x \notin A \textit{ and } t \geq t_0\}, \tag{10}$$

$$D^-(A_{t_0}) = \{(y,t) : d_F(y,x) > t_0 - t \textit{ for every } x \notin A \textit{ and } t \leq t_0\}. \tag{11}$$

Moreover, the Cauchy horizons can be described as

$$H^+(A_{t_0}) = \{(y,t) : \inf_{x \notin A} d_F(x,y) = t - t_0\} \tag{12}$$

$$H^-(A_{t_0}) = \{(y,t) : \inf_{x \notin A} d_F(y,x) = t_0 - t\}. \tag{13}$$

Proof. See [16, Proposition 4.7]. □

Last proposition can be used to study the differentiability of the Cauchy horizon in terms of the distance computed with the Fermat metric from a closed subset [16, Theorem 4.10] and vice versa (see [16, Sect. 5.4].

5 Causal Boundaries and Fermat Metrics

In general relativity, it is important to complete the spacetime with some kind of boundary. One way to obtain an intrinsic completion is using the causal structure. This completion has been largely discussed because of some topological pathologies, but it seems that after [21], the definition and the topology is now satisfactory. As the causal completion (or c-completion for short) depends only on the causal structure of the spacetime, it is expectable that, in conformally standard stationary spacetimes, can be computed in terms of the Fermat metric. Let us recall that the c-completion is constructed in strongly causal spacetimes by adding some ideal points to the spacetime in such a way that time-like curves always have an endpoint in the new space. This is done by identifying the points of the space with PIP's (resp. PIF's), namely, proper indecomposable past (resp. future) sets; in other words, every point $p \in M$ is identified with $I^-(p)$ and $I^+(p)$. Then, we add to the spacetime the TIP's (resp. TIF's), namely, terminal indecomposable past (resp. future) sets. Then, the future (resp. past) c-completion $\hat{\partial} M$ (resp. $\check{\partial} M$) is given by the TIP's (resp. TIF's). Moreover, $\hat{M} := M \cup \hat{\partial} M$ and $\check{M} := M \cup \check{\partial} M$. In order to obtain the causal completion of M, we must identify some TIP's and TIF's. This is done by means of the S-relation. Denote $\hat{M}_\emptyset = \hat{M} \cup \{\emptyset\}$ (resp. $\check{M}_\emptyset = \check{M} \cup \{\emptyset\}$). The S-relation is defined in $\hat{M}_\emptyset \times \check{M}_\emptyset$ as follows. If $(P,F) \in \hat{M} \times \check{M}$, then $P \sim_S F$ if and only if:

(i) F is included and a maximal indecomposable future set in $\uparrow P$ (the common future of P)
(ii) P is included and a maximal indecomposable past set in $\downarrow F$ (the common past of F).

Moreover, we also put

$$P \sim_S \emptyset, \qquad \emptyset \sim_S F.$$

In particular, the only S-relations between PIP's and PIF's are $I^+(p) \sim_S I^-(p)$. Then, the c-completion $\bar{M}$ is the quotient set $\hat{M}_\emptyset \times \check{M}_\emptyset / \sim_S$ endowed with the *chronological topology* (see Definition 2.2 and the paragraph below in [20]). We can identify $M \equiv \{(I^-(p), I^+(p)) : p \in M\}$ and define the *c-boundary* as $\partial M := \bar{M} \setminus M$. We say that the c-completion is simple as a point set when every TIP (resp. TIF) determines a unique pair in ∂M (for the definition of topologically simple see [20, Definition 2.4]).

Up to the completion of the Finsler manifold (S,F), there are several nonequivalent ways to do it. You can compute the forward (resp. backward) Cauchy boundary $\partial_C^+ S$ (resp. $\partial_C^- S$) by adding ideal points in such a way that you can always obtain the convergence of forward (resp. backward) Cauchy sequences. Then the forward (resp. backward) *Cauchy completion* is $S_C^+ := S \cup \partial_C^+ S$ (resp. $S_C^- := S \cup \partial_C^- S$). Moreover, denote $\partial_C^s S := \partial_C^+ S \cap \partial_C^- S$ and $S_C^s = S \cup \partial_C^s S$. The map

$$d_Q : S_C^+ \times (S_C^+ \cup S_C^-) \to [0, \infty]$$

defined by

$$d_Q([\{x_n\}],[\{y_m\}]) := \lim_n(\lim_m d_F(x_n,y_m))$$

is a quasi-distance (see [20, Propositions 3.25 and 3.32]).

The *Gromov completion* is obtained by considering the subset $\mathscr{L}_1(S,d_F)$ of d_F-Lipschitz functions on S [20, Definition 5.2] and $\mathscr{L}_1(S,d_F)_* = \mathscr{L}_1(S,d_F)/\mathbb{R}$ (two functions are related when they differ in a constant). Then define the maps

$$j^+ : S_C^+ \to \mathscr{L}_1(S,d_F), \quad x \to -d_x^+, \quad \text{where } d_x^+ = d_Q(\cdot,x),$$
$$j^- : S_C^- \to \mathscr{L}_1(S,d_F), \quad x \to +d_x^-, \quad \text{where } d_x^- = d_Q(x,\cdot),$$

which are injective (see [20, Proposition 5.7]). We can identify the points of S with the class of (minus) the distance function to (resp. from) the point, which is denoted by $j^+(S)$ (resp. $j^-(S)$). The forward (resp. backward) Gromov completion S_G^+ (resp. S_G^-) is the closure of S in $\mathscr{L}_1(S,d_F)_*$ considering the compact-open topology. Observe that this topology is equivalent to that of uniform convergence on compact subsets and to that of pointwise convergence.

Let $C^+(S)$ be the set of piecewise smooth curves $c : [\alpha,+\Omega) \to S$, $\Omega \leq \infty$, such that $F(\dot{c}) < 1$. For $c \in C^+(S)$, the associated (forward) Busemann function $b_c^+ : S \to (-\infty,\infty]$ is $b_c^+(\cdot) = \lim_{s\to\Omega}(s - d_F(\cdot,c(s)))$, which always exists because $s \mapsto s - d_F(\cdot,c(s))$ is increasing [20, Lemma 4.14]. Observe that b_c^+ is finite or infinite everywhere. Denote

$$B^+(S) := \{b_c^+ < \infty : c \in C^+(S)\}.$$

Moreover, if $\Omega < \infty$, then there exists some $\bar{x} \in S_C^+$ such that $b_c^+(x) = \Omega - d_F(x,\bar{x})$ for all $x \in S$ (here d_F is extended to S_C^+), and we denote

$$b_c^+ = d_p^+ := \Omega - d_F(\cdot,\bar{x}),$$

with $p = (\bar{x},\Omega) \in S_C^+ \times \mathbb{R}$. If $\Omega = \infty$, we say that b_c^+ is a properly Busemann function and we write

$$\mathscr{B}^+(S) := \{b_c^+ < \infty : c \in C^+(S), \Omega = \infty\}.$$

The *Busemann completion* as a point set is the quotient $S_B^+ := B^+(S)/\mathbb{R} \subset S_G^+$ and the (forward) *Busemann boundary* $\partial_B^+ S := S_B^+ \setminus S (\subset \partial_G^+ S)$. Furthermore, the (forward) *properly Busemann boundary* is defined as $\partial_{\mathscr{B}}^+ S := \mathscr{B}^+(S)/\mathbb{R}$. Then $\partial_B^+ S = j^+(\partial_C^+ S) \cup \partial_{\mathscr{B}}^+ S$. S_B^+ will be endowed with the chronological topology (see [20, Sect. 5.2.2]).

Recall that given a topological space T, the *forward cone* with base T is constructed as the quotient topological space $(T \times (-\infty,\infty])/\sim$, where the unique non-trivial identifications are $(x,\infty) \sim (x',\infty)$ for all $x,x' \in T$. Moreover, the class of (x,∞) is called the *apex* of the cone.

Finally, given a future-pointing time-like curve $\gamma : [\alpha,\Omega) \to M$, parameterized as $\gamma(t) = (c(t),t)$, observe that $I^-[\gamma] = \{(x',t') \in M : t' < b_c^+(x')\}$. Therefore the indecomposable past sets ($\neq M$) can be identified with the elements of $B^+(S)$. When $b_c^+ \equiv \infty$, then $I^-[\gamma] = M$, and it will be denoted with i^+. Given a d_F-Lipschitz function $f : S \to \mathbb{R}$, we define its past $P(f)$ (resp. future $F(f)$) as

$$P(f) := \{(x,t) \in M : t < f(x)\} \subset M$$

(resp. $F(f) := \{(x,t) \in M : t > -f(x)\} \subset M$). If $(P,F) \in \partial M$ with $\emptyset \neq P = P(b_c^+)$, we define the line over (P,F), denoted as $L(P,F)$, as

- If $F = \emptyset$ then $L(P,\emptyset) := \{(P',\emptyset) : P' = P(b_c^+ + k), k \in \mathbb{R}\}$,
- If $F \neq \emptyset$, it follows that $P = P(d_p^+)$ and $F = F(d_{p'}^-)$, where $p \in \partial_c^+ S \times \mathbb{R}$, $p' \in \partial_c^- S \times \mathbb{R}$ (see [20, Theorem 6.15]), and then

$$L(P,F) := \{(P',F') : P' = P(d_p^+ + k), F' = F(d_{p'}^- + k), k \in \mathbb{R}\}.$$

A dual definition is assumed for $(P,F) \in \partial M$ with $\emptyset \neq F = F(b_c^-)$.

Let us first describe the structure of the c-completion as a point set.

Theorem 5. *Let $(S \times \mathbb{R}, g)$ be a (conformally) standard stationary spacetime as in Eq. (3) and denote $M = S \times \mathbb{R}$. Then, the c-boundary ∂M has the following structure:*

(i) The future (resp. past) c-boundary $\hat{\partial} M$ (resp. $\check{\partial} M$) is naturally a point set cone with base $\partial_B^+ S$ (resp. $\partial_B^- S$) and apex i^+ (resp. i^-).
(ii) A pair $(P,F) \in \partial M$ with $P \neq \emptyset$ satisfies that $P = P(b_c^+)$ for some $c \in C^+(S)$ and:

(a) If $b_c^+ \equiv \infty$ then $P = M, F = \emptyset$.
(b) If $b_c^+ \in \mathscr{B}^+(S) (\equiv \partial_{\mathscr{B}}^+ S \times \mathbb{R})$, then $F = \emptyset$.
(c) If $b_c^+ \in B^+(S) \setminus \mathscr{B}^+(S)$, then $b_c^+ = d_p^+$ with $p = (x^+, \Omega^+) \in \partial_C^+ S \times \mathbb{R}$, $P = P(d_p^+)$ and $F \subset F(d_p^-)$. In this case, there are two exclusive possibilities:

(c1) Either $F = \emptyset$.
(c2) Or $F = F(d_{p'}^-)$ with $p' = (x^-, \Omega^-) \in \partial_C^- S \times \mathbb{R}$ and satisfying

$$\Omega^- - \Omega^+ = d_Q(x^+, x^-)$$

(in this case, p' is not necessarily unique).

Moreover, if $x^+ \in \partial_C^s S$, then $p' = p$, $\uparrow P = F(d_p^-)$ and P is univocally S-related with $F = F(d_p^-)$.

A dual result holds for pairs (P,F) with $F \neq \emptyset$. So, the total c-boundary is the disjoint union of lines $L(P,F)$.

When ∂M is simple as a point set, it is the quotient set $\hat{\partial} M \cup_d \check{\partial} M / \sim_S$ of the partial boundaries $\hat{\partial} M, \check{\partial} M$ under the S-relation.

Proof. See [20, Theorem 1.2]. □

Let us finally describe the causal and topological structures. As to the completions of (S,F), let us remark that in the description of the c-completion, we only need Busemann and Cauchy completions, while Gromov completion is useful to define the Busemann one. Observe that Gromov completion is a compact metrizable topological space and the Busemann one is T_1, sequentially compact but not necessarily Hausdorff. This is because, in S_B^+, the topology inherited from the Gromov completion is finer than the chronological topology (otherwise, the Busemann completion would not be sequentially compact). As a matter of fact, Busemann and Gromov completions coincide both as a point set and as a topological space when S_B^+ is Hausdorff.

Theorem 6. *Let $(S \times \mathbb{R}, g)$ be a (conformally) standard stationary spacetime as in Eq. (3) and denote $M = S \times \mathbb{R}$. Then, for each $(P,F) \in \partial M$, the line $L(P,F)$ is:*

(i) Time like if $P = P(d_p^+)$ and $F = F(d_p^-)$ for some $p \in \partial_C^s M \times \mathbb{R}$
(ii) Horismotic if either P or F are empty
(iii) Locally horismotic otherwise

(see [20, Definition 6.22]).

As to the topology of the c-completion:

(iv) If S_B^+ (resp. S_B^-) is Hausdorff, the future (resp. past) causal boundary has the structure of a (topological) cone with base $\partial_B^+ S$ (resp. $\partial_B^- S$) and apex i^+ (resp. i^-).
(v) If S_C^s is locally compact and d_Q^+ is a generalized distance, then $\overline{M}$ is simple, and so, it coincides with the quotient topological space $\hat{M} \cup_d \check{M}/ \sim_S$ of the partial completions $\hat{M}$ and $\check{M}$ under the S-relation.

Summarizing, if S_C^s is locally compact, d_Q is a generalized distance, and $S_B^\pm$ is Hausdorff; ∂M coincides with the quotient topological space $(\hat{\partial} M \cup_d \check{\partial} M)/ \sim_S$, where $\hat{\partial} M$ and $\check{\partial} M$ have the structure of cones with bases $\partial_B^+ S, \partial_B^- S$ and apexes i^+, i^-, respectively.

Proof. See [20, Theorem 1.2]. □

See the contribution by Flores and Herrera [19] to these proceedings for a more detailed study.

6 Existence of Light-Like Geodesics

The study of multiplicity of light-like geodesics between an event and a vertical line was the original scope of the use of Fermat metrics. For example, in [39], the authors use the shortening method applied to the Fermat metric to give some existence results. It is remarkable that in [39], the authors refer to the Fermat metric as a pseudo-Finsler metric and they are concerned about the local existence, uniqueness and regularity of minimizers of the length functional (see [39, Appendix A.1]).

Of course, this is because, in that moment, they were not aware of the fact that Randers metrics are fiberwise strongly convex. By the way, it seems that this fact is not collected in the classical books of Finsler geometry available at the time that [39] was published. This was done just two years later in [3, Sect. 11.1].

Once you know that Randers metrics are fiberwise strongly convex, the local existence and uniqueness of geodesics are guaranteed. Moreover, studying light-like geodesics between an event and a vertical line or ligth-like geodesics spatially closed in a conformally standard stationary spacetime is equivalent to studying the existence and multiplicity of geodesics between two points or closed geodesics of the Fermat metric, respectively (up to the case with boundary see [11] and [9, Proposition 4.9] in these proceedings). This can be done by applying the theories of Lyusternik–Schnirelmann and Morse to the energy functional of a Finsler manifold (M,F). In fact, you can consider the space of curves of Sobolev class H^1 on M. Recall that this space does not depend on the Riemannian metric that you fix on M. Thus, we fix an auxiliary Riemannian metric h on M. Moreover, fix a smooth submanifold N of $M \times M$ and consider the collection $\Lambda_N(M)$ of the curves $x : [0,1] \to M$, having H^1 regularity, that is, x is absolutely continuous and $\int_0^1 h(\dot{x},\dot{x})ds$ is finite, and with $(x(0),x(1)) \in N \subseteq M \times M$. Then, it is well known that $\Lambda_N(M)$ is a Hilbert manifold modeled on any of the equivalent Hilbert spaces of all the H^1-sections with endpoints in TN of the pulled back bundle x^*TM, with x a regular curve in $\Lambda_N(M)$. Let us observe that even when the strong convexity condition is available, we must pay some attention to the fact that F^2 is not even C^2 on the zero section unless F^2 is quadratic, that is, a Riemannian metric (see [55]).

Proposition 6. *A nonconstant curve $\gamma \in \Lambda_N(M)$ is a geodesic for the Finsler manifold (M,F) satisfying*

$$g_{\dot{\gamma}(0)}(V,\dot{\gamma}(0)) = g_{\dot{\gamma}(1)}(W,\dot{\gamma}(1)) \tag{14}$$

for any $(V,W) \in T_{(\gamma(0),\gamma(1))}N$ if and only if it is a (non-constant) critical point of the energy functional E_F on $\Lambda_N(M)$.

Proof. See, for example, [15, Proposition 2.1]. □

Moreover, recall that a functional J defined on a Banach manifold $(X, \|\cdot\|)$ satisfies the Palais–Smale condition if every sequence $\{x_n\}_{n\in\mathbb{N}}$ such that $\{J(x_n)\}_{n\in\mathbb{N}}$ is bounded and $\|dJ(x_n)\| \to 0$ contains a convergent subsequence. This condition is fundamental to apply the theories of Lyusternik–Schnirelmann and Morse, which study the relation between the number of critical points and the topology of the manifold. Palais–Smale is satisfied by the energy functional precisely when one of the equivalence conditions of the generalized Hopf–Rinow Theorem in 3 holds.

Theorem 7. *Let (M,F) be a Finsler manifold with $B_F^+(x,r) \cap B_F^-(x,r)$ relatively compact for every $x \in M$ and $r > 0$, and N, a closed submanifold on $M \times M$ such that the first or the second projection of N to M is compact, then E_F is a $C^{1,1}$ functional on $\Lambda_N(M)$ and it satisfies the Palais–Smale condition on $\Lambda_N(M)$.*

Proof. See [15, Theorem 3.1] or [43] and the comments before [16, Theorem 5.2]. □

Again, the most difficult part to prove Palais–Smale for the energy functional of a Finsler metric is the lack of differentiability of F^2 in the zero section. As F^2 is not C^2 on the zero section, we can only apply the mean value theorem to the derivatives of F^2 away from the zero section. Once Palais–Smale condition is available, we can apply Lyusternik–Schnirelmann theory to obtain the existence of infinitely many geodesics between two arbitrary points when the manifold is non-contractible (see [15, Proposition 3.1]). With a different approach, it is possible to prove the existence of only a finite number of geodesics between two nonconjugate points in the presence of a convex function for the Finsler metric [13, Theorem 2.4].

6.1 Morse Theory for Light-Like Geodesics

As to the Morse theory for the energy functional in the space of H^1 curves, the main difficulty is that E_F is not twice differentiable at a curve γ, even if γ is a geodesic, unless the restriction of F^2 to the geodesic is a quadratic function with respect to the velocities (see [1] and also [10]). As a consequence, Morse Lemma cannot be proved in the curves of class H^1 with the standard techniques. Even if Morse Theory works for $C^{1,1}$-functionals in Hilbert manifolds (see, e.g., [42, Chap. 8]), the Morse Lemma is essential to compute the critical groups in terms of the index of the critical point. In [14], this problem is circumvented using that the space of curves with C^1 regularity is a Banach manifold densely immersed in the Hilbert space of H^1 curves and E_F, restricted to the C^1 class, admits second differential in regular curves of C^1. To be more precise, consider the second differential of E_F in the space of C^1-curves, assume for simplicity that the kernel is trivial, and extend it by density to H^1. This gives a functional that it is represented by the identity plus a compact operator in a certain scalar product [14, Lemma 2]. Moreover, the restriction of this operator to the space of C^1-curves gives an invertible operator [14, Lemma 5], and then one can obtain a Morse Lemma for this restriction and the scalar product of H^1 [14, Theorem 7]. Finally, we show that the critical groups of the C^1-class coincide with those of H^1 using a classical result by Palais. As the geometrical index of a light-like geodesic coincides with the one of its projection as a Fermat geodesic [14, Theorem 13], the Morse relations for light-like geodesics in conformally standard stationary spacetimes follow.

Theorem 8. *Let $(S\times\mathbb{R},g)$ be a globally hyperbolic conformally standard stationary spacetime with g as in Eq. (3), $p=(p_0,t_0)\in S\times\mathbb{R}$ and $L_{q_0}=\{(q_0,s)\in S\times\mathbb{R}: s\in\mathbb{R}\}$. Assume that for each $s\in\mathbb{R}$ the points p and (q_0,s) are non-conjugate along every future-pointing light-like geodesic connecting them. Then there exists a formal series $Q(r)$ with coefficients in $\mathbb{N}\cup\{+\infty\}$ such that*

$$\sum_{z\in G_{p,L_{q_0}}} r^{\mu(z)} = P(r,\Lambda_{(p_0,q_0)}(S)) + (1+r)Q(r),$$

where $G_{p,L_{q_0}}$ is the set of all the future-pointing light-like geodesics connecting p to L_{q_0}, $\mu(z)$ is the number of conjugate points of z counted with multiplicity, and $P(r,\Lambda_{(p_0,q_0)}(S))$ is the Poincaré polynomial of $\Lambda_{(p_0,q_0)}(S)$.

Proof. See [14, Theorem 15]. □

Recall that the Gromoll–Meyer theorem ensures the existence of infinitely many geometrically distinct closed geodesics whenever $\limsup_{k\to\infty}\beta_k(\Lambda(M)) = +\infty$, where $\beta_k(\Lambda(M))$ are the Betti numbers of the loop space of M. Using the same hypothesis, in [12], it is obtained the existence of an infinite number of geometrically distinct geodesics joining two nonconjugate points p and q. By geometrically distinct, we mean that they do not come from the iterations of a finite number of closed geodesics that go through p and q (as in the round sphere).

Let us observe that even if the problem of existence of normal geodesics between two arbitrary submanifolds in a standard stationary spacetime cannot be reduced to a problem for the Fermat metric in general, in [5], the authors use completeness of the Fermat metric to prove a result of this type with some hypotheses in the submanifolds [5, Theorem 1.1].

6.2 t-Periodic Light-Like Geodesics and the Closed Geodesic Problem

Let us recall that a light-like geodesic $\gamma = (x,t):\mathbb{R}\to S\times\mathbb{R}$ in a standard stationary spacetime is said t-periodic if there exists $T\geq 0$ and $s_0>0$ such that x is periodic, that is, x and its derivatives coincide in 0 and s_0, $t(s_0)=t(0)+T$ and $\dot t(s_0)=\dot t(0)$. In this case, T is called the *universal period*. They are related with closed geodesics for the Fermat metric.

Proposition 7. *Let (M,g) be a conformally standard stationary spacetime as in Eq. (3). Then, $\gamma=(x,t):\mathbb{R}\to S\times\mathbb{R}$ is a t-periodic light-like geodesic if and only if $x:\mathbb{R}\to S$ is a closed geodesic of the Fermat metric.*

Proof. The implication to the right follows from Proposition 1. For the other one, first observe that $g(\dot\gamma,\partial_t)$ is constant. To see this, recall that as ∂_t is a conformal field, it satisfies

$$g(\nabla_V\partial_t,W)+g(\nabla_W\partial_t,V)=\lambda g(V,W) \tag{15}$$

for every $V,W\in\mathfrak{X}(M)$, where ∇ is the Levi–Civita connection of (M,g) and λ a smooth function from M to $\mathbb{R}$. Then, using that γ is a light-like geodesic and Eq. (15),

$$\frac{d}{ds}g(\dot\gamma,\partial_t)=g(\dot\gamma,\nabla_{\dot\gamma}\partial_t)=\frac{1}{2}\lambda g(\dot\gamma,\dot\gamma)=0.$$

Using again Proposition 1, we deduce that there exists $s_0 > 0$ such that $\gamma(s_0) = \gamma(0)$ and $\dot\gamma(s_0) = \mu\dot\gamma(0)$ for some $\mu > 0$, but the fact that $g(\dot\gamma, \partial_t)$ is a non null constant (it cannot be zero because $\dot\gamma$ is light-like and ∂_t time-like) implies that $\mu = 1$ as required. □

As to the closed geodesic problem for compact manifolds, most of the classical results for Riemannian metrics, such as Gromoll–Meyer and Bangert–Hingston theorems, which assume, respectively, that $\limsup_{k\to\infty} \beta_k(\Lambda(M)) = +\infty$ and that the fundamental group of the manifold is infinite abelian, are available in the Finslerian setting under the same topological hypothesis, obtaining the corresponding results of multiplicity for t-periodic light-like geodesics in conformally standard stationary spacetimes (see [8] and references therein). As an exception, there are Finsler metrics with a finite number of geometrically distinct closed geodesics, the so-called Katok metrics. Remarkably, these Finsler metrics are of Randers type, and they have constant flag curvature. Let us observe that the classification of Randers metrics of constant flag curvature has been obtained in [4] using the expression of a Randers metric as a Zermelo one, that is, a metric defined from a Riemannian metric g and a vector field W in a manifold M as

$$Z(v) = \sqrt{\frac{1}{\lambda}g(v,v) + \frac{1}{\lambda^2}g(v,W)^2} - \frac{1}{\lambda}g(v,W),$$

where $\lambda = 1 - g(W,W)$ must be positive. Indeed, (M,Z) has constant flag curvature if and only if W is a homothety and g has constant curvature [4]. We can then construct standard stationary spacetimes with compact orbit manifold S and a finite number of geometrically distinct t-periodic light rays (see [8, Propositions 3.1 and 3.4]).

6.3 *Alternative Functional to Energy*

Existence and multiplicity of Fermat geodesics can be studied by means of other functionals rather than the energy one. In [22], the authors use the functional defined as

$$J(x) = \sqrt{\int_0^1 h(\dot x, \dot x)\mathrm{d}s} + \int_0^1 \omega(\dot x)\mathrm{d}s$$

for every curve $x : [0,1] \to S$ of class H^1 with h as in Eq. (7). The advantage of this functional is that it is C^2 on geodesics. Its critical points are Fermat geodesics parameterized with h-constant speed (see also [38]). This functional has also been used in [23, 24] to obtain a result of genericity of stationary spacetimes without conjugate light-like geodesics between a fix event p and a fixed vertical line.

7 Further Applications

7.1 *Randers Metrics of Constant Flag Curvature and Stationary SpaceTimes*

Flag curvature plays a similar role in Finsler geometry as sectional curvature in the Riemannian setting, that is, it is an important invariant related to the behaviour of geodesics. Let us recall that Randers metrics with constant flag curvature have been classified in [4]. These metrics have already appeared in the context of Fermat metrics to provide examples of spacetimes with a finite number of geometrically distinct t-periodic light-like geodesics [8, Propositions 3.1 and 3.4]. Subsequently, these spacetimes were studied in [26].

Proposition 8. *Let $(S\times\mathbb{R},g)$ be a conformally standard stationary spacetime as in Eq. (3) whose Fermat metric is of constant flag curvature. Then $(S\times\mathbb{R},g)$ is locally conformally flat.*

Proof. See [26, Sect. II.E.2]. □

The converse of the last proposition is not true in general, because, for example, you can find a Randers metric of the form $\sqrt{g_0}+df$ with g_0 the Euclidean metric in $\mathbb{R}^n$ and $f:\mathbb{R}^n\to\mathbb{R}$ a smooth function, such that $\sqrt{g_0}+df$ does not have constant flag curvature (see [3, Sect. 3.9B]), and $\sqrt{g_0}+df$ is the Fermat metric associated to a certain splitting of Minkowski spacetime (see Proposition 2). Anyway, it is expectable, as commented in [26, Sect. II.E.2], that given a conformally flat stationary spacetime, you can find a space-like section having as a Fermat metric a Randers metric with constant flag curvature. Let us point out that in [26], the authors give several examples of Randers metrics coming from well-known stationary spacetimes. They also recall the relation between magnetic Lagrangians and Randers metrics.

7.2 *Time-Like Geodesics with Fixed Arrival Proper Time*

First of all, let us recall that the proper time of a time-like curve $\alpha:[a,b]\to M$ in a Lorentzian manifold (M,g) is defined as $\int_a^b\sqrt{-g(\dot\gamma,\dot\gamma)}\mathrm{d}s$. Let us also remark that existence of time-like geodesics with fixed arrival proper time between an event and a vertical line in a standard stationary spacetime can be reduced to existence of light-like geodesics in a one-dimensional higher standard stationary spacetime. Observe that, in this case, as time-like geodesics are not preserved by conformal changes, we cannot consider conformally standard stationary spacetimes as in Eq. (3), but standard stationary spacetimes $(S\times\mathbb{R},g)$ such that

$$g((v,\tau),(v,\tau))=g_0(v,v)+2\omega(v)\tau-\beta\tau^2,$$

in $(x,t) \in S \times \mathbb{R}$ for any $(v,\tau) \in T_xS \times \mathbb{R}$, where ω and g_0 are respectively a one-form and a Riemannian metric on S and β is a positive function on S. Then we can define the one-dimensional higher spacetime $(S \times \mathbb{R}^2, \eta)$, with η defined as

$$\eta((v,y,\tau),(v,y,\tau)) = g_0(v,v) + y^2 + 2\omega(v)\tau - \beta\tau^2,$$

in $(x,v,t) \in S \times \mathbb{R}^2$, where $(v,y,\tau) \in T_xS \times \mathbb{R}^2$. A curve from the event $(x_0,t_0) \in S \times \mathbb{R}$ to the line $L_{x_1} = \{(x_1,s) \in S \times \mathbb{R} : s \in \mathbb{R}\}$ is a time-like geodesic $\gamma = (x,t) : [0,1] \to S \times \mathbb{R}$ of $(S \times \mathbb{R}, g)$ with arrival proper time T if and only if $[0,1] \ni s \to (x(s),s,t(s)) \in S \times \mathbb{R}^2$ is a light-like geodesic of $(S \times \mathbb{R}^2, \eta)$ from the event $(x_0,0,t_0)$ to the line $\{(x_1,T,s) \in S \times \mathbb{R}^2 : s \in \mathbb{R}\}$. Moreover, the Fermat metric of this standard stationary spacetime is given as

$$\tilde{F}(v,y) = \sqrt{\frac{1}{\beta}g_0(v,v) + \frac{v^2}{\beta} + \frac{1}{\beta^2}\omega(v)^2} + \frac{1}{\beta}\omega(v),$$

in $(x,v) \in S \times \mathbb{R}$, where $(v,y) \in T_xS \times \mathbb{R}$. As completeness conditions for the original Fermat metric in Eq. (6), which in this case is expressed as

$$F(v) = \sqrt{\frac{1}{\beta}g_0(v,v) + \frac{1}{\beta^2}\omega(v)^2} + \frac{1}{\beta}\omega(v),$$

imply completeness conditions for $\tilde{F}$ (see the proof of [15, Proposition 4.2]), some multiplicity results [15, Proposition 4.2] and Morse relations [14, Theorem 18] are available when the spacetime is globally hyperbolic. Observe that in [11], the existence of such time-like geodesics under sharp conditions (weaker than global hyperbolicity) is obtained.

7.3 Conformal Maps and Almost Isometries

Another interesting relation between Fermat metrics and conformally standard stationary spacetimes occurs at the level of transformations (see [28]). As Fermat metrics remain invariant by conformal changes in the conformally stationary spacetime, we need to consider conformal maps in the spacetime. Moreover, as we want to project these maps into maps of the orbit manifold S, they have to preserve the conformal vector field K. Summing up, they have to be K-conformal maps, denoted by $\mathrm{Conf}_K(S \times \mathbb{R}, g)$, that is, they must preserve the metric up to a positive constant in every point and the conformal vector field K. As to general relativity, these maps are precisely those that preserve the causal structure and the observers along K. Their counterpart in Fermat metrics are the so-called *almost isometries*, which are maps $\varphi : S \to S$ such that $\varphi^*(F) = F + df$ for a certain smooth function

$f : S \to \mathbb{R}$ (here $*$ denotes the pullback operator). Let us denote by $\widetilde{\mathrm{Iso}}(S,F)$ the group of almost isometries of (S,F), which is a Lie group [28].

Theorem 9. *Let $\psi : S \times \mathbb{R} \to S \times \mathbb{R}$ be a K-conformal map of a conformally standard stationary spacetime as in Eq. (3). Then, there exist functions $\varphi : S \to S$ and $f : S \to \mathbb{R}$ such that $\psi(x,t) = \big(\varphi(x), t + f(x)\big)$ and φ is an almost isometry for the Fermat metric of $(S \times \mathbb{R}, g)$. Moreover, φ is a Riemannian isometry for the metric h in Eq. (7), and the map $\pi : \mathrm{Conf}_K(S \times \mathbb{R}, g) \to \widetilde{\mathrm{Iso}}(S,F)$, defined as $\pi(\psi) = \varphi$, is a Lie group homomorphism. The map can be projected to the quotient*

$$\bar{\pi} : \mathrm{Conf}_K(S \times \mathbb{R}, g)/\mathscr{K} \to \widetilde{\mathrm{Iso}}(S,F),$$

(where $\mathscr{K}$ is the subgroup generated by the flow of K) and gives an isomorphism of Lie groups.

Proof. See [28]. □

As a consequence of this relation, it follows a result of genericity of stationary spacetimes with discrete K-conformal group.

Corollary 1. *Given a manifold S, for a* generic *set of data (g_0, ω), the stationary metric $g = g(g_0, \omega)$ given in Eq. (4) on $S \times \mathbb{R}$ has discrete K-conformal group* $\mathrm{Conf}_K(S \times \mathbb{R}, g)$.

Proof. See [28]. □

Acknowledgements The author appreciates very useful suggestions given by the anonymous referees and would like to acknowledge the careful reading of the first version and the advices of Erasmo Caponio, Jose Luis Flores, Jonatan Herrera, and Miguel Sánchez.

The author is partially supported by Regional Junta de Andalucía Grant P09-FQM-4496, by MICINN project MTM2009-10418, and by Fundación Séneca project 04540/GERM/06.

References

1. Abbondandolo, A., Schwarz, M.: A smooth pseudo-gradient for the Lagrangian action functional. Adv. Nonlinear Stud. **9**(4), 597–623 (2009)
2. Abramowicz, M., Carter, B., Lasota, J.-P.: Optical reference geometry for stationary and static dynamics. Gen. Relativ. Gravit. **20**, 1172–1183 (1988)
3. Bao, D., Chern, S.-S., Shen, Z.: An Introduction to Riemann-Finsler geometry. Graduate Texts in Mathematics. Springer, New York (2000)
4. Bao, D., Robles, C., Shen, Z.: Zermelo navigation on Riemannian manifolds. J. Differ. Geom. **66**(3), 377–435 (2004)
5. Bartolo, R., Candela, A.M., Caponio, E.: Normal geodesics connecting two non-necessarily spacelike submanifolds in a stationary spacetime. Adv. Nonlinear Stud. **10**(4), 851–866 (2010)
6. Beem, J.K., Ehrlich, P.E., Easley, K.L.: Global Lorentzian geometry. Monographs and Textbooks in Pure and Applied Mathematics, vol. 202, 2nd edn. Marcel Dekker, New York (1996)

7. Bernal A.N., Sánchez, M.: On smooth Cauchy hypersurfaces and Geroch's splitting theorem. Comm. Math. Phys. **243**(3), 461–470 (2003)
8. Biliotti L., Javaloyes, M.Á.: t-periodic light rays in conformally stationary spacetimes via Finsler geometry. Houston J. Math. **37**(1), 127–146 (2011)
9. Caponio, E.: Infinitesimal and local convexity of a hypersurface in a semi-Riemannian manifold, these proceedings, Recent Trends in Lorentzian Geometry, Springer Proceedings in Mathematics & Statistics (2013)
10. Caponio, E.: The index of a geodesic in a Randers space and some remarks about the lack of regularity of the energy functional of a Finsler metric. Acta Math. Acad. Paedagog. Nyházi. (N.S.) **26**(2), 265–274 (2010)
11. Caponio, E., Germinario, A., Sánchez, M.: Geodesics on convex regions of stationary spacetimes and Finslerian Randers spaces. arXiv:1112.3892v1 [math.DG] (2011)
12. Caponio, E., Javaloyes, M.A.: A remark on the Morse Theorem about infinitely many geodesics between two points. arXiv:1105.3923v1 [math.DG] (2011)
13. Caponio, E., Javaloyes, M.Á., Masiello, A.: Finsler geodesics in the presence of a convex function and their applications. J. Phys. A **43**(13), 135207, 15 (2010)
14. Caponio, E., Javaloyes, M.Á., Masiello, A.: Morse theory of causal geodesics in a stationary spacetime via Morse theory of geodesics of a Finsler metric. Ann. Inst. H. Poincaré Anal. Non Linéaire **27**(3), 857–876 (2010)
15. Caponio, E., Javaloyes, M.Á., Masiello A.: On the energy functional on Finsler manifolds and applications to stationary spacetimes. Math. Ann. **351**(2), 365–392 (2011)
16. Caponio, E., Javaloyes, M.A., Sánchez, M.: On the interplay between Lorentzian Causality and Finsler metrics of Randers type. Rev. Mat. Iberoamericana **27**(3), 919–952 (2011)
17. Dirmeier, A., Plaue, M., Scherfner, M.: Growth conditions, Riemannian completeness and Lorentzian causality. J. Geom. Phys. **62**(3), 604–612 (2011)
18. Dowker J., Kennedy, G.: Finite temperature and boundary effects in static spacetimes. J. Phys. A: Math. Gen. **11**, 895–920 (1978)
19. Flores J.L., Herrera, J.: The c-boundary construction of spacetimes: application to stationary kerr spacetime, these proceedings, Recent Trends in Lorentzian Geometry, Springer Proceedings in Mathematics & Statistics (2013).
20. Flores, J.L., Herrera, J., Sánchez, M.: Gromov, Cauchy and causal boundaries for Riemannian, Finslerian and Lorentzian manifolds. arXiv:1011.1154v2 [math.DG], in Mem. Amer. Math. Soc. (to appear)
21. Flores, J.L., Herrera, J., Sánchez, M.: On the final definition of the causal boundary and its relation with the conformal boundary. Adv. Theor. Math. Phys. **15**(4), 991–1058 (2011)
22. Fortunato, D., Giannoni, F., Masiello, A.: A Fermat principle for stationary spacetimes and applications to light rays. J. Geom. Phys. **15**(2), 159–188 (1995)
23. Giambò, R., Giannoni, F., Piccione, P.: Genericity of nondegeneracy for light rays in stationary spacetimes. Comm. Math. Phys. **287**(3), 903–923 (2009)
24. Giambò, R., Javaloyes, M.A.: Addendum to: Genericity of nondegeneracy for light rays in stationary spacetimes. Comm. Math. Phys. **295**(1), 289–291 (2010)
25. Gibbons, G., Perry, M.: Black holes and thermal green functions. Proc. Roy. Soc. London, Ser. A **358**, 467–494 (1978)
26. Gibbons, G.W., Herdeiro, C.A.R., Warnick, C.M., Werner, M.C.: Stationary metrics and optical Zermelo-Randers-Finsler geometry. Phys. Rev. D **79**(4), 044022, 21 (2009)
27. Ingarden, R.S.: On the geometrically absolute optical representation in the electron microscope. Trav. Soc. Sci. Lett. Wrocław. Ser. B. (45), 60 (1957)
28. Javaloyes, M.A., Lichtenfelz, L., Piccione, P.: Almost isometries of non-reversible metrics with applications to stationary spacetimes. arXiv:1205.4539v1 [math.DG] (2012)
29. Javaloyes, M.A., Sánchez, M.: A note on the existence of standard splittings for conformally stationary spacetimes. Class. Quantum Grav. **25**(16), 168001, 7 (2008)
30. Javaloyes, M.A., Sánchez, M.: On the definition and examples of Finsler metrics. arXiv:1111.5066v2 [math.DG] (2011), in Ann. Scuola Norm. Sup. Pisa Cl. Sci. (to appear)

31. Kobayashi, S., Nomizu, K.: Foundations of differential geometry. Wiley Classics Library, vol. I. Wiley, New York (1996) Reprint of the 1963 original, A Wiley-Interscience Publication
32. Kovner, I.: Fermat principle in gravitational fields. Astrophys. J. **351**, 114–120 (1990)
33. Landau, L., Lifshitz, E.: The classical theory of fields, 2nd edn. Pergamon Press, Addison-Wesley, Oxford, Reading, MA (1962)
34. Levi-Civita, T.: La teoria di Einstein e il principio di Fermat. Nuovo Cimento **16**, 105–114 (1918)
35. Levi-Civita, T.: The Absolute Differential Calculus. Blackie & Son Limited, London (1927)
36. Lichnerowicz, A.: Théories relativistes de la gravitation et de l'électromagnétisme. Relativité générale et théories unitaires. Masson et Cie, Paris (1955)
37. Lichnerowicz, A., Thiry, Y.: Problèmes de calcul des variations liés à la dynamique classique et à la théorie unitaire du champ. C. R. Acad. Sci. Paris **224**, 529–531 (1947)
38. Masiello, A.: An alternative variational principle for geodesics of a Randers metric. Adv. Nonlinear Stud. **9**(4), 783–801 (2009)
39. Masiello, A., Piccione, P.: Shortening null geodesics in Lorentzian manifolds. Applications to closed light rays. Differ. Geom. Appl. **8**(1), 47–70 (1998)
40. Matsumoto, M.: On *C*-reducible Finsler spaces. Tensor (N.S.) **24**, 29–37 (1972) Commemoration volumes for Prof. Dr. Akitsugu Kawaguchi's seventieth birthday, Vol. I
41. Matsumoto, M.: On Finsler spaces with Randers' metric and special forms of important tensors. J. Math. Kyoto Univ. **14**, 477–498 (1974)
42. Mawhin, J., Willem, M.: Critical point theory and Hamiltonian systems. Applied Mathematical Sciences, vol. 74. Springer, New York (1989)
43. Mercuri, F.: The critical points theory for the closed geodesics problem. Math. Z. **156**(3), 231–245 (1977)
44. Minguzzi, E., Sánchez, M.: The causal hierarchy of spacetimes. In: Recent developments in pseudo-Riemannian geometry, ESI Lect. Math. Phys., pp. 299–358. Eur. Math. Soc., Zürich (2008)
45. Miron, R.: The geometry of Ingarden spaces. Rep. Math. Phys. **54**(2), 131–147 (2004)
46. O'Neill, B.: Semi-Riemannian geometry. Pure and Applied Mathematics, vol. 103. Academic [Harcourt Brace Jovanovich Publishers], New York (1983) With applications to relativity
47. Perlick, V.: On Fermat's principle in general relativity. I. The general case. Class. Quantum Grav. **7**, 1319–1331 (1990)
48. Perlick, V.: On Fermat's principle in general relativity. II. The conformally stationary case. Class. Quantum Grav. **7**(10), 1849–1867 (1990)
49. Perlick, V.: Fermat principle in Finsler spacetimes. Gen. Relat. Gravit. **38**(2), 365–380 (2006)
50. Pham, Q.: Inductions électromagnétiques en rélativité général et principe de Fermat. Arch. Ration. Mech. Anal. **1**, 54–80 (1957)
51. Randers, G.: On an asymmetrical metric in the fourspace of general relativity. hys. Rev. **59**(2), 195–199 (1941)
52. Sánchez, M.: Some remarks on causality theory and variational methods in Lorenzian manifolds. Conf. Semin. Mat. Univ. Bari (265), ii+12 (1997)
53. Shibata, C., Shimada, H., Azuma, M., Yasuda, H.: On Finsler spaces with Randers' metric. Tensor (N.S.) **31**(2), 219–226 (1977)
54. Synge, J.: An alternative treatment of Fermat's principle for a stationary gravitational field. Philos. Mag. J. Sci. **50**, 913–916 (1925)
55. Warner, F.: The conjugate locus of a Riemannian manifold. Amer. J. Math. **87**, 575–604 (1965)
56. Weyl, H.: Zur gravitationstheorie. Ann. Phys. (Berlin) **54**, 117–145 (1917)
57. Yasuda, H., Shimada, H.: On Randers spaces of scalar curvature. Rep. Math. Phys. **11**(3), 347–360 (1977)
58. Zaustinsky, E.: Spaces with non-symmetric distance. Mem. Amer. Math. Soc. **34**, 1–91 (1959)

Can We Make a Finsler Metric Complete by a Trivial Projective Change?

Vladimir S. Matveev

1 Statement of the Problem, Motivation and the Main Result

Let (M,F) be a connected Finsler manifold and $f : M \to \mathbb{R}$ be a function such that

$$F(x,v) + d_x f(v) > 0 \text{ for all } (x,v) \in TM \text{ with } v \neq 0. \tag{1}$$

By a *trivial projective change*, we understand the Finsler metric $F + df$.

It is customary in Finsler geometry to require the Finsler metric to be *strongly convex*, that is, the Hessian of the restriction of F^2 to $T_xM \setminus \{0\}$ is assumed to be positive definite for any $x \in M$. Our results do not require this assumption and are valid also for Finsler metrics that are merely positively homogeneous (i.e., $F(x,\lambda\xi) = \lambda F(x,\xi)$ for all $\lambda \geq 0$, $x \in M$, $\xi \in T_xM$) and convex (i.e., the set $\Omega_x := \{\xi \in T_xM \mid F(x,\xi) \leq 1\}$ is a convex body in $T_xM \equiv \mathbb{R}^{\dim(M)}$).

Let us note though that if the metric F is strongly or strictly convex, then the trivial projective change $F + df$ is also strongly or, respectively, strictly convex, cf. [11, Corollary 4.17].

The metric $F + df$ has the same unparameterized geodesics as F. Indeed, a forward-geodesic connecting two points $x, y \in M$ is an extremal of the forward-length functional

$$L_F^+(c) := \int_a^b F(c(t), \dot{c}(t))dt \tag{2}$$

V.S. Matveev (✉)
Institute of Mathematics, Friedrich-Schiller-Universität Jena, 07737 Jena, Germany
e-mail: vladimir.s.matveev@gmail.com

M. Sánchez et al. (eds.), *Recent Trends in Lorentzian Geometry*, Springer Proceedings in Mathematics & Statistics 26, DOI 10.1007/978-1-4614-4897-6_10,

in the set of all smooth curves $c : [a,b] \to M$ connecting x and y. Now, replacing F by $F + df$ in (2), we obtain

$$L^+_{F+df}(c) := \int_a^b \left[F(c(t),\dot c(t)) + d_{c(t)}f(\dot c(t))\right] dt = L^+_F(c) + f(y) - f(x). \quad (3)$$

We see that the difference $L^+_{F+df}(c) - L^+_F(c)$ is the constant $f(y) - f(x)$ so extremals of the functional (2) are extremals of the functional (3) and vice versa.

Analogously, for the backward-length

$$L^-_F := \int_a^b F(c(t), -\dot c(t))dt,$$

we obtain $L^-_{F+df}(c) - L^-_F(c) = f(x) - f(y)$ implying that the backward-geodesics of F and $F + df$ coincide.

Note that, though the unparameterized geodesics of F and $F + df$ coincide, the arc-length parameter of the geodesics and also the distance functions generated by the Finsler metrics do not coincide unless f is a constant. More precisely, the forward and backward distance functions

$$\mathrm{dist}^\pm_F(x,y) = \inf\{L^\pm_F(c) \mid c : [a,b] \to M \text{ with } c(a) = x\ ,\ c(b) = y\}$$

and the corresponding distance functions for $F + df$ are related by

$$\mathrm{dist}^\pm_{(F+df)}(x,y) = \mathrm{dist}^\pm_F(x,y) \pm (f(y) - f(x)). \quad (4)$$

The goal of this note is to answer the questions under what conditions one could make a Finsler metric simultaneously forward complete and backward complete by an appropriate trivial projective change. We will assume that all objects we consider in our paper are sufficiently smooth. The assumption that the metric F is smooth is very natural in view of our motivation (see Sect. 1.1 below). We will see that the restriction that the (searched) function f is smooth (which is also natural in view of the motivation) actually makes our proof more complicated: if we allow Lipschitz functions f (and in Remark 3 we explain why we may do it), the proof becomes shorter and does not require the Appendix where we proved that it is possible, for any $\varepsilon_1, \varepsilon_2 > 0$, to ε_1-approximate a 1-Lipschitz function by a $(1+\varepsilon_2)$-Lipschitz function, where the Lipschitz property is understood with respect to the (nonsymmetric) distance function coming from the metric F.

1.1 Motivation

This problem has been inspired by the mathematical relativity and Lorentz differential geometry. Following [5–8], see also references therein, we consider the (normalized, standard) stationary spacetime $(M^4 = \mathbb{R} \times S^3, G)$. Here S is a 3-dimensional manifold. The condition that the spacetime is normalized, standard, and stationary means that, in any local coordinate system (t, x^1, x^2, x^3) where t is the coordinate on $\mathbb{R}$ and x^1, x^2, x^3 are local coordinates on the three-manifold S, the metric G is given by the formula

$$\begin{aligned} G &= -dt^2 + 2\omega_i dx^i dt + g_{ij} dx^i dx^j \\ &= -(dt - \omega_i dx^i)(dt - \omega_j dx^j) + (g_{ij} + \omega_i \omega_j) dx^i dx^j, \end{aligned} \tag{5}$$

where $g = g(x)_{ij}$ $i, j = 1, \ldots, 3$, is a Riemannian metric on S and $\omega = \omega(x)_i$ is a 1-form on S.

Remark 1. Note that the above definition of the normalized, standard, stationary spacetime is not the usual one. Usually, one defines a standard stationary spacetime as the one which is distinguishing (i.e., any two different points must have different chronological futures and different chronological pasts) and which admits a complete time-like Killing vector field K. By [10], this is equivalent to the condition that M is isometric to a product $\mathbb{R} \times S$, where S is some (appropriate) space-like hypersurface and K, in the coordinate system corresponding to this decomposition, is the vector field $\frac{\partial}{\partial t}$. Then, the metric can be written locally as in (5); it is easy to check that ω_i and g_{ij} could be viewed as objects globally defined on S since for $\xi \in TS$ we have $\omega(\xi) = G\left(\frac{\partial}{\partial t}, \xi\right) = G(K, \xi)$ and g is simply the restriction of G to S.

Next, on S, we consider the (Randers) Finsler metric

$$F(x, \dot{x}) = \sqrt{(g(x)_{ij} + \omega(x)_i \omega(x)_j)\dot{x}^i \dot{x}^j} + \omega(x)_i \dot{x}^i. \tag{6}$$

As it was observed and actively studied in [7–9], this Finsler metric and the initial Lorentz metric G are closely related. In particular, for every light-like geodesic $\gamma(\tau) = (t(\tau), x^1(\tau), x^2(\tau), x^3(\tau))$ of G, its "projection" to S,that is, the curve $\tau \mapsto (x^1(\tau), x^2(\tau), x^3(\tau))$ on S is a (probably, reparameterized) geodesic of the Finsler metric (6). Moreover, the slice

$$\{0\} \times S = \{(0, x) \mid x \in S\} \tag{7}$$

is a Cauchy hypersurface of (M, G) if and only if the metric F is forward and backward complete, see [8, Theorem 4.4].

Note that it is possible to take another decomposition of M in the product of $\mathbb{R} \times S'$ such that the metric G written in the coordinates adapted to the new decomposition still has the form (5) with possibly different g and ω.

Indeed, consider another local coordinate systems (t', x^1, x^2, x^3) such that $t' = t + f(x^1, x^2, x^3)$ (and the coordinates x^1, x^2, x^3 are the same). Physically, this choice of the coordinates corresponds to the choice of another space-like slice: by the "old" slice we understand the 3-dimensional submanifold (7), and by the new one we understand $\{(f(x), x) \mid x \in S\}$.

In the new coordinates (t', x^1, x^2, x^3), in view of $dt' = dt - df$, the metric G reads

$$-\left(dt' - \underbrace{\left(\omega_i + \tfrac{\partial f}{\partial x^i}\right)}_{\omega'_i} dx^i\right)\left(dt' - \underbrace{\left(\omega_j + \tfrac{\partial f}{\partial x^j}\right)}_{\omega'_j} dx^j\right) + (g_{ij} + \omega_i\omega_j)\, dx^i dx^j. \quad (8)$$

We see that the Finsler metric (6) constructed by using the metric (8) is related to the initial metric (6) constructed by using the metric (5) by the formula $F' = F + df$, i.e., is the trivial projective change of the metric (6). It is easy to check that the slice $\{(f(x), x) \mid x \in S\}$ is space-like if and only if

$$\sqrt{(g_{ij} + \omega_i\omega_j)\dot{x}^i\dot{x}^j} + \omega_i\dot{x}^i + \tfrac{\partial f}{\partial x^i}\dot{x}^i \text{ is positive for all } \dot{x}^i \neq 0,$$

that is, if and only if f satisfies the condition from the definition of the trivial projective change with respect to the Finsler metric (6).

Thus, the question we are studying,that is, the existence of a trivial projective change of a Finsler metric such that the result is forward and backward complete is, in the special case when the metric is the Randers metric coming from the Lorentz metric (5) by formula (6), equivalent to the existence of the function f such that the corresponding slice $\{(f(x), x) \mid x \in S\}$ is a Cauchy hypersurface. Note that if such Cauchy hypersurface exists, then the spacetime is globally hyperbolic, see [2, 3] for details, and global hyperbolicity is an important condition to be studied in any spacetime.

It appears though that the special form of the metric F suggested by the motivation does not make (our version of) the answer simpler, so we give the answer for the general Finsler metrics. If the Finsler metric is a Randers one, that is, in the situation suggested by relativity, our main result which is Theorem 1 is a generalization of [8, Theorem 5.10]. Actually, [8, Theorem 5.10] and our main Theorem 1 restricted to the Randers metrics are very similar: the difference is that in [8, Theorem 5.10] one (essentially) assumes that the function $D^+ + D^-$ (see below) is proper for all choices of the point p as the initial point and we require this for one point p only. It is easy to see though (by using the triangle inequality and the fact that a closed subset of a compact set is compact) that if $D^+ + D^-$ is proper for one choice of p then it is proper for all choices of p; in view of this, our main Theorem 1 restricted to the Randers metrics is equivalent to [8, Theorem 5.10].

1.2 Main Result

We fix an arbitrary point $p \in M$ and consider the functions $D^+, D^- : M \to \mathbb{R}$ given by

$$D^\pm(x) := \mathrm{dist}^\pm(p, x).$$

Theorem 1. *F can be made forward and backward complete by a trivial projective change if and only if the function $D^+ + D^-$ is proper.*

Recall that a (continuous) function is *proper*, if the preimage of every compact set is compact or empty. Since the function $D^+ + D^-$ is nonnegative and $D^+(p) + D^-(p) = 0$, the function $D^+ + D^-$ is proper if and only if for every $c \in \mathbb{R}_{\geq 0}$ the set

$$\{x \in M \mid D^+(x) + D^-(x) \leq c\} \tag{9}$$

is compact.

2 Proof of Theorem 1

First observe that if the function $D^+ + D^-$ is proper then the function $\alpha_1 D^+ + \alpha_2 D^-$ is proper for arbitrary positive numbers α_i and that if the function $\alpha_1 D^+ + \alpha_2 D^-$ is proper for some positive numbers α_i, then the function $D^+ + D^-$ is proper. Indeed, the set

$$\{x \in M \mid \alpha_1 D^+(x) + \alpha_2 D^-(x) \leq c\} \tag{10}$$

is a (evidently, closed) subset of

$$\{x \in M \mid D^+(x) + D^-(x) \leq \frac{c}{\alpha_{\min}}\}, \ \text{ where } \alpha_{\min} = \min(\alpha_1, \alpha_2)$$

Then, if all the sets of the form (9) are compact, all sets of the form (10) are compact as well, implying that the function $\alpha_1 D^+(x) + \alpha_2 D^-(x)$ is proper.

Now, if the function $\alpha_1 D^+ + \alpha_2 D^-$ is proper, the set (9) is compact as a closed subset of

$$\{x \in M \mid \alpha_1 D^+ + \alpha_2 D^- \leq \alpha_{\max} c\}, \quad \text{where } \alpha_{\max} = \max(\alpha_1, \alpha_2),$$

implying that $D^+ + D^-$ is proper.

We will now show Theorem 1 in the direction "$\Longrightarrow$": we show that if the function $D^+ + D^-$ is not proper, then no projective change $F + df$ is complete. Let $R > 0$ be the number such that the set

$$B_R := \{x \in M \mid D^+(x) + D^-(x) \leq R\}$$

is not compact. Since in view of (4) for all $x \in M$ we evidently have

$$D^+(x) + D^-(x) = \text{dist}^+_{F+df}(p,x) + \text{dist}^-_{F+df}(p,x),$$

the set B_R coincides with the set

$$\{x \in M \mid \text{dist}^+_{F+df}(p,x) + \text{dist}^-_{F+df}(p,x) \leq R\}$$

Fig. 1 Triangles for the triangle inequalities

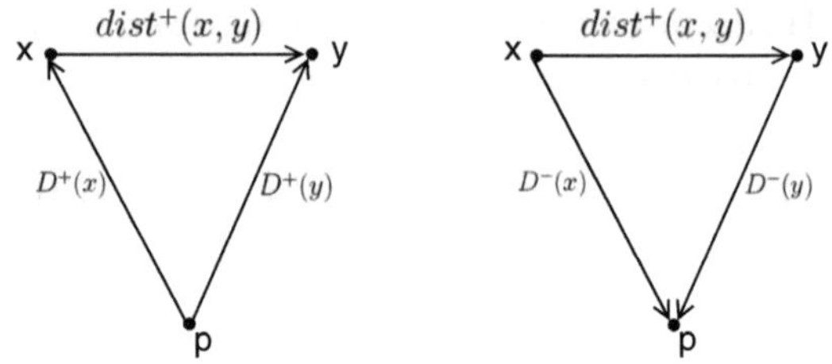

which is evidently a closed subset of the intersection

$$\{x \in M \mid \mathrm{dist}^{+}_{F+df}(p,x) \leq R\} \bigcap \{x \in M \mid \mathrm{dist}^{-}_{F+df}(p,x) \leq R\}. \tag{11}$$

Now, under the assumption that the metric $F+df$ is forward and backward complete, both subsets $\{x \in M \mid \mathrm{dist}^{\pm}_{F+df}(p,x) \leq R\}$ are compact: in the case when the Finsler $F+df$ metric is symmetric, this follows directly from [4, Proposition 2.5.22], and, actually, the proof of [4, Proposition 2.5.22] can be easily generalized to the situation when the Finsler metric is not symmetric. Then, (11) and all its closed subsets are compact which contradicts our assumption that B_R is not compact. Theorem is proved in one direction.

In order to prove it in the other direction, we consider the function

$$f : M \to \mathbb{R}, \;\; f(x) := \frac{D^{-}(x) - D^{+}(x)}{2}.$$

The function f is not a priori smooth; next we show that the function is 1-Lipshitz w.r.t. the distance dist^{+}, that is, for every $x, y \in M$ we have

$$\mathrm{dist}^{+}(x,y) \geq f(x) - f(y). \tag{12}$$

Indeed, consider the triangles in Fig. 1 and the corresponding triangle inequalities:

$$D^{+}(x) + \mathrm{dist}^{+}(x,y) \geq D^{+}(y)\ , \;\; D^{-}(y) + \mathrm{dist}^{+}(x,y) \geq D^{-}(x). \tag{13}$$

The sum of these inequalities is equivalent to (12)

Remark 2. As a consequence, we obtain that the function f is also 1-Lipschitz w.r.t. the symmetrized distance $\mathrm{dist}^{sym} := \mathrm{dist}^{+} + \mathrm{dist}^{-}$. Since, locally, in a sufficiently small neighborhood, we can evidently find a Euclidean structure such that the corresponding distance is not less than dist^{sym}, the function f is locally Lipschitz w.r.t. an Euclidean structure and is therefore differentiable at almost every point. Moreover, the restriction of the function to every smooth curve is a locally Lipschitz function, and the formula (3) remains valid though df is not everywhere defined.

Let us now take a smooth function $\tilde{f}$ on M such that

(a) $|\tilde{f}(x) - f(x)| \leq 1$ for all $x \in M$.
(b) $\tilde{f}(x)$ is 1.5-Lipschitz w.r.t. dist^{+}.

We will show the existence of such a function in the Appendix. Let us note that the proof of its existence essentially repeats the proof of the existence, for arbitrary $\varepsilon_1, \varepsilon_2 > 0$, of an ε_1-approximation of a 1-Lipschitz function by a smooth $(1+\varepsilon_2)$-Lipschitz function on the standard $\mathbb{R}^n$ with the standard metric. Note also that one of the main technical difficulties in approximating a 1-Lipschitz function by a smooth one appears implicitly in the solution of the Randers case provided in [8] by using spacetimes.

Now let us take the function $\frac{1}{2}\tilde{f}$ and consider $F + \frac{1}{2}d\tilde{f}$. This is a Finsler metric. Indeed, we need to check that $F(x,v) + \frac{1}{2}d_x\tilde{f}(v) > 0$ for all x and for all $v \neq 0$. In a local coordinate system in a neighborhood of x we consider the curve $t \mapsto x + t \cdot v$, $t \in [0, \varepsilon]$. From the definition (2), it follows that

$$F(x,v) = \lim_{t \to 0^+} \tfrac{1}{t}\mathrm{dist}^+(x, x+tv). \tag{14}$$

Indeed, we will work in our fixed coordinate system around x. Consider the Minkowski norm $F_0(\xi) = F(x,\xi)$. Since F is continuous, for every (small) $\varepsilon > 0$ there exists a sufficiently small neighborhood $U(x)$ of x such that the Minkowski - Finsler metric $(1+\varepsilon)F_0(\xi)$ bounds from above and the Minkowski - Finsler metric $(1-\varepsilon)F_0(\xi)$ bounds from below the restriction of the metric F to $U(x)$:

$$(1-\varepsilon)F_0(\xi) \le F(y,\xi) \le (1+\varepsilon)F_0(\xi)$$

for all $y \in U(x)$ and all $\xi \in T_yM$. Then, for sufficiently small t, we have

$$t(1-\varepsilon)F_0(v) \le \mathrm{dist}^+(x, x+tv) \le t(1+\varepsilon)F_0(v).$$

Making ε smaller and dividing the last formula by t, we obtain (14).

Now,

$$-d_x\tilde{f}(v) = \lim_{t\to 0^+} \tfrac{1}{t}(\tilde{f}(x) - \tilde{f}(x+tv)) \overset{(b)}{\le} \lim_{t\to 0^+} \tfrac{1.5}{t}\mathrm{dist}^+(x,x+tv) \overset{(14)}{=} 1.5F(x,v).$$

Then, since $F(x,v) > 0$ for all $v \neq 0$, we obtain $-\frac{1}{2}d_x\tilde{f}(v) < F(x,v)$ for all $v \neq 0$ implying that $F + \frac{1}{2}d\tilde{f}$ is a Finsler metric.

Let us now prove that the Finsler metric $F + \frac{1}{2}d\tilde{f}$ is forward and backward complete. It is sufficient to show that for every $r \in \mathbb{R}_{\ge 0}$ the balls

$$B_r^{\pm}(p) := \{x \in M \mid \mathrm{dist}^{\pm}_{F+\frac{1}{2}d\tilde{f}}(p,x) \le r\}$$

are compact. Indeed, any forward-Cauchy sequence lies in $B_r^+(p)$ for sufficiently large r. Then, if such balls are compact, they are complete implying that our forward-Cauchy sequence converges. Similar arguments show that if all balls $B_r^-(p)$ are compact, then the metric is backward complete.

Now, the $\left(F+\frac{1}{2}d\tilde{f}\right)$ -distance is given by

$$\begin{aligned}\operatorname{dist}^{+}_{F+\frac{1}{2}d\tilde{f}}(p,x) &= \operatorname{dist}^{+}_{F}(p,x)+\frac{1}{2}(\tilde{f}(x)-\tilde{f}(p))\\ &\geq D^{+}(x)+\tfrac{1}{2}(f(x)-f(p))-1\\ &= D^{+}(x)+\tfrac{1}{4}\left(D^{-}(x)-D^{+}(x)\right)-1\\ &= \tfrac{3}{4}D^{+}(x)+\tfrac{1}{4}D^{-}(x)-1.\end{aligned}$$

As we explained in the beginning of the proof, since the function $D^{+}+D^{-}$ is proper, the function $\frac{3}{4}D^{+}(x)+\frac{1}{4}D^{-}(x)-1$ is proper as well, implying that the function $\operatorname{dist}^{+}_{F+\frac{1}{2}d\tilde{f}}(p,x)$ is proper implying that the balls $B_r(p)$ are compact, so the metric $F+\frac{1}{2}d\tilde{f}$ is forward-complete. The proof that the metric is backward complete is similar. Theorem 1 is proved.

Remark 3. In the proof we constructed a **smooth** function $\tilde{f}$ such that the trivial projective change $F+\frac{1}{2}\tilde{f}$ is a forward and backward complete Finsler metric. If we do not require the smoothness, we can simply take the trivial projective change corresponding to the function $\frac{1}{2}f$. As we have shown above, the function is locally Lipschitz, so its differential is defined at almost every point, so the function $F+\frac{1}{2}df$ is defined almost everywhere. Moreover, for every curve c, the formula (3) gives us a well-defined length (because the restriction of a locally Lipschitz function to a smooth curve is locally Lipschitz) and the length in $F+\frac{1}{2}df$ is related to the length in F by the formula (4), so the (not everywhere defined) Finsler metric $F+\frac{1}{2}df$ generates a forward- and backward complete distance function.

Appendix: Smooth Approximation of Lipschitz Functions on Finsler Manifolds

Let (M,F) be a Finsler manifold. Assume that the function f is 1-Lipschitz w.r.t. to the distance function generated by F, that is for every $x,y \in M$ we have

$$\operatorname{dist}^{+}(x,y) \geq f(x)-f(y). \tag{15}$$

Our goal is to show that
for every $\varepsilon_1, \varepsilon_2 > 0$ there exists a smooth function $\tilde{f}$ such that

- $|\tilde{f}(x)-f(x)| < \varepsilon_1$ *for all x*
- $\tilde{f}$ *is $(1+\varepsilon_2)$-Lipschitz w.r.t. to the distance function generated by F*

The special cases of this statement, when the function F generates a Euclidean distance or is a Riemannian metric, are known: in the Euclidean case, this is a well known folklore, and in the Riemannian case, it was proved, for example, in [1].

Let us first do it in a small neighborhood of an arbitrary point p. We assume that the closure of the neighborhood is compact. We identify the neighborhood with a domain $U' \subseteq \mathbb{R}^n$, and take a small number $r > 0$ and an infinitely smooth positive function $\sigma : \mathbb{R}^n \to \mathbb{R}$ such that its support lies in the Eucledian r–ball, such that it is spherically symmetric with respect to 0 and such that the integral $\int_{\mathbb{R}^n} \sigma(\xi)d\xi = 1$. We denote by U the interior of the set of the points $\{x \in U' \mid \bar{B}_r(x) \in U'\}$, where $\bar{B}_r(x)$ is the Euclidean closed r–ball centered at x. The set U is open and, if r is sufficiently small, contains the point p.

Now, denote by $\tilde{f}_p$ the convolution of the function f with the function σ:

$$\tilde{f}_p(x) = \int_{\mathbb{R}^n} \sigma(x-\xi)f(\xi)d\xi.$$

The function $\tilde{f}_p$ is defined for all $x \in U$ and is smooth. Let us show that, if r is small enough, $|\tilde{f}(x) - f(x)| \leq \varepsilon_1$ for all x and $\tilde{f}$ is $(1+\varepsilon_2)$-Lipschitz.

Since the function f is Lipschitz, it is uniformly continuous on U, so for sufficiently small r we have $|f(x) - f(y)| < \varepsilon_1$ for all $x, y \in U$ such that $d(x,y) < r$. We consider

$$\begin{aligned} |\tilde{f}_p(x) - f(x)| &= \left| \int_{\mathbb{R}^n} \sigma(x-\xi)f(\xi)d\xi - \int_{\mathbb{R}^n} \sigma(x-\xi)f(x)d\xi \right| \\ &= \left| \int_{\mathbb{R}^n} \sigma(x-\xi)(f(\xi)-f(x))d\xi \right| \\ &\leq \int_{\mathbb{R}^n} \sigma(x-\xi)\varepsilon_1 d\xi = \varepsilon_1. \end{aligned}$$

Let us show that, for a sufficiently small r, the function $\tilde{f}_p$ is 1-Lipschitz. Since the function is smooth, it is sufficient to show that for every x and for every v the directional derivative of the function $\tilde{f}_p$ at the point x in the direction v is less than $(1+\varepsilon_2)F(x,v)$. Without loss of generality, we can think that $v = \frac{\partial}{\partial x^1}$.

We have

$$\begin{aligned} &\frac{\partial}{\partial x^1} \int_{\mathbb{R}^n} \sigma(x-\xi)f(\xi)d\xi^1 \wedge \cdots \wedge d\xi^n \\ &\overset{(1)}{=} \int_{\mathbb{R}^n} f(\xi)\frac{\partial}{\partial x^1}\sigma(x-\xi)d\xi^1 \wedge \cdots \wedge d\xi^n \\ &\overset{(2)}{=} -\int_{\mathbb{R}^n} f(\xi)\frac{\partial}{\partial \xi^1}\sigma(x-\xi)d\xi^1 \wedge \cdots \wedge d\xi^n \\ &\overset{(3)}{=} \int_{\mathbb{R}^n} \sigma(x-\xi)d_{\xi^1}f \wedge d\xi^2 \wedge \cdots \wedge d\xi^n \\ &\overset{(4)}{\leq} \int_{\mathbb{R}^n} \sigma(x-\xi)F\left(\xi, \tfrac{\partial}{\partial x^1}\right) d\xi^1 \wedge \cdots \wedge d\xi^n. \end{aligned}$$

Let us explain the equalities/inequalities in the formula above:

1. Here we used the standard formula of the differentiation of an integral depending on the parameter (in this case, x^1).
2. Here we used that $\frac{\partial}{\partial x^1}\sigma(x-\xi) = -\frac{\partial}{\partial \xi^1}\sigma(x-\xi)$.
3. Here we used integration by parts: $\int u dv = uv| - \int v du$. The role of the function v is played by the function $\sigma(x-\xi)$. The role of the function u is played by the function $f(x)$ considered as a function of one variable x^1. We denote by $d_{\xi^1} f$ the variation of f w.r.t. the variable x^1; since f is Lipschitz, it has bounded variation, so $d_{\xi^1} f$ is a well-defined measure.
   Now, for this choice of the functions u, v, the term $uv|$ disappears since the function $u = \sigma(x-\xi)$ has compact support, so we obtain $\int u dv = -\int v du$ which gives us (2).
4. Here we used that σ is nonnegative and that the measure $F\left(\xi, \frac{\partial}{\partial x^1}\right) d\xi^1$ is greater than $d_{\xi^1} f$ since the function f is 1-Lipschitz.

Now, since the Finsler function F is continuous, it is uniformly continuous on the unit spherical bundle S_1U, implying that, if r is sufficiently small, $F\left(\xi, \frac{\partial}{\partial x^1}\right)$ is $\varepsilon_2 \cdot F\left(x, \frac{\partial}{\partial x^1}\right)$ −close to $F\left(x, \frac{\partial}{\partial x^1}\right)$ for ξ that are r−close to x. Then,

$$\int_{\mathbb{R}^n} \sigma(x-\xi)F\left(\xi, \frac{\partial}{\partial x^1}\right) d\xi \leq \int_{\mathbb{R}^n} \sigma(x-\xi)(1+\varepsilon_2)F\left(x, \frac{\partial}{\partial x^1}\right) d\xi = (1+\varepsilon_2)F\left(x, \frac{\partial}{\partial x^1}\right)$$

implying that the v-derivative of the function $\tilde f$ is not greater than $(1+\varepsilon_2)F(x,v)$ implying that the function $\tilde f$ is $(1+\varepsilon_2)$-Lipschitz.

Thus, for every point p, we can choose a neighborhood U_p such that for every $\tilde\varepsilon_1 > 0, \tilde\varepsilon_2 > 0$ we can $\tilde\varepsilon_1$-approximate f in the neighborhood U_p by a $(1+\tilde\varepsilon_2)$-Lipschitz function $\tilde f_p$ on U_p. We take a locally finite cover U_p, $p \in P$ of M by such neighborhood and choose a smooth partition of unity μ_p corresponding to this cover. We think that the approximation functions $\tilde f_p$ are defined on the whole manifold (though it is not important what values the functions $\tilde f_p$ have on the points that do not lie in U_p since in all formulas below we will multiply $\tilde f_p$ by μ_p and all μ_p are zero outside of U_p).

Now, set

$$\tilde f := \sum_p \tilde f_p \cdot \mu_p. \tag{16}$$

The function $\tilde f$ is well defined since in a small neighborhood of every point x only finitely many terms of the sum are not zero, and is evidently smooth. Let us show that we can choose the numbers $\varepsilon_1(p), \varepsilon_2(p)$ for every U_p such that the function $\tilde f$ satisfies our requirements.

Suppose a point x lies in the intersection of k neighborhoods of the cover U_p, which we denote by $U_1, \dots, U_k$. We will denote by $\mu_1, \dots, \mu_k$ the corresponding elements of the partition of unity and by $\tilde f_1, \dots, \tilde f_k$ the correspondent approximation

$\tilde{f}_p$; we will show that there exist $\tilde{\varepsilon}_1, \tilde{\varepsilon}_2 > 0$ such that, if f_i are $\tilde{\varepsilon}_1$-approximations of the restriction of the functions f to U_i, then $\tilde{f}$ is an ε_1-approximation of the restriction of the functions f to $\bigcap_{i=1}^k U_i$.

Indeed,

$$\tilde{f}(x) - f(x) = \mu_1(x)(\tilde{f}_1(x) - f(x)) + \cdots + \mu_k(\tilde{f}_k(x) - f(x)) \leq \tilde{\varepsilon}_1.$$

Thus, for all $\tilde{\varepsilon}_1 < \varepsilon_1$, the function $\tilde{f}$ is indeed an ε_1-approximation of f.

Let us now show that for a sufficiently small $\tilde{\varepsilon}_1, \tilde{\varepsilon}_2 > 0$ the function $\tilde{f}$ is indeed $(1+\varepsilon_2)$-Lipschitz.

It is sufficient to prove that for every tangent vector v the directional derivative of $\tilde{f}$ in the direction v is less than or equal to $(1+\varepsilon_2)F(x,v)$. We take a point x such that it lies in the intersection of k elements of the cover which we again denote by $U_1, \ldots, U_k$. Then $\tilde{f}$ given by (16) is actually a finite sum

$$\tilde{f} = \mu_1 \tilde{f}_1 + \cdots + \mu_k \tilde{f}_k.$$

Without loss of generality, we can think that $v = \frac{\partial}{\partial x^1}$; we need to show that

$$\frac{\partial}{\partial x^1}(\mu_1 \tilde{f}_1 + \cdots + \mu_k \tilde{f}_k) \leq (1+\varepsilon_2) F\left(x, \frac{\partial}{\partial x^1}\right).$$

The left hand side of the above inequality is equal, in view of the equalities $\frac{\partial}{\partial x^1}\mu_1 + \cdots + \frac{\partial}{\partial x^1}\mu_k = \frac{\partial}{\partial x^1} 1 = 0$, to

$$\begin{aligned} &\frac{\partial}{\partial x^1}(\mu_1 \tilde{f}_1 + \cdots + \mu_k \tilde{f}_k) - \tilde{f} \cdot \left(\frac{\partial}{\partial x^1}\mu_1 + \cdots + \frac{\partial}{\partial x^1}\mu_k\right) \\ &= (\tilde{f}_1 - \tilde{f})\frac{\partial}{\partial x^1}\mu_1 + \cdots + (\tilde{f}_k - \tilde{f})\frac{\partial}{\partial x^1}\mu_k \\ &\quad \underbrace{+\mu_1 \frac{\partial}{\partial x^1}\tilde{f}_1 + \cdots + \mu_k \frac{\partial}{\partial x^1}\tilde{f}_k}_{\leq (1+\tilde{\varepsilon}_2) F\left(x, \frac{\partial}{\partial x^1}\right)}. \end{aligned}$$

Now, since the functions μ_i have bounded support, the derivatives $\frac{\partial}{\partial x^1}\mu_i$ are bounded, implying the sum above is less than $(1+\varepsilon_2)F\left(x, \frac{\partial}{\partial x^1}\right)$ for sufficiently small $\tilde{\varepsilon}_1, \tilde{\varepsilon}_2$ as we claimed.

Acknowledgements The work was started during the VI International Meeting on Lorentzian Geometry (Granada, September 6–9, 2011) and was initiated by the questions by E. Caponio, M.A. Javaloyes, and M. Sánchez; I thank them for this and for the stimulating discussions at the final stage of the preparation of this chapter, and Mike Scherfner for pointing out a misprint. I thank the anonymous referees for correcting Remark 1 and for many useful suggestions.

The standard proof of the existence of a smooth Lipschitz approximation of a Lipschitz function on $\mathbb{R}^n$ whose generalization for Finsler metrics is the main result of the appendix was explained to me by Yu. Burago, S. Ivanov, and A. Petrunin.

References

1. Azagra, D., Ferrera, J., Lopez-Mesas, F., Rangel, Y.: Smooth approximation of Lipschitz functions on Riemannian manifolds. J. Math. Anal. Appl. **326**, 1370–1378 (2007)
2. Bernal, A.N., Sánchez, M.: On smooth Cauchy hypersurfaces and Geroch's splitting theorem. Comm. Math. Phys. **243**(3), 461–470 (2003)
3. Bernal, A.N., Sánchez, M.: Smoothness of time functions and the metric splitting of globally hyperbolic spacetimes. Comm. Math. Phys. **257**(1), 43–50 (2005)
4. Burago, D., Burago, Yu., Ivanov, S.: A course in metric geometry. Graduate Studies in Mathematics, vol. 33, pp. xiv+415. AMS, Providence (2001)
5. Caponio, E., Germinario, A.V., Sánchez, M.: Geodesics on convex regions of stationary spacetimes and Finslerian Randers spaces. arXiv:1112.3892
6. Caponio, E., Javaloyes, M.A., Masiello, A.: Morse theory of causal geodesics in a stationary spacetime via Morse theory of geodesics of a Finsler metric. Annales de l'Institut Henri Poincare (C) Non Linear Analysis **27**, 857–876 (2010)
7. Caponio, E., Javaloyes, M.A., Masiello, A.: On the energy functional on Finsler manifolds and applications to stationary spacetimes. Math. Ann. **351**, 365–392 (2011). arXiv:math/0702323v4
8. Caponio, E., Javaloyes, M.A., Sánchez, M.: On the interplay between Lorentzian Causality and Finsler metrics of Randers type. Rev. Mat. Iberoamericana **27**(3), 919–952 (2011). arXiv:0903.3501
9. Dirmeier, A., Plaue, M., Scherfner, M.: Growth conditions, Riemannian completeness and Lorentzian causality. J. Geom. Phys. **62**(3), 604–612 (2012). doi:10.1016/j.geomphys.2011.04.017
10. Javaloyes, M.A., Sánchez, M.: A note on the existence of standard splittings for conformally stationary spacetimes. Classical Quant. Grav. **25**(16), 168001 (2008)
11. Javaloyes, M.A., Sánchez, M.: On the definition and examples of Finsler metrics. Ann. Scuola Norm. Sup. Pisa Cl. Sci. arXiv:1111.5066v2[math.DG] (2011)

The C-Boundary Construction of SpaceTimes: Application to Stationary Kerr SpaceTime

J.L. Flores and J. Herrera

1 Introduction

There are many different boundaries which can be attached to a space in order to complete it in certain sense. For instance, the Cauchy and Gromov completions provide two different completions for a Riemannian manifold. The choice of one over another will depend on the concrete problem which is being studied. One of the major problems in Lorentzian geometry is precisely to define a natural boundary for spacetimes which encodes relevant information on it, as its conformal structure and related topics (event horizons, singularities). This question is important because such a boundary would help to go deeper in the global causal structure of the spacetime, and would represent a "concrete place" where asymptotic conditions could be posed, among other benefits.

The aim of this chapter is twofold. On one hand, we are going to review some recent results about the c-boundary of spacetimes and other boundaries in differential geometry, completing some minor details which were left open in the original works (see, for instance, Example 1, Sect. 3.4 or the Appendix). On the other hand, we are going to apply these results to provide a precise description of the c-boundary of the stationary part of (slow rotating) Kerr spacetime. To overcome these objectives, this chapter is structured as follows.

In Sect. 2, we make a concise description of the final definition of the c-boundary of spacetimes, and its main properties, as was originally obtained in [8]. This

J.L. Flores
Departamento de Álgebra, Geometría y Topología, Facultad de Ciencias, Universidad de Málaga, Campus Teatinos, 29071 Málaga, Spain
e-mail: floresj@agt.cie.uma.es

J. Herrera (✉)
Departamento de Álgebra, Geometría y Topología, Facultad de Ciencias, Universidad de Málaga, Campus Teatinos, 29071 Málaga, Spain
e-mail: jherrera@uma.es

M. Sánchez et al. (eds.), *Recent Trends in Lorentzian Geometry*, Springer Proceedings in Mathematics & Statistics 26, DOI 10.1007/978-1-4614-4897-6_11,

description is overcome in four sections. In Sect. 2.1, we focus our attention on the point-set and chronological structures of the c-boundary. Then, in Sect. 2.2, we extend the construction to a topological level. The main properties of the c-boundary, which justify the choice of this construction, are stated in Sect. 2.3. Finally, in Sect. 2.4, we comment some relevant results about the relation between the c-boundary and the conformal one.

In Sect. 3, we review some boundary constructions for Finsler spaces, firstly obtained in [9]. After some brief preliminaries about Finsler elements in Sect. 3.1, in Sects. 3.2 and 3.3 we describe the Cauchy and Gromov completions for any Finsler manifold. In Sect. 3.4, the Busemann completion is analyzed. This construction is endowed with a topology derived from two operators, which significance is interpreted in Sect. 3.4.1. Finally, in Sect. 3.4.2 we study the topological properties of the Busemann completion.

In Sect. 4, following also [9] (and taking into account the construction in [4]), we apply the c-boundary construction introduced in Sect. 2 to the particular case of (standard) stationary spacetimes, obtaining that the c-boundary in this case is closely related to the Finslerian boundary constructions studied in Sect. 3.

In Sect. 5, we use the results obtained in previous section to the stationary part of Kerr spacetime.

Finally, we have also included an appendix (Sect. 6), which is a precise formulation of an example explained at a intuitive level in [9]. This example is of interest because it illustrates some remarkable differences, previously stated in Sect. 3.4.2, between the Busemann and Gromov completions.

2 C-Boundary of SpaceTimes

[1]By a spacetime we will understand a time-oriented connected smooth Lorentzian manifold (V,g) of signature $(-,+,\dots,+)$. A tangent vector $v \in T_pV$, $p \in V$ is called *time like* (resp. *light like*; *causal*) if $g(v,v)<0$ (resp. $g(v,v)=0$, $v\neq 0$; v is either time like or light like). A causal vector is called *future* or *past-directed* if it belongs to the future or past cone. Accordingly, a smooth curve $\gamma : I \to V$ (I real interval) is called time like, light like, causal and future or past-directed if so is $\dot{\gamma}(s)$ for all $s \in I$.

Two events $p,q \in V$ are *chronologically related* $p \ll q$ (resp. *causally related* $p \leq q$) if there exists some future-directed timelike (resp. either future-directed causal or constant) curve from p to q. If $p\neq q$ and $p \leq q$ but $p \not\ll q$, then p is said *horismotically related* to q. The *chronological past* (resp. *future*) of p, $I^-(p)$ (resp. $I^+(p)$) is defined as

$$I^-(p)=\{q\in V : q \ll p\} \qquad (\text{resp. } I^+(p)=\{q\in V : p \ll q\}).$$

[1]We will use typical background and terminology in Lorentzian geometry as in [2, 15, 16].

The chronological past $I^-(\gamma) = \cup_{s\in I} I^-(\gamma(s))$ (resp. future $I^+(\gamma) = \cup_{s\in I} I^+(\gamma(s))$) of future-directed (resp. past-directed) timelike curves $\gamma : I \to V$ will play a relevant role along the chapter.

Let V be a *strongly causal* spacetime, i.e. the topology of V is generated by the intersections between the chronological futures and pasts, $I^+(p) \cap I^-(q)$ for $p,q \in V$. The purpose of the c-boundary ∂V of V is to attach a boundary endpoint $P \in \partial V$ to any inextensible future-directed γ or past-directed ρ timelike curve. The basic idea is that the boundary points will be represented by the chronological past $I^-(\gamma)$ or future $I^+(\rho)$ of the curve, in such a way that some boundary points will be associated to both, some $I^-(\gamma)$ and some $I^+(\rho)$. In the next sections, we give precise definitions of the c-completion $\overline{V} := V \cup \partial V$ at all the natural levels, that is, as a point set, chronological, and topological levels.

2.1 Point Set and Chronological Levels

First, let us introduce the future and past (pre)completions $\hat{V}$, $\check{V}$, of any strongly causal spacetime V, which come from the original approach by Geroch, Kronheimer, and Penrose [10]. A non-empty subset $P \subset V$ is called a *past set* if $I^-(P) = P$. In this case, P is *decomposable* when $P = P_1 \cup P_2$ with P_i past sets such that $P_1 \neq P \neq P_2$, and *indecomposable past set (IP)* otherwise. The *future (pre)completion* $\hat{V}$ of V is defined as the set of all IPs. One can prove that if $P \in \hat{V}$, then either P is a *proper past set (PIP)*,that is, $P = I^-(p)$ for some $p \in V$, or P is a *terminal past set (TIP)*, so that $P = I^-(\gamma)$ for some inextensible future-directed timelike curve γ. The *future (pre)boundary* $\hat{\partial} V$ of V is defined as the set of all the TIPs. Therefore, the spacetime V, which is distinguishing, thus identifiable to the set of all the PIPs, coincides with $\hat{V} \backslash \hat{\partial} V$. Analogously, one defines the *indecomposable future sets (IF)*, which constitute the *past (pre)completion* $\check{V}$. Each IF is either *terminal (TIF)* or *proper (PIF)*. The set of TIFs is the *past (pre)boundary* $\check{\partial} V$.

Next, in order to define the (total) c-completion $\overline{V}$, denote by $\uparrow, \downarrow$ the common future and past, resp. (say, $\uparrow P = I^+(\{q \in V : p \ll q \;\; \forall p \in P\})$). Let $\hat{V}_\emptyset := \hat{V} \cup \{\emptyset\}, \check{V}_\emptyset := \check{V} \cup \{\emptyset\}$, and choose any $(P,F) \in (\hat{V}_\emptyset \times \check{V}_\emptyset) \backslash \{(\emptyset,\emptyset)\}$. We say that P is *S-related* to F if F is included and maximal in $\uparrow P$ and P is included and maximal in $\downarrow F$ (see [3, 18, 19]). By *maximal* we mean that no other $F' \in \check{V}_\emptyset$ (resp. $P' \in \hat{V}_\emptyset$) satisfies the stated property and includes strictly F (resp. P). In addition, if P (resp. F) does not satisfy previous condition for any F (resp. P), we also say that P is S-related to $\emptyset$ (resp. $\emptyset$ is S-related to F); notice that $\emptyset$ is never S-related to itself. Now, we define the *c-completion* $\overline{V}$ of V as (see [8, 14])

$$\overline{V} := \{(P,F) \in \hat{V}_\emptyset \times \check{V}_\emptyset : P \sim_S F\}. \tag{1}$$

In particular, for $p \in V$, we have that $I^-(p) \sim_S I^+(p)$ (recall [18]), and so, V can be regarded as the subset $\{(I^-(p), I^+(p)) : p \in V\} \subset \overline{V}$.

At this level, the c-completion has been defined only as a point set. However, it can be endowed with the following *chronological relation* (i.e. a binary relation which is transitive, anti reflexive, without isolates and chronologically separable, see [8, Definition 2.1]) that extends naturally the chronological relation $\ll$ on V:

$$(P,F)\overline{\ll}(P',F') \Longleftrightarrow F\cap P' \neq \emptyset$$

for all $(P,F),(P',F') \in \overline{V}$.

2.2 Topological Level

In order to define the topology on $\overline{V}$, first note that the following operator $\hat{L}$ determines the limit $x \in V$ of any sequence $\sigma = \{x_n\}_n \subset V$ as follows:

$$x \in \hat{L}(\sigma) \Leftrightarrow \begin{cases} y \ll x \Rightarrow y \ll x_n, \\ I^-(x) \subset P'(\in \hat{V}),\, I^-(x) \neq P' \Rightarrow \exists z \in P' : z \not\ll x_n \end{cases} \text{for large } n.$$

This operator suggests a natural way to introduce a topology for the future causal completion $\hat{V}$. In fact, re-define the operator $\hat{L}$, now acting on sequences $\sigma = \{P_n\}_n \subset \hat{V}$, as follows:

$$P \in \hat{L}(\sigma) \Leftrightarrow \begin{cases} P \subset LI(\{P_n\}) \\ P \text{ is a maximal IP in } LS(\{P_n\}), \end{cases} \tag{2}$$

where LI (resp. LS) denotes the point-set inferior limit (resp. superior limit) : That is, $\mathrm{LI}(\{A_n\}) := \cup_{n=1}^{\infty} \cap_{k=n}^{\infty} A_k$ (resp. $\mathrm{LS}(\{A_n\}) := \cap_{n=1}^{\infty} \cup_{k=n}^{\infty} A_k$). Then, one can check that a topology is defined on $\hat{V}$ as follows: C is closed iff $\hat{L}(\sigma) \in C$ for any sequence $\sigma \subset C$. This topology, called *future chronological topology*, satisfies that $P \in \hat{L}(\sigma)$ only if the sequence σ converges to P. The dual operator $\check{L}$ defines the analogous *past chronological topology* on $\check{V}$.

Finally, the topology on the completion $\overline{V}$ or *chronological topology* is introduced by means of a new operator L on sequences in $\overline{V}$ which defines the closed subsets and limits as above. Let $\sigma = \{(P_n,F_n)\} \subset \overline{V}(\subset \hat{V}_\emptyset \times \check{V}_\emptyset)$, then:

$$(P,F) \in L(\sigma) \Longleftrightarrow \begin{cases} \text{when } P \neq \emptyset,\; P \in \hat{L}(\{P_n\}) \\ \text{when } F \neq \emptyset,\; F \in \check{L}(\{F_n\}). \end{cases} \tag{3}$$

The chronological topology (as well as the future and past ones) extends the manifold topology of V. Moreover, it is sequential (i.e., any subset $C \subset \overline{V}$ which is sequentially closed, i.e., satisfying that all the limits of sequences in C remain in C, must be closed), but may be non-Hausdorff.

Remark 1. We remark the following assertions about the chronological topology:

(1) Clearly, if $(P,F) \in L(\{(P_n,F_n)\})$ then $\{(P_n,F_n)\}$ converges to (P,F). When the converse happens, L is called *of first order* (see the discussion [8, Sect. 3.6]).
(2) Given a pair $(P,F) \in \partial V$, any timelike curve defining P (or F) converges to (P,F) with the chronological topology (see [8, Theorem 3.27]).

These definitions for the c-boundary construction involve some particular subtleties, which are essentially associated to the following two facts: An IP (or IF) does not determine a unique pair in the c-boundary, and the topology does not always agree with the S-relation, in the sense that

$$P \in \hat{L}(P_n) \not\Leftrightarrow F \in \check{L}(F_n).$$

This makes natural to remark the following special cases:

Definition 1. A spacetime V has a c-completion $\overline{V}$ which is *simple as a point set* if each TIP (resp. each TIF) determines a unique pair in ∂V.

Moreover, the c-completion is *simple* if it is simple as a point set and also *topologically*, that is, $(P,F) \in L(P_n,F_n)$ holds when either $P \in \hat{L}(\{P_n\})$ or $F \in \check{L}(\{F_n\})$.

2.3 Properties of the C-Boundary

The following result justifies our choice of c-completion, since none of the previous choices in the literature satisfy all these properties simultaneously, [8, Theorem 3.27].

Theorem 1. *Under our definition of c-completion $\overline{V}$, one has:*

(a) (consistence with the original motivation:) each future (resp. past) timelike curve γ converges to some pair (P,F) with $P = I^-(\gamma)$ (resp. $F = I^+(\gamma)$).
(b) The c-boundary ∂V is a closed subset of the completion $\overline{V}$.
(c) $\overline{V}$ is a T_1 topological space.
(d) $\overline{V}$ is a sequential space.
(e) $I^{\pm}((P,F))$ (computed with $\overline{\ll}$) is open in $\overline{V}$.

As we have already commented, the c-completion $\overline{V}$ is not necessarily Hausdorff (see, for example, [6, Example 10.3]). Even more, the non-Hausdorff relation may link a boundary point with a point of the manifold (see [8, Footnote 6]): In fact, in $\mathbb{L}^3$ with the discs of radius 2 and centers $(1,0,-1)$ and $(-1,0,1)$ removed, the sequence of points $\{(0,1/n,0)\}_n$ converges to both, the origin $(0,0,0)$ and the pair (P,F), with $P = I^-((-1,0,-1))$, $F = I^+((1,0,1))$. The next example illustrates this same situation but now involving a boundary point of the form $(P',\emptyset)$.

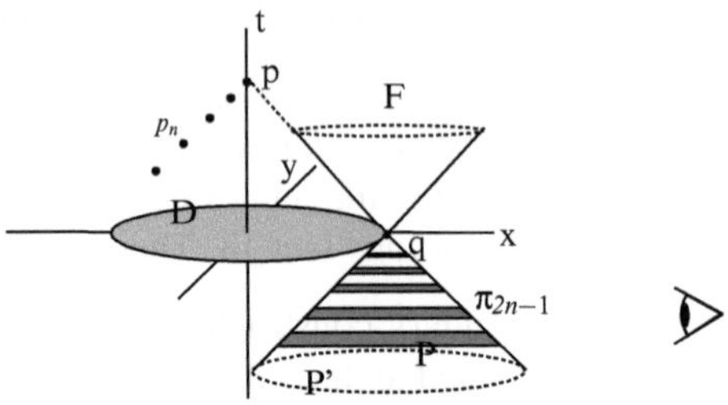

Fig. 1 Representation of the spacetime V, obtained by removing from $\mathbb{L}^3$ the *dashed regions*, which consist of the unitary disc D and the sheets π_{2n-1}. The figure also illustrates the sequence $\sigma = \{p_n\}_n$, which chronologically converges to both $p \in V$ and $(P', \emptyset) \in \partial V$

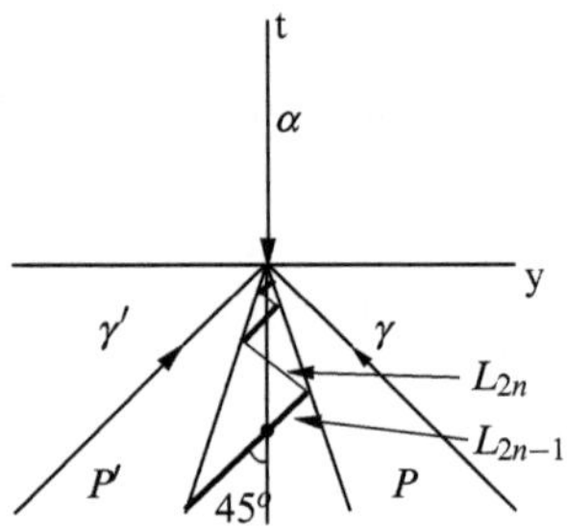

Fig. 2 Section of V at $x = x_0$, as seen from the "eye" in previous figure. The segments $L_n = \pi_n \cap \{x = x_0\}$ for any $x_0 \in \mathbb{R}$ are represented. The curves γ, γ', α will only appear when $x_0 = 1$

Example 1. Consider the following region V of Lorentz-Minkowski spacetime $\mathbb{L}^3$:

$$V = \mathbb{L}^3 \setminus (D \cup (\cup_{n=1}^{\infty}(\pi_{2n-1} \cap P)))$$

where $D = \{(t,x,y) : t = 0,\ x^2 + y^2 \leq 1\}$, $P = I^-(\gamma)$, $\gamma(s) = (s, 1, -s)$, $s \in [-1, 0)$, and $\{\pi_n\}_n$ is a sequence of 2-dimensional sheets of $\mathbb{L}^3$ determined by the sections $\pi_n \cap \{x = x_0\} = L_n$ for all $n \in \mathbb{N}$ and any $x_0 \in \mathbb{R}$ (see Figs. 1 and 2).

Notice that $P' = I^-(\gamma')$, $\gamma'(s) = (s, 1, s)$, $s \in [-1, 0)$ is strictly included in $P = I^-(\gamma)$, while the common future for both, P and P', is $F = I^+(\alpha)$, $\alpha(s) = (-s, 1, 0)$, $s \in [-1, 0)$. On the other hand, the common past of F is P. So, the boundary points associated to the removed point $q = (0, 1, 0) \in \mathbb{L}^3$ are $(P', \emptyset), (P, F) \in \partial V$.

Next, consider the sequence of points $\sigma = \{p_n\}_n \subset V$ with $p_n = (1, 0, -1/n)$ for all n. Clearly, σ converges to $p = (1, 0, 0)$ with the manifold topology, and so, also with the chronological topology. But, notice that $P' = I^-(\gamma') \in \hat{L}(\sigma)$, since it is included in $\mathrm{LI}(I^-(p_n))$ and is maximal in $\mathrm{LS}(I^-(p_n))$ (in fact, $P = I^-(\gamma)$ is not included in $\mathrm{LS}(I^-(p_n))$). So, the ideal point $(P', \emptyset) \in \partial V$ is also chronological limit of σ, and so, it is non-Hausdorff related to the point $p \in V$.

2.4 Relation with the Conformal Boundary

Consider an *envelopment* or open conformal embedding $i : V \hookrightarrow V_0$, and let us study under which conditions the conformal completion $\overline{V}_i := \overline{i(V)}$ and boundary $\partial_i V := \overline{i(V)} \backslash i(V)$ can be identified with the causal ones.

In order to relate the conformal and causal boundaries, one has to focus only on the *accessible (conformal) boundary* $\partial_i^* V$, that is, on those points of $\partial_i V$ which are reachable as endpoints of some timelike curve. If $\overline{V}_i$ is an embedded manifold with C^1 boundary, then all the points of $\partial_i V$ are accessible, i.e. $\partial_i^* V = \partial_i V$ (see [8, Proposition 4.20]). However, a point as the spacelike infinity i^0 in the standard conformal embedding of Lorentz-Minkowski space into Einstein static universe $\mathbb{L}^n \hookrightarrow \mathrm{ESU}^n$ is not accessible, that is, cannot correspond to any point of the c-boundary.

On the other hand, we will also require the envelopment i to be *chronologically complete*, that is: any inextensible (future or past-directed) timelike curve in V must have an endpoint in the conformal boundary. Obviously, otherwise the corresponding point in the causal boundary would not correspond to any point of the conformal one. If, for example, $\overline{V}_i^*$ is compact, this property is fulfilled (see [8, Proposition 4.8]). It is also fulfilled by $\mathbb{L}^n \hookrightarrow \mathrm{ESU}^n$ in spite of the fact that $\overline{V}_i^*$ is not compact in this case (as i^0 has been removed).

Summing up, if the envelopment i is chronologically complete, then the natural projections

$$\hat{\pi} : \hat{\partial} V \to \partial_i^* V, \quad P = I^-(\gamma) \mapsto \text{the limit of} \, \gamma$$

$$\check{\pi} : \check{\partial} V \to \partial_i^* V, \quad F = I^+(\rho) \mapsto \text{the limit of} \, \rho$$

are well-defined. Then, we say that the causal completion $\overline{V}$ and the (accessible) conformal one $\overline{V}_i^*$ are *identifiable* and write $\overline{V} \equiv \overline{V}_i^*$, when the map

$$\pi : \partial V \to \partial_i^* V, \qquad \pi((P,F)) = \begin{cases} \hat{\pi}(P) \text{ if } P \neq \emptyset \\ \check{\pi}(F) \text{ if } F \neq \emptyset \end{cases}$$

satisfies: (1) It is well defined (i.e., $(P,F) \in \partial V$ and $P \neq \emptyset \neq F$ implies $\hat{\pi}(P) = \check{\pi}(F)$), (2) π is bijective, and (3) its natural extension to the completions $\overline{\pi} : \overline{V} \to \overline{V}_i^*$ is both, a homeomorphism and a chronological isomorphism.

Even though there are simple examples where the conformal and causal boundaries differ, there are also simple conditions which ensure that they are identifiable. They are essentially fulfilled when the points of the conformal boundary are *regularly accessible*. We will not enter in this general notion here (see [8, Definition 4.14] for a precise definition), but only state this particular result, [8, Sect. 4.4.3]:

Theorem 2. *Assume that an envelopment is chronologically complete, and satisfies that $\overline{V}_i$ is a C^1 manifold with boundary (in particular, $\overline{V}_i = \overline{V}_i^*$). If, in addition, the boundary $\partial_i V$ has no timelike points, that is, the tangent hyperplane to each point* *of the boundary is either space like or light like, then the conformal and causal completions are identifiable* $\overline{V}_i^* \equiv \overline{V}$.

In particular, if $\partial_i^{\mathrm{asy}} V$ *denotes the set of asymptotic ideal points of the conformal boundary of an asymptotically flat spacetime* V, *the points in* $\partial_i^{\mathrm{asy}} V$ *are regularly accessible, and then, they correspond with a subset of the c-boundary* ∂V.

3 Boundaries for Finsler Spaces

3.1 Finsler Elements

Recall that, essentially, a Finsler metric F on a manifold M means to put smoothly *a positively homogeneous norm* at each $p \in M$, where positive homogeneity means that the equality $F(\lambda v) = |\lambda| F(v)$ for $v \in TM, \lambda \in \mathbb{R}$ (typical in the usual norms), is only ensured now when $\lambda > 0$. So, given such an F, one can define the *reverse Finsler metric*: $F^{\mathrm{rev}}(v) := F(-v)$. Any Finsler metric induces a map $d : M \times M \to \mathbb{R}$, where $d(x,y)$ is the infimum of the lengths of the curves $c : [a,b] \to M$ which start at x and end at y (by definition, length$(c) = \int_a^b F(\dot{c}(s))ds$). Such a d is a *generalized distance*, that is, it satisfies all the axioms of a distance except symmetry (i.e. d is a *quasi-distance*) and, additionally, the following condition: A sequence $\{x_n\} \subset M$ satisfies $d(x,x_n) \to 0$ iff $d(x_n,x) \to 0$. So, centered at any $x_0 \in M$, one can define the forward and backward open balls of radius $r \geq 0$ as $B^+(x_0,r) = \{x \in M : d(x_0,x) < r\}$ and $B^-(x_0,r) = \{x \in M : d(x,x_0) < r\}$, resp. Both types of balls generate the topology of M. Recall that the backward balls for d are the forward ones for the distance d^{rev}, defined as $d^{\mathrm{rev}}(x,y) = d(y,x)$ for all x,y. One can define also the *symmetrized distance* $d^s = (d + d^{\mathrm{rev}})/2$, which is a (true) distance, even though it cannot be obtained as a length space (i.e. as the infimum of lengths of connecting curves).

3.2 Cauchy Completions

For any generalized distance d (in particular, one derived from a Finsler metric F), one can define Cauchy sequences and completions. Namely, $\{x_n\} \subset M$ is *(forward) Cauchy* if for all $\varepsilon > 0$, there exists $n_0 \in \mathbb{N}$ such that $d(x_n,x_m) < \varepsilon$ whenever $n_0 \leq n \leq m$. Two Cauchy sequences $\sigma = \{x_n\}$, $\sigma' = \{x'_n\}$ are *related*, $\sigma \sim \sigma'$ if

$$\lim_n \lim_m d(x_n,x'_m) = \lim_n \lim_m d(x'_n,x_m) = 0.$$

The *backward* Cauchy sequence, and the corresponding relation, is defined analogously but using d^{rev}. Accordingly, one has two types (forward, backward) of Cauchy boundaries ($\partial_C^+ M$, $\partial_C^- M$) and of Cauchy completions (M_C^+, M_C^-), defined in the usual way. One can also consider the Cauchy boundary associated to the symmetrized distance d^s, or *symmetrized Cauchy boundary*, which satisfies $\partial_C^s M = \partial_C^- M \cap \partial_C^+ M$ (see [9, Proposition 3.22]). Of course, the *symmetrized Cauchy completion* is then defined as $M_C^s := M \cup \partial_C^s M$.

In the Riemannian case, the natural distance is extensible to a distance on the Cauchy completion. A generalized distance d on M can be also extended to M_C^+ (resp. M_C^-) in a natural way, that is,

$$d_Q([\{x_n\}],[\{y_n\}]) := \lim_n(\lim_m d(x_n,y_m))$$
$$(\text{resp. } d_Q^{\text{rev}}([\{x_n\}],[\{y_n\}]) := \lim_n(\lim_m d^{\text{rev}}(x_n,y_m))).$$

However, d_Q (resp. d_Q^{rev}) is *not* a generalized distance but only a quasi-distance. As a consequence, d_Q (resp. d_Q^{rev}) generates two (in general, different) topologies on M_C^+ (resp. M_C^-): one by using the corresponding forward balls and the other one by using the backward ones. The natural one, so that forward (resp. backward) Cauchy sequences converge to the point represented by its class in M_C^+ (resp. M_C^-), is generated by the backward (resp. forward) balls. On M_C^s, one can also define the extension $\overline{d}^s$ of d^s on M, which satisfies $\overline{d}^s = (d_Q + d_Q^{\text{rev}})/2$ on M_C^s. All the relevant properties of these constructions are summarized in [9, Theorem 1.1].

3.3 Gromov Completions

In the complete Riemannian case, Gromov developed a compactification by means of an embedding of the original space M into the space of Lipschitz functions [11]. In order to define the Gromov completion for a (possibly incomplete) Finsler manifold (M,F), one begins by defining an appropriate (nonsymmetric) notion of *Lipschitzian function*, namely: a function f on M is *Lipschitzian* if it satisfies $f(y) - f(x) \leq d(x,y)$ for all $x,y \in M$. Let $\mathscr{L}_1(M,d)$ be the set of all the Lipschitzian functions endowed with the pointwise topology, and let $\mathscr{L}_1(M,d)/\mathbb{R}$ its quotient by additive constants, i.e., where two functions are related if they differ only in an additive constant. Each point $x \in M$ determines the function $-d_x^+ \in \mathscr{L}_1(M,d)$ defined as $d_x^+(y) = d(y,x)$ for all $y \in M$. The class $[-d_x^+] \in \mathscr{L}_1(M,d)/\mathbb{R}$ characterizes univocally x. Moreover, the map

$$M \hookrightarrow \mathscr{L}_1(M,d)/\mathbb{R}, \qquad x \mapsto [-d_x^+]$$

is a topological embedding, so we can regard M as a subset of $\mathscr{L}_1(M,d)/\mathbb{R}$. Then, the forward Gromov's compactification M_G^+ is defined as the closure in $\mathscr{L}_1(M,d)/\mathbb{R}$ of the image of M by this embedding and the forward Gromov's boundary as $\partial_G^+ M = M_G^+ \setminus M$

Analogously, each point $x \in M$ determines the function $d_x^- \in \mathscr{L}_1(M,d)$ defined by $d_x^-(y) = d(x,y)$ for all $y \in M$. The class $[d_x^-] \in \mathscr{L}_1(M,d)/\mathbb{R}$, which characterizes univocally x, defines the topological embedding

$$M \hookrightarrow \mathscr{L}_1(M,d)/\mathbb{R}, \qquad x \mapsto [d_x^-],$$

which makes M a subset of $\mathscr{L}_1(M,d)/\mathbb{R}$. Consequently, the backward Gromov's compactification M_G^- is the closure in $\mathscr{L}_1(M,d)/\mathbb{R}$ of the image of M by this second embedding, and the backward Gromov's boundary is $\partial_G^- M = M_G^- \setminus M$.

Remark 2. Let us remark some of the main properties of this construction:

(1) $M_G^{\pm}$ is second countable, Hausdorff and compact.
(2) Even though all the points in $\partial_G^{\pm} M$ are the limit of sequences in the manifold M, some of them *may not be the limit of any curve in* M.
(3) Define the *forward/backward Cauchy Gromov boundary* $\partial_{CG}^{\pm} M$ as the set of points in $\partial_G^{\pm} M$ which are limits of some *bounded* sequence $\sigma \subset M$ (i.e., σ is included in some ball $B^{\pm}(x,r) := \{y \in M : d_x^{\mp}(y) < r\}$) and the *forward/backward proper Gromov boundary* as $\partial_{\mathscr{G}}^{\pm} M = \partial_G^{\pm} M \setminus \partial_{CG}^{\pm} M$. Clearly, the forward/backward Cauchy boundary $\partial_C^{\pm} M$ are included in $\partial_{CG}^{\pm} M$, and *this inclusion may be strict*—apart from the fact that may be not an embedding.

Summarizing, [9, Theorem 1.1] (see also [9, Sect. 5] for further details):

Theorem 3. *Let (M,F) be a (possibly incomplete) Finslerian manifold. Then, the forward/backward Gromov's completion $M_G^{\pm}$ is a compact, Hausdorff, and second countable space, and M is included in $M_G^{\pm}$ as an open dense subset. Moreover, the forward/backward Gromov's boundary $\partial_G^{\pm} M$ is the disjoint union of the forward/backward Cauchy-Gromov boundary $\partial_{CG}^{\pm} M$ and the forward/backward proper Gromov boundary $\partial_{\mathscr{G}}^{\pm} M$. The former includes—perhaps strictly—the forward/backward Cauchy boundary $\partial_C^{\pm} M$, and the inclusion $M_C^{\pm} \hookrightarrow M_G^{\pm}$ satisfies the following:*

(i) It is continuous if, and only if, the natural topology generated in $M_C^{\pm}$ by the d_Q-backward/forward balls is finer than the one generated by the d_Q-forward/backward balls. In particular, this happens if d_Q is a generalized distance.
(ii) If d_Q is a generalized distance and $\partial_C^{\pm} M$ is locally compact on $M_C^{\pm}$, the inclusion above is an embedding, and $\partial_C^{\pm} M$ coincides with $\partial_{CG}^{\pm} M$.

Remark 3. In the Riemannian case, d_Q is always a generalized distance so, according to point (i) in the previous theorem, the inclusion is always continuous. About (ii), there exist examples such that the Cauchy completion is not locally compact and where the inclusion is not an embedding (see [9, Fig. 3]).

3.4 Busemann Completion

The notion of Busemann function, successfully used in Eberlein and O'Neill's compactification [5], can be easily generalized to any Finsler manifold (M,F) as follows. Given any curve $c:[\alpha,\Omega^{+})\rightarrow M$, $\Omega^{+}\leq\infty$, with $F(\dot{c})\leq 1$, we define the *forward Busemann function associated to c* as:

$$b_c^{+}(x)=\lim_{s\rightarrow\Omega^{+}}(s-d(x,c(s))) \qquad \forall x\in M. \tag{4}$$

One can check that this limit always exists and, if it is ∞ at some point, then it is constantly equal to ∞, $b_c^{+}\equiv\infty$. Let $B^{+}(M)$ be the set of all finite-valued Busemann functions. It is easy to check that all the functions in $B^{+}(M)$ are 1-Lipschitz. So, in order to endow $B^{+}(M)$ with a topology, the most natural option is to consider $B^{+}(M)$ as a topological subspace of $\mathscr{L}_1(M,d)$, that is, endow $B^{+}(M)$ with the pointwise topology induced from $\mathscr{L}_1(M,d)$. However, as we will see in Sect. 4, there is a natural identification (10) between $B^{+}(M)\cup\{f\equiv\infty\}$ and the future causal completion $\hat{V}$ of a (standard) stationary spacetime $V=\mathbb{R}\times M$. As $\hat{V}$ was endowed with the future chronological topology (recall Sect. 2.2), we can define a topology on $B^{+}(M)\cup\{f\equiv\infty\}$ and thus on $B^{+}(M)$, just by requiring that the cited identification becomes a homeomorphism. This topology, called *forward chronological topology*, can be also defined directly, with no mention to the stationary spacetime, by using the following operator (in the sense of Sect. 2.2):

$$f\in\hat{L}(\sigma)\Longleftrightarrow\begin{cases}(a)f\leq\liminf_n f_n \text{ and}\\ (b)\forall g\in B^{+}(M) \text{ with } f\leq g\leq\limsup_n f_n, \text{ it is } g=f.\end{cases} \tag{5}$$

As usual, we can introduce "backward" definitions just by using the reversed elements F^{rev}, d^{rev} instead of F, d. So, consider any curve $c:[\alpha,\Omega^{-})\rightarrow M$, $\Omega^{-}\leq\infty$, such that $F^{\mathrm{rev}}(\dot{c})\leq 1$. Then, the *backward Busemann function* associated to c is defined as

$$b_c^{-}(x)=\lim_{s\rightarrow\Omega^{-}}(-s+d^{\mathrm{rev}}(x,c(s))) \qquad \forall x\in M.$$

Denote by $B^{-}(M)$ the space of backward Busemann functions which are finite. Then, by using the analogous identification (11) between $B^{-}(M)\cup\{f\equiv-\infty\}$ and the past causal completion $\check{V}$ of a (standard) stationary spacetime $V=\mathbb{R}\times M$, we define a topology on $B^{-}(M)$, called *backward chronological topology*, by requiring that it becomes a homeomorphism. In this case, the topology is determined by the operator:

$$f\in\check{L}(\sigma)\Longleftrightarrow\begin{cases}(a)f\geq\limsup_n f_n \text{ and}\\ (b)\forall g\in B^{-}(M) \text{ with } f\geq g\geq\liminf_n f_n, \text{ it is } g=f.\end{cases} \tag{6}$$

3.4.1 Interpretation of the Operators $\hat{L}$, $\check{L}$

[2]It is well known that the pointwise convergence topology presents an important limitation: Many natural spaces of functions are not closed for the pointwise limit operator. For instance, if we consider the space $C(M)$ of continuous real functions on a metric space (M,d), there exist sequences of continuous functions whose pointwise limit is not continuous and so does not belong to $C(M)$. A natural problem associated to this question is to define a topology on general spaces of real functions on M which, being free of this deficiency, is very similar to the pointwise convergence one.

Let $S(M)$ be any space of real functions on M, and let σ be a sequence of functions in $S(M)$. The operators defined in Eqs. (5) and (6) (with $B^{\pm}(M)$ replaced by $S(M)$ in condition (b)) provides the best possible approximation from below and above, resp., to the pointwise limit of σ. So, in this sense, the topologies associated to these operators can be interpreted as the best possible approximations (from below and above, resp.) to the pointwise convergence topology, which makes $S(M)$ close for these operators. The following result reinforces this interpretation.

Proposition 1. *Let $S(M)$ be any space of real functions on M:*

(1) A sequence $\sigma = \{f_n\} \subset S(M)$ converges pointwise to some $f \in S(M)$ iff $f \in \hat{L}(\sigma) \cap \check{L}(\sigma)$.

(2) The map $G : (S(M),\hat{L}) \to (G(S(M)),\check{L})$, $f \mapsto -f$, is a homeomorphism.

In particular, if (M,d) is a generalized metric space, the map G over $S(M) = \mathscr{L}_1(M,d)$ gives a homeomorphism between $(\mathscr{L}_1(M,d),\hat{L})$ and $(\mathscr{L}_1(M,d^{\mathrm{rev}}),\check{L})$.

(3) If (M,d) is a metric space then the spaces $(\mathscr{L}_1(M,d),\hat{L})$ and $(\mathscr{L}_1(M,d),\check{L})$ (resp. $(B^+(M),\hat{L})$ and $(B^-(M),\check{L})$) are homeomorphic.

Proof. (1) It directly follows from the following equivalences:

$$f \in \hat{L}(\sigma) \cap \check{L}(\sigma) \Leftrightarrow \limsup_n f_n \leq f \leq \liminf_n f_n \Leftrightarrow f = \liminf_n f_n = \limsup_n f_n.$$

(2) The first assertion is straightforward from the fact that

$$f \in \hat{L}(\{f_n\}) \Longleftrightarrow -f \in \check{L}(\{-f_n\}).$$

For the second assertion, just recall that $G(\mathscr{L}_1(M,d)) = \mathscr{L}_1(M,d^{\mathrm{rev}})$.

(3) In the symmetric case, $d = d^{\mathrm{rev}}$ and $G(B^+(M)) = B^-(M)$. So, the homeomorphism is obtained by restricting conveniently G. □

As condition (b) in the definition of limit operators $\hat{L}$, $\check{L}$ depends explicitly on the space of functions considered, the topology derived from them also depend on

[2]This section is the development of a seminal idea by Prof. Miguel Sánchez (see also [17]).

the space of functions. Even more, the larger the space of functions is, the more functions to test the maximal character in condition (b) will be, and it will be closer to condition $f = \limsup_n f_n$. This suggests that these topologies will coincide for spaces of functions large enough. In this sense, we establish the following result:

Proposition 2. *Let (M,d) be a generalized metric space and $S(M) = \mathscr{L}_1(M,d)$ the space of Lipschitz functions. A sequence $\sigma = \{f_n\} \subset \mathscr{L}_1(M,d)$ converges pointwise to $f \in \mathscr{L}_1(M,d)$ iff $f \in \hat{L}(\sigma)$ iff $f \in \check{L}(\sigma)$.*

Proof. We will only prove the equivalence between the pointwise convergence and the $\hat{L}$-convergence (as the equivalence between the pointwise convergence and the $\check{L}$-convergence is analogous).

The implication to the right is obvious. So, assume by contradiction the existence of some sequence $\sigma = \{f_n\} \subset \mathscr{L}_1(M,d)$ and $f \in \mathscr{L}_1(M,d)$ such that $f \in \hat{L}(\sigma)$, but σ does not converge pointwise to f. Then, there exists $x_0 \in M$ such that $\{f_n(x_0)\}$ does not converge to $f(x_0)$. Since $f(x_0) \leq \liminf_n f_n(x_0)$, there exists $\varepsilon_0 > 0$ and some subsequence $\{f_{n_k}\}_k$ of σ such that

$$f(x_0) + \varepsilon_0 < f_{n_k}(x_0) \qquad \forall k \in \mathbb{N}.$$

Take some $t_0 \in \mathbb{R}$ such that $f(x_0) < t_0 < f_{n_k}(x_0)$ for all k. Then,

$$d^+_{(t_0,x_0)}(\cdot) := t_0 - d(\cdot,x_0) < f_{n_k}(x_0) - d(\cdot,x_0) \leq f_{n_k}(\cdot).$$

Therefore, the function $h(\cdot) := \max\{f(\cdot), d^+_{(t_0,x_0)}(\cdot)\} \in \mathscr{L}_1(M,d)$ satisfies

$$f(x) \leq h(x) \leq f_{n_k}(x) \qquad \forall k \in \mathbb{N},$$

being the left inequality strict for $x = x_0$. This contradicts that $f \in \hat{L}(\sigma)$. □

3.4.2 Topology on the Busemann Completion

Consider the spaces $B^{\pm}(M)$ endowed with the forward and backward chronological topologies defined by $\hat{L}$, $\check{L}$, resp. Consider the quotient topological spaces $M_B^{\pm} = B^{\pm}(M)/\mathbb{R}$ obtained by identifying Busemann functions up to a constant. Then, M can be regarded as a topological subspace of $M_B^{\pm}$. So, $M_B^{\pm}$ is called the *Busemann completion* of M and $\partial_B^{\pm} M := M_B^{\pm} \backslash M$ the *Busemann boundary*. The theorems associated to these constructions can be summarized in the following result, [9, Theorem 1.1] (see also [9, Sect. 5] for further details).

Theorem 4. *Let $M_B^{\pm}$ be the Busemann completion of a Finsler manifold (M,F). Then:*

(1) $M_B^{\pm}$ is sequentially compact, and M is naturally embedded as an open dense subset in $M_B^{\pm}$.

(2) *All the points in* $\partial_B^{\pm}M$ *can be reached as limits of curves in* M.
(3) $M_C^{\pm}$ *is naturally included as a subset in* $M_B^{\pm}$, *and the inclusion is continuous if the natural topology generated in* $M_C^{\pm}$ *by the* d_Q*-backward/forward balls is finer than the one generated by the forward/backward ones (see Sect. 3.2). In particular, this happens if* d_Q *is a generalized distance.*
(4) $M_B^{\pm}$ *is the disjoint union of* M, *the Cauchy boundary* $\partial_C^{\pm}M$, *and the asymptotic Busemann boundary* $\partial_{\mathscr{B}}^{\pm}M$, *where:*

- $M_C^{\pm} = M \cup \partial_C^{\pm}M$ *corresponds to the subset of* $B^{\pm}(M)$ *constructed from curves* c *with* $\Omega^{\pm} < \infty$.
- $\partial_{\mathscr{B}}^{\pm}M$ *corresponds to the subset of* $B^{\pm}(M)$ *constructed from curves* c *with* $\Omega^{\pm} = \infty$.

(5) $M_B^{\pm}$ *is* T_1, *and any non-*T_2 *related points must lie in* $\partial_B^{\pm}M$.
(6) $M_B^{\pm} \hookrightarrow M_G^{\pm}$ *in a natural way, and the inverse of the inclusion is continuous (the topology on* $M_G^{\pm}$ *is finer).*
(7) $M_B^{\pm}$ *is homeomorphic to* $M_G^{\pm}$ *iff* $M_B^{\pm}$ *is Hausdorff. In particular, for Hadamard manifolds, any of previous conditions hold, and so,* $M_B^{\pm}$ *also coincides with Eberlein and O'Neill's compactification.*

Remark 4. (i) The possible non-Hausdorff character of $M_B^{\pm}$ alerts on the discrepancy between $M_B^{\pm}$ and $M_G^{\pm}$ (by the assertion (7)). This discrepancy comes from the fact that, in some sense, $M_B^{\pm}$ is a compactification of M by using (finite or asymptotic) directions; while, $M_G^{\pm}$ is a more analytical one. This points out a possible non-nice behavior of $M_G^{\pm}$ with respect to limits of curves in M: in order to ensure Hausdorffness, $M_G^{\pm}$ must include additional points which cannot be reached by curves in M.

The situation described above is illustrated in the appendix. In the first case ("one chimney"), Busemann and Gromov completions coincide, both including one boundary point associated to the chimney. In the second case ("two chimneys"), Busemann completion contains two boundary points associated to the corresponding chimneys, as expected. However, this completion is no longer Hausdorff, and so, it differs from the Gromov one. In fact, Gromov completion now includes a whole segment of additional boundary points, which cannot be reached by any curve in M.

(ii) The differences between M_B^{+} and M_B^{-} in the stationary case reflect asymmetries between the future and past causal boundaries, which did not appear in the static case. Moreover, the S-relation (inspired on the spacetime) allows to relate some points in M_C^{+} with M_C^{-} (and M_B^{+} with M_B^{-}).

Finally, the following proposition shows that the equivalence in Theorem 4 (7) between the topological spaces $M_B^{\pm}$ and $M_G^{\pm}$ can be extended to an equivalence between the chronological and the pointwise topology in $B^{\pm}(M)$. Even though the proof of this result follows from the arguments in [9, Sect. 5], we include a detailed proof here for the convenience of the reader. In this sense, let us recall first [9, Lemma 5.38].

Lemma 1. *$B^+(M)$ is ULL (i.e., for any sequence $\sigma \subset B^+(M)$, the set $\hat{L}(\sigma)$ has, at most, a single point) if, and only if, M_B^+ is Hausdorff.*

Proposition 3. *Consider (M,F) a Finsler manifold. The following statements are equivalent:*

(i) The pointwise and chronological topologies coincide in $B^\pm(M)$.
(ii) $B^\pm(M)$ endowed with the chronological topology is Hausdorff.
(iii) $M_B^\pm$ is Hausdorff.

Proof. The implication $(i) \Rightarrow (ii)$ is straightforward from the Hausdorff character of the pointwise topology. For $(ii) \Rightarrow (iii)$, observe that if $B^\pm(M)$ is Hausdorff, then it is also ULL (recall that Remark 1 (1) is true for any topology defined by a limit operator), and so, previous lemma applies.

For the implication $(iii) \Rightarrow (i)$, it suffices to prove the following equivalence under the Hausdorff hypothesis:

$$\sigma \text{ converges pointwise to } b_c^\pm \iff b_c^+ \in \hat{L}(\sigma) \text{ (resp. } b_c^- \in \check{L}(\sigma))$$

where $\sigma = \{b_{c_n}^\pm\} \subset B^\pm(M)$ and $b_c^\pm \in B^\pm(M)$. The implication to the right is general (it follows without the Hausdorff hypothesis, see [9, Proposition 5.29]), so we will focus in the left one. Assume that $b_c^+ \in \hat{L}(\sigma)$ (resp. $b_c^- \in \check{L}(\sigma)$). Then, σ converges to $b_c^\pm$ with the chronological topology, and thus, $\{[b_{c_n}^\pm]\}$ converges to $[b_c^\pm]$ in $M_B^\pm$. As $M_B^\pm$ is Hausdorff, Theorem 4 (7) ensures that $M_B^\pm = M_G^\pm$ both as point set and topologically. In particular, $\{[b_{c_n}^\pm]\}$ converges to $[b_c^\pm]$ in the topological space $M_G^\pm$, that is, there exists a sequence $\{t_n\}$ such that $\{b_{c_n}^\pm + t_n\}$ converges to $b_c^\pm$ with the pointwise topology. In order to conclude the proof, we have to show that $\{t_n\} \to 0$ and, so, that σ converges to $b_c^\pm$ pointwise. Observe that, if $\{t_n\}$ does not converge to 0, we can obtain a subsequence $\{t_{n_k}\} \subset \{t_n\}$ such that: either diverges, and so, $b_{c_n}^\pm(x_0)$ also diverges for all x_0, or it converges to $t_\infty \in \mathbb{R} \setminus \{0\}$. In any of previous cases, we obtain a contradiction with $b_c^+ \in \hat{L}(\sigma)$ (resp. $b_c^- \in \check{L}(\sigma)$), recall Eq. (5). Then, $\{t_n\} \to 0$ as claimed. □

4 C-Boundary for Stationary SpaceTimes

Now, we are in conditions to make a precise description of the c-boundary for any standard stationary spacetime (see [4] for further relations between Lorentzian causality and Finsler metrics). To this aim, first note that the conformal invariance of the c-boundary allows to consider the expression (7) below as the one of any standard stationary spacetime, without loss of generality:

$$V = (\mathbb{R} \times M, g_L = -dt^2 + \pi^*\omega \otimes dt + dt \otimes \pi^*\omega + \pi^* g), \tag{7}$$

where ω is a 1-form and g is a Riemannian metric, both on M, and $\pi : \mathbb{R} \times M \to M$ the natural projection.

The elements in Eq. (7) allows to construct the following Finsler metrics (of Randers type) on M:

$$\begin{aligned} F^{+}(v)(=F(v)) &= \sqrt{g(v,v)+\omega(v)^2}+\omega(v), \\ F^{-}(v)(=F^{\mathrm{rev}}(v)) &= \sqrt{g(v,v)+\omega(v)^2}-\omega(v). \end{aligned} \tag{8}$$

Denote by $d^{\pm}$ the associated distance to $F^{\pm}$, and by $d_{Q}^{\pm}$ the corresponding extensions to $M_{C}^{\pm}$. It is not difficult to check that a curve $\gamma(t)=(t,c(t)), t \in [\alpha,\Omega^{+})$ is time like and future directed iff $F^{+}(\dot{c})<1$. So, the chronological relation can be characterized as follows:

$$(t_0,x_0) \ll (t_1,x_1) \iff d^{+}(x_0,x_1) < t_1 - t_0$$

In order to compute an IP, notice that if $P = I^{-}(\gamma)$ for the future-directed timelike curve γ, then:

$$\begin{aligned} P &= \{(t_0,x_0) \in V : (t_0,x_0) \ll \gamma(t) \text{ for some } t \in [\alpha,\Omega^{+})\} \\ &= \{(t_0,x_0) \in V : t_0 < t - d^{+}(x_0,c(t)) \text{ for some } t \in [\alpha,\Omega^{+})\} \\ &= \{(t_0,x_0) \in V : t_0 < \lim_{t\to\Omega^{+}} (t - d^{+}(x_0,c(t)))\} \\ &= \{(t_0,x_0) \in V : t_0 < b_c^{+}(x_0)\}, \end{aligned} \tag{9}$$

where b_c^{+} is the Busemann function defined in Eq. (4). That is, there exists a natural identification between the future causal completion and the set of all the Busemann functions:

$$\begin{aligned} \hat{V} &\equiv B^{+}(M) \cup \{f \equiv \infty\} \\ &= V \cup (\mathbb{R} \times \partial_B^{+} M) \cup \{f \equiv \infty\}. \end{aligned} \tag{10}$$

Recall that $f \equiv \infty$ corresponds with the TIP $P=V$ obtained from the curve $\gamma(t) = (t,x_0), t \in [\alpha,\infty)$, for any $x_0 \in M$. Moreover, the PIPs are associated to Busemann functions b_c^{+} such that c converges (thus, $\Omega^{+}<\infty$), and the TIPs to b_c^{+} with non-converging c. As commented before, via the identification (10), the limit operator $\hat{L}$ defined in Eq. (2) takes the form (5). Of course, by proceeding analogously for the past c-boundary, we obtain

$$\begin{aligned} \check{V} &\equiv B^{-}(M) \cup \{f \equiv -\infty\} \\ &= V \cup (\mathbb{R} \times \partial_B^{-} M) \cup \{f \equiv -\infty\}, \end{aligned} \tag{11}$$

being the dual limit operator $\check{L}$ expressed as Eq. (6).

Regarding to the S-relation, in the static case (firstly analyzed in [1, 7, 12]) it becomes trivial, and the c-completion becomes simple as a point set. In the stationary case, however, the S-relation may be highly nontrivial. In fact, all the classical difficulties for the definition of the S-relation and, then, for obtaining a consistent c-boundary can be found in standard stationary spacetimes. So, we describe here the c-boundary of a stationary spacetime under some additional natural hypotheses, which imply that the c-completion is *simple*, according to Definition 1 (see [9, Theorem 1.2]).

Theorem 5. *Let V be a standard stationary spacetime [as in Eq. (7)], and assume that the associated Finsler manifolds $(M,F^{\pm})$ satisfy the following: (a) The quasi-distance d_Q^+ is a generalized distance, (b) the Cauchy completion M_C^s for the symmetrized distance $\overline{d}^s$ is locally compact, and (c) the Busemann completions $M_B^{\pm}$ are Hausdorff.*

Then, the preboundaries $\hat{\partial}V$ and $\check{\partial}V$ have the structures of cones with bases ∂_B^+M, ∂_B^-M and apexes i^+, i^-, resp., and the c-boundary ∂V coincides with the quotient topological space $(\hat{\partial}V \cup \check{\partial}V)/\sim_S$.

As a consequence, each point of $\partial_B^{\pm}M \setminus \partial_C^sM$ yields a line in ∂V (which is horismotic in a natural sense), starting at $i^{\pm}$, and each point in ∂_C^sM yields a single timelike line from i^- to i^+.

In the general (non-simple) case, the points in $\partial_C^+M\backslash\partial_C^sM$ may yield light-like (i.e., locally horismotic) curves starting at i^+, which may be eventually identified with the corresponding ones starting at i^- for $\partial_C^-M\backslash\partial_C^sM$.

5 Application to Stationary Kerr SpaceTime

Next, we are going to apply these results to study the c-boundary of the (standard) stationary region of Kerr spacetime, which is the standard relativistic model for the gravitational field of a rotating massive object. Here, the Kerr metric tensor will be written in terms of the time coordinate t on $\mathbb{R}$ and the spherical coordinates r, θ, φ on $\mathbb{R}^3$,that is, the so-called *Boyer-Lindquist* coordinates. In these coordinates, Kerr metric takes the form

$$g = -g_{tt}dt^2 + g_{rr}dr^2 + g_{\theta\theta}d\theta^2 + g_{\varphi\varphi}d\varphi^2 + 2g_{\varphi t}d\varphi dt \tag{12}$$

with

$$g_{rr} = \frac{\lambda(r,\theta)}{\Delta(r)} \quad g_{\varphi\varphi} = \left[r^2 + a^2 + \frac{2mra^2\sin^2\theta}{\lambda(r,\theta)}\right]\sin^2\theta \quad g_{tt} = 1 - \frac{2mr}{\lambda(r,\theta)}$$

$$g_{\theta\theta} = \lambda(r,\theta) \quad g_{\varphi t} = -\frac{2mra\sin^2\theta}{\lambda(r,\theta)} \tag{13}$$

and

$$\lambda(r,\theta) = r^2 + a^2\cos^2\theta \quad \text{and} \quad \Delta(r) = r^2 - 2mr + a^2. \tag{14}$$

Constants $m \geq 0$ and a represent the mass of the object and the angular momentum by unit of mass as measured from infinity, respectively. If the rotation is suppressed ($a = 0$), Kerr spacetime becomes Schwarzschild spacetime; if, in addition, the mass is removed ($m = 0$), it becomes (empty) Minkowski spacetime.

Fig. 3 Asymptotic c-boundary of stationary Kerr spacetime

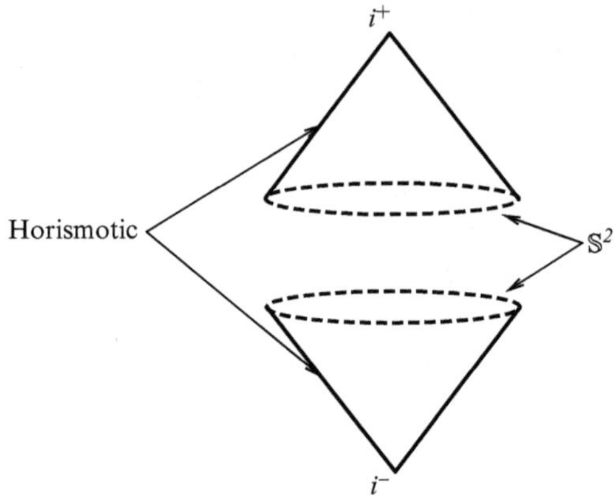

For simplicity, we consider *slow-rotating Kerr spacetime* $(0 < a^2 < m^2)$. We will focus our attention on the stationary region ($g_{tt} > 0$):

$$V = \mathbb{R} \times M, \quad \text{with } M = \{x \equiv (r, \theta, \varphi) : r > (r_\theta :=) m + \sqrt{m^2 - a^2 \cos^2\theta}\}. \quad (15)$$

As the c-boundary is conformal invariant, we are allowed to rewrite the metric as:

$$g = -dt^2 + \frac{g_{rr}}{g_{tt}} dr^2 + \frac{g_{\theta\theta}}{g_{tt}} d\theta^2 + \frac{g_{\varphi\varphi}}{g_{tt}} d\varphi^2 + 2\frac{g_{\varphi t}}{g_{tt}} d\varphi dt.$$

According to Sect. 4, the c-boundary of a stationary spacetime V can be described in terms of the Busemann boundaries $\partial_B^{\pm} M$, which consist of classes of finite Busemann functions associated to inextensible curves on M. Taking into account the geometry of M in Eq. (15), these curves $c : [\alpha, \Omega^+) \to M$, $c(s) = (r(s), \theta(s), \varphi(s))$ may present two alternative behaviors as $s \to \Omega^+$: either *asymptotic* if $r(s) \to \infty$ or *interior* if $c(s) \to (r_0, \theta_0, \varphi_0) \in \mathscr{S}$, where

$$\mathscr{S} = \left\{x \equiv (r, \theta, \varphi) : r = (r_\theta :=) m + \sqrt{m^2 - a^2 \cos^2\theta}\right\}.$$

This allows to distinguish two regions of the c-boundary, the *asymptotic* one, denoted by $\partial^{\text{asy}} V$, and the *interior* one, denoted by $\partial^{\text{int}} V$.

In order to study $\partial^{\text{asy}} V$, recall that Kerr spacetime V is asymptotically flat and the set $\partial_i^{\text{asy}} V$ of asymptotic ideal points of the conformal boundary has the structure of a double horismotic cone of basis $\mathbb{S}^2$ and apexes i^+, i^- (see, for example, [13, 20]). Thus, according to Theorem 2, one concludes that $\partial^{\text{asy}} V$ also presents this double-cone structure (Fig. 3).

So, we can focus on the study of the interior c-boundary $\partial^{\text{int}} V$. This will be done by applying Theorem 5. But, previously, we need to study in depth the Busemann and Cauchy completions $M_B^{\pm}$, $M_C^{\pm}$ of the Finsler manifolds $(M, F^{\pm})$, where $F^{\pm}$ are given in Eq. (8) with

$$\omega = \frac{g_{\varphi t}}{g_{tt}} d\varphi, \qquad g_0 = \frac{g_{rr}}{g_{tt}} dr^2 + \frac{g_{\theta\theta}}{g_{tt}} d\theta^2 + \frac{g_{\varphi\varphi}}{g_{tt}} d\varphi^2. \tag{16}$$

Next, we will denote by N and S the north and south poles of $\mathscr{S}$, resp.

Proposition 4. *There exists a homeomorphism between the symmetrized Cauchy completion M_C^s, endowed with the topology associated to $d_Q^{\pm}$, and $M \cup (\mathscr{S} \setminus \{N,S\})$, endowed with the extension d_0 of the standard distance on $M(\subset \mathbb{R}^3)$, such that the symmetrized Cauchy boundary $\partial_C^s M$, which coincides with $\partial_C^{\pm} M$, is identified with $\mathscr{S} \setminus \{N,S\}$.*

In particular, the quasi-distance d_Q^+ is a generalized distance, and M_C^s is locally compact for the symmetrized distance $\overline{d}^s$.

The proof of this proposition requires a previous technical lemma.

Lemma 2. *There exists some constant $k > 0$ such that $k\sqrt{h_0(v,v)} \leq F^{\pm}(v)$ for all $v \in TM$, where $h_0 = dr^2 + r^2 d\theta^2 + r^2 \sin^2\theta d\varphi^2$ is the standard metric on $M(\subset \mathbb{R}^3)$.*

Proof. We distinguish two cases according to the sign of $\pm\omega$. First, assume that $\pm\omega(v) \geq 0$ for $v \in TM$. From the expressions of the metric coefficients (13) and (14), one has:

$$g_{rr}/g_{tt} > 1, \qquad g_{\theta\theta}/g_{tt} \geq r^2, \qquad g_{\varphi\varphi}/g_{tt} \geq r^2 \sin^2\theta.$$

In fact, the second and third inequalities are trivial from the expressions of $g_{\theta\theta}$, $g_{\varphi\varphi}$ and the fact that $g_{tt} \leq 1$. For the first inequality, just note that $r > m > a \geq a|\sin\theta|$ implies $mr > a^2 \sin^2\theta$, and thus, $\lambda(r,\theta) > \Delta(r)$. Therefore:

$$\sqrt{h_0(v,v)} \leq \sqrt{g_0(v,v)} \leq \pm\omega(v) + \sqrt{\omega(v)^2 + g_0(v,v)} = F^{\pm}(v). \tag{17}$$

Next, assume that $\pm\omega(v) < 0$ for $v \equiv (\dot{r}, \dot{\theta}, \dot{\varphi}) \in TM$. There exists some constant $k_1 > 0$ such that

$$\mp\omega(v) g_{tt} = \frac{2mr|a|\sin^2\theta}{r^2 + a^2\cos^2\theta} |\dot{\varphi}| \leq k_1 r |\sin\theta| |\dot{\varphi}| \leq k_1 \sqrt{g_{\varphi\varphi}\dot{\varphi}^2} \leq k_1 \sqrt{g_0(v,v) g_{tt}}. \tag{18}$$

Therefore, from Eq. (18) and the fact that $g_{tt} \leq 1$, if we take $k := (k_1 + \sqrt{k_1^2 + 1})^{-1}$, the following expressions hold for all $v \in TM$:

$$\begin{aligned} F^{\pm}(v) &= \pm\omega(v) + \sqrt{\omega(v)^2 + g_0(v,v)} = \frac{g_0(v,v)}{\mp\omega(v) + \sqrt{\omega(v)^2 + g_0(v,v)}} \\ &= \frac{g_0(v,v)\sqrt{g_{tt}}}{\mp\omega(v)\sqrt{g_{tt}} + \sqrt{\omega(v)^2 g_{tt} + g_0(v,v) g_{tt}}} \geq \frac{g_0(v,v)\sqrt{g_{tt}}}{k_1\sqrt{g_0(v,v)} + \sqrt{k_1^2 g_0(v,v) + g_0(v,v) g_{tt}}} \\ &= \frac{\sqrt{g_0(v,v) g_{tt}}}{k_1 + \sqrt{k_1^2 + g_{tt}}} \geq k\sqrt{g_0(v,v) g_{tt}} \geq k\sqrt{h_0(v,v)}. \end{aligned}$$

□

Proof of Proposition 4. We are going to show that the map

$$H^{\pm}: M \cup (\mathscr{S} \setminus \{N,S\}) \to M_C^s \subset M_B^{\pm}, \qquad p \mapsto [b_p^{\pm}] \tag{19}$$

is a homeomorphism, where $b_p^{\pm} \equiv b_{c_p}^{\pm}$ and $c_p : [0, \Omega_p^{\pm}) \to M$ denotes the unitary reparametrization of $\overline{c}_p(s) = (r+1-s, \hat{p})$, $s \in [0,1)$, being $p = (r, \hat{p})$, $\hat{p} = (\theta, \varphi) \in \mathbb{S}^2$ and $r \geq r_\theta (= m + \sqrt{m^2 - a^2\cos^2\theta})$.

In order to show that $H^{\pm}$ is well defined, first note that $H^{\pm}|_M \equiv Id$. So, one only needs to check that $H^{\pm}$ is well-defined on $\mathscr{S} \setminus \{N,S\}$, or, equivalently, $\Omega_p^{\pm} < \infty$ for any $p \in \mathscr{S} \setminus \{N,S\}$. In fact, for any $p = (r_\theta, \hat{p})$, $\hat{p} \in \mathbb{S}^2 \setminus \{N,S\}$,

$$\begin{aligned}
\Omega_p^{\pm} &= \mathrm{length}_{\pm}(\overline{c}_p) = \int_0^1 F^{\pm}(\dot{\overline{c}}_p(s))ds = \int_0^1 \sqrt{\tfrac{g_{rr}(\overline{c}_p(s))}{g_{tt}(\overline{c}_p(s))}}ds \\
&= \int_0^1 \sqrt{\tfrac{(\overline{r}(s)^2+a^2\cos^2\theta)^2}{(\overline{r}(s)^2-2m\overline{r}(s)+a^2)(\overline{r}(s)^2-2m\overline{r}(s)+a^2\cos^2\theta)}}ds \qquad (\text{here, } \overline{r}(s) = r_\theta + 1 - s) \\
&= \int_{r_\theta}^{r_\theta+1} \sqrt{\tfrac{(r^2+a^2\cos^2\theta)^2}{(r^2-2mr+a^2)(r^2-2mr+a^2\cos^2\theta)}}dr.
\end{aligned} \tag{20}$$

Since $\hat{p} = (\theta, \varphi) \in \mathbb{S}^2 \setminus \{N,S\}$, it is $r_\theta > r_+ (> r_-)$, where $r_\pm$ are the roots of $\Delta(r)$, that is,

$$r_\pm = m \pm \sqrt{m^2 - a^2}.$$

Moreover,

$$r^2 - 2mr + a^2\cos^2\theta = (r - r_\theta)q_\theta(r), \quad \text{with } q_\theta(r) \geq \varepsilon_0 > 0 \text{ for all } r \in (r_\theta, r_\theta + 1].$$

Hence, there exists some constant $\Lambda > 0$ such that

$$\begin{aligned}
\Omega_p^{\pm} &= \int_{r_\theta}^{r_\theta+1} \sqrt{\tfrac{(r^2+a^2\cos^2\theta)^2}{(r^2-2mr+a^2)(r^2-2mr+a^2\cos^2\theta)}}dr = \int_{r_\theta}^{r_\theta+1} \sqrt{\tfrac{(r^2+a^2\cos^2\theta)^2}{(r-r_-)(r-r_+)(r-r_\theta)q_\theta(r)}}dr \\
&\leq \int_{r_\theta}^{r_\theta+1} \tfrac{\Lambda}{\sqrt{r-r_\theta}}dr < \infty.
\end{aligned}$$

Next, we are going to show that $H^{\pm}$ is injective, or, more precisely, $[b_p^{\pm}] \neq [b_{p'}^{\pm}]$ if $p \neq p'$, $p, p' \in M \cup (\mathscr{S} \setminus \{N,S\})$. From Lemma 2,

$$d_Q^{\pm}([b_p^{\pm}], [b_{p'}^{\pm}]) = \lim_{s,s'} d^{\pm}(c_p(s), c_{p'}(s')) \geq k \cdot \lim_{s,s'} d_0(c_p(s), c_{p'}(s')) = k \cdot d_0(p, p') > 0. \tag{21}$$

Therefore, $[b_p^{\pm}] \neq [b_{p'}^{\pm}] \in M_C^s$.

Let us prove now that $H^{\pm}$ is surjective. To this aim, it suffices to show that, for any inextensible curve c with finite $F^{\pm}$-length there exists some $p \in \mathscr{S} \setminus \{N,S\}$

such that $[b_p^{\pm}] = [b_c^{\pm}]$. From Lemma 2, any such a curve must have finite length when computed with h_0. Therefore, c must converge to some point $p = (r_\theta, \hat{p}) \in \mathscr{S}$. The rest of this paragraph is devoted to prove that, indeed, $0 < \theta < \pi$. Assume by contradiction that, say, $\theta = 0$, and let us arrive to the absurd $\text{length}_\pm(c) = \infty$. One can assume without restriction that $c(s) = (r(s), \theta(s), \varphi(s))$, $s \in [-\varepsilon, 0)$, with $r(s) = r_+ - s$ and $\theta(s) \to 0$ (otherwise, restrict the curve conveniently—this is not a problem because we are interested in a lower bound for the length). First, in order to bound $F^\pm(\dot{c}(s))$ observe that

$$\begin{aligned} \Delta(s) &= r(s)^2 - 2mr(s) + a^2 = (r(s) - r_+)(r(s) - r_-) = o(-s) \\ \lambda(s) - 2mr(s) &= q_{\theta(s)}(r(s))(r(s) - r_{\theta(s)}) \leq q_{\theta(s)}(r(s))(r(s) - r_+) = o(-s), \end{aligned} \tag{22}$$

where $o(f(s))$ denotes a term with dependence on s of order $f(s)$. On the other hand, since $c(s)$ must remain out of the ergosphere, necessarily

$$r(s) > m + \sqrt{m^2 - a^2 \cos^2 \theta(s)} \implies \sin^2 \theta(s) < \frac{\Delta(r(s))}{a^2} = o(-s). \tag{23}$$

Therefore, from Eq. (22)

$$\sqrt{\frac{g_{rr}(c(s))}{g_{tt}(c(s))}} = \frac{\lambda(s)}{\sqrt{\Delta(s)(\lambda(s) - 2mr(s))}} \geq o(1/-s). \tag{24}$$

Taking into account that

$$\begin{aligned} F^\pm(\dot{c}(s)) &\geq -|\omega(\dot{c}(s))| + \sqrt{\omega^2(\dot{c}(s)) + g_0(\dot{c}(s), \dot{c}(s))} \\ &\geq -\left|\tfrac{g_{\varphi t}(c(s))}{g_{tt}(c(s))}\dot{\varphi}(s)\right| + \sqrt{\tfrac{g_{\varphi t}(c(s))^2}{g_{tt}(c(s))^2}\dot{\varphi}(s)^2 + \tfrac{g_{rr}(c(s))}{g_{tt}(c(s))} + \tfrac{g_{\varphi\varphi}(c(s))}{g_{tt}(c(s))}\dot{\varphi}(s)^2}, \end{aligned} \tag{25}$$

we consider two possibilities for the parameter $s \in [-\varepsilon, 0)$:

1. If s satisfies $|g_{\varphi t}(c(s))\dot{\varphi}(s)| / -s \leq g_{rr}(c(s))$, then Eqs. (25) and (24) imply

$$F^\pm(\dot{c}(s)) \geq \frac{\frac{g_{rr}(c(s))}{g_{tt}(c(s))}}{\left|\frac{g_{\varphi t}(c(s))}{g_{tt}(c(s))}\dot{\varphi}(s)\right| + \sqrt{\frac{g_{\varphi t}(c(s))^2}{g_{tt}(c(s))^2}\dot{\varphi}(s)^2 + \frac{g_{rr}(c(s))}{g_{tt}(c(s))}}} \geq o(1/-s).$$

2. If s satisfies $|g_{\varphi t}(c(s))\dot{\varphi}(s)| / -s > g_{rr}(c(s))$, then Eqs. (13), (14) and (23) imply

$$|\dot{\varphi}(s)| > \frac{1}{o(-s)}, \quad \sqrt{\frac{g_{\varphi\varphi}(c(s))}{g_{tt}(c(s))}}|\dot{\varphi}(s)| > \frac{1}{o(-s)} \quad \text{and} \quad g_{\varphi\varphi}(s) = o(1) \cdot g_{\varphi t}(s).$$

And these inequalities joined to Eq. (25) imply

$$
\begin{aligned}
F^{\pm}(\dot{c}(s)) &\geq \frac{\frac{g_{\varphi\varphi}(c(s))}{g_{tt}(c(s))}\dot{\varphi}(s)^2}{\left|\frac{g_{\varphi t}(c(s))}{g_{tt}(c(s))}\dot{\varphi}(s)\right| + \sqrt{\frac{g_{\varphi t}(c(s))^2}{g_{tt}(c(s))^2}\dot{\varphi}(s)^2 + \frac{g_{rr}(c(s))}{g_{tt}(c(s))} + \frac{g_{\varphi\varphi}(c(s))}{g_{tt}(c(s))}\dot{\varphi}(s)^2}} \\
&\geq \frac{\frac{g_{\varphi\varphi}(c(s))}{g_{tt}(c(s))}\dot{\varphi}(s)^2}{\left|\frac{g_{\varphi t}(c(s))}{g_{tt}(c(s))}\dot{\varphi}(s)\right| + \sqrt{\frac{g_{\varphi t}(c(s))^2}{g_{tt}(c(s))^2}\dot{\varphi}(s)^2 + \left|\frac{g_{\varphi t}(c(s))}{g_{tt}(c(s))}\dot{\varphi}(s)\right| \cdot o(1/-s) + \frac{g_{\varphi\varphi}(c(s))}{g_{tt}(c(s))}\dot{\varphi}(s)^2}} \\
&\geq o(1) \cdot |\dot{\varphi}(s)| \cdot \min\{1, \sqrt{g_{\varphi\varphi}(c(s))/g_{tt}(c(s))}\} \geq o(1/-s).
\end{aligned}
$$

So, for any $s \in [-\varepsilon, 0)$, $F^{\pm}(\dot{c}(s)) \geq o(1/-s)$ for one particular o function working for all s, and we deduce the contradiction:

$$
\text{length}_{\pm}(c) \geq \int_{-\varepsilon}^{0} F^{\pm}(\dot{c}(s))ds \geq \int_{-\varepsilon}^{0} o(1/-s)ds = \infty.
$$

Summarizing, the surjectivity will follow if we show that $[b_c^{\pm}] = [b_p^{\pm}]$ whenever c converges to $p = (r_\theta, \hat{p}) \in \mathscr{S}$, with $\hat{p} = (\theta, \varphi) \in \mathbb{S}^2 \setminus \{N, S\}$. Since $[b_c^{\pm}], [b_p^{\pm}] \in \partial_C^s M$, this reduces to prove

$$
d_Q^{\pm}([b_c^{\pm}], [b_p^{\pm}]) = \lim_{s}\lim_{s'} d^{\pm}(c(s), c_p(s')) = 0. \tag{26}
$$

Assume $c : [0, \Omega^+) \to M$, $c(s) = (r(s), \theta(s), \varphi(s))$. For any $s \in [0, \Omega^+)$, define $c_2 : [s, \Omega^+) \to M$, $c_2(s_2) = (r_2(s_2), \theta_2(s_2), \varphi_2(s_2))$ with:

$$
\begin{aligned}
r_2(s_2) &= r_p(s') - r_\theta + r(s_2) \\
\theta_2(s_2) &= \theta(s_2) \\
\varphi_2(s_2) &= \varphi(s_2).
\end{aligned}
$$

Define also $c_1 : [0, s] \to M$ almost $F^{\pm}$-minimizing between $c_1(0) = c(s)$ and $c_1(s) = c_2(s)$. The concatenation $c_1 \cdot c_2$ gives a curve joining $c(s)$ with $c_p(s')$. Taking into account that c, c_p converge to the same point $p = (r_\theta, \hat{p})$ of $\mathscr{S}$, we have:

$$
\begin{aligned}
&c_1(s) = c_2(s) \to c(s) = c_1(0) \text{ as } s' \to \Omega^+ \\
&F^{\pm}(\dot{c}_2) \to F^{\pm}(\dot{c}) \text{ pointwise as } s' \to \Omega^+.
\end{aligned} \tag{27}
$$

From the first limit in Eq. (27) we deduce

$$
\lim_{s' \to \Omega^+} \text{length}_{\pm}(c_1) = 0. \tag{28}
$$

On the other hand, notice that $F^{\pm}(\dot{c}_2)$ is dominated by $\Lambda(F^+(\dot{c})+F^-(\dot{c}))$, for some $\Lambda > 0$, which satisfies

$$\int_s^{\Omega^+} \Lambda(F^+(\dot{c})+F^-(\dot{c})) = \Lambda \int_s^{\Omega^+} F^+(\dot{c}) + \Lambda \int_s^{\Omega^+} F^-(\dot{c}) < \infty. \qquad (29)$$

Therefore, from the second limit in Eq. (27) the Dominated Convergence Theorem can be applied, and so:

$$\lim_{s'\to\Omega^+} \mathrm{length}_{\pm}(c_2) = \lim_{s'\to\Omega^+} \int_s^{\Omega^+} F^{\pm}(\dot{c}_2) = \int_s^{\Omega^+} F^{\pm}(\dot{c}) = \mathrm{length}_{\pm}(c[s,\Omega^+]). \qquad (30)$$

Summarizing, from Eqs. (28) and (30) (Recall that $\mathrm{length}_{\pm}(c[s,\Omega^{\pm}]) < \infty$ since $[b_c^{\pm}] \in M_C^s$):

$$\lim_{s'\to\Omega^+} d^{\pm}(c(s),c_p(s')) \leq \lim_{s'\to\Omega^+} \mathrm{length}_{\pm}(c_1) + \lim_{s'\to\Omega^+} \mathrm{length}_{\pm}(c_2) = \mathrm{length}_{\pm}(c[s,\Omega^+]) < \infty,$$

and thus,

$$\lim_{s,s'\to\Omega^+} d^{\pm}(c(s),c_p(s')) \leq \lim_{s\to\Omega^+} \mathrm{length}_{\pm}(c[s,\Omega^+]) = 0.$$

Finally, it rests to show that $H^{\pm}$ and $(H^{\pm})^{-1}$ are continuous. For the continuity of $H^{\pm}$, assume that $\{p_n\} \subset M \cup (\mathscr{S} \setminus \{N,S\})$ converges to p in $M \cup (\mathscr{S} \setminus \{N,S\})$. As $M \cup (\mathscr{S} \cup \{N,S\})$ is a sequential space, it suffices to show that

$$\lim_n d_Q^{\pm}([b_{p_n}^{\pm}],[b_p^{\pm}]) = \lim_n (\lim_{ss'} d^{\pm}(c_{p_n}(s),c_p(s'))) = 0.$$

More precisely, we are going to show that for any $\varepsilon > 0$ there exists n_0 such that if $n \geq n_0$, then $\lim_{ss'} d^{\pm}(c_{p_n}(s),c_p(s')) < \varepsilon$. From the finiteness of the lengths of the curves c_p, c_{p_n}, there exists some m_0 big enough such that for any n,

$$\mathrm{length}_{\mp}(c_{p_n}[\Omega_{p_n}^{\pm} - 1/m_0, \Omega_{p_n}^{\pm}]) < \frac{\varepsilon}{3}, \quad \mathrm{length}_{\pm}(c_p[\Omega_p^{\mp} - 1/m_0, \Omega_p^{\mp}]) < \frac{\varepsilon}{3}.$$

On the other hand, there exists n_0 big enough such that for any $n \geq n_0$,

$$d^{\pm}(c_{p_n}(\Omega_{p_n}^{\pm} - 1/m_0), c_p(\Omega_p^{\pm} - 1/m_0)) < \frac{\varepsilon}{3}.$$

Therefore, if $n \geq n_0$, then

$$\begin{aligned}\lim_{ss'} d^{\pm}(c_{p_n}(s),c_p(s'))) &\leq \mathrm{length}_{\mp}(c_{p_n}[\Omega_{p_n}^{\pm} - 1/m_0, \Omega_{p_n}^{\pm}]) \\ &\quad + d^{\pm}(c_{p_n}(\Omega_{p_n}^{\pm} - 1/m_0), c_p(\Omega_p^{\pm} - 1/m_0)) \\ &\quad + \mathrm{length}_{\pm}(c_p[\Omega_p^{\mp} - 1/m_0, \Omega_p^{\mp}]) \\ &< \tfrac{\varepsilon}{3} + \tfrac{\varepsilon}{3} + \tfrac{\varepsilon}{3} = \varepsilon.\end{aligned}$$

For the continuity of $(H^{\pm})^{-1}$, assume that $\{[b^{\pm}_{p_n}]\}_n \subset M^s_C$ converges to $[b^{\pm}_p]$ in M^s_C. From Eq. (21),

$$0 \leq k \cdot d_0(p, p_n) \leq d^{\pm}_Q([b^{\pm}_{p_n}], [b^{\pm}_p]) \to 0, \quad \text{and thus,} \quad d_0(p, p_n) \to 0.$$

Hence, $\{p_n\}_n$ converges to p in $M \cup (\mathscr{S} \setminus \{N, S\})$. Therefore, as $M^s_C \subset M^{\pm}_B$ is a sequential space (Sect. 2.2), $(H^{\pm})^{-1}$ is continuous. □

Proposition 5. *The homeomorphism $H^{\pm}$ can be extended to a homeomorphism $\overline{H}^{\pm}$ between the Busemann completion[3] $M^{\pm}_B$ and $M \cup \mathscr{S}$, being $M^{\pm}_B \setminus M^s_C$ identified with $\{N, S\} \subset \mathscr{S}$.*

In particular, the Busemann completion $M^{\pm}_B$ is Hausdorff.

Proof. First, we are going to define the extension $\overline{H}^{\pm} : M \cup \mathscr{S} \to M^{\pm}_B$, and then, we are going to show that it is a homeomorphism.

Consider any sequence $\sigma \subset M$ converging to some pole of $\mathscr{S}$. Then, $H^{\pm}(\sigma)$ is a sequence in $M^s_C \subset M^{\pm}_B$. As $M^{\pm}_B$ is sequentially compact (recall Theorem 4 (1)), there exists some subsequence of $H^{\pm}(\sigma)$ which converges to some point $[b^{\pm}_c] \in M^{\pm}_B$, where c is some inextensible curve in M converging to some pole of $\mathscr{S}$. Therefore, from the symmetries of Kerr spacetime, we can ensure the existence of two inextendible curves c_N, c_S in M, converging to N, S, resp., with $c_S(s) = -c_N(s)$ for all s, and such that $[b^{\pm}_N], [b^{\pm}_S] \in M^{\pm}_B$, being $b^{\pm}_N = b^{\pm}_{c_N}$, $b^{\pm}_S = b^{\pm}_{c_S}$. Summarizing, we make the following definition:

$$\overline{H}^{\pm} : M \cup \mathscr{S} \to M^{\pm}_B, \qquad p \mapsto \overline{H}^{\pm}(p) = \begin{cases} H^{\pm}(p) & \text{if } p \in M \cup (\mathscr{S} \setminus \{N, S\}) \\ [b^{\pm}_N] & \text{if } p = N \\ [b^{\pm}_S] & \text{if } p = S. \end{cases}$$

Since $H^{\pm}$ is injective, $\mathrm{Im}(H^{\pm}) \subset M^s_C$, and $[b^{\pm}_N], [b^{\pm}_S] \in M^{\pm}_B \setminus M^s_C$, the injectivity of $\overline{H}^{\pm}$ will follow if we check that $[b^{\pm}_S] \neq [b^{\pm}_N]$. To this aim, first note that there exists some $\Lambda_0 > 0$ such that

$$\begin{aligned} &d^{\pm}(c_S(s), c_S(s')) - d^{\pm}(q, c_N(s')) \leq \Lambda_0, \quad \text{for } s(< s') \text{ big enough,} \\ &d^{\pm}(c_S(s), q) \to \infty \quad \text{as } s \to \infty, \end{aligned}$$

for any point q with $0 \leq \theta_q < \pi/2$ and $r_q = 2m$. Therefore, as any almost minimizing curve from $c_S(s)$ to $c_N(s')$ must pass through some point q as above, we can apply previous inequalities and deduce for s big enough:

$$\lim_s (b^{\pm}_S(c_S(s)) - b^{\pm}_N(c_S(s))) = \lim_{ss'} (d^{\pm}(c_S(s), c_N(s')) - d^{\pm}(c_S(s), c_S(s'))) = \infty. \tag{31}$$

[3]In this proposition, by the term "Busemann completion," we must understand the usual Busemann completion up to its asymptotic region,that is, those points of the Busemann boundary associated to curves with diverging radial component.

Analogously,

$$\lim_{s}(b_N^{\pm}(c_N(s)) - b_S^{\pm}(c_N(s))) = \lim_{ss'}(d^{\pm}(c_N(s), c_S(s')) - d^{\pm}(c_N(s), c_N(s'))) = \infty. \tag{32}$$

In particular, the difference $b_N^{\pm} - b_S^{\pm}$ cannot be a constant, and so, $[b_N^{\pm}] \neq [b_S^{\pm}]$.

Since $H^{\pm}$ is surjective, the surjectivity of $\overline{H}^{\pm}$ will follow if we check that, for any two curves c, c' with finite Busemann functions and $c(s), c'(s) \to N$, necessarily $[b_c^{\pm}] = [b_{c'}^{\pm}]$ (the proof for $c(s), c'(s) \to S$ is analogous). So, assume for example that $c(s), c'(s) \to N$. Since $[b_c^{\pm}], [b_{c'}^{\pm}] \in \partial_B^{\pm} M \setminus \partial_C^s M$, it suffices to prove the inequalities:

$$\begin{aligned} K_1^{\pm} &:= \lim_s (b_{c'}^{\pm}(c(s)) - b_c^{\pm}(c(s))) = \lim_{ss'}(d^{\pm}(c(s), c(s')) - d^{\pm}(c(s), c'(s'))) \geq 0 \\ K_2^{\pm} &:= \lim_s (b_c^{\pm}(c'(s)) - b_{c'}^{\pm}(c'(s))) = \lim_{ss'}(d^{\pm}(c'(s), c'(s')) - d^{\pm}(c'(s), c(s'))) \geq 0. \end{aligned} \tag{33}$$

In fact, from (see [7, Lemma 3.2]), if Eq. (33) holds then

$$b_{c'}^{\pm} - b_c^{\pm} - K_1^{\pm} \geq 0, \quad b_c^{\pm} - b_{c'}^{\pm} - K_2^{\pm} \geq 0. \tag{34}$$

Therefore,

$$b_c^{\pm} + K_1^{\pm} \leq b_{c'}^{\pm} \leq b_c^{\pm} - K_2^{\pm}, \text{ and thus, } \quad b_c^{\pm} = b_{c'}^{\pm}.$$

Assume $c, c' : [0, \infty) \to M$, $c(s) = (r(s), \theta(s), \varphi(s))$, $c'(s) = (r'(s), \theta'(s), \varphi'(s))$. For any $s \in [0, \infty)$, define $c_2 : [s, s'] \to M$, $c_2(s_2) = (r_2(s_2), \theta_2(s_2), \varphi_2(s_2))$ with:

$$\begin{aligned} r_2(s_2) &= r(s_2) + (r'(s') - r(s')) \\ \theta_2(s_2) &= \theta(s_2) + \tfrac{s_2 - s}{s' - s}(\theta'(s_2) - \theta(s_2)) \\ \varphi_2(s_2) &= \varphi(s_2) + \tfrac{s_2 - s}{s' - s}(\varphi'(s') - \varphi(s')). \end{aligned}$$

Define also $c_1 : [0, s] \to M$ almost $F^{\pm}$-minimizing between $c_1(0) = c(s)$ and $c_1(s) = c_2(s)$. The concatenation $\tilde{c} = c_1 \cdot c_2$ gives a curve joining $c(s)$ with $c'(s')$. Taking into account that c, c' converge to $N \in \mathscr{S}$ as $s \to \infty$, we have:

$$\begin{aligned} & c_1(s) = c_2(s) \to c(s) = c_1(0) \quad \text{as } s' \to \infty \\ & F^{\pm}(\dot{c}) - F^{\pm}(\dot{c}_2) \to 0 \text{ pointwise as } s' \to \infty. \end{aligned} \tag{35}$$

From the first limit in Eq. (35) we deduce $\lim_{s' \to \infty} \text{length}_{\pm}(c_1) = 0$, and so

$$\lim_{s' \to \infty} \text{length}_{\pm}(\tilde{c}[0, s']) = \lim_{s' \to \infty} \text{length}_{\pm}(c_2[s, s']). \tag{36}$$

On the other hand, notice that $F^{\pm}(\dot{c}) - F^{\pm}(\dot{c}_2)$ is dominated by the function $\Lambda(\tilde{F}^+(\dot{c}) + \tilde{F}^-(\dot{c}))$, for some $\Lambda > 0$, where $\tilde{F}^{\pm}(\cdot)$ is as $F^{\pm}(\cdot)$ but with the factor $\lambda - 2mr$ suppressed from the denominator of the g_{rr}-term. Moreover,

$$\int_s^{\infty} \Lambda(\tilde{F}^+(\dot{c}) + \tilde{F}^-(\dot{c})) = \Lambda \int_s^{\infty} \tilde{F}^+(\dot{c}) + \Lambda \int_s^{\infty} \tilde{F}^-(\dot{c}) < \infty. \tag{37}$$

Therefore, from Eq. (36) and the Dominated Convergence Theorem, we have:

$$\begin{aligned}\lim_{ss'}(d^{\pm}(c(s),c(s'))-d^{\pm}(c(s),c'(s'))) &= \lim_{ss'}(\text{length}_{\pm}(c[s,s'])-d^{\pm}(c(s),c'(s')))\\ &\geq \lim_{ss'}\left(\text{length}_{\pm}(c[s,s'])-\text{length}_{\pm}(\tilde{c}[0,s'])\right)\\ &= \lim_{ss'}(\text{length}_{\pm}(c[s,s'])-\text{length}_{\pm}(c_2[s,s']))\\ &= \lim_{ss'}\textstyle\int_s^{s'}(F^{\pm}(\dot{c})-F^{\pm}(\dot{c}_2))\\ &= \lim_s \textstyle\int_s^{\infty} 0 = 0.\end{aligned} \tag{38}$$

By interchanging the roles of c and c' in previous argument, we also deduce

$$\lim_{ss'}(d^{\pm}(c'(s),c'(s'))-d^{\pm}(c'(s),c(s')))\geq 0.$$

Next, we are going to prove that $\overline{H}^{\pm}$ is continuous. Since we already know that $H^{\pm}$ is continuous and $M\cup\mathscr{S}$ is sequential, it suffices to show that, for any sequence $\sigma\subset M\cup\mathscr{S}$ converging to N, the sequence $\overline{H}^{\pm}(\sigma)\subset M_B^{\pm}$ converges pointwise to $[b_N^{\pm}]$ (the case for S is analogous). Assume by contradiction that $\overline{H}^{\pm}(\sigma)$ does not converge pointwise to $[b_N^{\pm}]$. As $M_B^{\pm}$ is sequentially compact, there exists some subsequence $\overline{H}^{\pm}(\sigma')$ of $\overline{H}^{\pm}(\sigma)$ which converges to some $[b_c^{\pm}]\in M_B^{\pm}$ different from $[b_N^{\pm}]$. Moreover, as σ' converges to N, we can also assume that $c(s)$ converges to N. This is in contradiction with the proof of the surjectivity of $\overline{H}^{\pm}$, where the two curves to be compared are c and c_N.

Finally, we need to prove that $(\overline{H}^{\pm})^{-1}$ is continuous. Since $(H^{\pm})^{-1}$ is continuous and $M_B^{\pm}$ is sequential, it suffices to show that, for any sequence $\sigma\subset M\cup\mathscr{S}$ converging to $p\in M\cup\mathscr{S}$, the only limit of $\{\overline{H}^{\pm}(\sigma)\}$ in $M_B^{\pm}$ is $\overline{H}^{\pm}(p)$. In fact, this is a direct consequence of the fact that the convergence of $\{\overline{H}^{\pm}(\sigma)\}$ to $\overline{H}^{\pm}(p)$ is always pointwise (as showed in the proof of Proposition 4). □

Summarizing, from Propositions 4, 5 and Theorem 5 we obtain:

Theorem 6. *Let (V,g) be the stationary part of Kerr spacetime. Then, the asymptotic preboundary $\hat{\partial}^{\text{asy}}V$ (resp. $\check{\partial}^{\text{asy}}V$) has the structure of a cone with basis $\mathscr{S}$ and apex i^+ (resp. i^-). The asymptotic c-boundary $\partial^{\text{asy}}V$ coincides with the quotient topological space $(\hat{\partial}^{\text{asy}}V\cup\check{\partial}^{\text{asy}}V)/\sim_S$.*

Moreover, each pole $\{N,S\}\subset\mathscr{S}$ yields a horismotic line in $\partial^{\text{asy}}V$ starting at $i^{\pm}$, and each point in $\mathscr{S}\setminus\{N,S\}$ yields a single timelike line from i^- to i^+ (Fig. 4).

6 Appendix

Here, we rigorously describe an example, commented at an intuitive level in [9], which realizes some remarkable differences between the Busemann and Gromov completions.

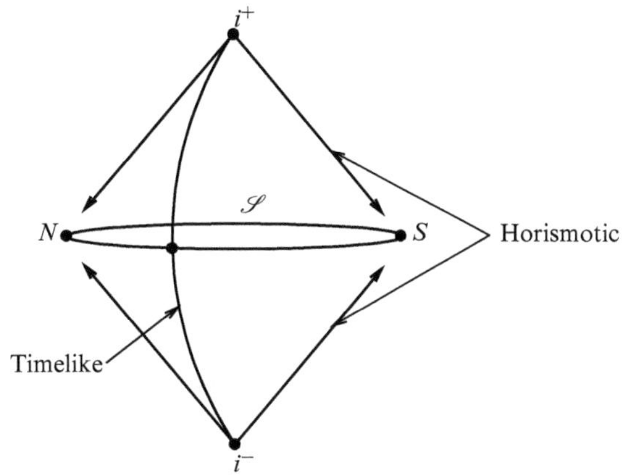

Fig. 4 Interior c-boundary of stationary Kerr spacetime. The ergosphere $\mathscr{S}$ has been deformed consistently with the causal character of the c-boundary

Consider any continuous function $\mu_a : \mathbb{R} \times (0,\infty) \to [0,1]$, $a \in \mathbb{R}$, satisfying

$$\mu_a(x,y) = \begin{cases} 0 & \text{if } |x-a| \geq \frac{1}{1+y}, \\ 1 & \text{if } x = a. \end{cases}$$

Given the curve $c_a : (0,\infty) \to \mathbb{R}^2, c_a(s) = (a,s)$, define the *chimney associated to* c_a as the region of $\mathbb{R} \times (0,\infty)$ given by:

$$\mathrm{Ch}_a = \left\{ (x,y) \in \mathbb{R} \times (0,\infty) : |x-a| < \frac{1}{1+y} \right\} \qquad \text{(gray region in Fig. 5).} \quad (39)$$

One Chimney

Consider the half-cylinder $M = ([-6,6]/R) \times (0,\infty)$, where the unique nontrivial relation is $-6R6$, and endow it with the Riemannian metric $g_a = dx^2 + e^{-\mu_a(x,y)y}dy^2$ for some $a \in (-6,6)$. Then, the Riemannian manifold (M, g_a) satisfies the following properties[4]:

(i) The inextensible curve $c_a : [0,\infty) \to M, c_a(s) = (a,s)$, $a \in [-6,6]/R$, has finite length, and thus, it defines a point z_a of the Cauchy boundary of (M, g_a). In fact,

$$\mathrm{length}(c_a) = \int_0^\infty \sqrt{g_a(\dot{c}_a(s), \dot{c}_a(s))}ds = \int_0^\infty e^{-s}ds = 1.$$

[4]For simplicity, the elements $[x] \in [-6,6]/R$ will be denoted by x.

(ii) The Cauchy boundary of (M, g_a) is formed by two connected components: one component naturally identifiable with $([-6,6]/R) \times \{0\}$ and the other one formed by the single point $\{z_a\}$.

If $c : [\alpha, \Omega) \to M, c(s) = (x(s), y(s))$ is any inextensible curve of finite length, then either $y(s) \to 0$ and $x(s)$ converges in $[-6,6]/R$, or $y(s) \to \infty$, $x(s) \to a$, and $c(s)$ converges to z_a. In fact, as length$(c) < \infty$, $x(s)$, $y(s)$ must converge (perhaps to ∞ in the case of $y(s)$). As $x(s)$ takes values in $[-6,6]/R$, necessarily $x(s)$ converges to some $x \in [-6,6]/R$. For $y(s)$, there are two possibilities: either $y(s) \to 0$ or $y(s) \to \infty$. In the second case, $x(s) \to a$, since, otherwise, $\mu_a(c(s)) \equiv 0$ for all s big enough; thus g_a becomes the Euclidean metric on $c(s)$, and, taking into account that $y(s) \to \infty$, necessarily length$(c(s)) = \infty$, a contradiction. Moreover:

$$\begin{aligned}\lim_{s\to\Omega} d(c(s), z_a) &\leq \lim_{s\to\Omega} \left(d(c(s), (a, y(s))) + d((a, y(s)), z_a)\right) \\ &= \lim_{s\to\Omega} d((x(s), y(s), (a, y(s))) + \lim_{s\to\Omega} d((a, y(s)), z_a) \\ &= 0.\end{aligned}$$

(iii) The diameter of (M, g_a) is finite, and the Cauchy, Busemann and Gromov completions of (M, g_a) coincide:

$$M_C = M_B = M_G.$$

In fact, for any pair of points $(x, y), (x', y') \in M$, consider the curve γ which joins (x, y) with (a, y) by means of a horizontal line, then (a, y) with (a, y') by a vertical line (thus, a segment of the curve c_a), and finally, (a, y') with (x', y') by a horizontal line. Obviously,

$$\text{length}(\gamma) \leq 6 + 6 + \text{length}(c_a) = 13.$$

For the last assertion, since the diameter of (M, g_a) is finite, if $c : [\alpha, \Omega) \to M$, $g_a(\dot{c}, \dot{c}) \leq 1$, is any curve defining a finite Busemann function, that is,

$$b_c(\cdot) = \lim_{s\to\Omega} (s - d(\cdot, c(s))) < \infty,$$

then Ω must be finite. Thus, also the length of c is finite, and so, $M_C = M_B$. As $M_B = M_C$ is Hausdorff, then $M_B = M_G$ (Theorem 4 (7)).

Two Chimneys

Next, consider the Riemannian manifold (M, g), formed by a half-cilinder M as before, but now endowed with the following metric:

$$g(= g_{-3,3}) = dx^2 + e^{-(\mu_{-3}(x,y) + \mu_3(x,y))y} dy^2.$$

(i) The Cauchy and Busemann completions of (M,g) coincide and are formed by three connected components: one component naturally identifiable with $([-6,6]/R) \times \{0\}$, and the other two being $\{z_{-3}\}$ and $\{z_3\}$.

This structure is deduced by arguing as in the case of one chimney. Note however that, in this case, we cannot say that these completions coincide also with the Gromov one. In order to study the Gromov boundary of (M,g), previously, we need to understand better the geometry of (M,g).

(ii) Estimation of the behavior of almost minimizing curves connecting (x_0,y_0) and (x,y) with $|y-y_0|$ big enough.

Taking into account the finiteness of the lengths of the vertical curves $c_{\pm 3}$ inside the chimneys $\mathrm{Ch}_{\pm 3}$, it is easy to deduce that any almost minimizing curve $\overline{\gamma}(s) = (\overline{x}(s), \overline{y}(s))$ connecting points $(x_0,y_0),(x,y)$, with $|y-y_0|$ big enough, must enter into some of these chimneys, then move inside the chimney until $\overline{y}(s)$ being close enough to y, and finally, reach the point (x,y) by an almost horizontal line. Observe that, if $\overline{s}^1 \in \mathbb{R}$ satisfies $\overline{\gamma}(\overline{s}^1) \in \mathrm{Ch}_{\pm 3}$ but $\overline{\gamma}(s) \notin \mathrm{Ch}_{\pm 3}$ for $s > \overline{s}^1$, we have that $|y - \overline{y}(\overline{s}^1)| \to 0$ when y diverges (see Fig. 5). In particular, any such a curve $\overline{\gamma}(s)$ can be modified for values of s bigger than $\overline{s}^1$ to provide a new curve $\gamma(s) = (x(s), y(s))$ such that: γ remains in the chimney also for $s \in [\overline{s}^1, s^1]$, where $s^1 > \overline{s}^1$ satisfies $y(s^1) = y$; γ reaches the point (x,y) by an horizontal line for $s \geq s^1$; and the difference between the lengths of γ and $\overline{\gamma}$ goes to zero as $y \to \infty$. As the chimneys $\mathrm{Ch}_{\pm 3}$ are getting narrower and narrower around the curves $c_{\pm 3}$ when the second coordinate increases, it is also

$$\lim_{y\to\infty} d(\gamma(s^1),(x,y)) = \lim_{y\to\infty} d((a,y),(x,y)) = d_0(a,x),$$

where d_0 denotes the standard distance in the circle $[-6,6]/R$. As we will be interested in estimating the distance between (x_0,y_0) and (x,y) as $y \to \infty$, we will assume without restriction that the almost minimizing curves between those points present the behavior described for γ.

(iii) For any $(x_0,y_0) \in M$ and any $\{(x_n,y_n)\}_n \subset M$ with $x_n \to x \in [-6,6]/R$ and $y_n \to \infty$, the following equality holds:

$$\lim_n d((x_0,y_0),(x_n,y_n)) = \lim_n d((x_0,y_0),(x,n)).$$

In fact, let $\gamma_n(s) = (x_n(s), y_n(s))$ be an almost minimizing curve from (x_0,y_0) to (x_n,y_n), which behavior for n big enough was described in previous item. For every n, denote by $m = m(n)$ the integer part of y_n (thus, $y_n \geq m$ for all n). Assume for example that, for infinitely many n, γ_n enters Ch_{-3}. Recalling the existence of some s_n^1 such that $y_n(s_n^1) = y_n$ (recall (ii)) and, so, taking into account the existence s_n^2 with $y_n(s_n^2) = m$, we can construct a curve β_n joining (x_0,y_0) with (x,m) in the following way: β_n joins (x_0,y_0) with $\gamma_n(s_n^2)$ by means of γ_n and $\gamma_n(s_n^2)$ with (x,m) by means of a distance-minimizing horizontal line.

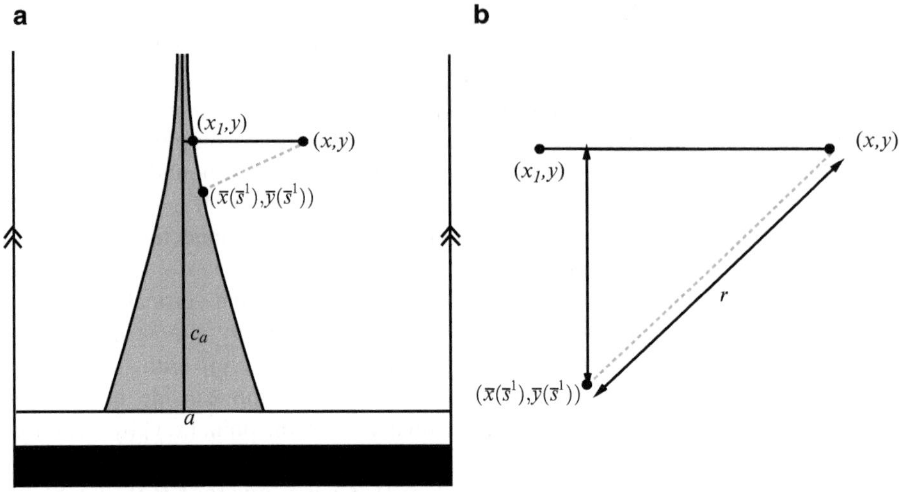

Fig. 5 Behaviour of the almost length-minimizing curves. Take the metric g_a as in the one chimney case and suppose that the *dotted line* above comes from an almost length-minimizing curve $\overline{\gamma}: I \to M, \overline{\gamma}(s) = (\overline{x}(s), \overline{y}(s))$ joining two points (x_0, y_0) and (x,y) (recall that, out of the *grey zone*, the metric g_a becomes Euclidean, and so, length-minimizing curves are lines). Consider $\overline{s}^1 \in I$ such that $\overline{\gamma}(\overline{s}^1) \in \mathrm{Ch_a}$ and $\overline{\gamma}(s) \notin \mathrm{Ch_a}$ for $s > \overline{s}^1$. As the diameter of (M,g) is bounded and the Euclidean distance r between $(\overline{x}(\overline{s}^1), \overline{y}(\overline{s}^1))$ and (x,y) satisfies $r^2 \geq (y - \overline{y}(\overline{s}^1))^2$, we deduce that $|y - \overline{y}(\overline{s}^1)|$ must be also bounded. In particular, if $y \to \infty$ then $\overline{y}(\overline{s}^1) \to \infty$. Therefore, as $(\overline{x}(\overline{s}^1), \overline{y}(\overline{s}^1))$ and (x_1, y) belong to the chimney $\mathrm{Ch_a}$ (see Definition 39), the terms $|x_1 - a|$ and $|x_1 - \overline{x}(\overline{s}^1)|$ converge to 0 when y diverges. So, it is not a restriction to assume that $r^2 \approx (x-a)^2 + (y - \overline{y}(\overline{s}^1))^2$ for y big enough. On the other hand, consider the curve σ formed by: a horizontal line from $(\overline{x}(\overline{s}^1), \overline{y}(\overline{s}^1))$ to $(a, \overline{y}(\overline{s}^1))$, a vertical line from $(a, \overline{y}(\overline{s}^1))$ to (a,y) and, finally, a horizontal line from (a,y) to (x,y). Observe that $\mathrm{length}(\sigma) = |\overline{x}(\overline{s}^1) - a| + |x - a| + \mathrm{length}(c_a\,|_{[\overline{y}(\overline{s}^1), y]})$. Thus, taking into account previous reasoning and the fact that $\mathrm{length}(c_a) < \infty$ (recall the definition of g_a), it is deduced that $\lim_{y \to \infty} \mathrm{length}(\sigma) = |x - a|$. Finally, as the *dotted line* comes from an almost length-minimizing curve, it is not restrictive to assume that $r \leq \mathrm{length}(\sigma)$, and so, we obtain that $|y - \overline{y}(\overline{s}^1)|$ converges to 0 when y diverges

Then,

$$\begin{aligned}\mathrm{length}(\beta_n) &= \mathrm{length}(\gamma_n|_{[0,s_n^2)}) + d(\gamma_n(s_n^2), (x,m)) \\ &= \mathrm{length}(\gamma_n|_{[0,s_n^1)}) + d(\gamma_n(s_n^2), (x,m)) - \mathrm{length}(\gamma_n|_{(s_n^2, s_n^1)}).\end{aligned}$$

So, taking the limit $n \to \infty$,

$$\begin{aligned}\lim_n d((x_0,y_0),(x,m))) &\leq \lim_n \mathrm{length}(\beta_n) \\ &\leq \lim_n \mathrm{length}(\gamma_n) \\ &\quad + \lim_n \left(d(\gamma_n(s_n^2),(x,m)) - d(\gamma_n(s_n^1),(x_n,y_n)) \right) \\ &\quad - \lim_n \mathrm{length}(\gamma_n|_{(s_n^2,s_n^1)}).\end{aligned}$$

From the almost minimizing character of γ_n,

$$\lim_n \text{length}(\gamma_n) = \lim_n d((x_0,y_0),(x_n,y_n)).$$

Moreover, as $\gamma_n(s_n^2)$ and $\gamma_n(s_n^1)$ lie in the chimney Ch_{-3} and both y_n and m diverge when $n \to \infty$ (recall that $m \equiv m(n)$ is the integer part of y_n), necessarily,

$$\lim_n \left(d(\gamma_n(s_n^2),(x,m)) - d(\gamma_n(s_n^1),(x_n,y_n))\right) = d_0(-3,x) - d_0(-3,x) = 0.$$

Finally, as $\gamma_n|_{(s_n^2,s_n^1)} \subset \text{Ch}_{-3}$ is almost minimizing and $\text{length}(c_{-3}) < \infty$,

$$\lim_n \text{length}(\gamma_n|_{(s_n^2,s_n^1)}) = 0.$$

Summarizing, we obtain

$$\lim_n d((x_0,y_0),(x,m)) \leq \lim_n d((x_0,y_0),(x_n,y_n)).$$

The other inequality is deduced analogously.

(iv) The Gromov boundary contains a region identifiable with subset $(-3,3) \subset \mathbb{R}$.
From the compactness of M_G, some subsequence of $\{[-d(\cdot,(x,n))]\}_n$, for any $x \in (-3,3)$, determines a point z_x in the Gromov boundary of (M,g). Moreover, item (iii) ensures that $z_x = \lim_n\{[-d(\cdot,(x_n,y_n))]\}_n$ for any sequence (x_n,y_n) with $x_n \to x$ and $y_n \to \infty$. On the other hand, assume for example that $x,x' \in (-3,3)$, $x < x'$. Taking into account the behavior of the almost minimizing curves described in (ii), one deduces

$$\lim_n \left(d(c_{-3}(1),(x',n)) - d(c_{-3}(1),(x,n))\right) > 0.$$

Analogously,

$$\lim_n \left(d(c_3(1),(x,n)) - d(c_3(1),(x',n))\right) > 0.$$

In conclusion, $\lim_n(d(\cdot,(x,n)) - d(\cdot,(x',n)))$ is not a constant, and so, the sequences $\{[-d(\cdot,(x,n))]\}_n$, $\{[-d(\cdot,(x',n))]\}_n$ determine different points $z_x \neq z_{x'}$ of the Gromov boundary.

(v) For any $0 \leq |\delta| \leq 3$ and any $a \in \{-3,3\}$, it is $z_{a+\delta} = z_{a-\delta}$.
In fact, consider the sequences $\{(a+\delta,n)\}_n$,$\{(a-\delta,n)\}_n$ in M. From our previous considerations, the value of $\lim_n d((x_0,y_0),(x,n))$ only depends on $|x_0-a|$ and $|x-a|$. So, if we consider the sequences $\{(a+\delta,n)\}_n$,$\{(a-\delta,n)\}_n$, then the limits

$$\lim_n d((x_0,y_0),(a+\delta,n)), \quad \lim_n d((x_0,y_0),(a-\delta,n))$$

will coincide, and so, $z_{a+\delta} = z_{a-\delta}$.

(vi) The space $\partial_G M \setminus \partial_B M$ can be identified with the set $\{z_x : x \in (-3,3)\}$.
In fact, any $z \in \partial_G M \setminus \partial_B M$ can be expressed as $z = \lim_n \{[-d(\cdot,(x_n,y_n))]\}$ for some sequence $\{(x_n,y_n)\}_n \subset M$ such that $x_n \not\to \pm 3$, $y_n \to \infty$. Then, $x_n \to x$ for some $x \in (-3,3)$, and thus, $z = z_x$.

Summarizing, the Gromov boundary is formed by two natural sets: the set of points which are limits of sequences with second component converging to 0, naturally identifiable with the circle $[-6,6]/R$, and the rest of points (i.e., when the second component converges to ∞), which are identifiable with the interval $[-3,3]$. Hence,

$$\partial_G M \equiv ([-6,6]/R) \cup [-3,3].$$

In particular, the Gromov and Busemann completions do not coincide, and, from Theorem 4 (7), the Busemann completion cannot be Hausdorff; for instance, the sequence $[-d_{(0,n)}]$ with $n \in \mathbb{N}$ converges to both, $[-d_{z_{-3}}], [-d_{z_3}] \in \partial_C M$.

Acknowledgements The authors acknowledge Prof. Miguel Sánchez the valuable supervision of this work, and the comments by the referee. The authors are partially supported by the Spanish MICINN Grant MTM2010-18099 and Regional J. Andalucía Grant P09-FQM-4496, both with FEDER funds.

References

1. Alaña, V., Flores, J.L.: The causal boundary of product spacetimes. Gen. Relat. Gravit. **39**(10), 1697–1718 (2007)
2. Beem, J.K., Ehrlich, P.E., Easley, K.L.: Global Lorentzian geometry. In: Monographs Textbooks Pure Appl. Math. vol. 202. Dekker, New York (1996)
3. Budic, R., Sachs, R.K.: Causal boundaries for general relativistic spacetimes. J. Math. Phys. **15**, 1302–1309 (1974)
4. Caponio, E., Javaloyes, M.A., Sánchez, M.: On the interplay between Lorentzian Causality and Finsler metrics of Randers type. Rev. Matem. Iberoamericana **27**, 919–952 (2011)
5. Eberlein, P., O'Neill, B.: Visibility manifolds. Pacific J. Math. **46**, 45–109 (1973)
6. Flores, J.L.: The Causal Boundary of spacetimes revisited. Commun. Math. Phys. **276**, 611–643 (2007)
7. Flores, J.L., Harris, S.G.: Topology of the causal boundary for standard static spacetimes. Class. Quant. Grav. **24**(5), 1211–1260 (2007)
8. Flores, J.L., Herrera, J., Sánchez, M.: On the final definition of the causal boundary and its relation with the conformal boundary. Adv. Theor. Math. Phys. **15**(4), 991–1058 (2011) Available at arXiv:1001.3270v3 [math-ph]
9. Flores, J.L., Herrera, J., Sánchez, M.: Gromov, Cauchy and causal boundaries for Riemannian, Finslerian and Lorentzian manifolds. Memoirs of the AMS, To appear. Available at arXiv:1011:1154v3 [math.DG]
10. Geroch, R.P., Kronheimer, E.H., Penrose, R.: Ideal points in spacetime. Proc. Roy. Soc. Lond. A **237**, 545–567 (1972)
11. Gromov, M.: Metric structures for Riemannian and non-Riemannian spaces. In: Progress in Mathematics, vol. 152. Birkhäuser, Boston (1999)
12. Harris, S.G.: Causal boundary for standard static spacetimes. Nonlinear Anal. **47**, 2971–2981 (2001)

13. Hawking, S.W., Ellis, G.F.R.: The Large Scale Structure of SpaceTime. Cambridge University, Cambridge (1973)
14. Marolf, D., Ross, S.R.: A new recipe for causal completions. Class. Quant. Grav. **20**, 4085–4117 (2003)
15. Minguzzi, E., Sánchez, M.: The causal hierarchy of spacetimes. In: Recent developments in pseudo-Riemannian Geometry, pp. 299–358 (2008). ESI Lect. in Math. Phys., European Mathematical Society Publishing House. (Available at gr-qc/0609119)
16. O'Neill, B.: Semi-Riemannian Geometry with Applications to Relativity. Academic, New York (1983)
17. Sánchez, M.: The causal boundary of a spacetime and its relation with the classical Gromov and Cauchy boundaries. Plenary contribution at the Int. Meeting on Differential Geometry, Córdoba, Nov. 15–17, 2010
18. Szabados, L.B.: Causal boundary for strongly causal spaces. Class. Quant. Grav. **5**, 121–134 (1988)
19. Szabados, L.B.: Causal boundary for strongly causal spacetimes: II. Class. Quant. Grav. **6**, 77–91 (1989)
20. Wald, R.M.: General Relativity. University of Chicago Press, Chicago (1984)

On the Isometry Group of Lorentz Manifolds

Leandro A. Lichtenfelz, Paolo Piccione, and Abdelghani Zeghib

1 Riemannian Versus Lorentzian Isometries

In Lorentzian geometry, one often profits from first studying what happens in the Riemannian setting and then examining the analogous situation in the Lorentzian setting, since the former is generally better understood. This approach is also the source of interesting comparisons, as we'll see. We begin by reviewing some properties of isometries.

If M is a smooth manifold, a *frame at* $x \in M$ is just a linear isomorphism $p : \mathbb{R}^n \to T_xM$. The set of all such frames has the structure of a $GL(n)$-principal bundle and will be denoted by $L(M)$.

A general metric g on M of index $0 \le k \le n$ determines a natural reduction of $L(M)$ by selecting, at each $x \in M$, those frames p satisfying

$$g_x(p(u), p(v)) = \langle u, v \rangle_k, \ \forall u, v \in \mathbb{R}^n, \tag{1}$$

where $\langle \cdot, \cdot \rangle_k$ denotes the canonical scalar product of index k on $\mathbb{R}^n$, and we denote this reduction by $L_g(M)$. This is known as the *g-orthogonal frame bundle*, and its elements are *orthogonal frames*. $L_g(M)$ is a $O(n,k)$-principal subbundle of $L(M)$. The isometry group $\mathrm{Iso}(M,g)$ is just the set of all diffeomorphisms of M which preserve the fibers of $L_g(M)$, that is, if p is an orthogonal frame at x and $\phi \in \mathrm{Iso}(M,g)$, then $d\phi(x) \circ p$ is an orthogonal frame at $\phi(x)$.

L.A. Lichtenfelz • P. Piccione (✉)
Departamento de Matemática, Universidade de São Paulo, Rua do Matão 1010, CEP 05508-900, São Paulo, SP Brazil
e-mail: silverratio@gmail.com; piccione.p@gmail.com

A. Zeghib
Unité de Mathématiques Pures et Appliquées École Normale Supérieure de Lyon allée d'Italie LYON Cedex 07 France
e-mail: abdelghani.zeghib@ens-lyon.fr

M. Sánchez et al. (eds.), *Recent Trends in Lorentzian Geometry*, Springer Proceedings in Mathematics & Statistics 26, DOI 10.1007/978-1-4614-4897-6_12,

We point out that $\text{Iso}(M,g)$ has a Lie group structure when considered with the compact-open topology. For Riemannian metrics, this has been established (long ago) in [6]. However, the techniques employed there do not generalize to semi-Riemannian metrics. Therefore, in the next section, we follow a construction outlined in [10], with some changes, which yields a general proof along with a detailed description of the underlying topology.

Recall that the isometry group of $(\mathbb{R}^n, g_0)$, where g_0 is the standard, flat Riemannian metric, is $\mathbb{R}^n \rtimes O(n)$, where $\mathbb{R}^n$ acts on itself by translations. The linear part of this group is $O(n)$, which is compact and has two connected components. Thus, if we compactify $\mathbb{R}^n$ by making a quotient relative to the action of some suitable subgroup of the translations, the new isometry group of this quotient will be compact.

It turns out that this compactness of the isometry group is characteristic of all compact Riemannian manifolds.

Theorem 1. *If* (M,g) *is a compact Riemannian manifold, then* $\text{Iso}(M,g)$ *is compact.*

Proof. In this case, $L_g(M)$ is compact since the base and the fibers (diffeomorphic to $O(n)$) are compact. Choose an arbitrary $p \in L_g(M)$ at $x \in M$, and consider the *orbit*

$$\mathcal{O}_p = \{d\phi(x) \circ p : \phi \in \text{Iso}(M,g)\}. \tag{2}$$

This is a closed subset of $L_g(M)$. Moreover, the map $\text{Iso}(M,g) \ni \phi \mapsto d\phi(x) \circ p$ is a homeomorphism, since isometries are determined by their value and derivative at any given point. □

In the flat Lorentzian case, the isometry group of $(\mathbb{R}^n, g_{LM})$, where g_{LM} is the flat Lorentz–Minkowski metric, is $\mathbb{R}^n \rtimes O(n,1)$. Again, $\mathbb{R}^n$ acts by translations, and $O(n,1)$ is the group

$$O(n,1) = \{T \in GL(n) : T^t L T = L\}, \tag{3}$$

where L is the diagonal matrix (l_{ij}) such that $l_{11} = -1$ and $l_{ii} = 1$ for $i > 1$. This time the linear part, $O(n,1)$, is not compact and has *four* connected components. Moreover, theorem 1 fails dramatically in the Lorentzian case, as the next example shows.

Example 1. Take $A \in \text{GL}(2,\mathbb{Z})$ hyperbolic (i.e., with real eigenvalues of absolute value different from 1). Then, A preserves a Lorentz form g on $\mathbb{R}^2$, having its eigendirections as isotropic directions. Moreover, A preserves the lattice $\mathbb{Z}^2$. Thus, A defines an isometry of the quotient $\mathbb{T}^2 = \mathbb{R}^2/\mathbb{Z}^2$ endowed with the Lorentz metric $\overline{g}$ induced by g. The powers A^n, $n \in \mathbb{Z}$, make an unbounded sequence in $\text{Iso}(\mathbb{T}^2, \overline{g})$, hence is non compact.

2 G-Structures

Definition 1. For a Lie subgroup $G \subseteq GL(n)$, a G-structure on M is a G-principal subbundle $\pi : B_G \to M$ of $L(M)$.

A useful characterization is this: given $p \in B_G$ and $g \in GL(n)$, we have $p \circ g \in B_G$ if and only if $g \in G$.

Example 2. When G is trivial, $G = \{1\}$, giving a G-structure on M is equivalent to choosing a single frame at each point of M.

Example 3. For $G = O(n,k)$, G-structures are the same as semi-Riemannian metrics of index k. In fact, if (M,g) is a semi-Riemannian manifold of index k, then $L_g(M)$ is the $O(n,k)$-structure. Conversely, if we are given any $O(n,k)$-structure $\pi : B_G \to M$, define, at each $x \in M$, the metric

$$g_x(u,v) = \langle p^{-1}(u), p^{-1}(v) \rangle_k, \ \forall u,v \in T_xM, \tag{4}$$

where p is *any* element of $\pi^{-1}(x)$. This is well defined: Any other $q \in \pi^{-1}(x)$ gives the same result, because their difference $q^{-1} \circ p$ belongs to $O(n,k)$.

Example 4. G-structures for $G = CO(n,k)$ are conformal structures of index k. They do not define a metric, but an equivalence class of metrics, all conformally related.

For other examples, see [10].

2.1 First-Order Structure Function

One of the most fundamental invariants of a G-structure is the so-called *first-order structure function*, which we now proceed to define. This function will also be used to construct certain *prolongations* of G-structures, that end up being important to build the Lie group structure of the group of automorphisms of certain G-structures.

Let $\pi : B_G \to M$ be a G-structure. There exists a canonical vector-valued 1-form θ on B_G, given by

$$\theta_p(X) = p^{-1} \circ d\pi_p(X), \ \forall X \in T_p(B_G), \ \forall p \in B_G. \tag{5}$$

Thus, the exterior derivative of θ is the vector-valued 2-form $d\theta : T(B_G) \wedge T(B_G) \to \mathbb{R}^n$. Recall that, being a fiber bundle, B_G has a distinguished family of subspaces $\mathcal{V}_p \subseteq T_p(B_G)$ for each $p \in B_G$, called the *vertical* space. The vertical space is usually defined by $\mathcal{V}_p = \ker(d\pi_p)$; here, it will be convenient to think of $\mathcal{V}_p$ as the linear map

$$\mathcal{V}_p \equiv d(\beta_p)_1 : \mathfrak{g} \to T_p(B_G), \tag{6}$$

where $\mathfrak{g}$ is the Lie algebra of G and $d(\beta_p)_1$ is the derivative of

$$\beta_p : G \to B_G \tag{7}$$

at $1 \in G$. The map β is just the action of G by right composition, $\beta_p(g) = p \circ g$. In this case, we have $\ker(d\pi_p) = \mathrm{Im}(\mathcal{V}_p)$.

A *horizontal complement* for $\mathcal{V}_p$, also called a *horizontal complement at* p, is a linear map $H_p : \mathbb{R}^n \to T_p(B_G)$ satisfying

$$\theta_p \circ H_p = 1, \tag{8}$$

where 1 is the identity map of $\mathbb{R}^n$. Clearly, if H_p is a horizontal complement for $\mathcal{V}_p$, then $\mathrm{Im}(H_p) \oplus \mathrm{Im}(\mathcal{V}_p) = T_p(B_G)$. The set of all horizontal complements at p, denoted by Hor_p, is an affine space in $\hom(\mathbb{R}^n, T_p(B_G))$. In fact, Hor_p is just a shift of $\hom(\mathbb{R}^n, \mathcal{V}_p)$, since the latter is the set of all linear maps $T : \mathbb{R}^n \to T_p(B_G)$ such that

$$\theta_p \circ T = 0. \tag{9}$$

Given $H \in \mathrm{Hor}_p$, consider the function

$$\begin{aligned} c_H &: \mathbb{R}^n \wedge \mathbb{R}^n \to \mathbb{R}^n \\ &u \wedge v \mapsto d\theta_p\left(H(u) \wedge H(v)\right). \end{aligned} \tag{10}$$

Let us investigate the dependence of c_H on H. Given $H_1, H_2 \in \mathrm{Hor}_p$, define

$$\begin{aligned} \mathcal{S}_{H_2,H_1} &: \mathbb{R}^n \to \mathcal{V}_p \\ &u \mapsto H_2(u) - H_1(u). \end{aligned} \tag{11}$$

Then, the following equation holds:

$$c_{H_2}(u \wedge v) - c_{H_1}(u \wedge v) = \mathcal{A}(\mathcal{S}_{H_2,H_1})(u \wedge v), \ \forall u, v \in \mathbb{R}^n, \tag{12}$$

where

$$\mathcal{A} : \hom(\mathbb{R}^n, \mathfrak{g}) \to \hom(\mathbb{R}^n \times \mathbb{R}^n, \mathbb{R}^n) \tag{13}$$

is the antisymmetrization operator, sending $T \in \hom(\mathbb{R}^n, \mathfrak{g})$ to

$$\begin{aligned} \mathcal{A}(T) &: \mathbb{R}^n \times \mathbb{R}^n \to \mathbb{R}^n \\ &u \wedge v \mapsto T(u)v - T(v)u. \end{aligned} \tag{14}$$

Hence, to eliminate the dependence of c_* on the horizontal complement, we define

$$\begin{aligned} c &: B_G \to \frac{\hom(\mathbb{R}^n \wedge \mathbb{R}^n, \mathbb{R}^n)}{\mathcal{A}(\hom(\mathbb{R}^n, \mathfrak{g}))} \\ &p \mapsto c_{H_p} + \mathcal{A}(\hom(\mathbb{R}^n, \mathfrak{g})), \end{aligned} \tag{15}$$

where $H_p \in \mathrm{Hor}_p$ is arbitrary.

Definition 2. The function c in Eq. (15) is the so-called *first-order structure function* of B_G.

Remark 1. When M is a Lie group itself, given a basis $\{X_1,\dots,X_n\}$ of T_1M, a natural parallelism can be defined on M by extending this basis to left-invariant vector fields. The structure function, c, of the corresponding G-structure gives the constants of structure of the Lie algebra $\mathfrak{m}$ of M with respect to the basis $\{X_1,\dots,X_n\}$, hence the name *structure function*.

Remark 2. c is an invariant of B_G. For some G-structures, however, this function is identically zero. This is the case for all $O(n,k)$-structures.

Remark 3. In some cases, c measures (obstruction to) integrability. For instance, when M is given an almost complex structure, c vanishes identically if and only if the Nijenhuis tensor does. By the celebrated theorem of Newlander-Nirenberg, this happens exactly when the almost complex structure is integrable; see [7].

2.2 *Prolongations*

Informally speaking, the *prolongations* of a Lie algebra $\mathfrak{g}$ corresponding to a given G-structure B_G on M are the spaces where the higher derivatives of automorphisms of B_G lie. This is most easily seen for G-structures on $\mathbb{R}^n$. We give the general definition, but we only prove what is strictly necessary for our purposes.

Choose $\mathcal{H} \subseteq \hom(\mathbb{R}^n \wedge \mathbb{R}^n, \mathbb{R}^n)$ such that

$$\mathcal{H} \oplus \mathcal{A}(\hom(\mathbb{R}^n, \mathfrak{g})) \simeq \hom(\mathbb{R}^n \wedge \mathbb{R}^n, \mathbb{R}^n). \tag{16}$$

This choice induces a family of special frames on B_G,

$$z : \mathbb{R}^n \times \mathfrak{g} \to T_p(B_G) \tag{17}$$

satisfying $z(A) = A$, for all $A \in \mathfrak{g}$ and $z(\mathbb{R}^n \times \{0\}) = H$, where H is such that $c_H \in \mathcal{H}$. These frames define a $G^{(1)}$-structure on B_G, called the *first prolongation of* B_G and denoted by $B_G^{(1)}$. Let us now see what the group $G^{(1)}$ looks like.

Consider the Lie algebra $\mathfrak{g}$ of G as a subspace of $\mathbb{R}^n \otimes (\mathbb{R}^n)^*$. Then, define the first prolongation of $\mathfrak{g}$ by

$$\mathfrak{g}^{(1)} = (\mathfrak{g} \otimes (\mathbb{R}^n)^*) \cap \left(\mathbb{R}^n \otimes \mathcal{S}^2(\mathbb{R}^n)^*\right), \tag{18}$$

where $\mathcal{S}^k$ denotes the symmetric k-tensors. Define, inductively, the ith prolongation of $\mathfrak{g}$ by

$$\mathfrak{g}^{(i)} = \left(\mathfrak{g}^{(i-1)} \otimes (\mathbb{R}^n)^*\right) \cap \left(\mathbb{R}^n \otimes \mathcal{S}^i(\mathbb{R}^n)^*\right). \tag{19}$$

Remark 4. It is often useful to think of $\mathfrak{g}^{(i)}$ as the space of multilinear symmetric maps $T : (\mathbb{R}^n)^{i+1} \to \mathbb{R}^n$ such that

$$\forall v_1, \ldots, v_i \in \mathbb{R}^n,\ v \mapsto T(v_1, \ldots, v_i, v) \in \mathfrak{g}. \tag{20}$$

We shall do so below, when working with some familiar groups.

Definition 3. The group $G^{(1)}$ consists of all the matrices of the form

$$\begin{bmatrix} 1_{\mathbb{R}^n} & 0 \\ T & 1_{\mathfrak{g}} \end{bmatrix},$$

where 1_* denotes the identity operator in the given space and $T \in \mathfrak{g}^{(1)}$. Because this is a subgroup of $GL(N)$ ($N = n + \dim(\mathfrak{g})$), it makes sense to define inductively $G^{(i+1)} = (G^{(i)})^{(1)}$, for $i \geq 1$.

Remark 5. In general, the Lie algebra of $G^{(i)}$ is just a representation of $\mathfrak{g}^{(i)}$. This means that if $\mathfrak{g}^{(i)} = \{0\}$ for some i, then $G^{(k)} = \{1\}$ for all $k \geq i$.

Definition 4. A G-structure is said to be of *finite type* if $\mathfrak{g}^{(i)} = \{0\}$ for some i. The smallest such i is called the *order* of the G-structure. As this definition depends only on the group G and not on the actual G-structure, the terminology carries over to the group itself.

Remark 6. Given any $k \in \mathbb{N}$, there exists $m \in \mathbb{N}$ and $G \subseteq GL(m)$ such that G is of finite type and has order k. We outline a construction of such an algebra for $k > 2$, since we already have examples for $k = 1, 2$. Given k, let $\{e_1, \ldots, e_{k+3}\}$ be the canonical basis of $\mathbb{R}^{k+3}$. Define $\mathfrak{g}_k$ to be the subalgebra of $\mathfrak{gl}(k+3)$ generated by $l_1, \ldots, l_k$, defined as

$$l_i = e_{i+3} \otimes e^1 + e_{i+2} \otimes e^2, \tag{21}$$

using upper indices for the dual basis. The first prolongation, $\mathfrak{g}_k^{(1)}$, will then have a basis consisting of the vectors $l_2^{(1)}, \ldots, l_k^{(1)}$, given by

$$l_i^{(1)} = e_{i+3} \otimes (e^1)^2 + e_{i+2} \otimes e^1 \otimes e^2 + e_{i+2} \otimes e^2 \otimes e^1 + e_{i+1} \otimes (e^2)^2. \tag{22}$$

Thus, $\mathfrak{g}_k^{(1)}$ has dimension $k-1$. It is easy to see from the above pattern that, in general, $\mathfrak{g}_k^{(j)}$ will have dimension $k-j$ for $0 \leq j \leq k$, and therefore $\mathfrak{g}_k$ has order exactly k.

3 Automorphism Groups

We now focus on the automorphism group of a G-structure. The theory we describe here works only for G-structures of *finite type*. Thus, it is important to know that some of the commonly encountered groups are of finite type. This is easy to establish.

Proposition 1. *The group $O(n,k)$ is of finite type and order* 1.

Proof. Let $u,v,w \in \mathbb{R}^n$ by any vectors and let $A \in \mathfrak{o}(n,k)^{(1)}$. Then,

$$\begin{aligned}\langle A(u,v),w\rangle_k &= \langle A(v,u),w\rangle_k = -\langle u,A(v,w)\rangle_k \\ &= -\langle u,A(w,v)\rangle_k = \langle A(w,u),v\rangle_k = \langle A(u,w),v\rangle_k \\ &= -\langle w,A(u,v)\rangle_k.\end{aligned} \tag{23}$$

This implies $A = 0$, which means that $O(n)^{(1)} = \{1\}$. □

Proposition 2. *For $n \geq 3$, $CO(n,k)$, the semi-Riemannian conformal group, is of finite type and order* 2.

Proof. First, we want to see that $\mathfrak{co}(n,k)^{(1)} \neq \{0\}$, so that the order of $CO(n,k)$ is not 1. Any $T \in \mathfrak{co}(n,k)^{(1)}$ can be realized as a bilinear map $T : \mathbb{R}^n \times \mathbb{R}^n \to \mathbb{R}^n$ such that

$$\langle T(u,v),w\rangle_k + \langle v,T(u,w)\rangle_k = \lambda_{T(u)}\langle v,w\rangle_k,\ \forall v,w \in \mathbb{R}^n. \tag{24}$$

Here, $\lambda_{T(u)}$ is the constant of conformality for the conformal map $T(u,*) : \mathbb{R}^n \to \mathbb{R}^n$. If we let u vary, we get a linear functional $\lambda_T : u \mapsto \lambda_{T(u)}$ associated with T. We use this functional to define the map

$$\begin{aligned}\Phi : \mathfrak{co}(n,k)^{(1)} &\to (\mathbb{R}^n)^* \\ T &\mapsto \lambda_T/2.\end{aligned} \tag{25}$$

This is actually a linear isomorphism; its inverse is given by

$$\begin{aligned}\Psi : (\mathbb{R}^n)^* &\to \mathfrak{co}(n,k)^{(1)} \\ u^* &\mapsto T_{u^*},\end{aligned} \tag{26}$$

where $T_{u^*} : \mathbb{R}^n \times \mathbb{R}^n \to \mathbb{R}^n$ is the bilinear map

$$T_{u^*}(v,w) = \langle u,v\rangle_k w + \langle u,w\rangle_k v - \langle v,w\rangle_k u,\ \forall v,w \in \mathbb{R}^n. \tag{27}$$

Thus, $\dim(\mathfrak{co}(n,k)^{(1)}) = n$, and in particular $\mathfrak{co}(n,k)^{(1)} \neq 0$. Now, given $T \in \mathfrak{co}(n,k)^{(2)}$, we have $T(u,v) \in \mathfrak{co}(n,k)$ for any $u,v \in \mathbb{R}^n$, which means that

$$\langle T(u,v)w,z\rangle_k + \langle w,T(u,v)z\rangle_k = \lambda_{T(u,v)}\langle w,z\rangle_k,\ \forall w,z \in \mathbb{R}^n. \tag{28}$$

The new map $\lambda_T : (v,w) \mapsto \lambda_{T(v,w)}$ is now bilinear, and if this map is zero, then Eq. (28) implies that T itself is zero. By the symmetry of λ_T, it suffices to show that $\lambda_{T(u,u)}$ is zero for all $u \in \mathbb{R}^n$. Choose any $u, v \in \mathbb{R}^n$ with $\{u,v\}$ orthonormal. Then,

$$\lambda_{T(u,u)} = \lambda_{T(u,u)}\langle v,v\rangle_k = 2\langle T(u,u)v,v\rangle_k = 2\langle T(u,v)u,v\rangle_k, \tag{29}$$

from the symmetry of T in the last two variables (recall that T is 3-linear; see remark 4). Since $T(u,v) \in \mathfrak{co}(n,k)$,

$$2\langle T(u,v)u,v\rangle_k = 2\lambda_{T(u,v)}\langle u,v\rangle_k - 2\langle u,T(u,v)v\rangle_k = -2\langle u,T(u,v)v\rangle_k. \tag{30}$$

Now, using the symmetry of T in the first and last variables,

$$-2\langle u,T(u,v)v\rangle_k = -2\langle u,T(v,v)u\rangle_k = \lambda_{T(v,v)}\langle u,u\rangle_k = -\lambda_{T(v,v)}. \tag{31}$$

Until this point, everything works even in dimension $n = 2$. If we now assume $n > 3$, given any $u \in \mathbb{R}^n$, $u \neq 0$, we may choose two other vectors v, w such that $\{u,v,w\}$ is orthonormal. Then, using the above calculations, we get

$$\lambda_{T(u,u)} = -\lambda_{T(v,v)} = \lambda_{T(w,w)} = -\lambda_{T(u,u)}, \tag{32}$$

so that λ_T is identically zero, as required. □

A more detailed discussion of this group along these lines can be found in [8].

Example 5. The group $Sp(n)$, consisting of all linear symplectomorphisms of $\mathbb{R}^n$ equipped with the canonical symplectic form, is of *infinite type*.

Back to generalities, we can now describe how the automorphism group of a G-structure is related to the automorphism groups of its prolongations.

Every diffeomorphism $\phi : M \to M$ generates a map

$$\begin{aligned} \phi_\star : L(M) &\to L(M) \\ p &\mapsto d\phi_{\pi(p)} \circ p, \end{aligned} \tag{33}$$

which is again a diffeomorphism and may or may not satisfy $\phi_\star(B_G) = B_G$. The automorphism group of B_G is the group

$$Aut_G(M) = \{\phi \in \mathrm{Diff}(M) : \phi_\star(B_G) = B_G\}. \tag{34}$$

Consider the sequence of principal bundles

$$M \overset{\pi}{\leftarrow} B_G \overset{\pi^1}{\longleftarrow} B_G^{(1)} \overset{\pi^2}{\longleftarrow} B_G^{(2)} \overset{\pi^3}{\longleftarrow} \dots \tag{35}$$

where each $B_G^{(i+1)}$ is the first prolongation of $B_G^{(i)}$. The interesting part is that, if $\phi \in Aut_G(M)$, then $\phi_\star \in Aut_{G^{(1)}}(B_G)$. The next theorem describes the situation more precisely.

Theorem 2. *The inclusions*

$$Aut_G(M) \hookrightarrow Aut_{G^{(1)}}(B_G) \hookrightarrow Aut_{G^{(2)}}(B_{G^{(1)}}) \hookrightarrow \dots \tag{36}$$

given by the map $\star$ are group monomorphisms. Moreover, each image is C^0-closed.

Proof. The main idea is to establish that

$$\star\,(Aut_G(M)) = \{\Phi \in Aut_{G^{(1)}}(B_G) : \Phi \text{ preserves fibres}\}. \tag{37}$$

This is not hard to do, and the rest of the proof is straightforward. □

Note that when $G = O(n,k)$, $B_G^{(1)}$ is already a 1-structure, that is, a parallelism. This yields an important conclusion, which we state separately as a corollary.

Corollary 1. *The isometry group of any semi-Riemannian manifold (M,g) can be realized as a C^0-closed subgroup of the automorphism group of a certain parallelism defined on $L_g(M)$.*

We now turn our attention to parallelisms.

3.1 The Automorphism Group of a Parallelism

Let M be a smooth, *connected* manifold with a complete parallelism $\{X_1, \dots, X_n\}$. Thus, $X_1, \dots, X_n$ are smooth vector fields such that $\{X_1(x), \dots, X_n(x)\}$ is a basis of T_xM at each $x \in M$. Let us denote simply by G the group of all automorphisms of the corresponding 1-structure. Hence, G consists of all maps $\phi \in \mathrm{Diff}(M)$ such that $d\phi_x(X_i(x)) = X_i(\phi(x))$, $i = 1, \dots, n$, for all $x \in M$. Finally, consider the action

$$\beta_x : G \to M \tag{38}$$

of G on M, given by $\beta_x(g) = g(x)$. Our main theorem, originally due to Kobayashi, is the following.

Theorem 3. *Given $x \in M$, the action $\beta_x : G \to M$ is a homeomorphism onto its image, the latter with the induced topology. Moreover, $\beta_x(G)$ is a closed submanifold of M, and the differentiable structure G inherits from it makes G a Lie group, now acting smoothly on M. This differentiable structure does not depend on the choice of x.*

Proof. The proof is rather long, so we sketch the three principal steps involved.

Step 1. Given $v = (\lambda_1, \ldots, \lambda_n) \in \mathbb{R}^n$, define the induced vector field

$$X_v(x) = \sum_{i=1}^{n} \lambda_i X_i(x). \tag{39}$$

Denote by $\exp(v)$ the time 1 flow of X_v, which is defined on a (possibly empty) open subset of M. Given $x, y \in M$, one can show that there always exists $v_1, \ldots, v_k$ such that

$$y = \exp(v_1) \circ \cdots \circ \exp(v_k)(x). \tag{40}$$

Using Eq. (40), it is easy to prove that G acts freely on M and that, given $x \in M$ and $g \in G$, if $\{g_i\}_{i\in\mathbb{N}}$ is a sequence on G such that $g_i(x) \to g(x)$, then $g_i \xrightarrow{C^0} g$.

Step 2. In order to show that $\beta_x(G)$ is closed, consider a sequence $\{g_i\}_{i\in\mathbb{N}}$ such that $g_i(x) \to q \in M$. Then, introduce a Riemannian metric $\hat{g}$ on M such that G becomes a subgroup of $\mathrm{Iso}(M, \hat{g})$ (this can be done, for instance, by declaring $\{X_1, \ldots, X_n\}$ to be an orthonormal basis). Since Riemannian isometries are equicontinuous, an application of Arzela-Ascoli yields a limit $g \in \mathrm{Iso}(M, \hat{g})$. This g actually belongs to G, since it will commute with the flows of the form $\exp(v)$ for $v \in \mathbb{R}^n$. It is then easy to see that $g(x) = q$, so that $\beta_x(G)$ is indeed closed.

Step 3. We first introduce the concept of *limit rays*. Let $\{x_i\}_{i\in\mathbb{N}}$ be a sequence in M converging to $x \in M$ such that $x_i \neq x$ for sufficiently large i. Given a local chart $\phi : U \subseteq M \to \mathbb{R}^n$ with $\phi(x) = 0$, we say that $\{x_i\}_{i\in\mathbb{N}}$ defines a *limit ray* at x if the sequence $\phi(x_i)/\|\phi(x_i)\|$, defined for large i, converges to some $u \in \mathbb{S}^{n-1}$. When this happens, we call the set $\{[d\phi_x]^{-1}(\lambda u) : \lambda > 0\} \subseteq T_xM$ the *limit ray* defined by the sequence $\{x_i\}_{i\in\mathbb{N}}$. Nothing in this definition depends on the specific coordinate chart used (as long as it is centered at x, evidently).

Consider the sets

$$\mathcal{D}_y = \{v \in T_yM : \exp(sv) \text{ exists and belongs to } \beta_y(G), \forall s \in \mathbb{R}\}, \tag{41}$$

for each $y \in M$. It is clear from the definition that if $v \in \mathcal{D}_y$, then $\lambda v \in \mathcal{D}_y$ for all $\lambda \in \mathbb{R}$. Moreover, one can show that $\mathcal{D}_y \backslash \{0\}$ equals exactly the set of all limit rays of all sequences $\{y_i\}_{i\in\mathbb{N}}$ in $\beta_y(G)$ converging to y (only those sequences that actually define a limit ray, of course). Then, given $v, w \in \mathcal{D}_y$ with $v + w \neq 0$, the flow

$$\begin{aligned} \eta : (-\varepsilon, \varepsilon) \times (-\varepsilon, \varepsilon) &\to M \\ (t, s) &\mapsto \exp(tw) \circ \exp(sv)(y), \end{aligned} \tag{42}$$

defined for small ε, is such that the sequence $y_i = \eta(1/i, 1/i)$, for large i, defines a limit ray containing $v + w$, thus showing that $\mathcal{D}_y$ is a vector space. Hence $\mathcal{D}$ is a distribution on M. Smoothness follows from the formula

$$\mathcal{D}_{\exp(v)(y)} = d(\exp(v))_y(\mathcal{D}_y), \tag{43}$$

for small v, which can be proved using limit rays. Finally, using the techniques outlined here and formula (40), it is possible to show that $\mathcal{D}$ is also involutive, which means that it is integrable.

Now the maximal, connected integral submanifold of $\mathcal{D}$ that intersect $\beta_x(G)$ are precisely the connected components of $\beta_x(G)$. Thus, $\beta_x(G)$ naturally inherits a differentiable structure from these submanifolds. Consequently, we may give G a differentiable structure, by pulling it back from $\beta_x(G)$ using β_x^{-1}. It is not hard to show, then, that this structure does not depend on x, and makes G a Lie group, acting smoothy on M. □

Now, let (M,g) be a semi-Riemannian manifold. As we have seen, $L_g(M)$ may be given a 1-structure, and we want to apply theorem 3 for this manifold. However, we do not know if $L_g(M)$ is connected, and this is crucial to the proof of theorem 3. Thus, we need to generalize the techniques above slightly. To do this, just observe that $L_g(M)$ has a finite number of connected components, say, $L_1,\dots,L_k$. A parallelism on $L_g(M)$ induces a parallelism for each one of these, whose automorphism group, $Aut_1(L_i)$, is a Lie group by theorem 3. Hence, the product

$$\mathbb{G}_0 = \prod_{i=1}^{k} Aut_1(L_i) \tag{44}$$

is also a Lie group with the compact-open topology. But $\mathbb{G}_0$ is a normal subgroup of the full automorphism group $Aut_1(L_g(M))$; thus, we may extend the differentiable structure (and the topology, at the same time) from $\mathbb{G}_0$ to $Aut_1(L_g(M))$ by declaring each map $\ell_g : G_0 \to gG_0$, $\ell_g(h) = gh$ to be a diffeomorphism. This leads to the main result of this section.

Corollary 2. *The isometry group* $\mathrm{Iso}(M,g)$ *is a Lie group, for any semi-Riemannian metric g on M.*

It is now clear that $\mathrm{Iso}(M,g)$ is a Lie group, but with the C^1 topology. This is because the map $\star$ used in theorem 2 takes derivatives. To show that the topology on $\mathrm{Iso}(M,g)$ coincides with the C^0 topology, we must prove that a sequence $\{\phi_i\}_{i\in\mathbb{N}}$ in $\mathrm{Iso}(M,g)$ converging C^0 to $\phi \in \mathrm{Iso}(M,g)$ actually converges C^1 to the same ϕ. To do this, consider the flows

$$\exp_x : U \subseteq \mathbb{R}^n \to M \tag{45}$$

given by $\exp_x(v) = \exp(v)(x)$ (in the notation of theorem 3). It's not hard to see that for each $x \in M$, there exists an open set $U \subseteq \mathbb{R}^n$ containing 0 such that Eq. (45) is defined. Now, every map from $\mathrm{Iso}(M,g)$ takes geodesics to geodesics, and then we have

$$\phi_n(\exp_x(v)) = \exp_{\phi_n(x)}\left(d(\phi_n)_x(v)\right),\ \forall v \in U. \tag{46}$$

This formula produces the equivalence between the topologies, as we wanted.

4 Groups Acting Isometrically on Compact Lorentz Manifolds

A systematic study of groups admitting an isometric action on Lorentzian or semi-Riemannian manifolds started in the 80s, by several authors, especially R. Zimmer, M. Gromov, and later N. Kowalsky, S. Adams, G. Stuck, and A. Zeghib. Let us recall some of the most important results, leading to the final classification due to Adams–Stuck and Zeghib in 1997.

Theorem 4 (Zimmer [12]). *Let (M,g) be a compact manifold, and let G be a connected Lie group acting by isometries locally faithfully on M. Then either of the two possibilities occur:*

- *G is locally isomorphic to $\mathrm{SL}(2,\mathbb{R}) \times K$, with K compact.*
- *G is* amenable *(equiv., compact modulo radical).*

Theorem 5 (Gromov [3]). *If (M,g) is a Lorentz manifold with finite volume, let G be a Lie group acting locally faithfully and isometrically on M. If the Zariski closure G^a of $\mathrm{Ad}(G) \subset \mathrm{Aut}(\mathfrak{g})$ has no proper co-compact normal algebraic subgroup, then:*

- *For almost every $m \in M$, the stabilizer $G_m \subset G$ is discrete.*
- *Either $\mathfrak{g} = \mathfrak{sl}(2,\mathbb{R}) \oplus$ abelian, or there exists a one-dimensional ideal $\mathfrak{I}_0 \subset \mathfrak{g}$ such that $\mathfrak{g}/\mathfrak{I}_0$ is abelian*

A breakthrough on the question of establishing the compactness of the isometry group of some classes of Lorentz manifolds was obtained by G. D'Ambra with the following result:

Theorem 6 (D'Ambra [2]). *If (M,g) is a compact, simply connected real-analytic Lorentz manifold, then $\mathrm{Iso}(M,g)$ is compact.*

In D'Ambra's theorem, it is important to observe that the result does not hold for general pseudo-Riemannian metrics of signature greater than 1. Also, the simple-connectedness assumption cannot be omitted in D'Ambra's theorem. On the other hand, it is an important open question whether one can extend D'Ambra's result to the C^∞-case:

Conjecture. *Compact simply connected Lorentzian manifolds have compact isometry group.*

A partial answer to the conjecture has been given in [9], see Corollary 4 below.

Let us recall that an action of a topological group G on a manifold M is said to be *proper* if for all compact subset $K \subset M$, the set $\{(g,x) \in G \times M : g \cdot x \in K\}$ is compact. Given a proper action, in particular, the stabilizer of every point is a compact subgroup of G.

Theorem 7 (Kowalsky [4]).

- *If G acts non properly on a compact Lorent manifold, and all G-stabilizers are discrete, then G is locally isomorphic to* $\mathrm{SL}(2,\mathbb{R})$.
- *In general, if G acts non properly, then G is locally isomorphic to either* $\mathrm{SO}(1,n)$ *or to* $\mathrm{SO}(2,n)$.

Finally, a complete classification of groups that act isometrically (locally faithfully) on compact Lorentz manifold was obtained simultaneously by Adams–Stuck and by Zeghib:

Theorem 8 (Adams and Stuck [1]; Zeghib [11]). *Let G be a connected, simply connected Lie group. The following are equivalent:*

- *G is the universal cover of identity connected component of the isometry group of some compact Lorentz manifold.*
- *G is isomorphic to $L \times K \times \mathbb{R}^d$, where K is compact and semisimple, $d \geq 0$, and L in the following list:*
  - $\widetilde{\mathrm{SL}(2,\mathbb{R})}$
  - *Heisenberg group* Heis^{2n+1}
  - *An* oscillator group.

Oscillator groups are characterized as the only (simply connected) solvable and non commutative Lie groups that admit a bi-invariant Lorentz metric and that possess a lattice, that is, a co-compact discrete subgroup [5]. More precisely, an oscillator group G is a semi-direct product $\mathbb{S}^1 \ltimes \mathrm{Heis}$, where Heis is a Heisenberg group (of some dimension $2d+1$). There are positivity condition on the eigenvalues of the automorphic $\mathbb{S}^1$ action on the Lie algebra $\mathfrak{heis}$ (ensuring the existence of a bi-invariant Lorentz metric), and arithmetic conditions on them (ensuring existence of a lattice).

5 Geometry of Manifolds with Non Compact Isometry Group

The geometry of compact Lorentz manifolds with non compact isometry group has been studied mostly by A. Zeghib and some collaborators. Basic references for the material exposed here are [9, 11].

The lack of compactness occurs in either one (not mutually exclusive) of the two situations: (a) the identity connected component of the isometry group is non compact and (b) the isometry group has infinitely many connected components. One has the following result:

Proposition 3 (Piccione and Zeghib [9]). *If (M,g) is a compact Lorentz manifold that admits a Killing vector field which is time like at some point, then (a) and (b) above are mutually exclusive.*

5.1 When the Identity Connected Component Is Not Compact

The geometric structure of compact Lorentz manifold whose isometry group contains a group which is locally isomorphic to an *oscillator group* or to $\mathrm{SL}(2,\mathbb{R})$ is studied in [11, Sect. 1.6]). A compact Lorentz manifold that admits a faithful isometric action of a group locally isomorphic to $\mathrm{SL}(2,\mathbb{R})$ has universal cover which is given by a warped product of the universal cover of $\mathrm{SL}(2,\mathbb{R})$, endowed with the bi-invariant Lorentz metric given by its Killing form and a Riemannian manifold. Every such manifold admits everywhere time-like Killing vector field, corresponding to the time-like vectors of the Lie algebra $\mathfrak{sl}(2,\mathbb{R})$.

Oscillator groups also admit a bi-invariant Lorentz metric and possess a lattice, that is, a co-compact discrete subgroup. From a Lorentz geometry viewpoint, it is useful to consider oscillator groups as objects completely similar to $\mathrm{SL}(2,\mathbb{R})$. Let us notice however, some differences (but with no incidence on our investigation here). First, of course, the bi-invariant Lorentz metrics on an oscillator group do not correspond to its Killing form, since this latter is degenerate (because the group is solvable). Another interesting fact is the nonuniqueness of these bi-invariant metrics but, surprisingly, their uniqueness up to automorphisms. In the $\mathrm{SL}(2,\mathbb{R})$-case, we have uniqueness up to a multiplicative constant. Also, we have essential uniqueness of lattices in an oscillator group, versus their abundance in $\mathrm{SL}(2,\mathbb{R})$.

Let us describe briefly the construction of Lorentz manifolds endowed with a faithful isometric G-action, where G is either $\mathrm{SL}(2,\mathbb{R})$ or an oscillator group. The construction starts by considering right quotients G/Γ, where Γ is a lattice of G. The G-left action is isometric exactly because the metric is bi-invariant. A slight generalization is obtained by considering a Riemannian manifold $(\widetilde{N},\widetilde{\mathbf{g}})$ and quotients of the direct metric product $X = \widetilde{N} \times G$ by a discrete subgroup Γ of $\mathrm{Iso}(\widetilde{N},\widetilde{\mathbf{g}}) \times G$. Observe here that since the isometry group of the Lorentz manifold X is $\mathrm{Iso}(\widetilde{N},\widetilde{\mathbf{g}}) \times (G \times G)$, it is possible to take a quotient by a subgroup Γ contained in this full group. The point is that we assumed G acting (on the left) on the quotient, and hence, G normalizes Γ; but since G is connected, it centralizes Γ. Therefore, only the right G factor in the full group remains (since the centralizer of the left action is exactly the right factor). Observe however that Γ does not necessarily split. Indeed, there are examples where Γ is discrete co-compact in $\mathrm{Iso}(\widetilde{N},\widetilde{\mathbf{g}}) \times G$, but its projection on each factor is dense!

Next, warped products yield a more general construction. Rather than a direct product metric $\widetilde{\mathbf{g}} \oplus \kappa$, one endows $\widetilde{N} \times G$ with a metric of the form $\widetilde{\mathbf{g}} \oplus w\kappa$, where w is a positive function on $\widetilde{N}$ and κ is the bi-invariant Lorentz metric on G. Such a metric passes to the quotient $(\widetilde{N} \times G)/\Gamma$ when the function w is Γ-invariant. Here, there is one difference between the case of $\mathrm{SL}(2,\mathbb{R})$ and the oscillator case. For $\mathrm{SL}(2,\mathbb{R})$, this is the more general construction, but in the oscillator case, some "mixing" between G and $\widetilde{N}$, and also a mixing of their metrics, is also possible; see [11, Sect. 1.2].

5.2 Amalgamated Products

Let us recall the notion of amalgamated product. Given any two manifolds X and Y carrying free (left) actions of the circle $\mathbb{S}^1$, then one can consider the diagonal action of $\mathbb{S}^1$ on the product $X \times Y$: $g(x,y) = (gx, gy)$ for all $g \in \mathbb{S}^1$, $x \in X$ and $y \in Y$. Let Z be the quotient $(X \times Y)/\mathbb{S}^1$ of this diagonal action. Assume that X is Lorentzian, Y is Riemannian, and the action of $\mathbb{S}^1$ in each manifold is isometric; one can define a natural Lorentzian structure on Z as follows. Let $A \in \mathfrak{X}(X)$ and $B \in \mathfrak{X}(Y)$ be smooth vector fields tangent to the fibers of the $\mathbb{S}^1$-action on X and on Y respectively; for $(x_0, y_0) \in X \times Y$, denote by $\big[(x_0, y_0)\big] \in Z$ the $\mathbb{S}^1$-orbit $\big\{(gx_0, gy_0) : g \in \mathbb{S}^1\big\}$. The subspace $T_{x_0}X \oplus B_{y_0}^\perp$ is complementary to the one-dimensional subspace spanned by $\big(A_{x_0}, B_{y_0}\big)$ in $T_{x_0}X \oplus T_{y_0}Y$. If we denote by $\mathfrak{q} : X \times Y \to Z$ the projection, then the linear map $\mathrm{d}\mathfrak{q}_{(x_0,y_0)} : T_{x_0}X \oplus T_{y_0}Y \to T_{[(x_0,y_0)]}Z \cong \big(T_{x_0}X \oplus T_{y_0}Y\big)/\mathbb{R}\cdot(A_{x_0}, B_{y_0})$ restricts to an isomorphism:

$$\mathrm{d}\mathfrak{q}_{(x_0,y_0)} : T_{x_0}X \oplus B_{y_0}^\perp \xrightarrow{\cong} T_{[(x_0,y_0)]}Z.$$

A Lorentzian metric can be defined on Z by requiring that such isomorphism be isometric; it is not hard to see that such a Lorentz structure on Z is well defined, that is, independent on the choice of (x_0, y_0) in the orbit $\big[(x_0, y_0)\big]$.

Using the long exact homotopy sequence for fibrations, one proves easily the following:

Proposition 4. *If X and Y are simply connected, then Z is simply connected. If the product of the fundamental groups $\pi_1(X) \times \pi_1(Y)$ is not a cyclic group, then Z is* not *simply connected.*

5.3 When the Isometry Group Has Infinitely Many Connected Components

The geometry of compact Lorentzian manifold whose isometry group has infinitely many connected components has been studied recently in [9], which is the main reference for the material presented here.

Let us consider the following situation. Let (M, g) be a compact Lorentz manifold, and let $G = \mathrm{Iso}(M, g)$ denote its full isometry group, $G_0 = \mathrm{Iso}_0(M, g)$ the identity connected component of G and $\mathfrak{g}$ its Lie algebra. We will consider the conjugacy action of G on $\mathfrak{g}$; we will say that this action is of *post-Riemannian type* if it preserves a positive–semidefinite inner product with degeneracy index (i.e., the dimension of the kernel) less than or equal to 1. Let us also denote by Γ the discrete part of G, that is, the quotient G/G_0

We make the following assumptions:

(a) G_0 is compact.
(b) The action of G_0 on M is not everywhere non-time-like, that is, there exists $x_0 \in M$ such that the orbit $G_0 \cdot x_0$ is time like.
(c) Γ is infinite.

Note that assumption (b) says that (M,g) possesses some Killing vector field which is time like at some point. Recall that a Lorentz manifold is said to be *stationary* when it admits an *everywhere* time-like Killing vector field. A first non-trivial fact is the following:

Proposition 5. *If the action of G on $\mathfrak{g}$ is not of post-Riemannian type, then assumption (b) holds.*

Example 6. A paradigmatic family of examples of the situation described above is given by flat tori: The dynamics and geometry of flat tori has the following flavor. Let q be a Lorentz form on $\mathbb{R}^k$; this induces a (flat) Lorentz metric on the torus $\mathbb{T}^k = \mathbb{R}^k/\mathbb{Z}^k$. The linear isometry group of $\mathbb{T}^k$ is $\mathrm{O}(q,\mathbb{Z}) = \mathrm{GL}(k,\mathbb{Z}) \cap \mathrm{O}(q)$, and its full isometry group is the semi-direct product $\mathrm{O}(q,\mathbb{Z}) \ltimes \mathbb{T}^k$.

The global and individual structure of $\mathrm{O}(q,\mathbb{Z})$ involves interesting geometric, arithmetic and dynamical interactions. For generic q, $\mathrm{O}(q,\mathbb{Z})$ is trivial. Nonetheless, if q is rational, i.e., if $q(x) = \sum a_{ij}x_ix_j$, where a_{ij} are rational numbers, then $\mathrm{O}(q,\mathbb{Z})$ is *big* in $\mathrm{O}(q)$; more precisely, by Harich-Chandra-Borel theorem, it is a lattice in $\mathrm{O}(q)$. When q is not rational, many intermediate situations are possible. It is a finite volume non-co-compact lattice in the case of the standard form $q_0 = -x_1^2 + x_2^2 + \cdots + x_k^2$ but can be co-compact for other forms. On the other hand, a given element $A \in \mathrm{O}(q_0,\mathbb{Z})$ could have complicated dynamics. For instance, if A is hyperbolic, then it has a leading simple eigenvalue. If furthermore A is irreducible, that is, it preserves no non-trivial sub-torus, then this eigenvalue is a Salem number. Conversely, any Salem number is the eigenvalue of such a hyperbolic $A \in \mathrm{O}(q,\mathbb{Z})$, for some rational Lorentz form q,

The main result proved in [9] says that under assumptions (a), (b), and (c) above, then (M,g) is *essentially* built up by tori. More precisely:

Theorem 9 (Piccione and Zeghib [9]). *Under assumptions (a), (b), and (c) above, then* $\mathrm{Iso}_0(M,g)$ *contains a torus $\mathbb{T} = \mathbb{T}^d$, endowed with a Lorentz form q, such that Γ is a subgroup of* $\mathrm{O}(q,\mathbb{Z})$.

Up to a finite cover, there is a new Lorentz metric g^{new} on M having a larger isometry group than g, such that the discrete factor Γ^{new} is given by $\mathrm{O}(q,\mathbb{Z})$ *(here* $\Gamma^{\mathrm{new}} = \mathrm{Iso}(M,g^{\mathrm{new}})/\mathrm{Iso}_0(M,g^{\mathrm{new}})$*).*

Geometrically, M is metric direct product $\mathbb{T} \times N$, where N is a compact Riemannian manifold or M is an amalgamated metric product $\mathbb{T} \times_{\mathbb{S}^1} L$, *where L is a light-like manifold with an isometric $\mathbb{S}^1$-action. The last possibility holds when Γ is a parabolic subgroup of* $\mathrm{O}(q)$.

We have two important corollaries. A first consequence of Theorem 9 is the following:

Corollary 3. *Assume that* (M, g) *is a compact Lorentzian manifold with infinite discrete factor* Γ*. If* (M, g) *has a* somewhere *time-like Killing vector field, then* (M, g) *has an* everywhere *time-like Killing vector field.*

Finally, using the structure results of compact Lorentz manifolds having non compact isometry group (Sect. 5.1, Theorem 9), Proposition 3 and Proposition 4, one obtains the following partial answer to the conjecture on the compactness of the isometry group:

Corollary 4. *Let* (M, g) *be a simply connected compact Lorentz manifold that admits a somewhere time-like Killing vector field. Then,* $\mathrm{Iso}(M, g)$ *is compact.*

References

1. Adams, S., Stuck, G.: The isometry group of a compact Lorentz manifold. I, II. Invent. Math. **129**(2), 239–261, 263–287 (1997)
2. D'Ambra, G.: Isometry groups of Lorentz manifolds. Invent. Math. **92**(3), 555–565 (1988)
3. Gromov, M.: Rigid transformations groups. Géométrie différentielle (Paris, 1986), 65–139, Travaux en Cours, 33, Hermann, Paris (1988)
4. Kowalsky, N.: Noncompact simple automorphism groups of Lorentz manifolds and other geometric manifolds. Ann. of Math. (2) **144**(3), 611–640 (1996)
5. Medina, A.; Revoy, P.: Les groupes oscillateurs et leurs réseaux. Manuscripta Math. **52**(1–3), 81–95 (1985)
6. Myers, S.B., Steenrod, N.: The group of isometries of a Riemannian manifold. Ann. of Math. **40**(400–416) (1939)
7. Newlander, A., Nirenberg, L.: Complex analytic coordinates in almost complex manifolds. Ann. of Math. (2) **(65)**, (1957)
8. Ogiue, K.: Theory of conformal connections. Kodai Math. Sem. Rep. **19**(193–224) (1967)
9. Piccione P., Zeghib, A.: On the isometry group and the geometric structure of compact stationary Lorentzian manifolds, preprint (2010) arXiv:1002.0814.
10. Sternberg, S.: Lectures on Differential Geometry. Prentice-Hall, Englewood Cliffs (1964)
11. Zeghib, A.: Sur les espaces-temps homogènes. The Epstein birthday schrift, pp. 551–576 (electronic), Geom. Topol. Monogr., vol. 1. Geom. Topol. Publ., Coventry (1998)
12. Zimmer, R. J.: On the automorphism group of a compact Lorentz manifold and other geometric manifolds. Invent. Math. **83**(3), 411–424 (1986)

Conformally Flat Homogeneous Lorentzian Manifolds

Kyoko Honda and Kazumi Tsukada

1 Introduction

We are interested in the classification problem of conformally flat homogeneous semi-Riemannian manifolds. Takagi [9] classified the Riemannian case. They are all symmetric spaces. In the previous paper [4], we classified three-dimensional conformally flat homogeneous Lorentzian manifolds. There are some examples which are not symmetric spaces. Calvaruso [1] also obtained the classification of three-dimensional conformally flat homogeneous Lorentzian manifolds independently and by different arguments. In this chapter we try to classify higher-dimensional conformally flat homogeneous Lorentzian manifolds.

Let M_q^n be an n-dimensional semi-Riemannian manifold equipped with the semi-Riemannian metric $\langle , \rangle$ of index q. We denote by ∇ the Levi-Civita connection of M and by R, Q and S the curvature tensor, the Ricci operator and the scalar curvature of M, respectively. Moreover we define a tensor field A of the symmetric endomorphisms on M by

$$A = \frac{1}{n-2}\left\{ Q - \frac{S}{2(n-1)} Id \right\}, \tag{1}$$

where Id denotes the identity transformation. Then it is known that M is conformally flat if and only if the curvature tensor R of M is related to A by the following equation for $n \geq 4$:

$$R(X,Y) = AX \wedge Y + X \wedge AY, \tag{2}$$

K. Honda • K. Tsukada (✉)
Department of Mathematics, Ochanomizu University, 2-1-1 Otsuka Bunkyo-ku Tokyo, 112–8610, Japan
e-mail: honda.kyoko@ocha.ac.jp; tsukada.kazumi@ocha.ac.jp

M. Sánchez et al. (eds.), *Recent Trends in Lorentzian Geometry*, Springer Proceedings in Mathematics & Statistics 26, DOI 10.1007/978-1-4614-4897-6_13,

where $X \wedge Y$ denotes the endomorphism defined by $(X \wedge Y)(Z) = \langle Y,Z\rangle X - \langle X,Z\rangle Y$. Moreover, the Eq. (2) implies

$$(\nabla_X A)Y = (\nabla_Y A)X. \tag{3}$$

Since the homogeneity of a semi-Riemannian metric does not imply the geodesic completeness, we think that a global classification is a difficult problem. In this chapter, our approach is a classification by local isometric classes. Due to the theory of infinitesimally homogeneous spaces by Singer [8], we will classify conformally flat homogeneous semi-Riemannian manifolds, investigating the forms of the symmetric endomorphism A and its covariant derivatives $\nabla A, \nabla^2 A, \cdots$. In Sect. 2, we recall the theory by Singer and apply it to our problem (Corollary 1).

From now on, we assume that M^n_q is an $n(\geq 4)$-dimensional conformally flat homogeneous semi-Riemannian manifold. First we study the form of the symmetric endomorphism A. Then evidently, the – possibly complex – eigenvalues of A and their algebraic multiplicities are constant on M. If the operator A is diagonalizable with real eigenvalues, it has at most two distinct eigenvalues. In this case, the classification is similar to that of the Riemannian case. That is, M^n_q is locally isometric to one of the following: (1)$M^n_q(k)$,; (2)$M^m_{q'}(k) \times M^{n-m}_{q-q'}(-k), k \neq 0, 2 \leq m \leq n-2$; and (3)$M^{n-1}_q(k) \times \mathbb{R}$ or $M^{n-1}_{q-1}(k) \times \mathbb{R}_1$, $k \neq 0$, where $M^m_{q'}(k)$ denotes a semi-Riemannian manifold of constant curvature k and index q' (Theorem 2.3 in [4]).

We consider the case in which A is not diagonalizable with real eigenvalues. In this chapter we study only the Lorentzian case. For such a case we have determined the form of A (Theorem 2.4 in [4]).

Theorem 1. *Let M^n_1 be an $n(\geq 4)$-dimensional conformally flat homogeneous Lorentzian manifold. If the linear operator A of M^n_1 defined by Eq. (1) is not diagonalizable with real eigenvalues, then A has exactly one of the following three forms:*

Case 1. $$a(e_1^* \otimes e_1 + e_2^* \otimes e_2) + b(e_1^* \otimes e_2 - e_2^* \otimes e_1) + \lambda \sum_{i=3}^{m} e_i^* \otimes e_i - \lambda \sum_{i=m+1}^{n} e_i^* \otimes e_i, \qquad b \neq 0,\ a^2 + b^2 = \lambda^2$$

Case 2. $$\lambda I_r - \lambda I_{n-r} + \varepsilon e_2^* \otimes e_1, \qquad \varepsilon = 1 \text{ or } \varepsilon = -1, \quad r \geq 2$$

Case 3. $$\lambda I_r - \lambda I_{n-r} + e_1^* \otimes e_3 + e_3^* \otimes e_2, \quad r \geq 3$$

where $I_r = \sum_{i=1}^{r} e_i^* \otimes e_i$ and $I_{n-r} = \sum_{i=r+1}^{n} e_i^* \otimes e_i$. Our expressions are those with respect to a semi-orthonormal basis $\langle e_1, e_2 \rangle = 1$, $\langle e_i, e_j \rangle = \delta_{ij}$ $(i, j \geq 3)$. We denote by $\{e_1^*, e_2^*, \cdots, e_n^*\}$ the dual basis of $\{e_1, e_2, \cdots, e_n\}$. In the expression above, if A has only one real eigenvalue λ , we delete $-\lambda$.

We try to classify conformally flat homogeneous Lorentzian manifolds for each case. For cases 1 and 2 in Theorem above, we obtain complete local classifications (Theorems 4, 5, 6 and 7). For case 3, we can show examples but cannot solve the classification problem at the present. We will discuss this case in the forthcoming paper.

After preliminaries in Sect. 2, we study case 1 in Sect. 3, case 2 with one eigenvalue λ in Sect. 4 and case 2 with two eigenvalues $\lambda, -\lambda$ in Sect. 5.

2 Preliminaries

In this section, we recall the theory of infinitesimally homogeneous spaces by Singer and study fundamental properties of the conullity operator C for the totally geodesic foliation induced from the eigenspace distribution of the operator A.

Let M be a semi-Riemannian manifold equipped with the semi-Riemannian metric $\langle\ ,\ \rangle$. A semi-Riemannian manifold M is said to be *homogeneous* if for any two points $p,q \in M$ there exists an isometry of M which maps p to q. On the other hand, it is called *locally homogeneous* if for any two points $p,q \in M$ there exist a neighbourhood U of p and a neighbourhood V of q and an isometry of U onto V which maps p to q. The notion of a curvature homogeneous space was introduced by Singer [8]. For a non-negative integer l, we consider the following condition:

$P(l)$: for every $p,q \in M$, there exists a linear isometry $\phi : T_pM \to T_qM$ such that

$$\phi^*(\nabla^i R)_q = (\nabla^i R)_p \quad i = 0,1,\ldots,l. \tag{4}$$

If M is locally homogeneous, then M obviously satisfies $P(l)$ for any l. A semi-Riemannian manifold which satisfies $P(l)$ is said to be *curvature homogeneous up to order l*.

We denote by $\mathfrak{so}(T_pM)$ the Lie algebra of the endomorphisms of T_pM which are skew-symmetric with respect to $\langle,\rangle$. For a non-negative integer l, we define a Lie subalgebra $\mathfrak{g}_l(p)$ of $\mathfrak{so}(T_pM)$ by

$$\mathfrak{g}_l(p) = \{X \in \mathfrak{so}(T_pM) \mid X \cdot (\nabla^i R)_p = 0, \quad i = 0,1,\ldots,l\ \}, \tag{5}$$

where X acts as a derivation on the tensor algebra on T_pM. Since $\mathfrak{g}_l(p) \supseteq \mathfrak{g}_{l+1}(p)$, there exists a first integer $s(p)$ such that $\mathfrak{g}_{s(p)}(p) = \mathfrak{g}_{s(p)+1}(p)$. Following Singer, we say that $(M,\langle,\rangle)$ is *infinitesimally homogeneous* if M satisfies $P(s(p)+1)$ for some point $p \in M$. If M is infinitesimally homogeneous, $s(q)$ does not depend on $q \in M$. We put $s_M = s(p)$ and call it *the Singer invariant* of an infinitesimally homogeneous semi-Riemannian manifold M. The remarkable result is the following:

Theorem 2 ([8]). *A connected infinitesimally homogeneous semi-Riemannian manifold is locally homogeneous.*

Singer proved the above theorem for the Riemannian case. On the other hand, another proof by Nomizu [6,7] using the theory of Killing generators is valid for the semi-Riemannian case.

The proof of Theorem 2 implies the following:

Theorem 3. *Let M and M' be two locally homogeneous semi-Riemannian manifolds and $p \in M$ and $p' \in M'$. Suppose that there exists a linear isometry $\phi : T_pM \to T_{p'}M'$ such that*

$$\phi^*(\nabla^i R')_{p'} = (\nabla^i R)_p \quad i = 0, 1, \dots, s_M + 1,$$

where s_M denotes the Singer invariant of M. Then there exists a local isometry φ of a neighbourhood of p onto a neighbourhood of p' which satisfies $\varphi(p) = p'$ and $\varphi_{*p} = \phi$.

Now, we consider conformally flat semi-Riemannian manifolds. By the Eq. (2), the curvature tensor R and its covariant derivatives $\nabla R, \cdots, \nabla^l R$ are determined by $A, \nabla A, \cdots, \nabla^l A$. Therefore, for the Lie subalgebra defined by Eq. (5), we have

$$\mathfrak{g}_l(p) = \left\{ X \in \mathfrak{so}(T_pM) \mid X \cdot (\nabla^i A)_p = 0, \quad i = 0, 1, \dots, l \right\}. \tag{6}$$

Applying Theorem 3, we obtain the following:

Corollary 1. *Let M and M' be two conformally flat locally homogeneous semi-Riemannian manifolds and $p \in M$ and $p' \in M'$. Suppose that there exists a linear isometry $\phi : T_pM \to T_{p'}M'$ such that*

$$\phi^*(\nabla^i A')_{p'} = (\nabla^i A)_p \quad i = 0, 1, \dots, s_M + 1.$$

Then there exists a local isometry φ of a neighbourhood of p onto a neighbourhood of p' which satisfies $\varphi(p) = p'$ and $\varphi_{*p} = \phi$.

In the remainder of this section, we assume that M^n is an $n(\geq 4)$-dimensional conformally flat homogeneous semi-Riemannian manifold. Let λ be a real eigenvalue of the operator A. We consider the case when the dimension of its eigenspace coincides with its algebraic multiplicity m. For each point $p \in M$, we define $T_\lambda(p) = \mathrm{Ker}(A - \lambda Id)$. Then the semi-Riemannian metric restricted to $T_\lambda(p)$ is non-degenerate and we have the orthogonal decomposition $T_pM = T_\lambda(p) + T_\lambda^\perp(p)$. By Eq. (3), we see that the eigenspace distribution T_λ is completely integrable and that its leaves are totally geodesic semi-Riemannian submanifolds in M. The orthogonal complement $T_\lambda^\perp$ is parallel with respect to the Levi-Civita connection ∇ along the leaves of T_λ. For this totally geodesic foliation, we define the conullity operator C as a smooth section of $\mathrm{Hom}(T_\lambda, \mathrm{End}(T_\lambda^\perp))$ (cf. Ferus [2]). We denote by $\pi : TM \to T_\lambda^\perp$ the orthogonal projection. Define a linear homomorphism C of $T_\lambda(p)$ into $\mathrm{End}(T_\lambda^\perp(p))$ by

$$C_u x = -\pi(\nabla_x U) \quad \text{for } x \in T_\lambda^\perp(p), u \in T_\lambda(p),$$

where U is a local smooth section of T_λ with $U_p = u$. At each point $p \in M$, we restrict a linear endomorphism $A - \lambda Id$ to $T_\lambda^\perp(p)$ and denote it by Φ_λ. Then Φ_λ is a linear isomorphism of $T_\lambda^\perp(p)$. We have the following identity at each point $p \in M$:

$$\nabla_u \Phi_\lambda = \Phi_\lambda C_u \quad \text{for } u \in T_\lambda(p) \tag{7}$$

(see the proof of Theorem 2.1 in [4]). Since the leaves of T_λ are totally geodesic, we have $R(x,v)v \in T_\lambda^\perp(p)$ for $x \in T_\lambda^\perp(p)$ and $v \in T_\lambda(p)$. For $v \in T_\lambda(p)$, we denote by $\widetilde{R}_v$ the linear endomorphism of $T_\lambda^\perp(p)$ defined by $x \mapsto R(x,v)v$ for $x \in T_\lambda^\perp(p)$. By Eq. (2), we have

$$\widetilde{R}_v = \langle v,v\rangle\,(A+\lambda Id)|_{T_\lambda^\perp(p)} = \langle v,v\rangle(\Phi_\lambda + 2\lambda Id) \quad \text{for } v \in T_\lambda(p). \tag{8}$$

Here, we denote by $(A+\lambda Id)|_{T_\lambda^\perp(p)}$ a linear endomorphism $A+\lambda Id$ restricted to $T_\lambda^\perp(p)$. Let γ be a geodesic in a leaf of T_λ. Then it is known that the following identity holds (cf. Ferus [2]):

$$\nabla_{\dot\gamma} C_{\dot\gamma} = C_{\dot\gamma}^2 + \widetilde{R}_{\dot\gamma}. \tag{9}$$

For a non-zero vector $v \in T_\lambda(p)$, we take a geodesic $\gamma : (-\varepsilon,\varepsilon) \to M$ such that $\gamma(0) = p$ and $\dot\gamma(0) = v$. Then γ is a curve in the leaf of T_λ through p. Let $\{e_1,\cdots,e_{n-m}\}$ be an orthonormal basis of $T_\lambda^\perp(p)$ and $\{E_1,\cdots,E_{n-m}\}$ be parallel frame field of $T_\lambda^\perp$ along γ such that $(E_i)_p = e_i (i = 1,\cdots,n-m)$. We express $\Phi_\lambda, \Phi_\lambda^{-1}, C_{\dot\gamma}$ and $\widetilde{R}_{\dot\gamma}$ as $(n-m)\times(n-m)$-matrices with respect to $\{E_1,\cdots,E_{n-m}\}$ and denote them by the same notations. Then, because of the homogeneity, there exists a smooth map $g : (-\varepsilon,\varepsilon) \to SO_o(n-m)$ such that $g(0) = I_{n-m}$ and

$$\Phi_\lambda(t) = g(t)\Phi_\lambda(0)g(t)^{-1}, \tag{10}$$

where I_{n-m} denotes the unit matrix of order $n-m$ and $SO_o(n-m)$ denotes the identity component of the Lie group of orthogonal transformations. Differentiating Φ_λ and $C_{\dot\gamma}$ along γ, by Eqs. (7) and (9), we have

$$\Phi_\lambda' = \Phi_\lambda C_{\dot\gamma}, \tag{11}$$

$$C_{\dot\gamma}' = C_{\dot\gamma}^2 + \widetilde{R}_{\dot\gamma}. \tag{12}$$

Lemma 1. *(1) tr* $C_v = 0$ *for any* $v \in T_\lambda(p)$.

(2) If the semi-Riemannian metric restricted to $T_\lambda(p)$ *is positive or negative definite and tr* $(A+\lambda Id)|_{T_\lambda^\perp(p)} \neq 0$ *, the homomorphism* $C : T_\lambda(p) \to End(T_\lambda^\perp(p))$ *is injective.*

(3) $\dim ImC \leq \dim \mathfrak{so}(T_\lambda^\perp(p))/\mathfrak{g}_0(\Phi_\lambda(p))$, *where* $\mathfrak{so}(T_\lambda^\perp(p))$ *denotes the Lie algebra of skew-symmetric endomorphisms of* $T_\lambda^\perp(p)$ *and* $\mathfrak{g}_0(\Phi_\lambda(p)) = \{X \in \mathfrak{so}(T_\lambda^\perp(p)) | X \cdot \Phi_\lambda(p) = 0\}$.

Proof. We use the notations above.

(1) We put $d(t) = \det \Phi_\lambda(t)$. Then $d(t)$ is a non-zero constant. Differentiating $d(t)$, we have $d'(t) = d(t)\mathrm{tr}\, C_{\dot\gamma}$. Since $d'(t) = 0$, we have $\mathrm{tr}\, C_{\dot\gamma} = 0$. In particular, $\mathrm{tr}\, C_v = 0$.
(2) Since $\mathrm{tr}\, C_{\dot\gamma} = 0$, we have $0 = \frac{d}{dt}\mathrm{tr}\, C_{\dot\gamma} = \mathrm{tr}\, \frac{d}{dt}C_{\dot\gamma} = \mathrm{tr}\, C_{\dot\gamma}^2 + \mathrm{tr}\, \widetilde{R}_{\dot\gamma}$. Therefore, we have $\mathrm{tr}\, C_v^2 = -\mathrm{tr}\, \widetilde{R}_v = -\langle v, v\rangle \mathrm{tr}\, (A + \lambda Id)|_{T_\lambda^\perp(p)}$. If the semi-Riemannian metric restricted to $T_\lambda(p)$ is positive or negative definite and $\mathrm{tr}\, (A + \lambda Id)|_{T_\lambda^\perp(p)} \neq 0$, $\mathrm{tr}\, C_v^2 \neq 0$ for a non-zero vector $v \in T_\lambda(p)$ and hence $C_v \neq 0$. This implies that $C : T_\lambda(p) \to \mathrm{End}(T_\lambda^\perp(p))$ is injective.
(3) Differentiating (10), we have $\Phi_\lambda' = g'(0) \cdot \Phi_\lambda(p)$. Therefore the image of the linear map $T_\lambda(p) \ni v \mapsto \tilde\nabla_v \Phi_\lambda$ is contained in $\{X \cdot \Phi_\lambda(p) \mid X \in \mathfrak{so}(T_\lambda^\perp(p))\}$. Since $\Phi_\lambda(p)$ is a linear isomorphism, by (7), we have $\dim \mathrm{Im} C \leq \dim \mathfrak{so}(T_\lambda^\perp(p))/\mathfrak{g}_0(\Phi_\lambda(p))$. □

By Eqs. (11) and (12), we have $(\Phi_\lambda^{-1})' = -C_{\dot\gamma}\Phi_\lambda^{-1}$ and $(\Phi_\lambda^{-1})'' = -\langle\dot\gamma, \dot\gamma\rangle(Id + 2\lambda\Phi_\lambda^{-1})$. Now, we assume that v is a unit vector of $T_\lambda(p)$. By the equation above, we have

$$(\Phi_\lambda^{-1})'' + 2\lambda\Phi_\lambda^{-1} + Id = 0. \tag{13}$$

If $\lambda \neq 0$, we put $Y(t) = \Phi_\lambda(t)^{-1} + \frac{1}{2\lambda}Id$. Then $Y'' + 2\lambda Y = 0$. Therefore, we obtain the explicit solutions:

$$Y(t) = \cos(\sqrt{2\lambda}\, t)Y(0) + \sin(\sqrt{2\lambda}\, t)\frac{1}{\sqrt{2\lambda}}Y'(0) \qquad \text{if } \lambda > 0, \tag{14}$$

$$Y(t) = \cosh(\sqrt{-2\lambda}\, t)Y(0) + \sinh(\sqrt{-2\lambda}\, t)\frac{1}{\sqrt{-2\lambda}}Y'(0) \qquad \text{if } \lambda < 0. \tag{15}$$

3 Case 1

First, we construct an example and will show that a conformally flat homogeneous Lorentzian manifold whose operator A has the form case 1 in Theorem 1 is locally isometric to this example.

Example 1. We define an indefinite inner product $\langle\ ,\ \rangle$ on $\mathbb{R}^m$ by

$$\langle x, y\rangle = {}^t x I_{1,m-1} y = -x_1y_1 + \sum_{i=2}^m x_iy_i \qquad \text{for } x, y \in \mathbb{R}^m,$$

where $I_{1,m-1}$ denotes the diagonal matrix whose entries are $-1,1,\cdots,1$. We denote by $SO_+(1,m-1)$ the identity component of the Lie group of orthogonal transformations of $(\mathbb{R}^m,\langle\ ,\ \rangle)$ and by K the product Lie group $SO_+(1,m-1)\times SO(2)$. We denote by $M(m,2;\mathbb{R})$ the linear space of real $m\times 2$ matrices. We define an inner product $(\ ,\ )$ on $M(m,2;\mathbb{R})$ by

$$(Z,W)=\mathrm{tr}({}^tZI_{1,m-1}W)=\langle z_1,w_1\rangle+\langle z_2,w_2\rangle,$$

for $Z=(z_1,z_2), W=(w_1,w_2)\in M(m,2;\mathbb{R})$, $z_i,w_i\in\mathbb{R}^m$. Then, this inner product $(\ ,\ )$ is non-degenerate and of index 2. We define the action of K on $M(m,2;\mathbb{R})$ by $(k_1,k_2)\cdot Z=k_1Zk_2^{-1}$ for $(k_1,k_2)\in K=SO_+(1,m-1)\times SO(2), Z\in M(m,2;\mathbb{R})$. Then, the inner product $(\ ,\ )$ is left invariant by the action of K. We put $Z_o=a(e_1,e_2)$ $(a>0)$, where $e_1={}^t(1,0,\cdots,0)$, and $e_2={}^t(0,1,\cdots,0)$. Then, Z_o is a light-like vector with respect to $(\ ,\ )$, i.e., $(Z_o,Z_o)=0$. We denote by M_a the K-orbit through Z_o. We induce the metric on M_a from the inner product $(\ ,\ )$ of $M(m,2;\mathbb{R})$. This is our exact example.

Proposition 1. *The manifold M_a is a $(2m-2)$-dimensional conformally flat Lorentzian manifold whose linear operator A has the form*

$$A=\frac{1}{2a^2}\left(e_1^*\otimes e_2-e_2^*\otimes e_1+\sum_{i=3}^{m}e_i^*\otimes e_i-\sum_{i=m+1}^{2m-2}e_i^*\otimes e_i\right),$$

relative to a semi-orthonormal basis $(e_1,e_2)=1,\ \ (e_i,e_j)=\delta_{ij}\ (i,j\geq 3)$.

Now we assume that M is an $n(\geq 4)$-dimensional conformally flat homogeneous Lorentzian manifold whose linear operator A has the form case 1 in Theorem 1. We use the same notations as in Sect. 2.

Lemma 2. *The linear operator A has two real eigenvalues λ and $-\lambda$.*

Proof. We assume that A has only one real eigenvalue $\lambda(\neq 0)$ and show a contradiction. The dimension of its eigenspace $T_\lambda(p)$ coincides with its algebraic multiplicity. Let C be the conullity operator associated with the eigenspace distribution T_λ. The Lorentzian metric restricted to $T_\lambda(p)$ is positive definite. We have $\mathrm{tr}\,(A+\lambda Id)|_{T_\lambda^\perp(p)}=2(a+\lambda)$. If $a+\lambda=0$, then $a^2=\lambda^2$ and hence $b=0$. It is contrary to our assumption $b\neq 0$. It follows that $\mathrm{tr}\,(A+\lambda Id)|_{T_\lambda^\perp(p)}\neq 0$. Therefore by Lemma 1 (2), the homomorphism $C: T_\lambda(p)\to \mathrm{End}(T_\lambda^\perp(p))$ is injective, and hence $\dim \mathrm{Im}C=\dim T_\lambda(p)=n-2\geq 2$. Since $\dim \mathfrak{so}(T_\lambda^\perp(p))=1$, by Lemma 1 (3), $\dim \mathrm{Im}C\leq 1$. It is a contradiction.

From now on, we assume that $\lambda>0$ and $\dim T_\lambda(p)=m_1>0$, $\dim T_{-\lambda}(p)=m_2=n-m_1-2>0$ and $b>0$ without loss of generality. Moreover, we assume that M is simply connected. Since our classification is on the equivalence by local isometries, the assumption of simply connectedness does not lose the generality.

For $\mathfrak{g}_0$ defined by Eq. (6), we have

$$\mathfrak{g}_0 = \left\{ \begin{pmatrix} 0 & & \\ & X' & \\ & & X'' \end{pmatrix} \mid X' \in \mathfrak{so}(m_1), X'' \in \mathfrak{so}(m_2) \right\}.$$

This, together with the simply connectedness of M, implies that there exist smooth light-like vector fields e_1 and e_2 with $\langle e_1, e_2 \rangle = 1$ such that $Ae_1 = ae_1 + be_2$ and $Ae_2 = -be_1 + ae_2$. Moreover the vector fields e_1 and e_2 are invariant by the action of the identity component $I_o(M)$ of the isometry group of M. We always use these vector fields e_1 and e_2 adapted to the linear operator A.

We will investigate properties of the conullity operators associated with T_λ and $T_{-\lambda}$, respectively. First we consider the conullity operator C associated with T_λ.

Lemma 3. *For a unit vector $v \in T_\lambda(p)$, we choose a suitable orthonormal basis $\{e_{m_1+3}, \cdots, e_n\}$ of $T_{-\lambda}(p)$ and can express the conullity operator C_v associated with the eigenspace distribution T_λ with respect to the semi-orthonormal basis $\{e_1, e_2, e_{m_1+3}, \cdots, e_n\}$ of $T_\lambda^\perp(p)$ as follows:*

$$C_v = \frac{b}{\sqrt{2\lambda}(\lambda - a)} \left\{ \frac{\lambda - a + b}{\sqrt{2}} e_1^* \otimes e_{m_1+3} + \frac{-\lambda + a + b}{\sqrt{2}} e_2^* \otimes e_{m_1+3} \right.$$
$$\left. - \sqrt{2}\lambda e_{m_1+3}^* \otimes e_1 + \sqrt{2}\lambda e_{m_1+3}^* \otimes e_2 \right\}.$$

Here we remark that by our assumption, $\lambda - a > 0$ and $\lambda - a + b > 0$.

Proof. At an arbitrary point $q \in M$, we take an orthonormal basis $\{e_{m_1+3}, \cdots, e_n\}$ of $T_{-\lambda}(q)$. As in Sect. 2, we define the linear endomorphisms $\Phi_\lambda = (A - \lambda Id)|_{T_\lambda^\perp(q)}$ and $Y = \Phi_\lambda^{-1} + \frac{1}{2\lambda} Id$ of $T_\lambda^\perp(q)$. Then, Y is expressed with respect to the semi-orthonormal basis $\{e_1, e_2, e_{m_1+3}, \cdots, e_n\}$ as follows:

$$Y = \tilde{b}(-e_1^* \otimes e_2 + e_2^* \otimes e_1), \qquad \tilde{b} = b/(2\lambda(\lambda - a)) > 0. \tag{16}$$

In particular, Y has the eigenvalues 0 with multiplicity $n - m_1 - 2$ and $\pm\sqrt{-1}\tilde{b}$ with multiplicity 1.

For a point $p \in M$, we denote by $M_\lambda(p)$ the leaf of T_λ through p. Then $M_\lambda(p)$ is a totally geodesic Riemannian submanifold in M. Since $M_\lambda(p)$ is a homogeneous Riemannian manifold, it is geodesically complete. For a unit vector $v \in T_\lambda(p)$, we take a geodesic $\gamma : \mathbb{R} \to M$ such that $\gamma(0) = p$ and $\dot{\gamma}(0) = v$. We apply the arguments of Sect. 2 to this case. Let $\{E_1, E_2, E_{m_1+3}, \cdots, E_n\}$ be parallel frame field of $T_\lambda^\perp$ along γ such that $(E_i)_p = e_i \, (i = 1, 2, m_1 + 3, \cdots, n)$. We express $Y = \Phi_\lambda^{-1} + 1/(2\lambda) Id$ as $(n - m_1) \times (n - m_1)$-matrices with respect to $\{E_1, E_2, E_{m_1+3}, \cdots, E_n\}$ along γ. Then by Eq. (14) we have

$$Y(t) = \cos(\sqrt{2\lambda}\, t)Y(0) + \sin(\sqrt{2\lambda}\, t)\frac{1}{\sqrt{2\lambda}}Y'(0).$$

For any $t \in \mathbb{R}$, $Y(t)$ has the same eigenvalues as $Y(0)$. In particular, putting $t = \pi/(2\sqrt{2\lambda})$, we see that $1/\sqrt{2\lambda}Y'(0)$ has the same eigenvalues as $Y(0)$. On the other hand, by (10) we have $Y(t) = g(t)Y(0)g(t)^{-1}$ for some smooth curve $g : \mathbb{R} \to SO_+(1, n-m_1-1)$ with $g(0) = I_{n-m_1}$. Then,

$$\frac{1}{\sqrt{2\lambda}}Y'(0) = \left[\frac{1}{\sqrt{2\lambda}}g'(0), Y(0)\right] = \tilde{b}X,$$

where

$$X = \begin{pmatrix} 0 & x_{21} & {}^t x_2 \\ x_{21} & 0 & {}^t x_1 \\ x_1 & x_2 & 0 \end{pmatrix}, \qquad x_1, x_2 \in \mathbb{R}^{n-m_1-2}. \tag{17}$$

Here the matrix X has the eigenvalues 0 with multiplicity $n - m_1 - 2$ and $\pm\sqrt{-1}$ with multiplicity 1. From these, it follows that x_1 and x_2 are non-zero vectors and are linearly dependent. Exchanging an orthonormal basis of $T_{-\lambda}(p)$ if necessary, we may assume that $x_1 = {}^t(c, 0, \cdots, 0)$ $(c > 0)$. By straightforward computations, we see that $x_{21} = 0$ and $x_2 = -1/(2c^2)x_1$. Moreover, $Y(t)$ has the eigenvalues 0 with multiplicity $n - m_1 - 2$ and $\pm\sqrt{-1}\tilde{b}$ with multiplicity 1 if and only if $c = 1/\sqrt{2}$. In particular,

$$Y'(0) = \frac{b}{2\sqrt{\lambda}(\lambda - a)}\left\{e_1^* \otimes e_{m_1+3} - e_2^* \otimes e_{m_1+3} - e_{m_1+3}^* \otimes e_1 + e_{m_1+3}^* \otimes e_2\right\}.$$

Since $\nabla_v \Phi_\lambda^{-1} = Y'(0)$ and by (7), $C_v = -(\nabla_v \Phi_\lambda^{-1})\Phi_\lambda$, we have our formula of C_v in this lemma. □

Lemma 4. *There exists a linear isometry ϕ of $T_\lambda(p)$ into $T_{-\lambda}(p)$ such that for any local section $V \in \Gamma(T_\lambda)$ around p*

$$(\nabla_{e_1} V)_{T_\lambda^\perp} = \frac{(\lambda - a + b)b}{2\sqrt{\lambda}(\lambda - a)}\phi(V_p), \quad (\nabla_{e_2} V)_{T_\lambda^\perp} = \frac{(-\lambda + a + b)b}{2\sqrt{\lambda}(\lambda - a)}\phi(V_p).$$

In particular, we have $\dim T_\lambda(p) \leq \dim T_{-\lambda}(p)$.

Proof. We define a linear map $\phi : T_\lambda(p) \to T_{-\lambda}(p)$ by

$$\phi(v) = -\frac{2\sqrt{\lambda}(\lambda - a)}{(\lambda - a + b)b}C_v(e_1). \tag{18}$$

Then, by Lemma 3 and the definition of the conullity operator, our formulas in this lemma hold.

Remark 1. Since the vector field e_1 is globally defined on M, the bundle homomorphism ϕ of T_λ into $T_{-\lambda}$ is also globally defined on M by Eq. (18) and it is invariant by the action of $I_o(M)$.

We investigate the conullity operator C associated with $T_{-\lambda}$ by the similar arguments to the proof of Lemma 3 and using Lemma 4 obtain the following:

Lemma 5. *For a unit vector $v \in T_{-\lambda}(p)$, we choose a suitable orthonormal basis $\{e_3, \cdots, e_{m_1+2}\}$ of $T_\lambda(p)$ and can express the conullity operator C_v associated with the eigenspace distribution $T_{-\lambda}$ with respect to the semi-orthonormal basis $\{e_1, e_2, e_3, \cdots, e_{m_1+2}\}$ of $T^{\perp}_{-\lambda}(p)$ as follows:*

$$C_v = \frac{b}{\sqrt{2\lambda}(\lambda+a)} \left\{ \frac{\lambda+a+b}{\sqrt{2}} e_1^* \otimes e_3 + \frac{\lambda+a-b}{\sqrt{2}} e_2^* \otimes e_3 + \sqrt{2}\lambda (e_3^* \otimes e_1 + e_3^* \otimes e_2) \right\}. \tag{19}$$

These arguments imply the following:

Proposition 2. *Let M be an $n(\geq 4)$-dimensional simply connected conformally flat homogeneous Lorentzian manifold whose linear operator A has the form case 1 in Theorem 1. Then, we have $a = 0$ and $b = \lambda$ for the eigenvalues of A and* $\dim T_\lambda = \dim T_{-\lambda}$. *In particular, the dimension of M is even. Moreover, there exists a vector bundle isomorphism ϕ of T_λ onto $T_{-\lambda}$ which satisfies the following:*

(1) ϕ is a linear isometry of $T_\lambda(p)$ onto $T_{-\lambda}(p)$ at any point $p \in M$.
(2) For any local section V of T_λ,

$$(\nabla_{e_1} V)_{T^{\perp}_{\lambda}} = \sqrt{\lambda}\,\phi(V), \quad \textit{and} \quad (\nabla_{e_2} V)_{T^{\perp}_{\lambda}} = 0$$

and for any local section W of $T_{-\lambda}$,

$$(\nabla_{e_1} W)_{T^{\perp}_{-\lambda}} = -\sqrt{\lambda}\,\phi^{-1}(W), \quad \textit{and} \quad (\nabla_{e_2} W)_{T^{\perp}_{-\lambda}} = 0.$$

(3) ϕ is invariant by the action of $I_o(M)$.

Proof. By Lemma 5, it follows that there exists a linear isometry ψ of $T_{-\lambda}(p)$ into $T_\lambda(p)$ such that for any local section $W \in \Gamma(T_{-\lambda})$ around p

$$(\nabla_{e_1} W)_{T^{\perp}_{-\lambda}} = \frac{(\lambda+a+b)b}{2\sqrt{\lambda}(\lambda+a)} \psi(W_p), \quad (\nabla_{e_2} W)_{T^{\perp}_{-\lambda}} = \frac{(\lambda+a-b)b}{2\sqrt{\lambda}(\lambda+a)} \psi(W_p).$$

In particular, we have $\dim T_\lambda(p) \geq \dim T_{-\lambda}(p)$. This, together with Lemma 4, implies that $\dim T_\lambda(p) = \dim T_{-\lambda}(p)$.

For any local sections $V, V' \in \Gamma(T_\lambda)$ around p ,

$$\langle \nabla_{e_1} V, \phi(V') \rangle = \frac{(\lambda - a + b)b}{2\sqrt{\lambda}(\lambda - a)} \langle \phi(V), \phi(V') \rangle = \frac{(\lambda - a + b)b}{2\sqrt{\lambda}(\lambda - a)} \langle V, V' \rangle$$

and

$$\langle V, \nabla_{e_1} \phi(V') \rangle = \frac{(\lambda + a + b)b}{2\sqrt{\lambda}(\lambda + a)} \langle V, \psi(\phi(V')) \rangle.$$

Since $\langle \nabla_{e_1} V, \phi(V') \rangle + \langle V, \nabla_{e_1} \phi(V') \rangle = 0$, we have

$$\frac{(\lambda + a + b)b}{2\sqrt{\lambda}(\lambda + a)} \psi \circ \phi = -\frac{(\lambda - a + b)b}{2\sqrt{\lambda}(\lambda - a)} Id \qquad \text{on } T_\lambda.$$

Since ψ and ϕ are linear isometries, we have $(\lambda + a + b)b/(2\sqrt{\lambda}(\lambda + a)) = (\lambda - a + b)b/(2\sqrt{\lambda}(\lambda - a))$ and $\psi = -\phi^{-1}$. By these it follows that $a = 0$ and $b = \lambda$. By Remark 1, the latter part of Proposition 2 holds. □

By Proposition 2, we may assume that $\dim T_\lambda = \dim T_{-\lambda} = m (\geq 1)$ and $n = \dim M = 2(m+1)$. For any point $p \in M$, we take local orthonormal frame field $\{e_3, \cdots, e_{m+2}\}$ of T_λ on a neighbourhood of p and put $e_{m+i} = \phi(e_i)$ $(i = 3, \cdots, m+2)$, where ϕ is the vector bundle isomorphism of T_λ onto $T_{-\lambda}$ given in Proposition 2. We call the semi-orthonormal frame field $\{e_1, e_2, e_3, \cdots, e_{m+2}, e_{m+3}, \cdots, e_n\}$ constructed as above *adapted frame field* to A and ϕ. If $\dim M = 4$, we have $\mathfrak{g}_0 = \{0\}$, where $\mathfrak{g}_0$ denotes the Lie algebra defined in Eq. (6). Therefore, M is a four-dimensional Lie group with a left invariant Lorentzian metric, and the Singer invariant S_M of M is equal to 0.

We denote by $\{\Gamma_{ij}^k\}$ $(i, j, k = 1, 2, \cdots, n)$ the connection functions,i.e., $\nabla_{e_i} e_j = \sum_{k=1}^n \Gamma_{ij}^k e_k$. Simply, we denote by Γ_k the matrix whose (i, j)-components are Γ_{kj}^i. We calculate the connection functions with respect to the adapted frame field $\{e_1, e_2, \cdots, e_n\}$ to A and ϕ. We write $X = Y \mod \mathfrak{g}_0$ if $X - Y \in \mathfrak{g}_0$ for $X, Y \in \mathfrak{so}(TM)$.

Lemma 6. *Using Eq. (3), we have the following:*

$$\begin{aligned} \Gamma_1 &= \sqrt{\lambda} \sum_{j=3}^{m+2} e_{m+j} \wedge e_j && \mod \mathfrak{g}_0 \\ \Gamma_2 &= 0 && \mod \mathfrak{g}_0 \\ \Gamma_i &= \sqrt{\lambda} \{ e_{m+i} \wedge e_1 + e_{m+i} \wedge e_2 \} && \mod \mathfrak{g}_0 \ (i = 3, \cdots, m+2) \\ \Gamma_{m+i} &= \sqrt{\lambda} \{ e_i \wedge e_1 - e_i \wedge e_2 \} && \mod \mathfrak{g}_0 \ (i = 3, \cdots, m+2) \end{aligned} \tag{20}$$

Calculating $\nabla_{e_i} A = [\Gamma_i, A]$, we have

$$
\begin{aligned}
\nabla_{e_1}A &= 2\lambda\sqrt{\lambda}\sum_{j=3}^{m+2}\left\{e_j^*\otimes e_{m+j}+e_{m+j}^*\otimes e_j\right\},\\
\nabla_{e_2}A &= 0, \\
\nabla_{e_i}A &= 2\lambda\sqrt{\lambda}\{e_1^*\otimes e_{m+i}+e_{m+i}^*\otimes e_2\}, && (i=3,\cdots,m+2)\\
\nabla_{e_{m+i}}A &= 2\lambda\sqrt{\lambda}\{e_1^*\otimes e_i+e_i^*\otimes e_2\}, && (i=3,\cdots,m+2)
\end{aligned}
\tag{21}
$$

Then for the Lie algebra $\mathfrak{g}_1$ in Eq. (6), we have

$$
\mathfrak{g}_1=\left\{\begin{pmatrix}0 & & \\ & X & \\ & & X\end{pmatrix} \mid X\in\mathfrak{so}(m)\right\}\simeq\mathfrak{so}(m).
$$

Hence, if $\dim M=n=2m+2\geq 6$, $\mathfrak{g}_0\simeq\mathfrak{so}(m)\oplus\mathfrak{so}(m), \mathfrak{g}_1\simeq\mathfrak{so}(m) and \mathfrak{g}_0\supsetneq\mathfrak{g}_1$. A semi-orthonormal basis adapted to A is said to be *adapted to A and ∇A* if ∇A is expressed as Eq. (21) with respect to the basis. A semi-orthonormal basis is adapted to A and ϕ if and only if it is adapted to A and ∇A.

Lemma 7. *Using the Eq. (2), we have*

$$
\Gamma_{a\,m+i}^{\;m+j}=\Gamma_{ai}^{j}\qquad a=1,\cdots,n,\quad i,j=3,\cdots,m+2
$$

with respect to adapted semi-orthonormal frame field to A and ∇A.

Proof. By Eq. (2), we have $R(e_1,e_i)e_1=-\lambda e_i\,(i=3,\cdots,m+2)$. On the other hand, using the connection, we have

$$
R(e_1,e_i)e_1=-\lambda e_i+\sqrt{\lambda}\sum_{j=3}^{m+2}\left(\Gamma_{1\,m+i}^{\;m+j}-\Gamma_{1i}^{j}\right)e_{m+j}
$$

and hence $\Gamma_{1\,m+i}^{\;m+j}=\Gamma_{1i}^{j}\ (i,j=3,\cdots,m+2)$. Similarly we obtain the other identities.

By Lemma 7, it follows that for adapted semi-orthonormal frame field to A and ∇A, the connection functions have the same form to Eq. (20) under $\operatorname{mod}\mathfrak{g}_1$, to which we refer as Eq. (20)'. We denote by P the one of the connected components of the bundle of adapted semi-orthonormal bases to A and ∇A over M. The structure group G_1 of P is the connected Lie subgroup of $SO_+(1,n-1)$ which corresponds to $\mathfrak{g}_1$. For each $u=(e_1,\cdots,e_n)\in P$ with the base point $p\in M$, we will define an element $\tilde{\Gamma}$ of $\mathrm{Hom}(\mathbb{R}^n,\mathfrak{so}(1,n-1)/\mathfrak{g}_1)$ as follows: we extend u to a local section $\tilde{u}=(\tilde{e}_1,\cdots,\tilde{e}_n)$ of P and denote by $\{\Gamma_{ij}^{k}\}(i,j,k=1,2,\cdots,n)$ the connection functions with respect to the local frame field $\{\tilde{e}_1,\cdots,\tilde{e}_n\}$. We view Γ_i as a $\mathfrak{so}(1,n-1)$-valued function for each i $(i=1,\cdots,n)$ and denote by μ the natural projection of $\mathfrak{so}(1,n-1)$ onto the quotient space $\mathfrak{so}(1,n-1)/\mathfrak{g}_1$. Identifying the tangent space T_pM with $\mathbb{R}^n$ by the basis $u=(e_1,\cdots,e_n)\in P$, we define $\tilde{\Gamma}(e_i)=\mu\circ\Gamma_i$ for each i $(i=1,\cdots,n)$. By Eq. (20)', it follows that $\tilde{\Gamma}$ is defined independently of the extensions and it is constant on P.

The Lie algebra $\mathfrak{g}_1$ naturally acts on the space $\mathrm{Hom}(\mathbb{R}^n,\mathfrak{so}(1,n-1)/\mathfrak{g}_1)$. By the straightforward computations, we obtain the following.

Lemma 8. *$\tilde{\Gamma}$ is invariant by the action of* $\mathfrak{g}_1$.

We calculate $\nabla^2 A$ with respect to adapted semi-orthonormal frame field to A and ∇A. Then, we have $\nabla_{e_i} \nabla A = \Gamma_i \cdot \nabla A$, where Γ_i acts as a derivation on ∇A. By Eqs. (20)' and (21), it follows that the expression of $\nabla^2 A$ with respect to the adapted basis is completely determined by the value of λ. Moreover, by Lemma 8, for $X \in \mathfrak{g}_1$, we have $X \cdot \nabla^2 A = 0$ and hence $\mathfrak{g}_1 = \mathfrak{g}_2$. Applying Corollary 1, we have the following.

Theorem 4. *Let M be an $n(\geq 4)$-dimensional conformally flat homogeneous Lorentzian manifold whose linear operator A has the form case 1 in Theorem 1.Then M is locally isometric to M_a $(a = 1/\sqrt{2|\lambda|})$ constructed in Example 1.*

4 Case 2 with One Eigenvalue λ

In this section, we study the case whose linear operator A has the form

$$A = \lambda I_n + \varepsilon e_2^* \otimes e_1. \tag{22}$$

First, we construct two examples and will show that a conformally flat homogeneous Lorentzian manifold whose linear operator has the form (22) is locally isometric to one of these examples (Theorem 5 and Theorem 6).

We construct two pairs of Lie algebras $\mathfrak{k}$ and their Lie subalgebras $\mathfrak{h}$.

Example 2. Let $\mathfrak{k}$ be the linear space with the bases E_i $(4 \leq i \leq n), F_{ij}$ $(4 \leq i < j \leq n)$ *and*, X_i $(1 \leq i \leq n)$. We define the bracket operation $[\ ,\ ]$ as follows:

$$\begin{array}{ll}
[E_i, E_j] = 0 & [F_{ij}, F_{kl}] = -\delta_{ik} F_{jl} + \delta_{jk} F_{il} \\
{[E_i, F_{jk}]} = \delta_{ij} E_k - \delta_{ik} E_j & \qquad\qquad + \delta_{il} F_{jk} - \delta_{jl} F_{ik} \\
{[E_i, X_1]} = 0 & [F_{ij}, X_1] = 0 \\
{[E_i, X_2]} = -X_i & [F_{ij}, X_2] = 0 \\
{[E_i, X_3]} = cE_i & [F_{ij}, X_3] = 0 \\
{[E_i, X_j]} = \delta_{ij} X_1 & [F_{ij}, X_k] = -\delta_{ik} X_j + \delta_{jk} X_i
\end{array}$$

$$\begin{array}{ll}
[X_1, X_2] = 0 & \\
{[X_1, X_3]} = 3cX_1 & \\
{[X_1, X_j]} = 0 & [X_3, X_j] = -2cX_j
\end{array}$$

$$\begin{array}{ll}
[X_2, X_3] = -\dfrac{\varepsilon}{2c} X_1 + cX_2 & [X_i, X_j] = 0 \\
{[X_2, X_j]} = 0 &
\end{array}$$

$$i, j, k, l \geq 4 \qquad c > 0 \qquad \varepsilon = 1 \text{ or } -1$$

Then, the bracket operation $[\ ,\ ]$ satisfies the Jacobi identity, and $\mathfrak{k}$ becomes a Lie algebra. Let $\mathfrak{h}$ be the linear subspace spanned by $\{E_i\ (4 \le i \le n) and, F_{ij}\ (4 \le i < j \le n)\}$. Then, $\mathfrak{h}$ is a Lie subalgebra of $\mathfrak{k}$.

Example 3. Let $\mathfrak{k}$ be the linear space with the bases $E_i\ (3 \le i \le n), F_{ij}\ (3 \le i < j \le n) and X_i\ (1 \le i \le n)$. We define the bracket operation $[\ ,\ ]$ as follows:

$$
\begin{array}{ll}
[E_i, E_j] = 0 & [F_{ij}, F_{kl}] = -\delta_{ik}F_{jl} + \delta_{jk}F_{il} \\
[E_i, F_{jk}] = \delta_{ij}E_k - \delta_{ik}E_j & \qquad\qquad +\delta_{il}F_{jk} - \delta_{jl}F_{ik} \\
[E_i, X_1] = 0 & [F_{ij}, X_1] = 0 \\
[E_i, X_2] = -X_i - cE_i & [F_{ij}, X_2] = 0 \\
[E_i, X_j] = \delta_{ij}X_1 & [F_{ij}, X_k] = -\delta_{ik}X_j + \delta_{jk}X_i \\
 & \\
[X_1, X_2] = -cX_1 & \\
[X_1, X_j] = 0 & [X_i, X_j] = 0 \\
 & \\
[X_2, X_j] = -\varepsilon E_j & \\
 & \\
i, j, k, l \ge 3 & c \in \mathbb{R} \qquad \varepsilon = 1 \text{ or } -1
\end{array}
$$

Then, $\mathfrak{k}$ becomes a Lie algebra and its linear subspace $\mathfrak{h}$ spanned by $\{E_i\ (3 \le i \le n), F_{ij}\ (3 \le i < j \le n)\}$ is a Lie subalgebra of $\mathfrak{k}$.

From these pairs $(\mathfrak{k}, \mathfrak{h})$ of Lie algebras and their Lie subalgebras, we construct homogeneous Lorentzian manifolds. Let K be a simply connected Lie group corresponding to $\mathfrak{k}$ and H be the connected Lie subgroup of K which corresponds to $\mathfrak{h}$. Then, H is a closed subgroup of K, and hence we obtain the homogeneous space $M = K/H$. Let $\pi : K \longrightarrow K/H = M$ be the projection. We put $\pi(H) = o$. The differential of π at the unit element $e \in K$ defines the projection $\pi : \mathfrak{k} \longrightarrow T_oM$, which gives the linear isomorphism of $\mathfrak{k}/\mathfrak{h}$ onto T_oM. Let $\mathfrak{p}$ be the subspace spanned by $\{X_i\ (1 \le i \le n)\}$, and we identify $\mathfrak{k}/\mathfrak{h}$ with $\mathfrak{p}$. Under this identification, we define an inner product on $\mathfrak{k}/\mathfrak{h}$ by

$$\langle X_1, X_2 \rangle = 1, \quad \langle X_i, X_j \rangle = \delta_{ij}\ (3 \le i, j \le n), \quad \text{otherwise } 0.$$

This inner product $\langle\ ,\ \rangle$ is invariant by the adjoint representation of $\mathfrak{h}$ on $\mathfrak{k}/\mathfrak{h}$. Since H is connected, it is an $Ad(H)$-invariant inner product. Therefore, we can define the K-invariant Lorentzian metric g on M such that

$$g(\pi(X), \pi(Y)) = \langle X, Y \rangle \qquad X, Y \in \mathfrak{p}$$

at $o \in M$. Thus, we obtain the homogeneous Lorentzian manifolds (M, g).

Using the theory of invariant connections in Chapter X in [5] , we compute the curvature tensor R of (M, g). Under the identification of T_oM with $\mathfrak{p}$, we define the linear operator A by

$$A = -2c^2 I_n + \varepsilon X_2^* \otimes X_1$$

for Example 2 and by

$$A = \varepsilon X_2^* \otimes X_1$$

for Example 3, respectively. Then the curvature tensor R of (M,g) is given by

$$R(X,Y) = AX \wedge Y + X \wedge AY \qquad X,Y \in T_oM \simeq \mathfrak{p}.$$

Therefore, (M,g) is a conformally flat homogeneous Lorentzian manifold whose linear operator A has the form (22) with $\lambda < 0$ for Example 2 and with $\lambda = 0$ for Example 3, respectively. Moreover, we have

$$\nabla_{X_i} A = 0 \quad (i = 1,3,\cdots,n), \quad \nabla_{X_2} A = 2\varepsilon c X_2^* \otimes X_1 \tag{23}$$

for Example 3.

Now, we assume that M is an $n(\geq 4)$-dimensional conformally flat homogeneous Lorentzian manifold whose linear operator A has the form (22) with respect to a semi-orthonormal basis $\{e_1,\cdots,e_n\}$. We call a semi-orthonormal basis with respect to which the linear operator A has the form (22) to be *adapted* to A. With respect to an adapted basis, we have

$$\mathfrak{g}_0 = \left\{ \begin{pmatrix} 0 & 0 & -{}^t x \\ 0 & 0 & 0 \\ 0 & x & X' \end{pmatrix} | x \in \mathbb{R}^{n-2},\ X' \in \mathfrak{so}(n-2) \right\}. \tag{24}$$

We define the subbundles T_0 and T_{00} of the tangent bundle TM by $T_0 = \mathrm{Ker}(A - \lambda Id)$ and $T_{00} = \mathrm{Im}(A - \lambda Id)$, respectively (cf. [3] Sect. 5). Then, T_0 is a subbundle of rank $n-1$ and T_{00} is a line bundle. These subbundles are invariant by the action of the identity component $I_o(M)$ of the isometry group of M. If $\{e_1,e_2,\cdots,e_n\}$ and $\{\tilde{e}_1,\tilde{e}_2,\cdots,\tilde{e}_n\}$ are adapted semi-orthonormal bases with respect to A, then we have $\tilde{e}_1 = \pm e_1$. Now, we assume that M is simply connected. Then, there exists the unique smooth section e_1 of T_{00} up to the sign such that at each point $\{e_1,e_2,\cdots,e_n\}$ is an adapted basis together with some tangent vectors $e_2,\cdots,e_n$. Moreover, the vector field e_1 and tensor field ∇e_1 are invariant by the action of $I_o(M)$.

We calculate the covariant derivative ∇A with respect to the adapted frame field to A and obtain

$$\nabla_{e_i} A = 2\varepsilon \Gamma_{i1}^{\ 1} e_2^* \otimes e_1 + \varepsilon \sum_{j=3}^{n} \Gamma_{i1}^{\ j} \left(e_2^* \otimes e_j + e_j^* \otimes e_1\right). \tag{25}$$

Lemma 9. *Using the Eq. (3), we have*

$$\begin{aligned} \Gamma_1 &= 0 && mod\ \mathfrak{g}_0 \\ \Gamma_2 &= a_1 e_1 \wedge e_2 + \textstyle\sum_{j=3}^n a_j e_j \wedge e_2 && mod\ \mathfrak{g}_0 \\ \Gamma_k &= \frac{1}{2} a_k e_1 \wedge e_2 && mod\ \mathfrak{g}_0 \quad (k = 3,\cdots,n), \end{aligned}$$

where we put $a_i = \Gamma_{21}^{\ i} \quad (i = 1,3,\cdots,n)$.

Since $\langle \nabla_X e_1, e_1 \rangle = 0$, $\mathrm{Im}\nabla e_1 \subset T_0$. We define a T_0/T_{00}-valued 1-form α by $\alpha(X) = \pi(\nabla_X e_1)$, where π denotes the projection of T_0 onto T_0/T_{00}. Then α is invariant by the action of $I_o(M)$. We consider two cases:

Case I $\alpha \not\equiv 0$, i.e., T_{00} is not parallel with respect to ∇,
Case II $\alpha \equiv 0$, i.e., T_{00} is parallel with respect to ∇.

First, we study Case I. By Lemma 9, it follows that $\mathrm{Ker}\alpha = T_0$. The following is easily seen.

Lemma 10. *At each point, there exists an adapted semi-orthonormal basis* $\{e_1, e_2, e_3, \cdots, e_n\}$ *to* A *which satisfies* $\nabla_{e_2} e_1 = a e_3$ *for some* $a (> 0)$. *Moreover, such* e_3 *is uniquely determined.*

We call a semi-orthonormal basis which satisfies Lemma 10 an *adapted basis* to A and ∇e_1. For such a basis, ∇e_1 has the form

$$\nabla e_1 = a e_2^* \otimes e_3 + \frac{1}{2} a e_3^* \otimes e_1 \qquad a > 0.$$

Since ∇e_1 is invariant by the action of $I_o(M)$, the coefficient a is constant on M. By the uniqueness of e_3 in Lemmma 10, we see that there exists the unique unit vector field e_3 such that at each point $\{e_1, e_2, e_3, \cdots, e_n\}$ is an adapted basis to A and ∇e_1 together with some tangent vectors $e_2, e_4, \cdots, e_n$. Moreover, the vector field e_3 and the tensor field ∇e_3 are invariant by the action of $I_o(M)$. With respect to an adapted semi-orthonormal basis to A and ∇e_1, ∇A has the following form:

$$\begin{aligned} &\nabla_{e_1} A = 0, && \nabla_{e_2} A = \varepsilon a (e_2^* \otimes e_3 + e_3^* \otimes e_1), \\ &\nabla_{e_3} A = \varepsilon a e_2^* \otimes e_1, && \nabla_{e_i} A = 0 \ (i \geq 4). \end{aligned} \tag{26}$$

Then, for the Lie algebra $\mathfrak{g}_1$ in Eq. (6), we have

$$\mathfrak{g}_1 = \{X \in \mathfrak{so}(T_oM) | X e_1 = X e_3 = 0\}.$$

Lemma 11. *Using Eq. (2), we have* $a^2 = -2\lambda$, *and in particular,* λ *is negative. Moreover,*

$$\begin{aligned} \Gamma_1 &= a e_3 \wedge e_1 && \textit{mod } \mathfrak{g}_1 \\ \Gamma_2 &= a e_3 \wedge e_2 - \frac{\varepsilon}{a} e_3 \wedge e_1 && \textit{mod } \mathfrak{g}_1 \\ \Gamma_3 &= \frac{1}{2} a e_1 \wedge e_2 && \textit{mod } \mathfrak{g}_1 \\ \Gamma_k &= a e_3 \wedge e_k && \textit{mod } \mathfrak{g}_1 \quad (k = 4, \cdots, n) \end{aligned}$$

Now, we proceed as in the proof of Theorem 4. We define a $\mathfrak{so}(1, n-1)/\mathfrak{g}_1$-valued function $\tilde{\Gamma}$ on $\mathbb{R}^n$ using connection functions Γ_i in Lemma 11. It is easily seen that $\tilde{\Gamma}$ is invariant by the action of $\mathfrak{g}_1$. Since $\nabla^2 A$ is determined by ∇A and $\tilde{\Gamma}$, we have $\mathfrak{g}_1 = \mathfrak{g}_2$. The constant a in Eq. (26) and Lemma 11 is given by $a = \sqrt{-2\lambda}$. Therefore the expressions of ∇A and $\nabla^2 A$ with respect to an adapted basis to A and ∇e_1 are completely determined by the value of λ (< 0) and ε $(= 1 \text{ or } -1)$. Applying Corollary 1, we have the following:

Theorem 5. *Let M be an $n(\geq 4)$-dimensional conformally flat homogeneous Lorentzian manifold whose linear operator A has the form (22). Moreover suppose that the line subbundle T_{00} is not parallel with respect to the Levi-Civita connection ∇. Then the constant λ in Eq. (22) is negative and M is locally isometric to the homogeneous Lorentzian manifold constructed in Example 2 with $c = \sqrt{-\lambda/2}$.*

Now, we consider Case II, that is, the line subbundle T_{00} is parallel. By Lemma 9, we have $\nabla e_1 = a e_2^* \otimes e_1$. Since ∇e_1 is invariant by the action of $I_o(M)$, the coefficient a is constant on M.

Lemma 12. *Using the Eq. (2), we have $\lambda = 0$.*

By Eq. (25), we obtain

$$\nabla_{e_i} A = 0 \quad (i = 1, 3, \cdots, n), \qquad \nabla_{e_2} A = 2\varepsilon a e_2^* \otimes e_1.$$

From these, it follows that $\mathfrak{g}_0 = \mathfrak{g}_1$. Comparing A and ∇A with those of Example 3, we have the following:

Theorem 6. *Let M be an $n(\geq 4)$-dimensional conformally flat homogeneous Lorentzian manifold whose linear operator A has the form (22). Moreover suppose that the line subbundle T_{00} is parallel with respect to the Levi-Civita connection ∇. Then, $\lambda = 0$, and M is locally isometric to the homogeneous Lorentzian manifold constructed in Example 3.*

5 Case 2 with Two Eigenvalues λ *and* $-\lambda$

In this section, we study the case whose linear operator A has the form

$$A = \lambda I_r - \lambda I_{n-r} + \varepsilon e_2^* \otimes e_1 \qquad (\lambda \neq 0), \tag{27}$$

where $I_r = \sum_{i=1}^{r} e_i^* \otimes e_i$ and $I_{n-r} = \sum_{i=r+1}^{n} e_i^* \otimes e_i$, $1 < r < n$. We will show that such case does not occur (Theorem 7).

Let M be an $n(\geq 4)$-dimensional conformally flat homogeneous Lorentzian manifold whose linear operator has the form (27). As in Sect. 4, there exists the unique light-like vector field e_1 up to sign such that at each point of M, $\{e_1, e_2, \cdots, e_n\}$ is an adapted semi-orthonormal basis together with tangent vectors $e_2, \cdots, e_n$.

Lemma 13. *Using the Eq. (3), we have*

$$\Gamma_{11}^{1} = \Gamma_{11}^{j} = \Gamma_{11}^{k} = \Gamma_{12}^{k} = \Gamma_{1j}^{k} = 0$$

$$\Gamma_{21}^{k} = 0, \quad \Gamma_{22}^{k} = \frac{\varepsilon}{\lambda}\Gamma_{k1}^{1}, \quad \Gamma_{2j}^{k} = \frac{\varepsilon}{2\lambda}\Gamma_{k1}^{j}$$

$$\Gamma_{i1}^{1} = \frac{1}{2}\Gamma_{21}^{i}, \quad \Gamma_{i1}^{j} = \Gamma_{i1}^{k} = 0, \quad \Gamma_{i2}^{k} = \frac{\varepsilon}{2\lambda}\Gamma_{k1}^{i}, \quad \Gamma_{ij}^{k} = 0$$

$$\Gamma_{k1}^{\ell} = \Gamma_{k2}^{\ell} = \Gamma_{ki}^{\ell} = 0 \qquad (3 \leq i, j \leq r,\ r+1 \leq k, \ell \leq n).$$

Lemma 14. *In Eq. (27), we have $r \geq 3$.*

Proof. Suppose that $r = 2$. Then, there exists the unique pair of smooth light-like vector fields e_1 and e_2 up to sign such that $Ae_1 = \lambda e_1$, $Ae_2 = \varepsilon e_1 + \lambda e_2 and \langle e_1, e_2 \rangle = 1$. Moreover, the vector fields e_1 and e_2 are invariant by the action of $I_o(M)$. By Lemma 13, we have $\nabla_{e_1} e_1 = 0$, $\nabla_{e_1} e_2 = 0 and\ \nabla_{e_2} e_1 = a e_1$. Since e_1 and e_2 are invariant by the action of $I_o(M)$, the coefficient a is constant on M. Then, we have $R(e_1, e_2)e_1 = 0$. On the other hand, by Eq. (2), $R(e_1, e_2)e_1 = 2\lambda e_1$, which implies $\lambda = 0$. This is contrary to our assumption $\lambda \neq 0$. □

At each point $p \in M$, we define $T_\lambda(p) = \mathrm{Ker}(A - \lambda I_d)$ and $T_{-\lambda}(p) = \mathrm{Ker}(A + \lambda I_d)$. Then, the Lorentzian metric restricted to $T_{-\lambda}(p)$ is positive definite. Let C be the conullity operator associated with the eigenspace distribution $T_{-\lambda}$. Then by Lemma 13, C is expressed with respect to an adapted semi-orthonormal basis as follows:

$$C_{e_k} = \frac{\varepsilon}{\lambda}\Gamma_{k1}^{1} e_2^* \otimes e_1 + \frac{\varepsilon}{2\lambda}\sum_{j=3}^{r}\Gamma_{k1}^{j}(e_2^* \otimes e_j + e_j^* \otimes e_1) \quad k = r+1, \cdots, n. \tag{28}$$

The linear endomorphism Y of $T_{-\lambda}^{\perp}$ defined in Sect. 2 is given by

$$Y = -\frac{\varepsilon}{4\lambda^2} e_2^* \otimes e_1. \tag{29}$$

For Y, we make similar arguments as in Sect. 3. For a point $p \in M$, we denote by $M_{-\lambda}(p)$ the leaf of $T_{-\lambda}$ through p. For a unit vector $v \in T_{-\lambda}(p)$, we take a unit speed geodesic $\gamma : \mathbb{R} \to M_{-\lambda}(p)$ such that $\gamma(0) = p$ and $\dot{\gamma}(0) = v$. Let $\{e_1, e_2, \cdots, e_r\}$ be an adapted semi-orthonormal basis of $T_{-\lambda}^{\perp}(p)$. We denote by $\{E_1, E_2, \cdots, E_r\}$ the parallel frame field of $T_{-\lambda}^{\perp}$ along γ such that $E_i(0) = e_i$ $(1 \leq i \leq r)$. We express Y as $r \times r$-matrices with respect to $\{E_1, E_2, \cdots, E_r\}$ along γ. Then, by Eqs. (14) and (15), we have

$$Y(t) = \cos(\sqrt{-2\lambda t})Y(0) + \sin(\sqrt{-2\lambda t})\frac{1}{\sqrt{-2\lambda}}Y'(0) \qquad \text{for } -\lambda > 0$$

$$Y(t) = \cosh(\sqrt{2\lambda t})Y(0) + \sinh(\sqrt{2\lambda t})\frac{1}{\sqrt{2\lambda}}Y'(0) \qquad \text{for } -\lambda < 0.$$

First, we consider the case when $-\lambda > 0$. By the same argument as in Sect. 3, there exsits an $r \times r$-matrix

$$X = \begin{pmatrix} x_{11} & 0 & -{}^t x_2 \\ 0 & -x_{11} & -{}^t x_1 \\ x_1 & x_2 & X' \end{pmatrix}, \qquad x_1, x_2 \in \mathbb{R}^{r-2}, X' \in \mathfrak{so}(r-2)$$

such that

$$\frac{Y'(0)}{\sqrt{-2\lambda}} = [X, Y(0)] = -\frac{\varepsilon}{4\lambda^2}\begin{pmatrix} 0 & 2x_{11} & {}^t x_1 \\ 0 & 0 & 0 \\ 0 & x_1 & 0 \end{pmatrix}.$$

Lemma 15. *The case $-\lambda > 0$ does not occur.*

Proof. Since the rank of $Y(t) = 1$, we have $x_{i1} = 0$ $(i = 3, \cdots, r)$. There exists some $t \in \mathbb{R}$ such that $\cos(\sqrt{-2\lambda t}) + 2x_{11}\sin(\sqrt{-2\lambda t}) = 0$. This implies $Y(t) = 0$. It is a contradiction. □

From now on, we assume that $-\lambda < 0$. By the same argument as the proof of Lemma 15, we see that $Y'(0) = be_2^* \otimes e_1$ for some $b \in \mathbb{R}$. Since $\nabla_v \Phi_{-\lambda}^{-1} = Y'(0)$ and $C_v = -\left(\nabla_v \Phi_{-\lambda}^{-1}\right)\Phi_{-\lambda} = -2\lambda be_2^* \otimes e_1$, comparing with Eq. (28), we have $\Gamma_{k1}^j = 0$ $(j = 3, \cdots, r,\ k = r+1, \cdots, n)$. Hence, by Lemma 13, $\Gamma_{2j}^k = \Gamma_{j2}^k = \varepsilon/(2\lambda)\Gamma_{k1}^j = 0$. By Lemma 13, we have $\mathrm{Im}\nabla e_1 \subset T_\lambda$. We define a line subbundle T_{00} of TM by $T_{00} = \mathrm{Im}(A - \lambda I_d)|_{T_{-\lambda}^\perp} = \mathbb{R}e_1$ and a T_λ/T_{00}-valued 1-form α by $\alpha(X) = \pi(\nabla_X e_1)$, where π denotes the projection of T_λ onto T_λ/T_{00}. Then, α is invariant by the action of $I_o(M)$. Moreover, we have $\alpha \not\equiv 0$. In fact, if $\alpha \equiv 0$, the line subbundle T_{00} is parallel with respect to the Levi-Civita connection ∇. On the other hand, by Eq. (2), $R(e_2, e_3)e_1 = -2\lambda e_3$ $(\lambda \neq 0)$, which is a contradiction. Therefore $\alpha \not\equiv 0$. Moreover, we have Ker $\alpha = T_\lambda + T_{-\lambda}$. By an argument similar to the one used in Lemma 10, we can prove the following.

Lemma 16. *At each point $p \in M$, there exists an adapted semi-orthonormal basis $\{e_1, e_2, e_3, \cdots, e_r\}$ of $T_{-\lambda}^\perp$ to A such that $\nabla_{e_2} e_1 = ae_3$ for some $a (> 0)$. Moreover, such e_3 is uniquely determined.*

By Lemma 16, we obtain the unit vector field e_3 and it is invariant by the action of $I_o(M)$. By this, it follows that the coefficient a in Lemma 16 is constant on M.

Lemma 17. *Using Eq. (2), we have $a^2 = -2\lambda$. In particular, $-\lambda > 0$, and this is contrary to our assumption $-\lambda < 0$.*

Proof. By Lemma 16, we have $R(e_2, e_3)e_1 = a^2 e_3 - a\nabla_{e_3} e_3 + a/2\ \Gamma_{32}^3 e_1$, and hence $\langle R(e_2, e_3)e_1, e_3 \rangle = a^2$. On the other hand, by Eq. (2), $R(e_2, e_3)e_1 = -2\lambda e_3$. Therefore, we have $a^2 = -2\lambda$. □

Consequently, we obtain the following.

Theorem 7. *There is no $n(\geq 4)$-dimensional conformally flat homogeneous Lorentzian manifold whose linear operator A has the form (27).*

Acknowledgements The second author is partially supported by Grants-in-Aid for Scientific Research No. 20540067.

References

1. Calvaruso, G.: Einstein-like metrics on three-dimensional homogeneous Lorentzian manifolds. Geom. Dedicata **127**, 99–119 (2007)
2. Ferus, D.: Totally geodesic foliations. Math. Ann. **188**, 313–316 (1970)
3. Honda, K., Tsukada, K.: Conformally flat semi-Riemannian manifolds with nilpotent Ricci operators and affine differential geometry. Ann. Global Anal. Geom. **25**, 253–275 (2004)
4. Honda, K., Tsukada, K.: Three-dimensional conformally flat homogeneous Lorentzian manifolds. J. Phys. A: Math. Theor. **40**, 831–851 (2007)
5. Kobayashi, S., Nomizu, K.: Foundations of Differential Geometry, vol. II. Interscience Publishers, New York (1969)
6. Nomizu, K.: On local and global existence of Killing vector fields. Ann. Math. **72**, 105–120 (1960)
7. Nomizu, K.: Sur les algèbres de Lie de générateurs de Killing et l'homogénéité d'une variété riemannienne. Osaka Math. J. **14**, 45–51 (1962)
8. Singer, I.M.: Infinitesimally homogeneous spaces. Comm. Pure Appl. Math. **13**, 685–697 (1960)
9. Takagi, H.: Conformally flat Riemannian manifolds admitting a transitive group of isometries I, II. Tohoku Math. J. **27**, 103–110(I), 445–451(II) (1975)

Polar Actions on Symmetric Spaces

José Carlos Díaz-Ramos

1 Introduction

An isometric action is said to be polar if there exists a submanifold (called section) of the manifold acted upon that intersects all the orbits of the action orthogonally. If the section is flat, then the action is called hyperpolar. A particular type of hyperpolar action is a cohomogeneity one action, in which case the section is a geodesic. Polar actions can be viewed as a way to generalize polar coordinates of the Euclidean space, and sections are sometimes regarded as sets of canonical forms.

The purpose of this chapter is to review the basic facts about polar and hyperpolar actions. We will be especially interested in classification results. We will also comment on the history of polar actions and provide a comprehensive bibliography on the topic.

The results presented here are purely Riemannian. Many difficulties arise in the Lorentzian setting. Remark 1 briefly reviews some of the problems in this situation.

The chapter is organized as follows. In Sect. 2 we introduce the basic definitions and notation, give some elementary properties and examples of polar actions, and comment on the history of this problem. Then in Sect. 3 we review some basic facts regarding the theory of symmetric spaces. In Sect. 3.1 we focus on symmetric spaces of noncompact type as a more detailed description will be needed in Sect. 4.3. Section 4 is the main part of the chapter. We present the classification results known so far for polar actions. We start with Euclidean spaces (Sect. 4.1). Then, we consider symmetric spaces of compact type, focusing first on rank one symmetric spaces (Sect. 4.2.1) and then on higher rank (Sect. 4.2.2). Finally, we study the noncompact case in Sect. 4.3.

J.C. Díaz-Ramos (✉)
University of Santiago de Compostela, Santiago de Compostela, Spain
e-mail: josecarlos.diaz@usc.es

M. Sánchez et al. (eds.), *Recent Trends in Lorentzian Geometry*, Springer Proceedings in Mathematics & Statistics 26, DOI 10.1007/978-1-4614-4897-6_14,

The author has been supported by a Marie-Curie European Reintegration Grant (PERG04-GA-2008-239162) and by projects MTM2009-07756 and INCITE 09 207151PR (Spain).

2 Polar and Hyperpolar Actions

Let M be a connected Riemannian manifold with Riemannian metric $\langle \cdot, \cdot \rangle$. An isometry of M is a diffeomorphism $\phi \colon M \to M$ such that $\langle \phi_{*p} v, \phi_{*p} w \rangle = \langle v, w \rangle$ for any v, $w \in T_pM$ and any $p \in M$, and where ϕ_{*p} denotes the differential of ϕ at p. It is easy to see that the set $I(M)$ of all isometries of M is a group under composition of maps. Moreover, it was proved in [41] that $I(M)$ can be given a differentiable structure under which it becomes a Lie group.

A Killing vector field is a vector field V on M that satisfies $\mathscr{L}_V \langle \cdot, \cdot \rangle = 0$, where $\mathscr{L}$ stands for the Lie derivative. Let X be an element of the Lie algebra $\mathfrak{i}(M)$ of $I(M)$. Then $\operatorname{Exp}(tX)$ induces a one-parameter family of isometries on M, where Exp denotes the Lie exponential map. The vector field X^* defined by $X^*_p = \frac{d}{dt}_{|t=0} \operatorname{Exp}(tX)(p)$ is a Killing vector filed on M. This definition establishes an anti-isomorphism of Lie algebras between $\mathfrak{i}(M)$ and the module of Killing vector fields endowed with the usual Lie bracket.

A connected Lie group G acts isometrically on M if there exists a smooth map $G \times M \to M$, $(g, p) \mapsto g \cdot p$, such that $e \cdot p = p$, $(gh) \cdot p = g \cdot (h \cdot p)$, for all $g, h \in G$ and $p \in M$, and the map $M \to M$, $p \mapsto g \cdot p$ is an isometry for all $g \in G$. (Here, e is the identity element of G.) We will denote by $G \cdot p = \{g \cdot p \in M : g \in G\}$ the orbit of G through p, and by $G_p = \{g \in G : g \cdot p = p\}$ the isotropy group of G at p. The isometric action is said to be transitive if $G \cdot p = M$ for some $p \in M$, and it is said to be effective if $g \cdot p = p$ for all $p \in M$ implies that $g = e$. If an isometric action is effective, the group G can be regarded as a subgroup of $I(M)$, and this way, the action is just the natural action $g \cdot p = g(p)$. We say that G acts simply transitively if for each pair $p, q \in M$, there exists a unique $g \in G$ such that $q = g \cdot p$. In this case, G and M are diffeomorphic.

An action of a Lie group G on a manifold M is called proper [42] if the map $G \times M \to M \times M$, $(g, p) \mapsto (p, g \cdot p)$ is proper, that is, the inverse image of a compact subset of $M \times M$ is compact in $G \times M$. For effective isometric actions, this condition can be simplified: if $G \subset I(M)$ is a subgroup, then G acts properly on M if and only if G is closed in $I(M)$ [19, 53]. If G acts on M properly then the set of orbits M/G is a Hausdorff space with the quotient topology, the orbits of G are closed and embedded in M, and the isotropy groups are compact. Throughout this chapter we will always assume that actions are proper.

It follows from [40] that it is possible to define a partial ordering on the set of orbits. Two orbits $G \cdot p$ and $G \cdot q$ are said to be of the same orbit type if the isotropy groups G_p and G_q are conjugate in G. The orbit type of $G \cdot p$ is said to be smaller than the orbit type of $G \cdot q$ if G_q is conjugate in G to some subgroup of G_p. It can be shown that there exists a largest orbit type, and an orbit belonging to this orbit type is called a principal orbit. An orbit of the same dimension as a principal orbit but not principal is called exceptional, and the orbits that have strictly smaller dimension than the principal orbits are called singular. The cohomogeneity of an action is by definition the codimension of a principal orbit.

An isometric action on a manifold induces two important isometric actions on Euclidean spaces. Let G act isometrically on M and take $p \in M$. We denote by $T_p(G \cdot p)$ the tangent space of the orbit through p, and by $\nu_p(G \cdot p)$ its normal space. Since G_p fixes p, its differential acts on T_pM, and since the action is isometric, it leaves the tangent space $T_p(G \cdot p)$ and the normal space $\nu_p(G \cdot p)$ to the orbit through p invariant. The corresponding actions

$$\begin{aligned} G_p \times T_p(G \cdot p) &\to T_p(G \cdot p) \\ (g, v) &\mapsto g_{*p}(v) \end{aligned} \qquad \text{and} \qquad \begin{aligned} G_p \times \nu_p(G \cdot p) &\to \nu_p(G \cdot p) \\ (g, \xi) &\mapsto g_{*p}(\xi) \end{aligned}$$

are called the *isotropy representation* and the *slice representation* at p, respectively.

We say that two isometric actions of G on a manifold M are conjugate if there exists an isometry $\phi \colon M \to M$ such that $\phi(g \cdot p) = g \cdot \phi(p)$ for each $g \in G$ and each $p \in M$. Let G and H act on the manifold M. We say that the actions of G and H are *orbit equivalent* if there exists an isometry $\phi \colon M \to M$ that maps the orbits of G to the orbits of H, that is, $\phi(G \cdot p) = H \cdot \phi(p)$ for all $p \in M$. (Note that we are always assuming that the groups acting are connected, and thus orbits are also connected.) In this chapter we will be interested in the orbits of isometric actions, and thus, it is orbit equivalence the concept that will be more relevant.

Let G be a closed subgroup of the isometry group of M. We say that the action of G on M is *polar* if there exists an embedded submanifold Σ of M such that:

1. Σ intersects all the orbits of G.
2. $T_p(G \cdot p)$ and $T_p\Sigma$ are orthogonal for each $p \in \Sigma$.

The submanifold Σ is called a *section* of this polar action. Furthermore, this action is said to be *hyperpolar* if Σ is a flat Riemannian manifold.

It is obvious from the definition that the polarity of an action depends solely on the orbit structure. On the other hand, all sections are conjugate in M by the group action. We also have (see, e.g., [3]):

Proposition 1. *For a polar action on a Riemannian manifold we have:*

1. *Sections are totally geodesic.*
2. *The normal bundle of a principal orbit of a polar action is flat and has trivial normal holonomy.*

The above result implies that a polar action on a Euclidean space is hyperpolar, and hence both concepts are equivalent in this space. However, as we will see later, these two concepts are not equivalent in general. In fact, there are polar actions that are not hyperpolar in any symmetric space of rank one [44].

Remark 1. Although the definition of polar action seems to be possible in the Lorentzian (or pseudo-Riemannian) setting, some difficulties arise. Many interesting actions on Lorentzian manifolds are not proper, and hence it is possible that the orbits are not closed, the isometry groups are not compact, and the orbit space is not Hausdorff. For example, the action of $SO(1,n)$ on $\mathbb{R}^{1,n}$, the Minkowski space, is not proper: it has orbits that are not closed (the light cones) where the metric tensor is degenerate. Moreover, some orbits are Riemannian and others are Lorentzian. Obviously, a connected submanifold with a nondegenerate metric cannot change signature, so it will be difficult in general to find a section for an action on a Lorentzian manifold, and the definition of polar action would need to be adapted somehow.

The study of isometric actions on Lorentzian manifolds is nevertheless an interesting topic. See, for example, the survey paper [18] where nonproper actions of Lie groups are investigated, or the papers [1, 17] where the isometry group of compact Lorentzian manifold is studied. Another relevant reference is [24], where the authors show that a polar orthogonal representation of a real reductive algebraic group has the same closed orbits as the isotropy representation of a pseudo-Riemannian symmetric space. This paper studies the so-called polar representations introduced by Dadok and Kac in [16].

We begin with some basic examples. More complicated ones will be supplied in Sect. 4.

Example 1. 1. Any trivial or transitive action is polar. A section is the whole manifold or any point, respectively.

2. If G acts on M polarly with section Σ and $p \in \Sigma$, then the slice representation $G_p \times \nu_p(G \cdot p) \to \nu_p(G \cdot p)$ is polar with section $T_p\Sigma$.

   The converse does not hold: the action of $\mathbb{R}$ on $\mathbb{R}^3$ given by

$$t \cdot (x,y,z) = \begin{pmatrix} \cos(t) & -\sin(t) & 0 \\ \sin(t) & \cos(t) & 0 \\ 0 & 0 & 1 \end{pmatrix} \begin{pmatrix} x \\ y \\ z \end{pmatrix} + \begin{pmatrix} 0 \\ 0 \\ t \end{pmatrix},$$

   is not polar (orbits are helices), but the slice representations are trivial, and hence polar.

3. A cohomogeneity one action is hyperpolar. Any geodesic intersecting one orbit orthogonally must intersect all the orbits orthogonally.
4. The action of $SO(2)$ on $\mathbb{R}^2$ is polar and any line through the origin is a section. The action of $SO(3)$ on $\mathbb{R}^3$ is polar, and any line through the origin is a section of the $SO(3)$-action. The action of $SO(2) \times \mathbb{R}$ is polar; its orbits are coaxial cylinders and their axis, and any line that intersects this axis orthogonally is a section of the action. Thus, polar, spherical, and cylindrical coordinates are in a way particular cases of polar actions. Indeed, the motivation for the name 'polar' comes precisely from the fact that polar actions generalize polar coordinates [15].
5. Sections of polar actions are sometimes regarded as sets of canonical forms [43]. The following example justifies this claim. Let us consider the vector space $\mathscr{S}_n$ of

symmetric $n \times n$ matrices with trace zero over the real numbers. The orthogonal group $SO(n)$ acts on $\mathscr{S}_n$ isometrically by conjugation, that is, $g \cdot A = gAg^{-1}$. This action is polar, and the set $\mathscr{D}_n$ of diagonal matrices with trace zero is a section of this action. Thus, the fact that $\mathscr{D}_n$ is a section implies the classical result that asserts that any symmetric matrix can be diagonalized by congruency with real eigenvalues. The set $\mathscr{D}_n$ corresponds to the Jordan canonical forms of $\mathscr{S}_n$. Incidentally, this action is exactly the isotropy representation of the symmetric space $SL(n,\mathbb{R})/SO(n)$.

Throughout this chapter we will consider polar actions that are neither trivial nor transitive. Whenever a classification result is presented we will always disregard these two cases even if not mentioned explicitly.

Hyperpolar actions were first considered by Conlon [14], although he did not use this terminology in his paper. Assume that G acts on M isometrically. We say that the action of G on M is variationally complete [13] if for any orbit $G \cdot p$ of G, and any geodesic γ orthogonal to $G \cdot p$, we have that any $(G \cdot p)$-Jacobi vector field along γ that is tangent to some other orbit of G is the restriction to γ of a Killing vector field induced by the action of G. Recall that if N is a submanifold of M, an N-Jacobi vector field is a Jacobi vector field corresponding to a geodesic variation of geodesics intersecting N perpendicularly. Conlon proved that a hyperpolar action by a compact Lie group K on M is variationally complete, although he called sections K-transversal domains. It was subsequently proved by Lytchak and Thorbergsson [39] that in complete Riemannian manifolds of nonnegative sectional curvature an isometric action is variationally complete if and only if it is hyperpolar. See also [25] for an earlier result on symmetric spaces of compact type.

The aim of this chapter is to address the following question: *given a connected complete Riemannian manifold M, classify polar actions on M up to orbit equivalence*. Obviously, this problem makes sense only in spaces with high degree of symmetry as, in general, a Riemannian manifold might not have nontrivial isometries at all. Hence, in Sect. 4, we will consider the classification problem in Riemannian symmetric spaces, which are a kind of homogeneous manifolds with large isometry group.

3 Symmetric Spaces

Symmetric spaces are ubiquitous in the study of polar actions. As we argued before, it is natural to consider the classification in these spaces due to their large symmetry, but they also appear when studying simpler spaces. See, for example, Sect. 4.1 or 4.2.1. Symmetric spaces were studied and classified by Cartan, and are interesting in Differential Geometry, Topology, Lie Group Theory, and Geometric Analysis. We begin with a quick review on this type of spaces. See [27] or [38] for more information.

Let M be a Riemannian manifold of dimension n with metric $\langle\cdot,\cdot\rangle$. Let $o \in M$. Take $r > 0$ sufficiently small so that normal coordinates are defined on the open ball $B_r(o)$. We define the local geodesic symmetry at o as the map $s_o\colon B_r(o) \to B_r(o)$ given by $s_o(\exp_o(tv)) = \exp_o(-tv)$ for $t \in \mathbb{R}$ and $v \in T_oM$, where TM denotes the tangent bundle of M. Note that, in general, this map is defined only locally. A Riemannian manifold M is said to be locally symmetric if at each point there is a ball such that the corresponding local geodesic symmetry is a local isometry. A locally symmetric space is characterized by the fact that $\nabla R = 0$, where ∇ is the Levi-Civita connection of M. A connected Riemannian manifold M is called a *symmetric space* if each local geodesic symmetry can be extended to a global isometry $s_o\colon M \to M$. Since isometries are characterized by their differential at a point, this is equivalent to saying that for each point $o \in M$ there is an involutive isometry of M such that o is an isolated fixed point of that isometry; this involutive isometry turns out to be s_o and we call it the symmetry of M at o.

If M is a connected, complete, locally symmetric space, then its Riemannian universal covering is a symmetric space. In particular, every locally symmetric space is locally isometric to a symmetric space. Moreover, every symmetric space is complete and homogeneous. However, there are complete locally symmetric spaces that are not symmetric, even not homogeneous (for example compact Riemann surfaces of genus $g \geq 2$).

Now we give a more algebraic description of symmetric spaces. Denote by G the connected component of the identity of $I(M)$ and by $\mathfrak{g}$ the Lie algebra of G. Let $o \in M$ and s_o the geodesic symmetry at o. Define K as the isotropy group of G at o, that is, $K = G_o$. The coset space G/K is diffeomorphic to M by means of the map $\Phi\colon G/K \to M$, $gK \mapsto g(o)$. If $\langle\cdot,\cdot\rangle$ denotes the metric obtained by pulling back the metric of M, then Φ becomes an isometry and the metric $\langle\cdot,\cdot\rangle$ is G-invariant, that is, the map $gK \mapsto hgK$ is an isometry for any $h \in G$. It can be shown that the Riemannian connection on G/K is the same for all G-invariant Riemannian metrics on G/K (e.g. [27, Chapter IV, Corollary 4.3]). The map $\sigma\colon G \to G$, $g \mapsto s_o g s_o$ is an involutive automorphism of G, and $G_\sigma^0 \subset K \subset G_\sigma$, where $G_\sigma = \{g \in G : \sigma(g) = g\}$, and G_σ^0 is the connected component of the identity of G_σ. Let θ be the differential of σ at the identity. The Lie algebra of K is given by $\mathfrak{k} = \{X \in \mathfrak{g} : \theta(X) = X\}$, and we define $\mathfrak{p} = \{X \in \mathfrak{g} : \theta(X) = -X\}$. The space $\mathfrak{p}$ may be identified with T_oM using the map Φ and taking into account that $\mathfrak{p}$ is a complementary subspace to $\mathfrak{k}$ in $\mathfrak{g}$. Thus, $\mathfrak{p}$ inherits an inner product from T_oM which turns out to be $\mathrm{Ad}(K)$-invariant. Moreover, we have the Lie bracket relations $[\mathfrak{k},\mathfrak{k}] \subset \mathfrak{k}$, $[\mathfrak{k},\mathfrak{p}] \subset \mathfrak{p}$, and $[\mathfrak{p},\mathfrak{p}] \subset \mathfrak{k}$. The decomposition $\mathfrak{g} = \mathfrak{k} \oplus \mathfrak{p}$ is called the *Cartan decomposition* of $\mathfrak{g}$ with respect to the involution θ (or the point $o \in M$), and θ is called the Cartan involution. The pair (G,K) so defined is an effective Riemannian symmetric pair, and it can be shown that there is a one-to-one correspondence between symmetric spaces and effective Riemannian symmetric pairs. The notion of Riemannian symmetric pair can also be defined if we drop the assumption that G acts effectively on M, see [27]. Actually, there is consensus to call a group G the isometry group of a Riemannian manifold if G is the isometry group modulo some finite group.

The rank of a symmetric space is by definition the dimension of a maximal abelian subspace of $\mathfrak{p}$, or equivalently, the dimension of a maximal flat, totally geodesic submanifold of M.

Let M be a symmetric space and $\tilde{M}$ its universal covering. Then the De Rham theorem says that $\tilde{M}$ can be decomposed as $\tilde{M} = \tilde{M}_0 \times \tilde{M}_1 \times \cdots \times \tilde{M}_k$. Here $\tilde{M}_0$ is the Euclidean factor, that is, $\tilde{M}_0$ is locally isometric to a Euclidean space, and each $\tilde{M}_i$, $i > 0$, is a simply connected irreducible symmetric space. A semisimple symmetric space is one for which the Euclidean factor of its universal covering space has dimension zero. In this case, the Lie algebra of the isometry group of $\tilde{M}$ is semisimple. A semisimple symmetric space is said to be of *compact type* if all the De Rham factors of its universal covering are compact. It is said to be of *noncompact type* if all the De Rham factors of its universal covering are non-Euclidean and noncompact. Again, the Lie algebra of the isometry group of a symmetric space of compact (resp. noncompact) type is compact (resp. noncompact). By definition, an irreducible symmetric space must be one of this three: flat, of compact type, or of noncompact type.

There is a duality between symmetric spaces of compact and noncompact type which we describe now. Assume (G,K) is a Riemannian symmetric pair such that $M = G/K$ is simply connected. We have the Cartan decomposition $\mathfrak{g} = \mathfrak{k} \oplus \mathfrak{p}$ as defined above. We consider the Lie subalgebra $\mathfrak{g}^* = \mathfrak{k} \oplus i\mathfrak{p}$ of the complexification $\mathfrak{g}^{\mathbb{C}}$ of $\mathfrak{g}$, and where i is the imaginary unit. Then, if G^* is the connected real Lie subgroup of $G^{\mathbb{C}}$ with Lie algebra $\mathfrak{g}^*$, we have that G^*/K is a simply connected Riemannian symmetric space. If G/K is of compact type, then G^*/K is of noncompact type, and if G/K is of noncompact type, then G^*/K is of compact type. This procedure establishes a one-to-one correspondence between simply connected symmetric spaces of compact and noncompact type, which is called duality. Some problems on Riemannian symmetric spaces (we will see that not all of them) can be addressed using duality. For example, the very classification of symmetric spaces can be performed in this way.

The action of K on T_oM is called the *isotropy representation* of the symmetric space $M = G/K$. By identifying T_oM with $\mathfrak{p}$, it follows that any maximal abelian subspace of $\mathfrak{p}$ is a section of this representation, and thus, the isotropy representation of a symmetric space is polar.

3.1 Riemannian Symmetric Spaces of Noncompact Type

The discussion above will be enough for Sects. 4.1 and 4.2, but we have to provide further information on the noncompact case for Sect. 4.3. We follow [27] for the theory of symmetric spaces and [32] for the theory of semisimple Lie algebras. We also follow the notation in [8].

Let M be a connected Riemannian symmetric space of noncompact type with dimension n and rank r. It is well known that M is a Hadamard manifold and therefore diffeomorphic to $\mathbb{R}^n$. Write M as $M = G/K$ as before. Now it can be shown that the Lie algebra of G is semisimple and of noncompact type, and K is a maximal compact subgroup of G. Let B be the Killing form of $\mathfrak{g}$. If $\mathfrak{g} = \mathfrak{k} \oplus \mathfrak{p}$ is the Cartan decomposition of $\mathfrak{g}$ with respect to $o \in M$ and θ is the corresponding Cartan involution, we can define a positive definite inner product on $\mathfrak{g}$ by $\langle X, Y \rangle = -B(X, \theta Y)$ for all $X, Y \in \mathfrak{g}$.

We now fix a maximal abelian subspace $\mathfrak{a} \subset \mathfrak{p}$ and denote by $\mathfrak{a}^*$ the dual space of $\mathfrak{a}$. For each $\lambda \in \mathfrak{a}^*$ we define $\mathfrak{g}_\lambda = \{X \in \mathfrak{g} : \mathrm{ad}(H)X = \lambda(H)X \text{ for all } H \in \mathfrak{a}\}$. We say that a nonzero $\lambda \in \mathfrak{a}^*$ is a restricted root if $\mathfrak{g}_\lambda \neq 0$, and we denote by Σ the set of all restricted roots. Since $\mathfrak{a}$ is abelian, $\mathrm{ad}(\mathfrak{a})$ is a commuting family of self-adjoint linear transformations of $\mathfrak{g}$. This implies that the subset $\Sigma \subset \mathfrak{a}^*$ of all restricted roots is nonempty, finite and $\mathfrak{g} = \mathfrak{g}_0 \oplus (\bigoplus_{\lambda \in \Sigma} \mathfrak{g}_\lambda)$ is an orthogonal direct sum called the restricted root space decomposition of $\mathfrak{g}$ determined by $\mathfrak{a}$. Here, $\mathfrak{g}_0 = \mathfrak{k}_0 \oplus \mathfrak{a}$, where $\mathfrak{k}_0 = Z_{\mathfrak{k}}(\mathfrak{a})$ is the centralizer of $\mathfrak{a}$ in $\mathfrak{k}$. We introduce an ordering in Σ and denote by Σ^+ the resulting set of positive roots. We denote by $\Lambda = \{\alpha_1, \ldots, \alpha_r\}$ the set of simple roots of Σ^+, that is, each root $\lambda \in \Sigma$ can be written as $\lambda = \sum_{i=1}^{r} c_i \alpha_i$ where all the c_i are integers, and they are all nonpositive or nonnegative depending on whether the root is negative or positive. The inner product on $\mathfrak{p}$ induces an inner product on $\mathfrak{a}$, which in turn induces an inner product on $\mathfrak{a}^*$. Given $\lambda \in \Sigma$ we can define its metric dual $H_\lambda \in \mathfrak{a}$ as $\langle H_\lambda, H \rangle = \lambda(H)$ for each $H \in \mathfrak{a}$. It can be shown that two simple roots are orthogonal if they are not connected in the so-called Dynkin diagram of Σ.

The subspace $\mathfrak{n} = \bigoplus_{\lambda \in \Sigma^+} \mathfrak{g}_\lambda$ of $\mathfrak{g}$ is a nilpotent subalgebra of $\mathfrak{g}$, and $\mathfrak{a} \oplus \mathfrak{n}$ is a solvable subalgebra of $\mathfrak{g}$ with $[\mathfrak{a} \oplus \mathfrak{n}, \mathfrak{a} \oplus \mathfrak{n}] = \mathfrak{n}$. We can write $\mathfrak{g}$ as the direct sum of vector subspaces $\mathfrak{g} = \mathfrak{k} \oplus \mathfrak{a} \oplus \mathfrak{n}$, the so-called Iwasawa decomposition of $\mathfrak{g}$. Let A, N and AN be the connected subgroups of G with Lie algebra $\mathfrak{a}$, $\mathfrak{n}$, and $\mathfrak{a} \oplus \mathfrak{n}$, respectively. All these subgroups are simply connected and G is diffeomorphic to the product $K \times A \times N$. Moreover, the simply connected, solvable Lie group AN acts simply transitively on M. We can then equip AN with a left-invariant Riemannian metric so that M and AN become isometric.

4 Classification of Polar Actions

In this section, we present some of the classification results of polar actions in symmetric spaces. The classification problem was initiated by Dadok [15], who classified polar actions by compact Lie groups on Euclidean spaces (see Sect. 4.1). In [26], the authors state the interest of classifying (hyper-)polar actions on symmetric spaces of compact type. We present the basic results in Sect. 4.2. The analogous problem in symmetric spaces of noncompact type seems to be more difficult (see Sect. 4.3). There are other interesting problems related to polar actions. For example, the classification of polar actions in Hilbert spaces [48]. Indeed, Hilbert spaces arise naturally in the study of homogeneous submanifolds of compact symmetric spaces, so the infinite dimensional case deserves attention. However, in this chapter we will restrict ourselves to finite dimensional manifolds.

It is also interesting to point out the relation between polar actions and isoparametric submanifolds and their generalizations. Some of the classification problems can be stated in much greater generality for singular Riemannian foliations, isoparametric or equifocal submanifolds. We refer to [49–51] for further information.

4.1 Euclidean Spaces

Let us denote by $\mathbb{R}^n$ the n-dimensional Euclidean space. Since sections are totally geodesic, it follows that any polar action is hyperpolar, so both concepts become equivalent. Dadok addressed the problem of classifying polar actions in Euclidean spaces, although he focused his attention in actions by compact Lie groups. However, the general case does not introduce much complexity:

Theorem 1. *Assume that H acts polarly on the Euclidean space $\mathbb{R}^n$. Then, the action of H on $\mathbb{R}^n$ is orbit equivalent to the action of $\mathbb{R}^k \times K$, where $\mathbb{R}^k$ acts on $\mathbb{R}^n$ by translations, and K is a compact Lie group acting polarly on $\mathbb{R}^{n-k}$, the orthogonal complement of $\mathbb{R}^k$.*

Thus, from now on, we will assume that H is a compact Lie group acting polarly on $\mathbb{R}^n$. Actually, we may also assume that the action is irreducible, as the next theorem shows [15]. This theorem is usually quoted in the literature as an important step in other classifications.

Theorem 2. *Assume that a compact group H acts polarly on a Euclidean space V, and that the action is reducible, that is, V can be decomposed as $V = V_1 \oplus V_2$, and H leaves V_1 and V_2 invariant. Then:*

1. *The group H acts on V_1 and on V_2 polarly; a section for the action of H on V is of the form $\Sigma = \Sigma_1 \oplus \Sigma_2$, where Σ_i is a section of the polar action of H on V_i, for $i = 1, 2$.*
2. *The action of H is orbit equivalent to a product action $H_1 \times H_2$ on $V_1 \oplus V_2$, where H_i acts polarly on V_i, $i = 1, 2$.*

Dadok's main result is the following:

Theorem 3. *Let H be a compact Lie group acting polarly on a Euclidean space. Then, the action of H is orbit equivalent to the isotropy representation of a Riemannian symmetric space.*

Since Riemannian symmetric spaces are classified (see e.g., [27]) and their isotropy representations are known, this concludes the classification of polar actions on Euclidean spaces. Dadok's proof is based on the classification of irreducible representations. A more conceptual proof of this theorem can be found in [22]. Not any polar action on a Euclidean space is conjugate to the isotropy representation of a Riemannian symmetric space. In fact [23] gives a list of irreducible representations that have the same orbits as the isotropy representation of a Riemannian symmetric space. See also [2] where reducible polar representations that are not products of irreducible ones are studied.

4.2 Symmetric Spaces of Compact Type

After Dadok's result, the classification was attempted in symmetric spaces of compact type as the paper [26] suggests. The rank one case is particularly interesting as many examples of polar, non-hyperpolar actions arise. We will review these results in the next subsection and concentrate in higher rank in Sect. 4.2.2.

4.2.1 Symmetric Spaces of Rank One

By definition of rank of a symmetric space, it is clear that a hyperpolar action on a symmetric space of rank one must be of cohomogeneity one, as the dimension of the section, which is totally geodesic and flat, cannot exceed the rank. Cohomogeneity one actions on spheres were classified by Takagi and Takahashi [47] following the work of Hsiang and Lawson [29], to whom the classification is normally merited. Takagi also derived the classification in complex projective spaces [46], while the classification of cohomogeneity one actions on quaternionic projective spaces and the Cayley projective plane is due to Iwata [30, 31]. The next remarkable result, Theorem 4, on the one hand, includes the previous classifications and on the other hand, provides examples of polar actions in symmetric spaces of rank one which are not hyperpolar.

The following theorem was proved by Podestà and Thorbergsson in [44]. We start by giving the main result and comment on it later.

Theorem 4. *We have:*

1. *If H acts polarly on a sphere S^n, then the action of H is, up to orbit equivalence, induced by the isotropy representation of a Riemannian symmetric space.*
2. *If H acts polarly on a complex projective space $\mathbb{C}P^n$, then the action of H is, up to orbit equivalence, induced by the isotropy representation of a Hermitian symmetric space.*
3. *If H acts polarly on a quaternionic projective space $\mathbb{H}P^n$, then the action of H is, up to orbit equivalence, induced by isotropy representations of products of k quaternionic-Kähler symmetric spaces, where at least $k-1$ have rank one.*
4. *Any polar action on the Cayley hyperbolic plane $\mathbb{O}P^2$ is conjugate to the action of one of the following groups: $Sp(1)\cdot Sp(3)$, $T^1\cdot Sp(3)$, $Sp(3)$, $Spin(9)$, $Spin(8)$, $T^1\cdot Spin(7)$, $SU(2)\cdot SU(4)$, or $SU(3)\cdot SU(3)$.*

The classification in spheres is actually very simple. A closed subgroup H of $SO(n)$, the connected component of the identity of the isometry group of S^n, is compact and hence its action can be extended to $\mathbb{R}^n$ with a fixed point at the origin.

If the original action on S^n is polar with section Σ, it is easy to see that the affine subspace generated by Σ is a section of the H action on $\mathbb{R}^n$. Conversely, a polar action on a Euclidean space by a compact group has a fixed point (by Cartan's fixed point theorem) and hence leaves the spheres centered at this point invariant. A section on one of these spheres is obtained just by intersecting the original section with the sphere. Hence, the classification in S^n follows from Dadok's result (Theorem 3), and there is a one-to-one correspondence (up to translation of the fixed point) between polar actions on spheres and polar actions by compact groups on Euclidean spaces.

Actually, [44] focuses on the last three cases. The first result proved there is the fact that sections must be totally real, and from there, they conclude that a section must be some totally geodesic real projective space. Our goal in the rest of this section is to explain how the examples of Theorem 4 are constructed.

We start with the complex projective space $\mathbb{C}P^n$. A Hermitian symmetric space $M = G/K$ is a Riemannian symmetric space endowed with a complex structure J invariant under the geodesic symmetries. Assume that K is the isotropy group at o. Then the tangent space at o has a complex structure J_o, which commutes with the elements of $\mathrm{Ad}(K)$, and which turns $T_oM \cong \mathfrak{p}$ into a complex vector space $\mathbb{C}^{n+1}$. By Dadok's result, the isotropy representation $K \times \mathfrak{p} \to \mathfrak{p}$ is polar with totally real section $\mathfrak{a}$, a maximal abelian subspace, and by the previous discussion, it induces a polar action on the unit sphere S^{2n+1} of $\mathbb{C}^{n+1}$. A complex projective space can be defined as $\mathbb{C}P^n = S^{2n+1}/S^1$, and since J_o is $\mathrm{Ad}(K)$-invariant, the action of K on $\mathfrak{p}$ descends to an isometric action on $\mathbb{C}P^n$. Using the fact that $\mathfrak{a}$ is real, it is not difficult to see that $(\mathfrak{a} \cap S^{2n+1})/S^1$ is a section of the induced action on $\mathbb{C}P^n$. The second item of Theorem 4 implies that any polar action on $\mathbb{C}P^n$ can be obtained, up to orbit equivalence, in this way.

Case 3 of Theorem 4 is similar to the previous case, although same difficulties arise in the description. The main difference stems from the fact that a reducible Hermitian symmetric space is the product of irreducible Hermitian symmetric spaces. The analogous statement does not hold for quaternionic-Kähler symmetric spaces, and this is the reason why a cartesian product appears explicitly in case 3. In order to handle the problem in $\mathbb{H}P^n$, we start with a quaternionic-Kähler symmetric space G/K. The tangent space $T_oM \cong \mathfrak{p}$ is now a quaternionic vector space $\mathbb{H}^{n+1}$, and arguing as before, one can conclude that the isotropy representation of K on $\mathfrak{p}$ descends to a polar action on $\mathbb{H}P^n$. Unlike in the complex case, there are more examples than these ones. Let (G_i, K_i), $i = 1, \ldots, k$, be quaternionic-Kähler symmetric pairs with corresponding Cartan decompositions $\mathfrak{g}_i = \mathfrak{k}_i \oplus \mathfrak{p}_i$. Then $K_i = H_i \cdot Sp(1)$, where H_i and $Sp(1)$ are normal subgroups of K_i. It follows that $H = H_1 \times \cdots \times H_k \times Sp(1)$ acts polarly on $\mathfrak{p} = \mathfrak{p}_1 \oplus \cdots \oplus \mathfrak{p}_k$, where $\mathfrak{p}$ has the product quaternionic structure, and $Sp(1)$ acts by right multiplication on $\mathfrak{p}$. Since $\mathfrak{p}$ is a quaternionic vector space we can write $\mathfrak{p} = \mathbb{H}^{n+1}$, and it follows that the H action on $\mathfrak{p}$ descends to an isometric action on the corresponding $\mathbb{H}P^n$. If at most one of the factors G_i/K_i has rank greater than one, the corresponding action is polar. Theorem 4 states that any polar action on $\mathbb{H}P^n$ can be obtained, up to orbit equivalence, in this way.

Finally, for Case 4 of Theorem 4, which corresponds to the Cayley projective plane $\mathbb{O}P^2 = F_4/Spin(9)$, the authors follow a different procedure. They disregard the cohomogeneity one case, which had already been solved by Iwata. This produces the examples $Sp(1)\cdot Sp(3), T^1\cdot Sp(3), Sp(3), Spin(9)$, or $Spin(8)$. Then, the authors consider polar actions by groups contained in the isotropy group $Spin(9)$. They show that these actions are polar if and only if their slice representation is polar (this result was stated in greater generality in [21]). All these actions have a fixed point and are conjugate to the action of the groups $Spin(9), Spin(8), T^1\cdot Spin(7)$, or $SU(2)\cdot SU(4)$. A completely different argument omitted here is used to show that the action of $SU(3)\cdot SU(3)$, which is a maximal subgroup of the exceptional group F_4, is polar on $\mathbb{O}P^2$.

4.2.2 Symmetric Spaces of Higher Rank

The classification problem for hyperpolar actions on symmetric spaces of compact type was posed in [26]. Kollross solved this problem for irreducible symmetric spaces of compact type in [33], which in particular implies the classification of cohomogeneity one actions on these spaces. We describe this classification result in what follows.

An irreducible symmetric space of compact type can be represented as $M = G/K$ as described in Sect. 3, where G is connected and compact. Now there are two possibilities: either G is simple, or $M = (G\times G)/\Delta G$, where G is simple and ΔG is the diagonal subgroup of $G\times G$. In the latter case, M is isometric to a simple group with a bi-invariant metric (i.e., $(G\times G)/\Delta G \cong G$).

We now introduce examples of hyperpolar actions. Let H and K be closed subgroups of a connected compact simple Lie group G, such that (G,K) and (G,H) are symmetric pairs (as described in Section 3). Then, the following actions are hyperpolar: the action $H\times G/K \to G/K$, $(h,gK)\mapsto hgK$, the action $K\times G/H \to G/H$, $(k,gH)\mapsto kgH$, and the action $(H\times K)\times G\to G$, $((h,k),g)\mapsto hgk^{-1}$. These actions are called *Hermann actions* [28].

Indeed we have

Proposition 2. *Let (G,K) be a symmetric pair and H a closed subgroup of G. Then, the action of H on G/K is hyperpolar if and only if the action of $H\times K$ on G is hyperpolar.*

The author goes on to argue that in order to classify hyperpolar actions on irreducible symmetric spaces of compact type it suffices to classify hyperpolar actions on compact simple Lie groups. This classification is given by the following theorem [33]:

Theorem 5. *A hyperpolar action on an irreducible symmetric space of compact type is orbit equivalent to a Hermann action or has cohomogeneity one.*

There are three types of hyperpolar actions:

1. *Hermann actions.*
2. *Cohomogeneity one actions induced by isotropy representations of symmetric spaces of rank two:*
3. *Seven exceptions of cohomogeneity one. They correspond to the action of* $H \times K$ *on* G*, where the triples* (H, G, K) *are given by*

H	G_2	G_2	U_3	$Spin_9$	Sp_1Sp_n	SU_3	SU_3
G	SO_7	SO_7	SO_7	SO_{16}	SO_{4n}	G_2	G_2
K	$SO_4 \times SO_3$	G_2	G_2	$SO_{14} \times SO_2$	$SO_{4n-2} \times SO_2$	SO_4	SU_3

Hermann actions are determined by pairs of symmetric pairs, and thus their classification follows from the classification of symmetric pairs. See [27]. The cohomogeneity one actions induced by the isotropy representations of symmetric spaces of rank two are also easy to derive. In fact, these appear in [47], as cohomogeneity one actions on spheres correspond exactly to isotropy representations of symmetric spaces of rank two. See [33, Theorem A] for a more explicit description of these examples.

Since the classification of hyperpolar actions on irreducible symmetric spaces was known after the work of Kollross [33], and the classification of polar actions on symmetric spaces of compact type and rank one was derived by Podestà and Thorbergsson in [44], the next natural step was to address this problem in symmetric spaces of higher rank.

An isometric action on a Kähler manifold is coisotropic if and only if the normal spaces to its principal orbits are totally real. Polar actions on irreducible symmetric spaces of compact type are coisotropic [45], and it turns out that it is possible to classify coisotropic actions on this spaces [11, 12]. It follows from this classification that polar actions on irreducible Hermitian symmetric spaces of rank greater than one are hyperpolar. This led Biliotti [11] to formulate the following conjecture: a polar action on an irreducible symmetric space of compact type and higher rank is hyperpolar. Kollross answered this question in the affirmative for symmetric spaces with simple isometry group (i.e, $M = G/K$, with G connected, compact, and simple) [34], and for the exceptional simple Lie groups [35]. The final step was given by Kollross and Lytchak [37], based on a result by the second author that states that polar singular Riemannian foliations of irreducible compact symmetric spaces of higher rank are hyperpolar if the codimension of the foliation is at least three. The answer to Biliotti's conjecture is affirmative:

Theorem 6. *A polar action on an irreducible symmetric space of compact type and rank higher than one is hyperpolar.*

As a final comment in this section, it is worthwhile to mention that the classification of (hyper-)polar actions on symmetric spaces of compact type cannot be obtained from the corresponding classification on the irreducible ones. This is a problem that remains open.

4.3 Symmetric Spaces of Noncompact Type

Since the classification of polar actions on irreducible symmetric spaces of compact type is known, the first approach that one may consider is to use the duality between symmetric spaces of compact and noncompact type. However, as we will see later duality cannot be applied in general. For example, it was proved in [44] that polar actions on irreducible compact symmetric spaces have singular orbits. However, horocycle foliations (which will be described later) induce foliations, all whose orbits are principal. Another possible argument that we will present later is the fact that in the noncompact case, there are polar actions that are not hyperpolar even in higher rank.

Indeed, the problem in the noncompact case seems to be much more involved. There are not so many spaces where classifications (even partial ones) are known. In what follows, we will try to summarize the main results obtained so far.

The first basic case is the real hyperbolic space. The classification of polar actions in $\mathbb{R}H^n = SO(1,n)/SO(n)$ follows from a work by Wu [52]:

Theorem 7. *Let H act polarly on $\mathbb{R}H^n$. Then, the action of H is orbit equivalent to:*

1. *The action of $SO(1,k)\times K$, where $k\in\{0,\dots,n-1\}$, and K is a compact group acting polarly on $\mathbb{R}^{n-k}$.*
2. *The action of $N\times K$, where N is the nilpotent part of the Iwasawa decomposition of $SO(1,k)$, $k\in\{2,\dots,n\}$, and K is a compact group acting polarly on $\mathbb{R}^{n-k}$.*

Hence, Theorem 7 states that either there is a totally geodesic $\mathbb{R}H^k$ (which might be a point) as an orbit, and the rest of the orbits are obtained by a polar action on the normal space to that orbit, or a totally geodesic $\mathbb{R}H^k$ is left invariant, one takes the horosphere foliation in that $\mathbb{R}H^k$, and the rest of the orbits are obtained by a polar action on the normal space to that totally geodesic $\mathbb{R}H^k$. As a particular case, N, the nilpotent part of the Iwasawa decomposition of $SO(1,n)$, acts polarly (even with cohomogeneity one) on $\mathbb{R}H^n$. The set of orbits of this action is known as the horosphere foliation, and each orbit is called a horosphere. All horospheres are congruent to each other, intrinsically they are Euclidean spaces, and extrinsically they are umbilical inside $\mathbb{R}H^n$. The horosphere foliation does not have a counterpart in spheres and hence cannot be obtained using duality. Additionally, note that, as was the case in spheres, there are polar non-hyperpolar actions on $\mathbb{R}H^n$.

There is one more symmetric space of noncompact type where a full classification of polar actions is known: the complex hyperbolic plane $\mathbb{C}H^2$. The root space decomposition of $\mathbb{C}H^n$ (see Sect. 3.1) can be written as $\mathfrak{g}=\mathfrak{g}_{-2\alpha}\oplus\mathfrak{g}_{-\alpha}\oplus\mathfrak{g}_0\oplus\mathfrak{g}_\alpha\oplus\mathfrak{g}_{2\alpha}$, where $\mathfrak{a}$ and $\mathfrak{g}_{2\alpha}$ are one-dimensional, and $\mathfrak{g}_\alpha$ is a complex vector space. In the case of $\mathbb{C}H^2$, $\mathfrak{g}_\alpha$ has complex dimension one, and we denote by $\mathfrak{g}_\alpha^{\mathbb{R}}$ a real form of $\mathfrak{g}_\alpha$, that is, a real one-dimensional linear subspace of $\mathfrak{g}_\alpha$. We have [7]:

Theorem 8. *A polar action on $\mathbb{C}H^2$ is orbit equivalent to the action of a group $H\subset SU(1,2)$ whose Lie algebra is one of the following:*

1. *Actions of cohomogeneity one - the section Σ is a totally geodesic real hyperbolic line $\mathbb{R}H^1 \subset \mathbb{C}H^2$:*

   (a) $\mathfrak{h} = \mathfrak{k} = \mathfrak{s}(\mathfrak{u}(1) \oplus \mathfrak{u}(2)) \cong \mathfrak{u}(2)$; *the orbits are a point and geodesic spheres centered at that point.*

   (b) $\mathfrak{h} = \mathfrak{g}_{-2\alpha} \oplus \mathfrak{g}_0 \oplus \mathfrak{g}_{2\alpha} = \mathfrak{s}(\mathfrak{u}(1,1) \oplus \mathfrak{u}(1)) \cong \mathfrak{u}(1,1)$; *the orbits are a totally geodesic complex hyperbolic line $\mathbb{C}H^1 \subset \mathbb{C}H^2$ and the tubes around $\mathbb{C}H^1$.*

   (c) $\mathfrak{h} = \theta(\mathfrak{g}_\alpha^{\mathbb{R}}) \oplus \mathfrak{a} \oplus \mathfrak{g}_\alpha^{\mathbb{R}} \cong \mathfrak{so}(1,2)$; *the orbits are a totally geodesic real hyperbolic plane $\mathbb{R}H^2 \subset \mathbb{C}H^2$ and the tubes around $\mathbb{R}H^2$.*

   (d) $\mathfrak{h} = \mathfrak{k}_0 \oplus \mathfrak{g}_\alpha \oplus \mathfrak{g}_{2\alpha}$; *the orbits form a foliation of $\mathbb{C}H^2$ by horospheres.*

   (e) $\mathfrak{h} = \mathfrak{a} \oplus \mathfrak{g}_\alpha^{\mathbb{R}} \oplus \mathfrak{g}_{2\alpha}$; *the orbits form a foliation of $\mathbb{C}H^2$, one of its leaves is the minimal ruled real hypersurface of $\mathbb{C}H^2$ generated by a real horocycle in $\mathbb{C}H^2$, and the other leaves are the equidistant hypersurfaces.*

2. *Actions of cohomogeneity two - the section Σ is a totally geodesic real hyperbolic plane $\mathbb{R}H^2 \subset \mathbb{C}H^2$:*

   (a) $\mathfrak{h} = \mathfrak{k} \cap (\mathfrak{g}_{-2\alpha} \oplus \mathfrak{g}_0 \oplus \mathfrak{g}_{2\alpha}) = \mathfrak{s}(\mathfrak{u}(1) \oplus \mathfrak{u}(1) \oplus \mathfrak{u}(1)) \cong \mathfrak{u}(1) \oplus \mathfrak{u}(1)$; *the orbits are obtained by intersecting the orbits of the two cohomogeneity one actions* (a) *and* (b) *in* (1)*: the action has one fixed point o, and on each distance sphere centered at o the orbits are two circles as singular orbits and two-dimensional tori as principal orbits.*

   (b) $\mathfrak{h} = \mathfrak{g}_0$; *the action leaves a totally geodesic $\mathbb{C}H^1 \subset \mathbb{C}H^2$ invariant. On this $\mathbb{C}H^1$, the action induces a foliation by a totally geodesic real hyperbolic line $\mathbb{R}H^1 \subset \mathbb{C}H^1$ and its equidistant curves in $\mathbb{C}H^1$. The other orbits are two-dimensional cylinders whose axis is one of the curves in that $\mathbb{C}H^1$.*

   (c) $\mathfrak{h} = \mathfrak{k}_0 \oplus \mathfrak{g}_{2\alpha}$; *the orbits are obtained by intersecting the orbits of the two cohomogeneity one actions* (b) *and* (d) *in* (1)*: the action leaves a horosphere foliation invariant, and on each horosphere, the orbits consist of a complex horocycle and the tubes around it.*

   (d) $\mathfrak{h} = \mathfrak{g}_\alpha^{\mathbb{R}} \oplus \mathfrak{g}_{2\alpha}$; *the orbits are obtained by intersecting the orbits of the two cohomogeneity one actions* (d) *and* (e) *in* (1)*: the action leaves a horosphere foliation invariant, and on each horosphere the action induces a foliation for which the minimally embedded Euclidean plane $\mathbb{R}^2$ and its equidistant surfaces are the leaves.*

It follows from the classification that there are exactly four families of polar actions that are not hyperpolar (of cohomogeneity one). Cohomogeneity one actions on $\mathbb{C}H^2$ had already been classified, as we will see below, but a geometric proof in terms of constant principal curvatures can be found in [4].

To the author's knowledge, there are no more general classifications of polar actions on symmetric spaces of noncompact type. We now deal with the hyperpolar case.

Berndt and Tamaru [9] addressed the problem of classifying cohomogeneity one actions on noncompact symmetric spaces of rank one. Note that in rank one, an isometric action is hyperpolar if and only if it is of cohomogeneity one.

The classification of cohomogeneity one actions on complex hyperbolic spaces is as follows (see [5] for a study of the orbits of these actions):

Theorem 9. *A cohomogeneity one action on $\mathbb{C}H^n = SU(1,n)/S(U(1)\times U(n))$ is orbit equivalent to the action of one of the following groups:*

1. *$S(U(1,k)\times U(n-k))$, $k\in\{0,\dots,n-1\}$; the singular orbit is a totally geodesic $\mathbb{C}H^k$ (maybe a point), and the rest of the orbits are tubes around it.*
2. *$SO^0(1,n)$ (the connected component of the identity of $SO(1,n)$); the singular orbit is a totally geodesic $\mathbb{R}H^n$ and the rest of the orbits are tubes around it.*
3. *The nilpotent part of the Iwasawa decomposition of $SU(1,n)$; the orbits form a foliation called the horosphere foliation of $\mathbb{C}H^n$.*
4. *$N_K^0(S)S$; here, S is the connected subgroup of $SU(1,n)$ whose Lie algebra is $\mathfrak{s}=\mathfrak{a}\oplus\mathfrak{w}\oplus\mathfrak{g}_{2\alpha}$, where $\mathfrak{w}$ is a subspace of $\mathfrak{g}_\alpha\cong\mathbb{C}^{n-1}$ such that $\mathfrak{g}_\alpha\ominus\mathfrak{w}$, the orthogonal complement of $\mathfrak{w}$ in $\mathfrak{g}_\alpha$, has constant Kähler angle $\varphi\in(0,\pi/2]$, and $N_K^0(S)$ is the connected component of the identity of the normalizer of S in K.*

Recall that a real subspace V of a complex vector space $\mathbb{C}^n$ has constant Kähler angle $\varphi\in[0,\pi/2]$ if for any unit vector $v\in V$, the angle between Jv and V is φ. It follows that in case 4 of Theorem 9, there are uncountably many examples.

Berndt and Tamaru also obtained the classification in the Cayley hyperbolic plane [9]:

Theorem 10. *A cohomogeneity one action on the Cayley hyperbolic plane $\mathbb{O}H^2 = F_4^{-20}/Spin(9)$ is orbit equivalent to the action of one of the following groups:*

1. *$Spin(9)$; orbits are a point and geodesic spheres centered at that point.*
2. *$Spin(1,8)$; there is a totally geodesic $\mathbb{O}H^1$ and the rest of the orbits are tubes around it.*
3. *$Sp(1)\cdot Sp(1,2)$; there is a totally geodesic $\mathbb{H}H^2$ and the rest of the orbits are tubes around it.*
4. *The nilpotent part of the Iwasawa decomposition of F_4^{-20}, which gives the horosphere foliation.*
5. *$N_K^0(S)S$; here, S is the connected subgroup of F_4^{-20} whose Lie algebra is $\mathfrak{s}=\mathfrak{a}\oplus\mathfrak{w}\oplus\mathfrak{g}_{2\alpha}$, where $\mathfrak{w}$ is a subspace of $\mathfrak{g}_\alpha\cong\mathbb{O}$ such that $\mathfrak{g}_\alpha\ominus\mathfrak{w}$, the orthogonal complement of $\mathfrak{w}$ in $\mathfrak{g}_\alpha$, has dimension $k\in\{1,2,3,4,6,7\}$, and $N_K^0(S)$ is the connected component of the identity of the normalizer of S in K. There is exactly one congruency class for $k\in\{1,2,3,6,7\}$, and for $k=4$, the set of congruency classes is parametrized by $[0,1]$.*

In [9], the authors also studied cohomogeneity one actions in quaternionic hyperbolic spaces $\mathbb{H}H^n$, but they could only get a classification for $\mathbb{H}H^2$. For $n\geq 3$ they reduced the classification to a problem in linear algebra and gave possible candidates for the examples in the classification. However, the problem seems to be rather involved in view of the new examples constructed in [20]. For symmetric spaces of higher rank, a full classification of cohomogeneity one actions is not yet known although much progress has been done in [10].

We will finish this section by providing some partial classifications of polar and hyperpolar actions on symmetric spaces of noncompact type.

As we argued at the beginning of this section, duality cannot be applied to derive a classification of polar actions on symmetric spaces of noncompact type from the corresponding classification in spaces of compact type. However, in certain situations it is possible to use this tool.

The first result in that direction was obtained in [21], where the authors classified polar actions by compact Lie groups in symmetric spaces of noncompact type. The result can be summarized as follows. Let $M = G/K$ be an irreducible symmetric space of noncompact type, and assume that H is a compact Lie group that acts polarly on M. Then, Cartan's fixed point theorem ensures that H has a fixed point, which we can assume, by homogeneity, is $o \in M$ in the notation of Section 3. Since K is maximal compact we have $H \subset K$. Its Lie subalgebra $\mathfrak{h}$ splits nicely with respect to the Cartan decomposition $\mathfrak{h} = \mathfrak{k} \oplus \mathfrak{p}$ (indeed, $\mathfrak{h} \subset \mathfrak{k}$), and hence duality may be applied. Therefore, if M has rank greater than one, the action of H is orbit equivalent to the action of K on M, the isotropy action. If M has rank one, then the action of H on M is (hyper-)polar if and only if the action of H on its compact dual $M^* = G^*/K$ is (hyper-)polar. It can actually be shown that polar actions by compact Lie groups on $\mathbb{R}H^n$, $\mathbb{C}H^n$, and $\mathbb{H}H^n$, are in correspondence with polar actions on $\mathbb{R}P^{n-1}$, $\mathbb{C}P^{n-1}$, $\mathbb{H}P^{n-1}$, respectively. To prove this statement, one uses duality to go from $\mathbb{F}H^n$ to $\mathbb{F}P^n$, where $\mathbb{F} \in \{\mathbb{R}, \mathbb{C}, \mathbb{H}\}$, and then uses the fact that a polar action on $\mathbb{F}P^n$ with a fixed point leaves a totally geodesic $\mathbb{F}P^{n-1}$ invariant where the action is still polar.

Kollross went further with this approach in [36]. In order to present his result, we need some definitions. Let $\mathfrak{g}$ be a complex semisimple Lie algebra. Then, $\mathfrak{g}$ is isomorphic to $\operatorname{ad}(\mathfrak{g})$, a subalgebra of $\mathfrak{gl}(\mathfrak{g})$. A subalgebra $\mathfrak{h}$ of $\mathfrak{g}$ is said to be algebraic if it is the Lie algebra of an algebraic subgroup H, that is, H is an affine algebraic variety for which the map $H \times H \to H$, $(g,h) \mapsto gh^{-1}$, is polynomial. On the other hand, a subalgebra $\mathfrak{h}$ of $\mathfrak{g}$ is a reductive subalgebra if $\mathfrak{h} = \mathfrak{z}(\mathfrak{h}) \oplus \mathfrak{h}'$, where $\mathfrak{z}(\mathfrak{h})$ is the center of $\mathfrak{h}$, $\mathfrak{h}' = [\mathfrak{h}, \mathfrak{h}]$ is the derived Lie algebra of $\mathfrak{h}$, and the elements of $\mathfrak{z}(\mathfrak{h})$ are semisimple in $\mathfrak{g}$. If $\mathfrak{g}$ is a real Lie algebra, by definition $\mathfrak{h}$ is a reductive subalgebra of $\mathfrak{g}$ if $\mathfrak{h}^{\mathbb{C}}$, the complexification of $\mathfrak{h}$, is reductive in $\mathfrak{g}^{\mathbb{C}}$. Kollross proved:

Theorem 11. *Let $M = G/K$ be a symmetric space of noncompact type, and H a Lie subgroup of G whose Lie algebra is an algebraic reductive subalgebra of $\mathfrak{g}$. Then, up to conjugation, $\mathfrak{h} = (\mathfrak{k} \cap \mathfrak{h}) \oplus (\mathfrak{p} \cap \mathfrak{h})$, with respect to the Cartan decomposition $\mathfrak{g} = \mathfrak{k} \oplus \mathfrak{p}$. Moreover, $\mathfrak{h}^* = (\mathfrak{k} \cap \mathfrak{h}) \oplus i(\mathfrak{p} \cap \mathfrak{h})$ is a subalgebra of $\mathfrak{g}^*$, the compact dual of $\mathfrak{g}$, and if H^* denotes the corresponding connected subgroup of G^* whose Lie algebra is $\mathfrak{h}^*$, then the action of H on $M = G/K$ is (hyper-)polar if and only if the action of H^* on $M^* = G^*/K$ is (hyper-)polar.*

In particular, one can derive the classification of polar actions by reductive algebraic subgroups of the isometry group on irreducible symmetric spaces of noncompact type from the classification of polar actions on irreducible compact symmetric spaces.

As mentioned earlier in this section, a polar action on a symmetric space of compact type always has singular orbits. This is not true in the noncompact setting as horosphere foliations show. Motivated by this fact Berndt, Tamaru, and the author studied hyperpolar actions on symmetric spaces that have no singular orbits [8]. We call these homogeneous hyperpolar foliations. We have (see Sect. 3.1 for the notation):

Theorem 12. *Let $M = G/K$ be a symmetric space of noncompact type with a root space decomposition $\mathfrak{g} = \mathfrak{g}_0 \oplus (\bigoplus_{\lambda\in\Sigma} \mathfrak{g}_\lambda)$, with respect to a Cartan decomposition $\mathfrak{g} = \mathfrak{k} \oplus \mathfrak{p}$, and a maximal abelian subspace $\mathfrak{a}$ of $\mathfrak{p}$. If H induces a homogeneous hyperpolar foliation on M, then the action of H is orbit equivalent to the action of the subgroup $S_{V,\Phi}$ of G whose Lie algebra is*

$$\mathfrak{s}_{V,\Phi} = V \oplus (\mathfrak{n} \ominus \ell_\Phi),$$

where Φ is a subset of orthogonal simple roots of Λ, $\ell_\Phi = \bigoplus_{\alpha\in\Phi} \ell_\alpha$, with ℓ_α any line in $\mathfrak{g}_\alpha$, and V is a subspace of $\mathfrak{a}$ containing $\bigoplus_{\alpha\in\Phi} \mathbb{R}H_\alpha$.

Note that different choices of $\ell_\alpha \subset \mathfrak{g}_\alpha$ give rise to congruent actions. There is a nice description of these foliations in terms of cohomogeneity one actions on symmetric spaces of rank one and the theory of parabolic subalgebras that we omit here. We refer to the original paper [8] for more details.

Incidentally, if N_Φ is the nilpotent part of the Langlands decomposition of the parabolic subalgebra of $\mathfrak{g}$ associated with a nonempty $\Phi \subset \Lambda$, and V is a linear subspace of $\mathfrak{a} \ominus (\bigoplus_{\alpha\in\Phi} \mathbb{R}H_\alpha)$, then the action of $V \times N_\Phi$ on M is polar but not hyperpolar (see [8, Proposition 4.2]). This is another reason why we cannot expect duality to provide a classification of hyperpolar actions on symmetric spaces on noncompact type, as in the irreducible compact setting, polar actions are always hyperpolar.

This result can be improved for homogeneous polar foliations in complex hyperbolic spaces [6]:

Theorem 13. *If H is a group inducing a homogeneous polar foliation on $\mathbb{C}H^n$, then the action of H is orbit equivalent to the action of the subgroup $S_{V,\mathfrak{w}}$ of $SU(1,n)$ whose Lie algebra is*

$$\mathfrak{s}_{V,\mathfrak{w}} = V \oplus (\mathfrak{g}_\alpha \ominus \mathfrak{w}) \oplus \mathfrak{g}_{2\alpha},$$

where V is a subspace of $\mathfrak{a}$ (hence 0 or $\mathfrak{a}$), and $\mathfrak{w}$ is a real subspace of $\mathfrak{g}_\alpha \cong \mathbb{C}^{n-1}$.

It is also shown that there are exactly $2n+1$ congruency classes of these actions because the actions of $S_{V,\mathfrak{w}}$ and $S_{V,\mathfrak{w}'}$ are congruent if and only if $\mathfrak{w}$ and $\mathfrak{w}'$ have the same dimension.

References

1. Adams, S., Stuck, G.: The isometry group of a compact Lorentz manifold I. Invent. Math. **129**, 239–261 (1997)
2. Bergmann, I.: Reducible polar representations. Manuscripta Math. **104**, 309–324 (2001)
3. Berndt, J., Console, S., Olmos, C.: Submanifolds and holonomy. Chapman & Hall/CRC Research Notes in Mathematics, vol. 434, Chapman & Hall/CRC, Boca Raton, FL (2003)
4. Berndt, J., Díaz-Ramos, J.C.: Real hypersurfaces with constant principal curvatures in the complex hyperbolic plane. Proc. Amer. Math. Soc. **135**, 3349–3357 (2007)
5. Berndt, J., Díaz-Ramos, J.C.: Homogeneous hypersurfaces in complex hyperbolic spaces. Geom. Dedicata **138**, 129–150 (2009)
6. Berndt, J., Díaz-Ramos, J. C.: Homogeneous polar foliations of complex hyperbolic spaces. Comm. Anal. Geom. **20**(3), 435–454 (2012)
7. Berndt, J., Díaz-Ramos, J.C.: Polar actions on the complex hyperbolic plane. Ann. Glob. Anal. Geom. DOI 10.1007/s10455-012-9335-z
8. Berndt, J., Díaz-Ramos, J.C., Tamaru, H.: Hyperpolar homogeneous foliations on symmetric spaces of noncompact type. J. Differential Geom. **86**, 191–235 (2010)
9. Berndt, J., Tamaru, H.: Cohomogeneity one actions on noncompact symmetric spaces of rank one. Trans. Amer. Math. Soc. **359**, 3425–3438 (2007)
10. Berndt, J., Tamaru, H.: Cohomogeneity one actions on symmetric spaces of noncompact type. To appear in J. Reine Angew. Math. Preprint arXiv:1006.1980 [math.DG].
11. Biliotti, L.: Coisotropic and polar actions on compact irreducible Hermitian symmetric spaces. Trans. Amer. Math. Soc. **358**, 3003–3022 (2006)
12. Biliotti, L., Gori, A.: Coisotropic and polar actions on complex Grassmannians. Trans. Amer. Math. Soc. **357**, 1731–1751 (2005)
13. Bott, R., Samelson, H.: Applications of the theory of Morse to symmetric spaces. Amer. J. Math. **80**, 964–1029 (1958)
14. Conlon, L.: Variational completeness and K-transversal domains. J. Differ. Geom. **5**, 135–147 (1971)
15. Dadok, J.: Polar coordinates induced by actions of compact Lie groups. Trans. Amer. Math. Soc. **288**, 125–137 (1985)
16. Dadok, J., Kac, V.: Polar representations. J. Algebra **92**, 504–524 (1985)
17. D'Ambra, G.: Isometry groups of Lorentz manifolds. Invent. Math. **92**, 555–565 (1988)
18. D'Ambra, G., Gromov, M.: Lectures on transformation groups: geometry and dynamics. Surveys in differential geometry, Cambridge, MA, (1990), pp. 19–111, Lehigh University Bethlehem, PA (1991)
19. Díaz-Ramos, J.C.: Proper isometric actions. arXiv:0811.0547v1 [math.DG]
20. Díaz-Ramos, J.C., Domínguez-Vázquez, M.: Isoparametric hypersurfaces in Damek-Ricci spaces. arXiv:1111.0264v1 [math.DG]
21. Díaz-Ramos, J.C., Kollross, A.: Polar actions with a fixed point. Differ. Geom. Appl. **29**, 20–25 (2011)
22. Eschenburg, J.-H., Heintze, E.: Polar representations and symmetric spaces. J. Reine Angew. Math. **507**, 93–106 (1999)
23. Eschenburg, J.-H., Heintze, E.: On the classification of polar representations. Math. Z. **232**, 391–398 (1999)
24. Geatti, L., Gorodski, C.: Polar orthogonal representations of real reductive algebraic groups. J. Algebra **320**, 3036–3061 (2008)
25. Gorodski, C., Thorbergsson, G.: Variationally complete actions on compact symmetric spaces. J. Differ. Geom. **62**, 39–48 (2002)
26. Heintze, E., Palais, R.S., Terng, C.-L., Thorbergsson, G.: Hyperpolar actions on symmetric spaces. Geometry, topology, and physics, Conf. Proc. Lecture Notes Geom. Topology, IV, pp. 214–245. Int. Press, Cambridge (1995)

27. Helgason, S.: Differential geometry, Lie groups, and symmetric spaces. Corrected reprint of the 1978 original, Graduate Studies in Mathematics, vol. 34. American Mathematical Society, Providence (2001)
28. Hermann, R.: Variational completeness for compact symmetric spaces, Proc. Amer. Math. Soc. **11**, 544–546 (1960)
29. Hsiang, W.-Y., Lawson Jr., H.B.: Minimal submanifolds of low cohomogeneity. J. Differ. Geom. **5**, 1–38 (1971)
30. Iwata, K.: Classification of compact transformation groups on cohomology quaternion projective spaces with codimension one orbits. Osaka J. Math. **15**, 475–508 (1978)
31. Iwata, K.: Compact transformation groups on rational cohomology Cayley projective planes. Tôhoku Math. J. **33**(2), 429–442 (1981)
32. Knapp, A.W.: Lie Groups Beyond an Introduction. Birkhäuser, Boston (1996)
33. Kollross, A.: A classification of hyperpolar and cohomogeneity one actions. Trans. Amer. Math. Soc. **354**, 571–612 (2002)
34. Kollross, A.: Polar actions on symmetric spaces. J. Differ. Geom. **77**, 425–482 (2007)
35. Kollross, A.: Low cohomogeneity and polar actions on exceptional compact Lie groups. Transform. Groups **14**, 387–415 (2009)
36. Kollross, A.: Duality of symmetric spaces and polar actions. J. Lie Theory **21**, 961–986 (2011)
37. Kollross, A., Lytchak, A.: Polar actions on symmetric spaces of higher rank. arXiv:1108.3256v2 [math.DG]
38. Loos, O.: Symmetric spaces. I: General Theory. II: Compact spaces and classification. W.A. Benjamin (1969)
39. Lytchak, A., Thorbergsson, G.: Variationally complete actions on nonnegatively curved manifolds. Illinois J. Math. **51**, 605–615 (2007)
40. Montgomery, D., Yang, C.T.: The existence of a slice. Ann. Math. **65**(2), 108–116 (1957)
41. Myers, S.B., Steenrod, N.E.: The group of isometries of a Riemannian manifold. Ann. Math. **40**(2), 400–416 (1939)
42. Palais, R.S., On the existence of slices for actions of non-compact Lie groups. Ann. Math. **73**(2), 295–323 (1961)
43. Palais, R.S., Terng, C.-L.: A general theory of canonical forms. Trans. Amer. Math. Soc. **300**, 771–789 (1987)
44. Podestà, F., Thorbergsson, G.: Polar actions on rank-one symmetric spaces. J. Differ. Geom. **53**, 131–175 (1999)
45. Podestà, F., Thorbergsson, G.: Polar and coisotropic actions on Kähler manifolds. Trans. Amer. Math. Soc. **354**, 1759–1781 (2002)
46. Takagi, R.: On homogeneous real hypersurfaces in a complex projective space. Osaka J. Math. **10**, 495–506 (1973)
47. Takagi, R., Takahashi, T.: On the principal curvatures of homogeneous hypersurfaces in a sphere. Differential Geometry (in Honor of Kentaro Yano), pp. 469–481. Kinokuniya, Tokyo (1972)
48. Terng, C.-L.: Polar actions on Hilbert space, J. Geom. Anal. **5**, 129–150 (1995)
49. Thorbergsson, G.: A survey on isoparametric hypersurfaces and their generalizations. Handbook of Differential Geometry, vol. I, pp. 963–995. North-Holland, Amsterdam (2000)
50. Thorbergsson, G.: Transformation groups and submanifold geometry. Rendiconti di Matematica, Serie VII, **25**, 1–16 (2005)
51. Thorbergsson, G.: Singular Riemannian foliations and isoparametric submanifolds. Milan J. Math. **78**, 355–370 (2010)
52. Wu, B.: Isoparametric submanifolds of hyperbolic spaces. Trans. Amer. Math. Soc. **331**, 609–626 (1992)
53. Yau, S.T.: Remarks on the group of isometries of a Riemannian manifold. Topology **16**, 239–247 (1977)

(para)-Kähler Weyl Structures

P. Gilkey and S. Nikčević

1 Introduction

Let ∇ be a torsion free connection on a pseudo-Riemannian manifold (M,g) of even dimension $m = 2\bar{m} \geq 4$. The triple (M,g,∇) is said to be a *Weyl structure* if there exists a smooth 1-form ϕ so that $\nabla g = -2\phi \otimes g$. Such a geometric structure was introduced by Weyl [37] in an attempt to unify gravity with electromagnetism. Although this approach failed for physical reasons, these geometries are still studied for their intrinsic interest [2, 10, 21, 27, 28]; they also appear in the mathematical physics literature [12, 20, 26]. Weyl geometry is relevant to submanifold geometry [25] and to contact geometry [15]. The pseudo-Riemannian setting also is important [1, 24, 32] as are para-complex geometries [11, 13]. See also [9, 14, 22, 29, 30] for related results. The literature in the field is vast, and we can only give a flavor of it for reasons of brevity. We shall be primarily interested in the Hermitian setting. However, since there are applications to higher signature geometry, we include the pseudo-Hermitian context as well; similarly, we treat para-Hermitian geometries as they can be studied with little additional effort.

Section 1.1 of the Introduction deals with the real setting. In Theorem 1, we recall the basic theorems of geometric realizability for affine, Riemannian, and Weyl curvature models and in Theorem 2 provide various characterizations of the notion of a trivial Weyl structure. Section 1.2 treats the (para)-Kähler setting. In Theorem 3, we recall geometric realizibility results for (para)-Kähler affine and (para)-Kähler Riemannian curvature models. Theorem 4 presents results in the geometric setting

P. Gilkey (✉)
Mathematics Department, University of Oregon, Eugene, OR 97403, USA
e-mail: gilkey@uoregon.edu

S. Nikčević
Mathematical Institute, Sanu, Knez Mihailova 36, p.p. 367, 11001 Belgrade, Serbia
e-mail: stanan@mi.sanu.ac.rs

M. Sánchez et al. (eds.), *Recent Trends in Lorentzian Geometry*, Springer Proceedings in Mathematics & Statistics 26, DOI 10.1007/978-1-4614-4897-6_15,

for (para)-Kähler Weyl manifolds. Theorem 5 is one of the two main results of this chapter: every (para)-Kähler curvature model is geometrically realizable. The proof of Theorem 5 relies on a curvature decomposition result; the second main result of the chapter, Theorem 6, discusses the space of (para)-Kähler Weyl algebraic curvature tensors.

1.1 Riemannian, Affine, and Weyl Geometry

Let $(V, \langle\cdot,\cdot\rangle)$ be an inner product space of signature (p,q) and dimension $m = p+q$; an inner product of signature $(0,4)$ is positive definite. A 4-tensor $A \in \otimes^4 V^*$ is said to be a *Riemannian algebraic curvature tensor* if A satisfies the symmetries of the Riemann curvature tensor, namely:

$$A(x,y,z,w) + A(y,x,z,w) = 0\,, \tag{1}$$

$$A(x,y,z,w) + A(y,z,x,w) + A(z,x,y,w) = 0\,, \tag{2}$$

$$A(x,y,z,w) = A(w,z,x,y)\,. \tag{3}$$

Let $\mathfrak{R}(V)$ be the subspace of $\otimes^4 V^*$ which consists of all tensors satisfying these relations. We say that a triple $\mathscr{R} := (V, \langle\cdot,\cdot\rangle, A)$ is a *Riemannian curvature model* if $A \in \mathfrak{R}(V)$. One says that $\mathscr{R}$ is *geometrically realizable* by a pseudo-Riemannian manifold if there is a point P of some pseudo-Riemannian manifold (M,g) and if there is an isomorphism $\Phi : V \to T_P M$ so

$$\Phi^* g_P = \langle\cdot,\cdot\rangle \quad \text{and} \quad \Phi^* R^g_P = A,$$

where R^g is the curvature tensor of the Levi–Civita connection ∇^g on M.

Affine differential geometry extends Riemannian geometry. A pair (M, ∇) is said to be an *affine manifold* if ∇ is a torsion-free connection on the tangent bundle TM. The curvature R^∇ of the connection ∇ then satisfies the identities of Eqs. (1) and (2) but need no longer satisfy Eq. (3); if $A \in \otimes^4 V^*$, one says A is an *affine algebraic curvature tensor* if A satisfies Eqs. (1) and (2) and one lets $\mathfrak{A}(V)$ be the set of all such tensors. Strictly speaking in affine geometry, the addition of an inner product is an additional structure that must be dealt with separately. This plays no role in our discussion since we have in fact prescribed an inner product. We refer to [17] for further details of this point for the interested reader. Note that the corresponding *curvature operator* $\hat{A}$ and the curvature tensor A are related by the identity

$$\langle \hat{A}(x,y)z, w\rangle = A(x,y,z,w).$$

The pair $\mathscr{A} := (V, A)$ is said to be an *affine curvature model* if $A \in \mathfrak{A}(V)$; such an $\mathscr{A}$ is said to be *geometrically realizable by an affine manifold* if there is a point P of some affine manifold (M, ∇) and if there is an isomorphism $\Phi : V \to T_P M$ so that $\Phi^* R^\nabla_P = A$.

Weyl geometry is in a sense midway between Riemannian geometry and affine geometry. A triple (M,g,∇) is said to be a *Weyl manifold* if (M,g) is a pseudo-Riemannian manifold, if (M,∇) is an affine manifold, and if there exists a smooth 1-form ϕ on M so that the structures are related by the equation:

$$\nabla g = -2\phi \otimes g. \tag{4}$$

Define the *Ricci-tensor* $\rho = \rho_\nabla$ and the *alternating Ricci tensor* $\rho_a = \rho_{a,\nabla}$ by

$$\begin{aligned}\rho(x,y) &:= \mathrm{Tr}(z \to R(z,x)y),\\ \rho_a(x,y) &:= \frac{1}{2}\{\rho(x,y) - \rho(y,x)\}.\end{aligned}$$

There is an additional curvature symmetry which pertains in Weyl geometry (see, e.g. the discussion in [17]):

$$R(x,y,z,w) + R(x,y,w,z) = -\frac{4}{m}\rho_a(x,y)g(z,w). \tag{5}$$

The defining 1-form ϕ is related to the curvature by the equation:

$$d\phi = -\frac{2}{m}\rho_a\,. \tag{6}$$

Let $\mathfrak{W}(V) \subset \otimes^4(V^*)$ be space of 4-tensors satisfying Eqs. (1), (2), and (5); these are the Weyl algebraic curvature tensors. If $A \in \mathfrak{R}$, then $\rho_a = 0$ and $A(x,y,z,w) + A(x,y,w,z) = 0$. Consequently:

$$\mathfrak{R}(V) \subset \mathfrak{W}(V) \subset \mathfrak{A}(V)\,.$$

A triple $\mathscr{W} := (V,\langle\cdot,\cdot\rangle,A)$ is said to be a *Weyl curvature model* if $A \in \mathfrak{W}(V)$. The notion of geometric realizability is defined analogously in this setting.

We refer to [8,16,17] for the proof of the following result; the first two assertions are, of course, well known:

Theorem 1. *1. Every Riemannian curvature model is geometrically realizable by a pseudo-Riemannian manifold.*
2. Every affine curvature model is geometrically realizable by an affine manifold.
3. Every Weyl curvature model is geometrically realizable by a Weyl manifold.

Weyl geometry is a conformal theory; if $g_1 = e^{2f}g$ is conformally equivalent to g and if (M,g,∇) is a Weyl manifold, then (M,g_1,∇) is again a Weyl manifold with associated 1-form ϕ_1 given by $\phi_1 = \phi - df$. One has the following well known result characterizing trivial Weyl structures (see, e.g., [17]):

Theorem 2. *Let (M,g,∇) be a Weyl manifold with $H^1(M;\mathbb{R}) = 0$. The following assertions are equivalent, and if any is satisfied, then the Weyl structure is said to be trivial.*

1. $d\phi = 0$
2. $\nabla = \nabla^{g_1}$ *for some conformally equivalent metric* g_1
3. $R^\nabla \in \mathfrak{R}$

1.2 Kähler Geometry

We now pass from the real to the (para)-complex setting. Let V be a real vector space of even dimension $m = 2\bar{m}$. A complex structure on V is an endomorphism J_- of V so $J_-^2 = -\operatorname{Id}$. Similarly, a para-complex structure on V is an endomorphism J_+ of V so $J_+^2 = \operatorname{Id}$ and $\operatorname{Tr}(J_+) = 0$; this trace-free condition is automatic in the complex setting but must be imposed in the para-complex setting. It is convenient to introduce the notation $J_\pm$ in order to have a common formulation in both contexts although we shall never be considering both structures simultaneously. In the geometric setting, $(M, J_\pm)$ is said to be an *almost (para)-complex manifold* if $J_\pm$ is a smooth endomorphism of the tangent bundle so that $(T_PM, J_\pm)$ is a (para)-complex structure for every $P \in M$. The almost (para)-complex structure $J_\pm$ is said to be *integrable* and the pair $(M, J_\pm)$ is said to be a *(para)-complex manifold* if there are coordinate charts $(x^1, y^1, \dots, x^{\bar{m}}, y^{\bar{m}})$ covering M so that:

$$J_\pm\left\{\frac{\partial}{\partial x_i}\right\} = \frac{\partial}{\partial y_i} \quad \text{and} \quad J_\pm\left\{\frac{\partial}{\partial y_i}\right\} = \pm\frac{\partial}{\partial x_i} \quad \text{for} \quad 1 \le i \le \bar{m}. \tag{7}$$

If $(M, J_\pm)$ is an almost (para)-complex manifold and if ∇ is a torsion free connection on M, then $(M, J_\pm, \nabla)$ is said to be a *Kähler affine manifold* if $\nabla J_\pm = 0$; this assumption then implies that $J_\pm$ is integrable. The curvature satisfies an extra symmetry in this setting:

$$R(x,y,z,w) = \mp R(x,y,J_\pm z, J_\pm w). \tag{8}$$

A (para)-complex pseudo-Riemannian manifold $(M, g, J_\pm)$ is said to be a *(para)-Kähler Hermitian* manifold if $J_\pm^* g = \mp g$ and $\nabla^g J_\pm = 0$. Finally, a (para)-complex-Riemannian Weyl manifold $(M, g, J_\pm, \nabla)$ is said to be a *(para)-Kähler Weyl* manifold if $\nabla J_\pm = 0$.

We now pass to the algebraic context. Define the space of (para)-Kähler tensors $\mathfrak{K}_\pm$, the space of (para)-Kähler affine algebraic curvature tensors $\mathfrak{K}_{\pm,\mathfrak{A}}$, the space of (para)-Kähler Riemannian algebraic curvature tensors $\mathfrak{K}_{\pm,\mathfrak{R}}$, and the space of (para)-Kähler Weyl algebraic curvature tensors $\mathfrak{K}_{\pm,\mathfrak{W}}$ by setting, respectively:

$$\mathfrak{K}_\pm := \{A \in \otimes^4 V^* : A(x,y,z,w) = \mp A(x,y,J_\pm z, J_\pm w)\},$$
$$\mathfrak{K}_{\pm,\mathfrak{A}} := \mathfrak{K}_\pm \cap \mathfrak{A}, \quad \mathfrak{K}_{\pm,\mathfrak{R}} := \mathfrak{K}_\pm \cap \mathfrak{R}, \quad \mathfrak{K}_{\pm,\mathfrak{W}} := \mathfrak{K}_\pm \cap \mathfrak{W}.$$

A triple $\mathscr{K}_\mathscr{A} = (V, J_\pm, A)$ is said to be a *(para)-Kähler affine curvature model* if $(V, J_\pm)$ is (para)-complex and if $A \in \mathfrak{K}_{\pm,\mathfrak{A}}$. A quadruple $\mathscr{K}_\mathscr{R} = (V, \langle\cdot,\cdot\rangle, J_\pm, A)$

is said to be a *(para)-Kähler Hermitian curvature model* if $J_\pm^*\langle\cdot,\cdot\rangle = \mp\langle\cdot,\cdot\rangle$ and if $A \in \mathfrak{K}_{\pm,\mathfrak{R}}$. A quadruple $\mathscr{K}_\mathscr{W} = (V,\langle\cdot,\cdot\rangle,J_\pm,A)$ is said to be a *(para)-Kähler Weyl curvature model* if $J_\pm^*\langle\cdot,\cdot\rangle = \mp\langle\cdot,\cdot\rangle$ and if $A \in \mathfrak{K}_{\pm,\mathfrak{W}}$.

Let $\langle\cdot,\cdot\rangle$ have signature (p,q); if $p=0$, then $\langle\cdot,\cdot\rangle$ is positive definite, while if $q=0$, then $\langle\cdot,\cdot\rangle$ is negative definite. In the para-complex setting, $p=q$ so $\langle\cdot,\cdot\rangle$ is necessarily indefinite. In the complex setting, p and q must both be even; we emphasize that we do not assume necessarily that the inner product is positive definite. We refer to [4] for the proof of Assertion (1) and to [3] for the proof of Assertion (2) in the following result:

Theorem 3. *1. Every (para)-Kähler affine curvature model is geometrically realizable by a (para)-Kähler affine manifold.*
2. Every (para)-Kähler Hermitian curvature model is geometrically realizable by a (para)-Kähler Hermitian manifold.

The (para)-Kähler form $\Omega_\pm$ is defined by the identity:

$$\Omega_\pm(x,y) := g(x,J_\pm y)\,.$$

Let δ be the co-derivative. We refer to [31, 35, 36] for the proof of Assertion (1) in the following result in the positive definite setting—the generalization to the indefinite setting is immediate. We refer to [23] for the proof of Assertion (2) in the Riemannian setting—the extension to the general setting is immediate:

Theorem 4. *1. Let $m \geq 6$. If $(M,g,J_\pm,\nabla)$ is a (para)-Kähler Weyl manifold, then the associated Weyl structure is trivial, i.e., locally there is a conformally equivalent metric g_1 so that $(M,g_1,J_\pm)$ is Kähler and so that $\nabla = \nabla^{g_1}$.*
2. Every (para)-Hermitian manifold of dimension 4 admits a unique (para)-Kähler Weyl structure defined by taking $\phi = \pm\frac{1}{2}J_\pm^*\delta\Omega_\pm$.

The following theorem is the first main result of this chapter:

Theorem 5. *Every (para)-Kähler Weyl curvature model is geometrically realizable by a (para)-Kähler Weyl manifold.*

Curvature decompositions play a central role in modern differential geometry. The following theorem is the second main result of this chapter and will play a central role in the proof of Theorem 5:

Theorem 6. *Let $(V,\langle\cdot,\cdot\rangle,J_\pm)$ be a (para)-Hermitian vector space.*

1. *If $m \geq 6$, then $\mathfrak{K}_{\pm,\mathfrak{W}} = \mathfrak{K}_{\pm,\mathfrak{R}}$.*
2. *If $m = 4$, then $\mathfrak{K}_{\pm,\mathfrak{W}} = \mathfrak{K}_{\pm,\mathfrak{R}} \oplus L^2_{0,\mp}$ where*

$$\rho_a : L^2_{0,\mp} \xrightarrow{\approx} \Lambda^2_{0,\mp} := \{\Phi \in \Lambda^2(V^*) : \Phi \perp \Omega_\pm \quad \textit{and} \quad J_\pm^*\Phi = \mp\Phi\}\,.$$

Theorem 6 is one of the facts about 4-dimensional geometry that distinguishes it from the higher-dimensional setting; the module $L^2_{0,\mp}$ provides additional curvature possibilities if $m = 4$.

Curvature decompositions are fundamental in establishing geometrical realizability results. For example, we can use Theorem 6(1) to establish Theorem 4(1) as follows. Suppose that $(M, g, J_\pm, \nabla)$ is a (para)-Kähler Weyl manifold of dimension $m \geq 6$. By Theorem 6, $R^\nabla \in \mathfrak{K}_{\pm,\mathfrak{R}} \subset \mathfrak{R}$. By Theorem 2, there is a locally conformally equivalent metric g_1 so that $\nabla = \nabla^{g_1}$; g_1 is globally defined if $H^1(M;\mathbb{R}) = 0$.

Here is a brief outline to the remainder of this chapter. In Sect. 2, we review well-known previous results concerning curvature decompositions that we shall need. Theorem 6 is established in Sect. 3 and Theorem 5 is established in Sect. 4.

2 Curvature Decompositions

In Sect. 2.1, the structure groups $\mathscr{O}$, $\mathscr{U}_\pm$, and $\mathscr{U}_\pm^\star$ will be defined and the fundamental facts needed from representation theory will be established. In Sect. 2.2, results of Singer and Thorpe [33] giving the decomposition of $\mathfrak{R}$ and results of Higa [18, 19] giving the decomposition of $\mathfrak{W}$ as an $\mathscr{O}$-module will be presented. In Sect. 2.3, the Tricerri–Vanhecke decomposition [34] of the space of Riemannian algebraic curvature tensors $\mathfrak{R}$ and the space of Kähler algebraic curvature tensors $\mathfrak{K}_{\pm,\mathfrak{R}}$ as $\mathscr{U}_\pm^\star$ modules will be outlined; this will rise to the decomposition of the space of Weyl algebraic curvature tensors $\mathfrak{W}$ as a $\mathscr{U}_\pm^\star$ module. As we shall not need the decomposition of $\mathfrak{K}_{\pm,\mathfrak{A}}$ as a $\mathscr{U}_\pm^\star$ module, we shall omit this decomposition and instead refer to the discussion in [5].

2.1 Representation Theory

Let $(V, \langle\cdot,\cdot\rangle)$ be an inner product space. The orthogonal group $\mathscr{O}$ is the subgroup of all invertible linear transformations of V preserving the inner product. If $(V, \langle\cdot,\cdot\rangle, J_\pm)$ is a (para)-Hermitian vector space, define:

$$\mathscr{U}_\pm := \{T \in \mathscr{O} : TJ_\pm = J_\pm T\},$$
$$\mathscr{U}_\pm^\star := \{T \in \mathscr{O} : TJ_\pm = J_\pm T \text{ or } TJ_\pm = -J_\pm T\}.$$

It is convenient to work with the $\mathbb{Z}_2$ extensions $\mathscr{U}_\pm^\star$ as we may then interchange the roles of $J_\pm$ and $-J_\pm$. Let χ be the $\mathbb{Z}_2$ valued character of $\mathscr{U}_\pm^\star$ so that

$$J_\pm T = \chi(T) T J_\pm \quad \text{and} \quad T^*\Omega_\pm = \chi(T)\Omega_\pm \quad \text{for} \quad T \in \mathscr{U}_\pm^\star.$$

By an abuse of notation, we identify χ with the associated 1-dimensional module. We can extend $\langle\cdot,\cdot\rangle$ to a natural nondegenerate inner product on $\otimes^k V$ and $\otimes^k V^*$. The following observation is fundamental in the subject:

Lemma 1. *Let $G \in \{\mathscr{O}, \mathscr{U}_-, \mathscr{U}_-^\star, \mathscr{U}_+^\star\}$ and let ξ be a G-submodule of $\otimes^k V^*$. Then, the restriction of the inner product on $\otimes^k V^*$ to ξ is nondegenerate.*

Proof. Let $\{e_i\}$ be an orthonormal basis for V and let $\{e^i\}$ be the associated dual basis for V^*. If $I = (i_1, \ldots, i_k)$ is a multi-index, set $e^I = e^{i_1} \otimes \ldots \otimes e^{i_k}$. Then:

$$(e^I, e^J) := \langle e^{i_1}, e^{j_1} \rangle \cdots \langle e^{i_k}, e^{j_k} \rangle = \left\{ \begin{array}{c} 0 \ \text{if}\, I \neq J \\ \pm 1 \ \text{if}\, I = J \end{array} \right\}. \tag{9}$$

Let $Te_i = \langle e_i, e_i \rangle \cdot e_i$ define an element $T \in \mathscr{O}$. Suppose that ξ is an $\mathscr{O}$ invariant subspace of $\otimes^k V^*$. Decompose $\xi = \xi_+ \oplus \xi_-$ and decompose $\otimes^k V^* = W_+ \oplus W_-$ into the ± 1 eigenspaces of T. Since $T \in \mathscr{O}$, these decompositions are orthogonal direct sums. By Eq. (9), W_+ is space like and W_- is time like. Since $\xi_\pm \subset W_\pm$, ξ_+ is space like and ξ_- is time like; the Lemma now follows in this special case. If $G = \mathscr{U}_-$ or if $G = \mathscr{U}_-^\star$, then we can choose the orthonormal basis so that

$$J_- e_{2\nu-1} = e_{2\nu} \quad \text{and} \quad J_- e_{2\nu} = -e_{2\nu-1}.$$

Since $J_-^* \langle \cdot, \cdot \rangle = \langle \cdot, \cdot \rangle$, $J_- T = T J_-$. Thus $T \in G$ and the same argument pertains. Finally suppose $G = \mathscr{U}_+^\star$. We can choose the basis so

$$J_+ e_{2\nu-1} = e_{2\nu} \quad \text{and} \quad J_+ e_{2\nu} = e_{2\nu-1},$$

where $e_{2\nu-1}$ is space like and $e_{2\nu}$ is time like. We now have $T \in \mathscr{U}_+^\star - \mathscr{U}_+$. □

We note that Lemma 1 fails for the group $G = \mathscr{U}_+$. For example, let $V_\pm$ be the ± 1 eigenspaces of J_+; then $J_\pm V_\pm = V_\pm$ and $V_\pm$ is totally isotropic. We can combine Lemma 1 with same arguments as used in the positive definite setting to establish the following result; we omit details in the interests of brevity:

Lemma 2. *Let $G \in \{\mathscr{O}, \mathscr{U}_-, \mathscr{U}_-^\star, \mathscr{U}_+^\star\}$ and let ξ be a G-submodule of $\otimes^k V^*$.*

1. *There is an orthogonal direct sum decomposition of $\xi = \xi_1 \oplus \ldots \oplus \xi_k$ into irreducible G-submodules of ξ. The multiplicity with which a given irreducible G-module η appears in ξ is independent of the particular decomposition which is chosen. If ξ_1 appears with multiplicity 1 in the decomposition of ξ and if η is any G-submodule of ξ, then either $\xi_1 \subset \eta$ or $\xi_1 \perp \eta$.*
2. *If $\xi_1 \to \xi \to \xi_2$ is a short exact sequence of G-modules, then ξ is isomorphic to $\xi_1 \oplus \xi_2$ as a G-module.*

We can illustrate Lemma 2 as follows. Decompose

$$\otimes^2 V^* = \Lambda^2(V^*) \oplus S^2(V^*)$$

as the direct sum of the alternating and the symmetric bilinear forms. We can further decompose $\Lambda^2(V^*) = \Lambda_\pm^2 \oplus \chi \oplus \Lambda_{0,\mp}^2$ and $S^2(V^*) = S_\pm^2 \oplus \mathbb{1} \oplus S_{0,\mp}^2$ where

$$\begin{aligned}
&\Lambda^2_\pm := \{\omega \in \Lambda^2 : J^*_\pm \omega = \pm\omega\}, && \chi := \Omega_\pm \cdot \mathbb{R},\\
&\Lambda^2_{0,\mp} := \{\omega \in \Lambda^2 : J^*_\pm \omega = \mp\omega,\ \omega \perp \Omega_\pm\},\\
&S^2_\pm := \{\theta \in S^2 : J^*_\pm \theta = \pm\theta\}, && 1\!\!1 := \langle\cdot,\cdot\rangle \cdot \mathbb{R},\\
&S^2_{0,\mp} := \{\theta \in S^2 : J^*_\pm \theta = \mp\theta,\ \theta \perp \langle\cdot,\cdot\rangle\}.
\end{aligned}$$

Lemma 3. *Let $(V,\langle\cdot,\cdot\rangle,J_\pm)$ be a (para)-Hermitian vector space. We have the following decomposition of $\Lambda^2(V^*)$, $S^2(V^*)$, and $\otimes^2 V^*$ into inequivalent and irreducible $\mathcal{U}^\star_\pm$ modules:*

$$\Lambda^2(V^*) = \Lambda^2_\pm \oplus \chi \oplus \Lambda^2_{0,\mp}, \qquad S^2(V^*) = S^2_\pm \oplus 1\!\!1 \oplus S^2_{0,\mp},$$
$$\otimes^2 V^* = \Lambda^2_\pm \oplus \chi \oplus \Lambda^2_{0,\mp} \oplus S^2_\pm \oplus 1\!\!1 \oplus S^2_{0,\mp}.$$

We note that $\Lambda^2_{0,\mp}$ and $S^2_{0,\mp}$ are isomorphic $U_\pm$ modules, that $\Lambda^2_{0,\mp}$ is isomorphic to $S^2_{0,\mp} \otimes \chi$ as a $U^\star_\pm$ module, and that Λ^2_+ is not an irreducible U_+ module. We complete our discussion of elementary representation theory with the following diagonalization result (see, e.g., the discussion in [6]):

Lemma 4. *If ξ is a nontrivial proper $\mathcal{U}^\star_\pm$ submodule of $\Lambda^2_\pm \oplus \Lambda^2_\pm$, then there exists $(a,b) \neq (0,0)$ so $\xi = \xi(a,b) := \{(a\theta, b\theta)\}_{\theta\in\Lambda^2_\pm} \subset \Lambda^2_\pm \oplus \Lambda^2_\pm$.*

2.2 The Singer–Thorpe and the Higa Decompositions

We now examine the $\mathcal{O}$-module structure of $\mathfrak{R}$ and $\mathfrak{W}$. Let

$$S^2_0 := \{\theta \in S^2 : \theta \perp \langle\cdot,\cdot\rangle\} \quad \text{and} \quad \mathfrak{C} := \ker\{\rho\} \cap \mathfrak{R}$$

be the $\mathcal{O}$ modules of trace-free symmetric 2-tensors and *Weyl conformal curvature tensors*, respectively. We refer to Singer and Thorpe [33] for the proof of Assertion (1) and to Higa [18, 19] for the proof of Assertion (2) in the following result:

Theorem 7. *Let $n \geq 4$.*

1. *We may decompose $\mathfrak{R} = 1\!\!1 \oplus S^2_0 \oplus \mathfrak{C}$ as the orthogonal direct sum of irreducible and inequivalent $\mathcal{O}$ modules.*
2. *We may decompose $\mathfrak{W} = 1\!\!1 \oplus S^2_0 \oplus \mathfrak{C} \oplus \mathfrak{P}$ as the orthogonal direct sum of irreducible and inequivalent $\mathcal{O}$ modules. Here ρ_a provides an $\mathcal{O}$ module isomorphism from $\mathfrak{P}$ to Λ^2 with the inverse embedding $\Xi : \Lambda^2 \xrightarrow{\approx} \mathfrak{P} \subset \mathfrak{W}$ given by:*

$$\begin{aligned}
\Xi(\psi)(x,y,z,w) := {}& 2\psi(x,y)\langle z,w\rangle + \psi(x,z)\langle y,w\rangle - \psi(y,z)\langle x,w\rangle\\
&- \psi(x,w)\langle y,z\rangle + \psi(y,w)\langle x,z\rangle. \qquad (10)
\end{aligned}$$

2.3 The Tricerri–Vanhecke Decompositions

The following decompositions of $\mathfrak{R}$ and $\mathfrak{K}_{\pm,\mathfrak{R}}$ as $\mathcal{U}_-$ modules was given by Tricerri and Vanhecke [34] in the positive definite setting; they extend easily to the more general context [3, 4]. The decomposition of $\mathfrak{W}$ as a $\mathcal{U}_\pm^\star$ module then follows from Lemma 3 and Theorem 7.

Theorem 8. *Let $(V,\langle\cdot,\cdot\rangle,J_\pm)$ be a (para)-Hermitian vector space. We have the following decompositions of $\mathfrak{R}$, $\mathfrak{K}_{\pm,\mathfrak{R}}$, and $\mathfrak{W}$ as $\mathcal{U}_\pm^\star$ modules:*

$$\begin{aligned}\mathfrak{R} &= W_{\pm,1}\oplus\ldots\oplus W_{\pm,10},\\ \mathfrak{K}_{\pm,\mathfrak{R}} &= W_{\pm,1}\oplus W_{\pm,2}\oplus W_{\pm,3},\\ \mathfrak{W} &= W_{\pm,1}\oplus\ldots\oplus W_{\pm,13}.\end{aligned} \tag{11}$$

If $n=4$, we omit the modules $\{W_{\pm,5},W_{\pm,6},W_{\pm,10}\}$. If $n=6$, we omit the module $W_{\pm,6}$. The decomposition of Eq. (11) is then into irreducible $\mathcal{U}_\pm^\star$ modules. We have $\mathcal{U}_\pm^\star$ module isomorphisms:

$$W_{\pm,1}\approx W_{\pm,4}\approx \mathbb{1},\quad W_{\pm,2}\approx W_{\pm,5}\approx S^2_{0,\mp},\quad W_{\pm,9}\approx W_{\pm,13}\approx \Lambda^2_\pm, \tag{12}$$

$$W_{\pm,8}\approx S^2_\pm,\quad W_{\pm,11}\approx\chi,\quad W_{\pm,12}\approx\Lambda^2_{0,\mp}. \tag{13}$$

With exception of the isomorphisms described in Eq. (12), these are inequivalent $\mathcal{U}_\pm^\star$ modules. The isomorphism Ψ from $\Lambda^2_\pm$ to $W_{\pm,9}$ is given by setting

$$\begin{aligned}\Psi(\psi)(x,y,z,w) := {}& 2\langle x,J_\pm y\rangle\psi(z,J_\pm w)+2\langle z,J_\pm w\rangle\psi(x,J_\pm y)\\ &+\langle x,J_\pm z\rangle\psi(y,J_\pm w)+\langle y,J_\pm w\rangle\psi(x,J_\pm z)\\ &-\langle x,J_\pm w\rangle\psi(y,J_\pm z)-\langle y,J_\pm z\rangle\psi(x,J_\pm w).\end{aligned} \tag{14}$$

It is worth describing some of these in a bit more detail. Let $\{e_i\}$ be a basis for V. Set $\varepsilon_{ij}:=\langle e_i,e_j\rangle$. Define $\rho_{J_\pm}(x,y):=\varepsilon^{il}A(e_i,x,J_\pm y,J_\pm e_l)$. We then have:

$$\begin{aligned}W_{\pm,7} &= \{A\in\mathfrak{R}: A(J_\pm x,y,z,w)=A(x,y,J_\pm z,w)\},\\ W_{\pm,3} &= \mathfrak{K}_{\pm,\mathfrak{R}}\cap\ker(\rho),\\ W_{\pm,6} &= \{A\in\mathfrak{R}: J_\pm^*A=A\}\cap\{\mathfrak{K}_{\pm,\mathfrak{R}}\}^\perp\cap\{W_{\pm,7}\}^\perp\cap\ker(\rho\oplus\rho_{J_\pm}),\\ W_{\pm,10} &= \{A\in\mathfrak{R}: J_\pm^*A=-A\}\cap\ker(\rho\oplus\rho_{J_\pm}).\end{aligned}$$

3 The Proof of Theorem 6

If η is an irreducible $\mathscr{U}_\pm^\star$ module and if ξ is a submodule of $\otimes^4 V^*$, let $n_\eta(\xi)$ be the multiplicity with which η appears in the decomposition of ξ given in Lemma 2; note that $W_{\pm,4} \approx W_{\pm,1}$ and $W_{\pm,2} \approx W_{\pm,5}$. We apply Theorem 8. If η is isomorphic to $W_{i,\pm}$ for $i \in \{1,2,3,4,5,6,7,8,10\}$, then $n_\eta(\Lambda^2) = 0$ so

$$n_\eta(\mathfrak{K}_{\pm,\mathfrak{W}}) = n_\eta(\mathfrak{K}_{\pm,\mathfrak{R}}) = \left\{ \begin{array}{ll} 1 & \text{if } i = 1,2,3,4,5 \\ 0 & \text{if } i = 6,7,8,10 \end{array} \right\}.$$

Thus only the multiplicities of the representations $\{\chi, \Lambda^2_{0,\mp}, \Lambda^2_\pm\}$ are at issue.

3.1 The Module $\chi = \Xi\,(\Omega_\pm)$ for $m \geq 4$

Let $\{e_i\}$ be an orthonormal basis for V with $J_\pm e_{2i-1} = e_{2i}$ and $J_\pm e_{2i} = \pm e_{2i-1}$. Let $\varepsilon_{ij} := \langle e_i, e_j \rangle$. We use Eq. (10) to see:

$$\begin{aligned} \Xi(\Omega_\pm)(e_1,e_4,e_3,e_1) &= -\langle e_4, J_\pm e_3\rangle\langle e_1,e_1\rangle = -\varepsilon_{11}\varepsilon_{44}, \\ \mp\Xi(\Omega_\pm)(e_1,e_4,J_\pm e_3,J_\pm e_1) &= \pm\langle e_1, J_\pm J_\pm e_1\rangle\langle e_4, J_\pm e_3\rangle = \varepsilon_{11}\varepsilon_{44}. \end{aligned}$$

Thus $\Xi(\Omega_\pm)$ does not satisfy the Kähler identity given in Eq. (8). Consequently, $n_\chi(\mathfrak{K}_{\pm,\mathfrak{W}}) = 0$.

3.2 The Module $\mathrm{W}_{\pm,12} = \Xi\,(\Lambda^2_{0,\mp})$ for $m \geq 6$

Set $\psi_{0,\pm} := e^1 \otimes e^2 - e^2 \otimes e^1 - \varepsilon_{11}\varepsilon_{33}\{e^3 \otimes e^4 - e^4 \otimes e^3\}$. Clearly $\psi_{0,\pm} \perp \Omega_\pm$. Since $J_\pm^* \psi_{0,\pm} = \mp\psi_{0,\pm}$, $\psi_{0,\pm} \in \Lambda^2_{0,\mp}$. By Eq. (10):

$$\begin{aligned} \Xi(\psi_{0,\pm})(e_5,e_1,e_2,e_5) &= -\psi_{0,\pm}(e_1,e_2)\langle e_5,e_5\rangle = -\varepsilon_{55}, \\ \mp\Xi(\psi_{0,\pm})(e_5,e_1,J_\pm e_2,J_\pm e_5) &= 0. \end{aligned}$$

Consequently $\Xi(\psi_{0,\pm})$ does not satisfy the Kähler identity and we conclude that $n_{\Lambda^2_{0,\mp}}(\mathfrak{K}_{\pm,\mathfrak{W}}) = 0$ if $m \geq 6$.

3.3 The Module $\Lambda^2_{0,\mp}$ if $m = 4$

The argument given above in Sect. 3.2 does not, of course, pertain if $m = 4$ since we can not examine $\Xi(\psi_{0,\pm})(e_5,e_1,e_2,e_5)$. Let $\eta = \Lambda^2_{0,\mp}$. As noted above,

$n_\eta(\mathfrak{K}_{\pm,\mathfrak{W}}) \leq 1$. Thus if we can exhibit a non-trivial element of $W_{\pm,12} \cap \mathfrak{K}_{\pm,\mathfrak{W}}$, we will have $n_\eta(\mathfrak{K}_{\pm,\mathfrak{W}}) = 1$. We work in the positive definite setting for the moment to simplify the argument. Let

$$\psi_{0,+} := e^1 \otimes e^2 - e^2 \otimes e^1 - e^3 \otimes e^4 + e^4 \otimes e^3,$$
$$\langle \cdot, \cdot \rangle := e^1 \otimes e^1 + e^2 \otimes e^2 + e^3 \otimes e^3 + e^4 \otimes e^4.$$

Decompose $A := \Xi(\psi_{0,+}) = A_1 + A_2 + A_3 + A_4 + A_5$ using the notation of Eq. 10) where

$$A_1(x,y,z,w) := 2\psi_{0,+}(x,y)\langle z,w\rangle, \qquad A_2(x,y,z,w) := \psi_{0,+}(x,z)\langle y,w\rangle,$$
$$A_3(x,y,z,w) := -\psi_{0,+}(y,z)\langle x,w\rangle, \qquad A_4(x,y,z,w) := -\psi_{0,+}(x,w)\langle y,z\rangle,$$
$$A_5(x,y,z,w) := \psi_{0,+}(y,w)\langle x,z\rangle .$$

As a short hand, we set $e^{ijkl} := e^i \otimes e^j \otimes e^k \otimes e^l$. We may then express:

$$\begin{aligned}
A_1 &= 2e^{1211} + 2e^{1222} + 2e^{1233} + 2e^{1244} - 2e^{2111} - 2e^{2122} - 2e^{2133} - 2e^{2144} \\
&\quad - 2e^{3411} - 2e^{3422} - 2e^{3433} - 2e^{3444} + 2e^{4311} + 2e^{4322} + 2e^{4333} + 2e^{4344}, \\
A_2 &= e^{1121} + e^{1222} + e^{1323} + e^{1424} - e^{2111} - e^{2212} - e^{2313} - e^{2414} \\
&\quad - e^{3141} - e^{3242} - e^{3343} - e^{3444} + e^{4131} + e^{4232} + e^{4333} + e^{4434}, \\
A_3 &= -e^{1121} - e^{2122} - e^{3123} - e^{4124} + e^{1211} + e^{2212} + e^{3213} + e^{4214} \\
&\quad + e^{1341} + e^{2342} + e^{3343} + e^{4344} - e^{1431} - e^{2432} - e^{3433} - e^{4434}, \\
A_4 &= -e^{1112} - e^{1222} - e^{1332} - e^{1442} + e^{2111} + e^{2221} + e^{2331} + e^{2441} \\
&\quad + e^{3114} + e^{3224} + e^{3334} + e^{3444} - e^{4113} - e^{4223} - e^{4333} - e^{4443}, \\
A_5 &= e^{1112} + e^{2122} + e^{3132} + e^{4142} - e^{1211} - e^{2221} - e^{3231} - e^{4241} \\
&\quad - e^{1314} - e^{2324} - e^{3334} - e^{4344} + e^{1413} + e^{2423} + e^{3433} + e^{4443}.
\end{aligned}$$

We may ignore the terms in A_1 as these belong to $\mathfrak{K}_+$. The remaining terms yield a tensor which is anti-symmetric both in the first two and in the last two indices. Thus, automatically terms of the form e^{**12} or e^{**34} will belong to $\mathfrak{K}_+$ and can be ignored. Using the $\mathbb{Z}_2$ symmetry, we may consider terms e^{ijkl} where $i < j$ and $k < l$. We establish the Kähler identity and show that $n_\eta(\mathfrak{K}_{+,\mathfrak{W}}) = 1$ if $m = 4$ in the positive definite setting by examining the following crucial terms:

Term	Coeff.	Term	Coeff.
e^{1323}	$A_2 = 1$	e^{1314}	$A_5 = -1$
e^{1424}	$A_2 = 1$	e^{1413}	$A_5 = 1$
e^{2313}	$A_2 = -1$	e^{2324}	$A_5 = -1$
e^{2414}	$A_2 = -1$	e^{2423}	$A_5 = 1$

We now complexify and let $W := V \otimes_{\mathbb{R}} \mathbb{C}$. Extend $\langle\cdot,\cdot\rangle$, J_-, and A to be complex bilinear, complex linear, and complex multi-linear, respectively. Let:

$$V_{2,2} := \mathrm{Span}_{\mathbb{R}}\left\{\sqrt{-1}e_1, \sqrt{-1}e_2, e_3, e_4\right\}.$$

Then, $(\langle\cdot,\cdot\rangle, J_-)$ restricts to a pseudo-Hermitian almost complex structure on $V_{2,2}$ of signature $(2,2)$. Note that

$$\mathrm{Re}(A|_{V_{2,2}}) \in W_{\pm,12}(V_{2,2}) \cap \mathfrak{K}_{-,\mathfrak{W}}(V_{2,2}),$$
$$\mathrm{Im}(A|_{V_{2,2}}) \in W_{\pm,12}(V_{2,2}) \cap \mathfrak{K}_{-,\mathfrak{W}}(V_{2,2}).$$

Since $A|_{V_{2,2}} \neq 0$, at least one of these tensors is nontrivial and the desired conclusion follows for neutral signature $(2,2)$; a similar argument applied to

$$V_{4,0} := \mathrm{Span}_{\mathbb{R}}\left\{\sqrt{-1}e_1, \sqrt{-1}e_2, \sqrt{-1}e_3, \sqrt{-1}e_4\right\}$$

establishes the desired result in signature $(4,0)$ (which is the negative definite setting). Finally, by considering

$$U_{2,2} := \mathrm{Span}_{\mathbb{R}}\left\{e_1, \sqrt{-1}e_2, e_3, \sqrt{-1}e_4\right\}$$

and $J_+ := \sqrt{-1}J_-$, we can construct an example in the para-complex setting.

3.4 The Module $\Lambda^2_\pm$ if $m \geq 6$

Let $\eta = \Lambda^2_\pm$. Then $W_{\pm,9} \oplus W_{\pm,13} \approx 2 \cdot \eta$. We adopt the notation of Eq. (10) and of Eq. (14). For $(a,b) \neq (0,0)$, let

$$\xi(a,b) := \mathrm{Range}\{a\Xi + b\Psi\} \subset W_{\pm,9} \oplus W_{\pm,13}.$$

By Lemma 4, every non-trivial proper submodule of $W_{\pm,9} \oplus W_{\pm,13}$ is isomorphic to $\xi(a,b)$ for some $(a,b) \neq 0$. We suppose $\xi(a,b) \subset \mathfrak{K}_{\pm,\mathfrak{W}}$ and thus

$$(a\Xi + b\Psi)\psi \in \mathfrak{K}_{\pm,\mathfrak{W}} \quad \text{for all} \quad \psi_\pm \in \Lambda^2_\pm.$$

Set $\psi_\pm := e^1 \otimes e^3 - e^3 \otimes e^1 \pm e^2 \otimes e^4 \mp e^4 \otimes e^2$. Then $J^*_\pm \psi_\pm = \pm\psi_\pm$ so $\psi_\pm \in \Lambda^2_\pm$. We show that $b = 0$ by checking:

$$a\Xi(\psi_\pm)(e_5, e_6, e_1, e_4) = 0,$$
$$\mp a\Xi(\psi_\pm)(e_5, e_6, J_\pm e_1, J_\pm e_4) = 0,$$
$$b\Psi(\psi_\pm)(e_5, e_6, e_1, e_4) = 2b\langle e_5, J_\pm e_6\rangle \psi_\pm(e_1, J_\pm e_4) = 2b\varepsilon_{55},$$
$$\mp b\Psi(\psi_\pm)(e_5, e_6, J_\pm e_1, J_\pm e_4) = \mp 2b\langle e_5, J_\pm e_6\rangle \psi_\pm(J_\pm e_1, J_\pm J_\pm e_4) = -2b\varepsilon_{55}.$$

We show that $a = 0$ and complete the proof of Theorem 6 if $m \geq 6$ by checking:

$$\begin{aligned} a\Xi(\psi_{\pm})(e_5,e_1,e_3,e_5) &= -a\psi_{\pm}(e_1,e_3)\langle e_5,e_5\rangle = -a\varepsilon_{55}, \\ \mp a\Xi(\psi_{\pm})(e_5,e_1,e_4,e_6) &= 0. \end{aligned}$$

3.5 The Module $\Lambda^2_{\pm}$ if $m = 4$

Again, the argument given in Sect. 3.4 is not available if $m = 4$ since, for example, we can not examine (e_5,e_1,e_4,e_6). Let $\eta = \Lambda^2_{\pm}$. Again, we first work in the positive definite setting. Since $n_\eta(\mathfrak{K}_{+,\mathfrak{R}}) = 0$, to show $n_\eta(\mathfrak{K}_{+,\mathfrak{W}}) = 1$, it suffices to construct a suitable element of $\mathfrak{K}_{+,\mathfrak{W}}$. Let

$$\begin{aligned} \psi_- &:= e^1 \otimes e^3 - e^3 \otimes e^1 - e^2 \otimes e^4 + e^4 \otimes e^2 \in \Lambda^2_-, \\ \langle \cdot,\cdot \rangle &:= e^1 \otimes e^1 + e^2 \otimes e^2 + e^3 \otimes e^3 + e^4 \otimes e^4 . \end{aligned}$$

Adopt the notation of Eq. (10) to decompose $\Xi(\psi_-) = F + G + H + J + K$ where

$$\begin{aligned} &F(x,y,z,w) := 2\psi_-(x,y)\langle z,w\rangle, \ G(x,y,z,w) := \psi_-(x,z)\langle y,w\rangle, \\ &H(x,y,z,w) := -\psi_-(y,z)\langle x,w\rangle, \ J(x,y,z,w) := -\psi_-(x,w)\langle y,z\rangle, \\ &K(x,y,z,w) := \psi_-(y,w)\langle x,z\rangle . \end{aligned}$$

We compute:

$$\begin{aligned} F = &+2e^{1311} + 2e^{1322} + 2e^{1333} + 2e^{1344} - 2e^{3111} - 2e^{3122} - 2e^{3133} - 2e^{3144} \\ &+ 2e^{4211} + 2e^{4222} + 2e^{4233} + 2e^{4244} - 2e^{2411} - 2e^{2422} - 2e^{2433} - 2e^{2444}, \\ G = &+e^{1131} + e^{1232} + e^{1333} + e^{1434} - e^{3111} - e^{3212} - e^{3313} - e^{3414} \\ &+ e^{4121} + e^{4222} + e^{4323} + e^{4424} - e^{2141} - e^{2242} - e^{2343} - e^{2444}, \\ H = &-e^{1131} - e^{2132} - e^{3133} - e^{4134} + e^{1311} + e^{2312} + e^{3313} + e^{4314} \\ &- e^{1421} - e^{2422} - e^{3423} - e^{4424} + e^{1241} + e^{2242} + e^{3243} + e^{4244}, \\ J = &-e^{1113} - e^{1223} - e^{1333} - e^{1443} + e^{3111} + e^{3221} + e^{3331} + e^{3441} \\ &- e^{4112} - e^{4222} - e^{4332} - e^{4442} + e^{2114} + e^{2224} + e^{2334} + e^{2444}, \\ K = &+e^{1113} + e^{2123} + e^{3133} + e^{4143} - e^{1311} - e^{2321} - e^{3331} - e^{4341} \\ &+ e^{1412} + e^{2422} + e^{3432} + e^{4442} - e^{1214} - e^{2224} - e^{3234} - e^{4244}. \end{aligned}$$

Next, we examine the role of Ψ. Set $\tilde{\varepsilon}(x,y) := \langle x, Jy\rangle$ and $\tilde{\psi}_-(x,y) := \psi_-(x,Jy)$. We expand $\Psi(\psi_-) = R+S+T+U+V+W$ where

$$\begin{aligned}&R(x,y,z,w) := 2\tilde{\varepsilon}(x,y)\tilde{\psi}_-(z,w),\ S(x,y,z,w) := 2\tilde{\varepsilon}(z,w)\tilde{\psi}_-(x,y),\\ &T(x,y,z,w) := \tilde{\varepsilon}(x,z)\tilde{\psi}_-(y,w),\ U(x,y,z,w) := \tilde{\varepsilon}(y,w)\tilde{\psi}_-(x,z),\\ &V(x,y,z,w) := -\tilde{\varepsilon}(x,w)\tilde{\psi}_-(y,z),\ W(x,y,z,w) := -\tilde{\varepsilon}(y,z)\tilde{\psi}_-(x,w).\end{aligned}$$

We compute

$$\begin{aligned}\tilde{\varepsilon} &= -e^1\otimes e^2 + e^2\otimes e^1 - e^3\otimes e^4 + e^4\otimes e^3,\\ \tilde{\psi}_- &= -e^1\otimes e^4 + e^4\otimes e^1 - e^2\otimes e^3 + e^3\otimes e^2,\\ R &= -2e^{1241} + 2e^{2141} - 2e^{3441} + 2e^{4341} + 2e^{1214} - 2e^{2114} + 2e^{3414} - 2e^{4314}\\ &\quad -2e^{1232} + 2e^{2132} - 2e^{3432} + 2e^{4332} + 2e^{1223} - 2e^{2123} + 2e^{3423} - 2e^{4323},\\ S &= -2e^{4112} + 2e^{4121} - 2e^{4134} + 2e^{4143} + 2e^{1412} - 2e^{1421} + 2e^{1434} - 2e^{1443}\\ &\quad -2e^{3212} + 2e^{3221} - 2e^{3234} + 2e^{3243} + 2e^{2312} - 2e^{2321} + 2e^{2334} - 2e^{2343},\\ T &= -e^{1421} + e^{2411} - e^{3441} + e^{4431} + e^{1124} - e^{2114} + e^{3144} - e^{4134}\\ &\quad -e^{1322} + e^{2312} - e^{3342} + e^{4332} + e^{1223} - e^{2213} + e^{3243} - e^{4233},\\ U &= -e^{4112} + e^{4211} - e^{4314} + e^{4413} + e^{1142} - e^{1241} + e^{1344} - e^{1443}\\ &\quad -e^{3122} + e^{3221} - e^{3324} + e^{3423} + e^{2132} - e^{2231} + e^{2334} - e^{2433},\\ V &= e^{1412} - e^{2411} + e^{3414} - e^{4413} - e^{1142} + e^{2141} - e^{3144} + e^{4143}\\ &\quad +e^{1322} - e^{2321} + e^{3324} - e^{4323} - e^{1232} + e^{2231} - e^{3234} + e^{4233},\\ W &= e^{4121} - e^{4211} + e^{4341} - e^{4431} - e^{1124} + e^{1214} - e^{1344} + e^{1434}\\ &\quad +e^{3122} - e^{3212} + e^{3342} - e^{3432} - e^{2123} + e^{2213} - e^{2343} + e^{2433}.\end{aligned}$$

We may ignore the F and the S terms as these belong to $\mathfrak{K}_+$. The remaining terms yield a tensor which is antisymmetric in the first indices and antisymmetric in the last indices. Thus, automatically things of the form e^{**12} or e^{**34} belong to $\mathfrak{K}_+$ and don't need to be worried about. Thus the only terms which matter are the following:

Term	Coef	Coef	Coef	Contribution
e^{1223}	$J=-1$	$R=2$	$T=1$	$-a+3b$
e^{1214}	$K=-1$	$R=2$	$W=1$	$-a+3b$
e^{3423}	$H=-1$	$R=2$	$U=1$	$-a+3b$
e^{3414}	$G=-1$	$R=2$	$V=1$	$-a+3b$

Thus we must have $-a+3b=0$ so we may take $a=3$ and $b=1$. This completes the proof in signature $(0,4)$; the remaining cases are handled using the same techniques used in Sect. 3.3.

4 The Proof of Theorem 5

Adopt the notation of Eq. (7). Fix a bilinear form $\varepsilon = (\varepsilon_{ij})$ on $\mathbb{R}^m$ which is $\pm$invariant under $J_\pm$. Let "$\circ$" denote symmetric tensor product. Let $\theta \in S^2_\mp \otimes S^2$. We form the germ of a pseudo-Riemannian metric which is $\pm$-invariant under the action of $J_\pm$ by setting:

$$g = \varepsilon + \theta_{ijkl} x^k x^l dx^i \circ dx^j\,;$$

g is a (para)-Hermitian metric on a neighborhood $\mathscr{O}$ of 0 in $\mathbb{R}^m$. By Theorem 4(2), there is a unique Weyl connection $\nabla = \nabla(\theta)$ so that $(\mathscr{O}, J_\pm, g, \nabla)$ is a (para)-Kähler Weyl manifold. Let $\Theta(\theta) := R^\nabla(0)$; Θ defines an equivariant linear map

$$\Theta : S^2_\mp \otimes S^2 \to \mathfrak{K}_{\pm,\mathfrak{W}}\,.$$

To show that Θ is surjective and complete the proof of Theorem 5, we must show:

$$n_\eta(\mathrm{Range}(\Theta)) = 1 \quad \text{for} \quad \eta \in \left\{\mathbb{1}, S^2_{0,\pm}, W_{\pm,3}, \Lambda^2_{0,\mp}, \Lambda^2_\pm\right\}.$$

4.1 The Representations $W_{\pm,i}$ for $i = 1, 2, 3$

Let $R^g(0)$ be the curvature of the Levi–Civita connection at the origin. The map $\mathscr{L} : \theta \to R^g(0)$ is a linear function of θ given by:

$$(\mathscr{L}\theta)(x,y,z,w) := \theta(x,z,y,w) + \theta(y,w,x,z) - \theta(x,w,y,z) - \theta(y,z,x,w)\,.$$

We set $A := \mathscr{L}(\theta)$. Similarly the map $K_\pm : \Theta \to d\Omega^g$ is a linear map which takes $S^2_\mp(V^*) \otimes S^2(V^*)$ to $\Lambda^3(V^*) \otimes V^*$. It is given by:

$$\{(K_\pm\Theta)(x,y,z)\}(w) := \Theta(x,J_\pm y,z,w) + \Theta(y,J_\pm z,x,w) + \Theta(z,J_\pm x,y,w)\,.$$

This shows that $\ker(K_\pm)$ is invariant under the action of $\mathscr{U}^\star_\pm$. Clearly $\theta \in \ker(K_\pm)$ if and only if g_θ is a Kähler metric. On $\ker(K_\pm)$, we have $\Theta = \mathscr{L}$ since $\phi = 0$. Thus:

$$\mathscr{L} : \ker(K_\pm) \to W_{1,\pm} \oplus W_{2,\pm} \oplus W_{3\pm}\,.$$

Take

$$\Theta = \frac{1}{2}(e^1 \otimes e^1 \mp e^2 \otimes e^2) \otimes (e^1 \otimes e^1 + e^2 \otimes e^2)$$

so that the metric has the form

$$g_\Theta = \varepsilon + \frac{1}{2}(u_1^2 + u_2^2)(du_1^2 \mp du_2^2)\,.$$

The metric g_Θ is Kähler since it takes the form $M_2 \times \mathbb{C}$ where M_2 is a Riemann surface. Thus $\Theta \in \ker(K_\pm)$. Furthermore, the only non-zero curvature components of the curvature tensor $A = R^g(0)$ at the origin, up to the usual $\mathbb{Z}_2$ symmetries, are given by

$$A(e_1, e_2, e_2, e_1) = 1\,.$$

The symmetric Ricci tensor $\rho_s(x,y) := \frac{1}{2}(\rho(x,y) + \rho(y,x))$ defines a map from $\mathfrak{K}_{\pm,\mathfrak{W}}$ to S^2. We have

$$\rho_s(e_i, e_j) = \left\{ \begin{array}{cl} \varepsilon_{22} & \text{if } i = j = 1, \\ \varepsilon_{11} & \text{if } i = j = 2, \\ 0 & \text{otherwise} \end{array} \right\}.$$

Since ρ_s is neither a multiple of $\langle\cdot,\cdot\rangle$ nor is ρ_s perpendicular to $\langle\cdot,\cdot\rangle$, ρ_s has components both in $\mathfrak{U}$ and in $S^2_{0,\mp}$. Consequently

$$W_{\pm,1} \oplus W_{\pm,2} \subset \mathscr{L}(K_\pm).$$

Let $S \in S^2_\mp$. Following [34], define:

$$\begin{aligned} S_1(x,y,z,w) &:= \langle x,z\rangle S(y,w) + \langle y,w\rangle S(x,z) \\ &\quad - \langle x,w\rangle S(y,z) - \langle y,z\rangle S(x,w) \\ S_2(x,y,z,w) &:= 2\langle x,J_\pm y\rangle S(z,J_\pm w) + 2\langle z,J_\pm w\rangle S(x,J_\pm y) \\ &\quad + \langle x,J_\pm z\rangle S(y,J_\pm w) + \langle y,J_\pm w\rangle S(x,J_\pm z) \\ &\quad - \langle x,J_\pm w\rangle S(y,J_\pm z) - \langle y,J_\pm z\rangle S(x,J_\pm w). \end{aligned}$$

Then the map $\Sigma : S \to S_1 \mp S_2$ splits ρ_s modulo a suitable normalizing constant.[1] We have:

$$\Sigma(\rho_s)(e_1, e_3, e_3, e_1) = -\varepsilon_{33}\varepsilon_{22}$$

and thus $\Sigma(\rho_s)$ is not a multiple of R so R has a non-zero component in $W_{\pm,3}$ and

$$W_{\pm,3} \subset \mathscr{L}(K_\pm)\,.$$

[1] This result was established in the positive definite setting; it extends easily to the general context.

4.2 The Representations $\Lambda^2_\pm$ and $\Lambda^2_{0,\mp}$

The alternating part of the Ricci tensor, ρ_a provides a map from $\mathfrak{K}_{\pm,\mathfrak{W}}$ to Λ^2. If we can show $\rho_a\Theta$ is a surjective map to $\Lambda^2_{0,\mp}\oplus\Lambda^2_\pm$, it will follow from Lemma 2 that $n_\eta(\mathfrak{K}_{\pm,\mathfrak{W}})\geq 1$ which will complete the proof. We have that ϕ is a multiple of $J^*_\pm\delta\Omega_\pm$ and that $d\phi$ is a multiple of ρ_a. Thus it will suffice to give an example where $dJ^*_\pm\delta\Omega_\pm$ has components in both $\Lambda^2_{0,\mp}$ and $\Lambda^2_\pm$. Suppose $f(x)=x_1x_3$. Let:

$$ds^2 := \varepsilon_{11}e^{2f(x_1,x_3)}(dx^1\otimes dx^1 \mp dx^2\otimes dx^2)+\varepsilon_{22}(dx^3\otimes dx^3\mp dx^4\otimes dx^4)\,.$$

We have [7]:

$$\begin{aligned}(\nabla^g\Omega_\pm)(\partial_{x_i},\partial_{x_j};\partial_{x_k}) &= \frac{1}{2}\Big\{g(\partial_{x_i},\partial_{x_k};J_\pm\partial_{x_j})-g(\partial_{x_j},\partial_{x_k};J_\pm\partial_{x_i})\\ &\qquad +g(J_\pm\partial_{x_i},\partial_{x_k};\partial_{x_j})-g(J_\pm\partial_{x_j},\partial_{x_k};\partial_{x_i})\Big\}\,.\end{aligned}$$

This permits us to compute that:

$$(\nabla^g\Omega_\pm)(\partial_{x_1},\partial_{x_3};\partial_{x_k})=\left\{\begin{array}{rl}\mp\varepsilon_{11}e^{2f}\partial_{x_3}f & \text{if } k=2\\ 0 & \text{if } k\neq 2\end{array}\right\}.$$

The covariant derivative of the Kähler form has the symmetries [7]:

$$\begin{aligned}(\nabla^g\Omega_\pm)(x,y;z) &= -(\nabla^g\Omega_\pm)(y,x;z)=\pm(\nabla^g\Omega_\pm)(J_\pm x,J_\pm y;z)\\ &= \mp(\nabla^g\Omega_\pm)(x,J_\pm y;J_\pm z)\,.\end{aligned}$$

It now follows that the non-zero components of $\nabla^g\Omega_\pm$ are given, up to the $\mathbb{Z}_2$ symmetry in the first components, by:

$$\begin{aligned}(\nabla^g\Omega_\pm)(\partial_{x_1},\partial_{x_3};\partial_{x_2}) &= \mp\varepsilon_{11}e^{2f}\partial_{x_3}f,\\ (\nabla^g\Omega_\pm)(\partial_{x_1},\partial_{x_4};\partial_{x_1}) &= \pm\varepsilon_{11}e^{2f}\partial_{x_3}f,\\ (\nabla^g\Omega_\pm)(\partial_{x_2},\partial_{x_4};\partial_{x_2}) &= -\varepsilon_{11}e^{2f}\partial_{x_3}f,\\ (\nabla^g\Omega_\pm)(\partial_{x_2},\partial_{x_3};\partial_{x_1}) &= \pm\varepsilon_{11}e^{2f}\partial_{x_3}f\,.\end{aligned}$$

This then implies

$$\begin{aligned}J^*_\pm\delta\Omega_\pm &= 2\mp\partial_{x_3}f\cdot dx^3,\\ dJ^*_\pm\delta\Omega_\pm &= 2\mp\partial_{x_1}\partial_{x_3}f\cdot dx^1\wedge dx^3\,.\end{aligned}$$

This has components in both $\Lambda^2_{0,\mp}$ and in $\Lambda^2_\pm$. The desired result now follows.

Acknowledgements Research of both authors supported by project MTM2009-07756 (Spain), by project 174012 (Serbia), and by DFG PI 158/4-6 (Germany). It is a pleasure to acknowledge useful suggestions from the Referees.

References

1. Aguilar, J., Romero, C.: Inducing the cosmological constant from five-dimensional Weyl space. Found. Phys. **39**, 1205–1216 (2009)
2. Alexandrov, B., Ivanov, S.: Weyl structures with positive Ricci tensor. Differ. Geom. Appl. **18**, 343–350 (2003)
3. Brozos-Vázquez, M., Gilkey, P., Merino, E.: Geometric realizations of Kaehler and of para-Kaehler curvature models. Int. J. Geom. Methods Mod. Phys. **7**, 505–515 (2010)
4. Brozos-Vázquez, M., Gilkey, P., Nikčević, S.: Geometric realizations of affine Kaehler curvature models. Results Math. **59**, 507–521 (2011)
5. Brozos-Vázquez, M., Gilkey, P., Nikčević, S.: The structure of the space of affine Kähler curvature tensors as a complex module. Int. J. Geom. Methods Mod. Phys. **8**, 1849–1868 (2011)
6. Brozos-Vázquez, M., Gilkey, P., Nikčević, S.: Geometric Realizations of Curvature. Imperial College Press, London (2012)
7. Brozos-Vázquez, M., García-Río, E., Gilkey, P., Hervella, L.: Geometric realizability of covariant derivative Kähler tensors for almost pseudo-Hermitian and almost para-Hermitian manifolds. Ann. Mat. Pura Appl. **191**, 487–502 (2012) http://arxiv.org/abs/1012.4964
8. Blažić, N., Gilkey, P., Nikčević, S., Simon, U.: Algebraic theory of affine curvature tensors. Archivum Mathematicum, Masaryk University (Brno, Czech Republic) ISSN 0044–8753, tomus **42**, supplement: Proceedings of the 26th Winter School Geometry and Physics 2006 (SRNI), 147–168 (2006)
9. Calderbank, D., Pedersen, H.: Self dual spaces with complex structures, Einstein-Weyl geometry and geodesics. Ann. I. Fourier **50**, 921–963 (2000)
10. Calderbank, D., Tod, P.: Einstein metrics, hypercomplex structures and the Toda field equation. Differ. Geom. Appl. **14**, 199–208 (2001)
11. Dunajski, M.: Paraconformal geometry of nth-order ODEs, and exotic holonomy in dimension four. J. Geom. Phys. **56**, 1790–1809 (2006)
12. Dunajski M., Tod, P.: Einstein–Weyl spaces and dispersionless Kadomtsev–Petviashvili equation from Painleve I and II. Phys. Lett. A **303**, 253–264 (2002)
13. Fatibene, L., Ferraris, M., Francaviglia, M., Mercadante, S.: Further extended theories of graviatation part II. Int. J. Geom. Methods Mod. Phys. **7**, 899–906 (2010)
14. Ganchev, G., Ivanov, S.: Semi-symmetric W-metric connections and the W-conformal group. God. Sofij. Univ. Fak. Mat. Inform. **81**, 181–193 (1994)
15. Ghosh, A.: Einstein–Weyl structures on contact metric manifolds. Ann. Glob. Anal. Geom. **35**, 431–441 (2009)
16. Gilkey, P.: The Geometry of Curvature Homogeneous Pseudo-Riemannian Manifolds. Imperial College Press, London (2007)
17. Gilkey, P., Nikčević, S., Simon U.: Geometric realizations, curvature decompositions, and Weyl manifolds. J. Geom. Phys. **61**, 270–275 (2011)
18. Higa, T.: Weyl manifolds and Einstein–Weyl manifolds. Comm. Math. Univ. St. Pauli **42**, 143–160 (1993)
19. Higa, T.: Curvature tensors and curvature conditions in Weyl geometry. Comm. Math. Univ. St. Pauli **43**, 139–153 (1994)
20. Itoh, M.: Affine locally symmetric structures and finiteness theorems for Einstein–Weyl manifolds. Tokyo J. Math. **23**, 37–49 (2000)

21. Ivanov, S.: Geometry of quaternionic Kähler connections with torsion. J. Geom. Phys. **41**, 235–257 (2002)
22. Kalafat, M.: Geometric invariant theory and Einstein-Weyl geometry. Expo. Math. **29**, 220–230 (2011)
23. Kokarev, G. Kotschick, D.: Fibrations and fundamental groups of Kähler-Weyl manifolds. Proc. Am. Math. Soc. **138**, 997–1010 (2010)
24. Matsuzoe, H.: Geometry of semi-Weyl manifolds and Weyl manifolds. Kyushu J. Math. **55**, 107–117 (2001)
25. Matzeu, P.: Submanifolds of Weyl flat manifolds. Monatsh. Math. **136**, 297–311 (2002)
26. Miritzis, J.: Isotropic cosmologies in Weyl geometry. Classical Quant. Grav. **21**, 3043–3055 (2004)
27. Nurowski, P.: Differential equations and conformal structures. J. Geom. Phys. **55**, 19–49 (2005)
28. Ozdeger, A: Conformal and generalized concircular mappings of Einstein–Weyl manifolds. Acta Math. Sci. Ser. B Engl. Ed. **30**, 1739–1745 (2010)
29. Pedersen, H. Swann, A.: Riemannian submersions, four manifolds, and Einstein–Weyl geometry. Proc. London Math. Soc. **66**, 381–399 (1991)
30. Pedersen, H., Tod, K.: Three-dimensional Einstein–Weyl geometry. Adv. Math. **97**, 74–109 (1993)
31. Pedersen, H., Poon, Y., Swann, A.: The Einstein-Weyl equations in complex and quaternionic geometry. Diff. Geo. Appl. **3**, 309–321 (1993)
32. Scholz, E.: Cosmological spacetimes balanced by a Weyl geometric scale covariant scalar field. Found. Phys. **39**, 45–72 (2009)
33. Singer, I.M., Thorpe, J.A.: The curvature of 4-dimensional Einstein spaces. 1969 Global Analysis (Papers in Honor of K. Kodaira), pp. 355–365. University Tokyo Press, Tokyo (1969)
34. Tricerri, F. Vanhecke, L.: Curvature tensors on almost Hermitian manifolds. Trans. Amer. Math. Soc. **267**, 365–397 (1981)
35. Vaisman, I.: Generalized Hopf manifolds. Geom. Dedicata **13**, 231–255 (1982)
36. Vaisman, I.: A survey of generalized Hopf manifolds. Differential Geometry on Homogeneous Spaces. Proceedings of Conference Torino, Italy, Rendiconti del Seminario Matematico. Editrice Rend. Semin. Mat. Torino, Fasc. Spec. pp. 205–221 (1983)
37. Weyl, H.: SpaceTime-Matter. Dover, New York (1922)

MIX
Papier aus verantwortungsvollen Quellen
Paper from responsible sources
FSC® C105338

If you have any concerns about our products,
you can contact us on
ProductSafety@springernature.com

In case Publisher is established outside the EU,
the EU authorized representative is:
Springer Nature Customer Service Center GmbH
Europaplatz 3, 69115 Heidelberg, Germany

Printed by Libri Plureos GmbH
in Hamburg, Germany